DOUBT

a history

DOUBT

a history

The Great Doubters and
Their Legacy of Innovation
from Socrates and Jesus to
Thomas Jefferson and Emily Dickinson

JENNIFER MICHAEL HECHT

HarperSanFrancisco
A Division of HarperCollins*Publishers*

Raphael, *School of Athens;* Stanza della Segnatura, Stanza di Rafaello, Vatican Palace, Vatican State. Photo credit: Eric Lessing / Art Resource, NY.

HarperCollins books may be purchased for educational, business, or sales promotional use. For information please write: Special Markets Department, HarperCollins Publishers, Inc., 10 East 53rd Street, New York, NY 10022.

HarperCollins Web site: http://www.harpercollins.com

HarperCollins®, 📖 ®, and HarperSanFrancisco™ are trademarks of HarperCollins Publishers, Inc.

FIRST EDITION

Designed by Kris Tobiassen

Library of Congress Cataloging-in-Publication Data is available on request.

ISBN 0–06–009772–8

03 04 05 06 07 ❖/RRD 10 9 8 7 6 5 4 3 2 1

For my parents

Contents

INTRODUCTION

Doubt Is No Shadow

A Quiz and a Guide to the Question

L ike belief, doubt takes a lot of different forms, from ancient Skepticism to modern scientific empiricism, from doubt in many gods to doubt in one God, to doubt that recreates and enlivens faith and doubt that is really disbelief. There are also celebrations of the state of doubt itself, from Socratic questioning to Zen koans; there is the sigh of the world-weary, the distracted hum of the scientist, and the rant of the victimized. Yet with all this conceptual difference there is a narrative to tell here: doubters in every century have made use of that which came before. At other times, great notions of doubt have been reinvented in relative isolation from the original and in fascinating new forms. This is a study of religious doubt, all over the world, from the beginning of recorded history to the present day. The story builds and does so in the same erratic, wildly creative way that the history of belief does. Once we see it as its own story, rather than as a mere collection of shadows on the history of belief, a whole new drama appears and new archetypes begin to come into focus. Without having the doubt story sketched out as such, it's hard to see how patterns of questioning have mirrored certain types of social change, for instance, and hard to identify doubt's most enduring themes. There are saints of doubt, martyrs of atheism, and sages of happy disbelief who have not been lined up as such, made visible by their relationships across time, and given the context of their story.

Issues of belief and doubt tend to get into some very partisan ruts. Atheists tend to see believers as naïve and dependent. Believers tend to see atheists as having abandoned themselves to meaninglessness, amorality, and pain. To shake off these and other modern habits will take exercise. It may be useful to begin by taking one's own pulse on a handful of questions—a quiz—intended both to vitalize the issues by pulling them apart a bit and to help situate some readers among their peers. Answer Yes, No, or Not Sure.

The Scale of Doubt Quiz

1. Do you believe that a particular religious tradition holds accurate knowledge of the ultimate nature of reality and the purpose of human life?

2. Do you believe that some thinking being consciously made the universe?

3. Is there an identifiable force coursing through the universe, holding it together, or uniting all life-forms?

4. Could prayer be in any way effective, that is, do you believe that such a being or force (as posited above) could ever be responsive to your thoughts or words?

5. Do you believe this being or force can think or speak?

6. Do you believe this being has a memory or can make plans?

7. Does this force sometimes take a human form?

8. Do you believe that the thinking part or animating force of a human being continues to exist after the body has died?

9. Do you believe that any part of a human being survives death, elsewhere or here on earth?

10. Do you believe that feelings about things should be admitted as evidence in establishing reality?

11. Do you believe that love and inner feelings of morality suggest that there is a world beyond that of biology, social patterns, and accident— i.e., a realm of higher meaning?

12. Do you believe that the world is not completely knowable by science?

13. If someone were to say, "The universe is nothing but an accidental pile of stuff, jostling around with no rhyme nor reason, and all life on earth is but a tiny, utterly inconsequential speck of nothing, in a corner of

space, existing in the blink of an eye never to be judged, noticed, or remembered," would you say, "Now that's going a bit far, that's a bit wrongheaded"?

If you answered No to all these questions, you're a hard-core atheist and of a certain variety: a rationalist materialist. If you said No to the first seven, but then had a few Yes answers, you're still an atheist, but you may have what I will call a pious relationship to the universe. If your answers to the first seven questions contained at least two Not Sure answers, you're an agnostic. If you answered Yes to some of the questions, you still might be an atheist or an agnostic, though not of the materialist variety. If you answered Yes to nine or more, you are a believer. But more than providing titles for various states of mind, the questions above may serve to demonstrate common clusters of opinion.

In the Eastern hemisphere of the planet, we find powerful and extraordinarily popular religions that did not posit a God or gods. In the West most religious doubt must be categorized as oppositional: in recorded history, belief in God or gods has been the norm and those who questioned or rejected the idea generally did so under at least some constraint. Of course, in every tradition of theistic belief there are records of questioning, doubt, and disbelief. In fact, the great religious texts are all a terrific jumble of affirmation and denial, and the greatest of them record valiant efforts to reconcile these impulses: in the Hebrew Bible, Job rants at God, and Augustine, the early Church Father, tears at his hair in his *Confessions,* beset by doubts. Whether you are a nonbeliever, or you belong to a religion without God, or you are a believer troubled by dark nights of the soul, we are all part of the same discussion. This is because, whatever our position may be, we all have the same contradictory information to work with. Sometimes it feels like there is a God or ultimate certainty, and it would be a great comfort if such a thing existed and we knew the answers to life's ultimate mysteries: who or what created the universe and why; what is human life for; what happens when I die? But there is no universally compelling, empirical, or philosophical evidence for the existence of God, a purposeful universe, or life after death.

Some people may be tone-deaf to the idea of evidence, some may be tone-deaf to the *feeling* that there is a higher power—we must forgive them each their failing. But there is also a tradition by which both sides refuse to engage the interesting questions: believers refuse to consider the reasonableness of doubt, and nonbelievers refuse to consider the feeling of faith.

Believers value the sense of mystery human beings can feel when they look inward or beyond; nonbelievers value the ability to map out the world by rational proofs. Yet there is a kind of mutual blindness, as if personal affiliation with one camp or another means more than does interest in the truth. These refusals to consider the opposing viewpoint are in some ways the result of recent history, a still-warm turf war between science and religion that got out of hand. A little historical context does the most to counteract this, but before launching into the narrative, I offer two interpretive ideas: "A Great Schism" and "Patterns of Doubt." These discussions are distinctly open to the doubting interpretation. This is, after all, doubt's story.

A GREAT SCHISM

Great believers and great doubters seem like opposites, but they are more similar to each other than to the mass of relatively disinterested or acquiescent men and women. This is because they are both awake to the fact that we live between two divergent realities: On one side, there is a world in our heads—and in our lives, so long as we are not contradicted by death and disaster—and that is a world of reason and plans, love, and purpose. On the other side, there is the world beyond our human life—an equally real world in which there is no sign of caring or value, planning or judgment, love, or joy. We live in a meaning-rupture because we are human and the universe is not.

Great doubters, like great believers, have been people occupied with this problem, trying to figure out whether the universe actually has a hidden version of humanness, or whether humanness is the error and people would be better off weaning themselves from their sense of narrative, justice, and love, thereby solving the schism by becoming more like the universe in which they are stuck. Cosmology can be stunning in this context. It is meaningful to get to your wedding on time, to do well in the marathon for which you have been training, to not spill coffee on your favorite shirt. But if we take a few steps back from the planet Earth and from our tiny moment in history, we see a very different picture: the Earth is a ball of water and dirt swarming with creatures, living and dying, passing in and out of existence, shifting around the continents. A few steps further back and we see planets coming into being, stars being born and dying, galaxies swarming in clusters across billions of years. The Earth blips into existence, life appears and swarms, and the Earth blips out of existence. From this perspective, the importance of a favorite shirt, a finish in the next marathon, and even whether you show up at your wedding—all of this begins to seem inconsequential. Concentrating on the macro-picture of reality is enough to make you sit down on a park bench and never get up again. When you face this schism in meaning, the idea that the universe has an agenda can get you off the park bench and back to your life.

Also in terrific contrast to the universe, human beings have a seemingly innate notion of what is fair. Yet as John F. Kennedy famously put it, "Life is unfair." We are indignant when things are not fair and yet there is little evidence of fairness in the world outside our heads. Unbelievably painful things happen, sometimes for no apparent reason and with no justification. The question of value is part of this. We make sense of things in life, all day every day, by sorting the important from the unimportant (the phone is ringing, the desk is dusty, the baby is falling), but the larger universe seems devoid of these calculations: a stray bullet strikes the generous and beloved mathematician and spares his gnarly little dog. Unstick yourself from our local human time and place, and it is hard to imagine that human values have any real meaning. Poets have often described the oddness of considering a dead emperor, or the skull of a genius: human power in life has no translation in death, or in the greater universe.

A related but not identical rupture has to do with the fact of answers in general. We have an almost violent desire to understand things, and our brains seem to take the whole of life as a great puzzle. Puzzles in the human world usually have solutions. We spend our entire lives working on an intriguing mystery, and we do not have any reason to expect ever to be presented with a solution, or even that there is one. The French philosopher Gabriel Marcel wrote about the difference between problems and mysteries, as did the great, offbeat student of Buddhism, Alan Watts. Both pointed out that problems must be solved but mysteries are to be enjoyed unsolved—and that we will be happier if we regard the universe and existence itself as mysteries. More commonly, the world strikes human beings as something to be figured out, and comes with no solution. Consciousness itself seems missing in the wider universe, and the human heart seems quite out of place. There is a serious weirdness to the mind, thinking amid the vast unthinking world.

Another huge difference between our human world and the universe as we know it is that, within the human world, as Bob Dylan sings, "Everybody's got to serve somebody." We are all inferior to someone in some areas. In the universe, we human beings are the only ones talking and the only ones articulating any answers. The universe is more powerful than we, but when it comes to demonstration of sentience and will, we find ourselves in the uncomfortable position of being the smartest, most powerful creatures around. There is no one to help us. Thus there is a rupture between daily life, in which individuals are rarely the highest authority, and the larger picture, the macro-reality of humankind, in which we as a group are the authority on everything.

Again, faced with two contradictory truths—that of the human world and that of the universe—religious virtuosos have all suggested some kind of reconciliation. They all say the schism is illusory, either because the universe is really possessed of human attributes and only looks chaotic, uncaring, and without direction, or because our sense of meaning is ridiculous, and we ought to train ourselves away from our willfulness and our struggle to invent, succeed, and sustain. Most prophets, preachers, and seers have made use of both these ideas. On the one hand, when the religious virtuoso tells us to focus our attention on infinity and eternity, and tells us that the importance of our concerns is illusory, he or she seeks to rouse us from our waking dream; to teach us to concentrate on the incomprehensible mysteries of our situation. Along these same lines, people who are driven to speak to such issues often insist that dedicated truth seekers must physically absent themselves from the human contest, living instead in seclusion, doing meditative exercises to strip away belief in human purpose, plans, and meaning. On the other hand, when the religious virtuoso tells us that God has meanings and purposes that we merely do not understand, he or she suggests that we trust the sense of justice and narrative that we have in our heads, and that we claim the meaninglessness of the universe to be the illusion. It is a comfort; it sends us back into our lives of meaning and purpose with the sense that meaning and purpose reign throughout the human experience and the universe at large. Again, most of us walk the line.

The sage tries to help contemplative people hold both of these thoughts in their minds. Jesus supported the idea that God created the world with purpose and care—an example of a preacher reading human-type meaning into the universe. But he also said to give up daily-life contests, habits, and even family bonds, to learn to see them as meaningless—an example of a preacher imposing the nonhumanness of the universe onto daily life. The Hebrew Bible says that vengeance is the Lord's, meaning the fairness that human beings crave really does exist in the world outside our heads. But the Hebrew Bible also says that the race is not to the swift, nor the battle to the strong, that time and chance happen to them all. Almost all important religious figures and texts make both of these impositions (meaninglessness on the human world, meaning on the world beyond human control), because the chief issue of religion is the breach between these two worlds.

Religion is not the only discipline to address these concerns, but it is the one by which human beings have attempted to integrate these two realities through practices and emotions as well as ideas. Writing philosophy and reading philosophy are practices, but the text of philosophy is primarily concerned with ideas that can be articulated, not instructions on how to arrive personally at an awareness of what cannot be articulated. Also, a lot of philosophy is concerned with other things. This is especially true today as linguistic analysis and symbolic logic dominate the field. Moreover, much philosophy is inaccessible to many people. The arts are also concerned with the ruptures of human existence, they are full of ideas, and they are practices; indeed, when one performs them or attends great performances, these arts are very close to religious experience in their effect. Still, there is something about religion that is more completely centered on contemplating the rupture—perhaps it is because no end product (canvas, performance, or text) is expected or construed as the central point of the adventure. With religion, the point of the exercise is enlightenment; it is to teach us to live, well and wide awake, in our strange place between meaning and meaninglessness. Great doubters are concerned with this same area: they seek to understand the schism between humanness and the universe, and they very frequently do it through acts—rituals, meditations, life choices—as well as ideas.

The great doubters and believers have been preoccupied with another great schism: the one between what human beings are and what we wish we were, what we do and what we understand. That we love, and that love, among other possibilities, brings forth life, is very strange. We cannot say it is inexplicable, and yet, when it happens (either true love, or conception, or both) we stand amazed. Love can drastically alter a rational person's worldview. The birth of a child can bring extraordinarily religious feelings—because it is such a good thing, but also because it makes no real sense. Where did this miniature human being come from? Technically, we made it out of nine months' worth of French toast, salad, and lamb chops. Technically, our bodies hold tiny little instructions for how to build human eyes, a language center in a human brain, and a human spirit—fussy, joyful, or otherwise. But how strange that such a thing as *fussy* exists and is created thusly.

The fact that the human heart so often disagrees with and disobeys the human brain also seems to demand explanation. We feel "possessed" when we love someone we did not intend to, and when we are in great heights of artistic creation, and when we are acting with unusual honor or surprising deceit. In a similar way, human beings find it difficult to credit themselves with owning the virtues, since we lose touch with them so often. It would seem rational that any creature capable of feeling, contemplating, and praising kindness would in fact be extraordinarily kind, but we are not. We may strive for true altruism, pure love, and total clarity, yet we cannot possess these ultimate virtues; for some, this suggests that the ultimate virtues exist elsewhere.

The terms that we use to define God tend to be descriptions of the ruptures between human beings and the universe: meaning, purpose, infinity, and eternity. The terms that we use to describe the personality of God tend to be descriptions of the ruptures between our real selves and our potential selves: honesty, kindness, love, and compassion. Great doubters have been as profoundly invested in these questions as have great believers, and they have offered a bounty of answers, addressing not only what we might believe, but also how we might achieve this belief through study and practice, and how we ought to live. Without God to answer the question of virtue, some have taken on extraordinary codes of morality themselves, as the only way left to solve the breach between what we are and what we wish to be. The history of doubt is not only a history of the denial of God; it is also a history of those who have grappled with the religious questions and found the possibility of other answers.

PATTERNS OF DOUBT

Doubting the existence of God or some ultimate power or divinely mandated code is a very private experience, but it has everything to do with the larger community, and we can describe some loose relationships between certain types of communities and certain kinds of religious doubt. We start where belief starts: in a relatively isolated group of people, concerned with a very local religious world. In this locally oriented and homogeneous culture, religion and science are essentially the same thing, or are at least fully compatible—early on in ancient Greece, for example. Where everyone seems to believe the same thing, doubt is calm: when scientists or philosophers begin to question religious lore, they do so from within the religion, merely trying to get it all correct. The best religious minds help to question the specifics without hostility to the old version of things. Over time, vibrant details of the cult may fade from attention without much breach or upheaval. What was understood as history and science is increasingly seen as allegory.

Even when the walls of this bedrock-belief culture are worn down to the point of being out of sight, they still effortlessly hold the place together. For a while, the citizens' very personalities are held together by the massively stable and integrated culture, such that they do not fracture and become self-reflexive to the point of distress. Doubters who develop here tend to be more interested in what they have found than what they have lost. These figures are not howling in the abyss of the night; they're out there measuring the stars. They love thinking about the logic of the machine of the world: they're impressed with it and impressed with themselves for figuring it out. Generally speaking, marveling at the mechanism is regarded as a sufficient replacement for faith.

The second model is a heterogeneous or cosmopolitan culture—now, our average citizen belongs to a particular group within the community or is even from some other place. By peaceful trade or hostile clash or general upheaval, interaction between small groups has led to one big group. Alexander the Great mixed the Greeks and the Persians, but this is also a shift one person can make alone: from a village in the old country to the streets of New York City. The Hellenistic Age that Alexander started was one of the great cosmopolitan worlds, as was the Roman Peace, the golden age of Baghdad in the Middle Ages, the Tang dynasty in China, Europe during the Renaissance, and our own whirling modernism. They all experienced a massive mixing of peoples and cultures, and they all produced terrific cosmopolitan doubt. If my ostensibly universal God demands rest on a different day than your ostensibly universal God, we are both going to notice the glitch and wonder who's got it right, if anyone. So difference alone leads to a more questioning, critical attitude toward received truths, i.e., truths that have tradition as their primary proof or source of authority. But it is more than that: the heterogeneous society results from, and leads to, a shakeup of cultural constraints, so that eventually nothing feels unified and integrated. You speak a different language at home; or you have moved several times; they teach relativism in school; technology has proceeded beyond your skills; you raise your children differently than your parents did. You seek counsel from competing experts.

Yes, you are more likely to lose your faith here, but even more important, when you lose your faith here you are much more alienated, because
you were already a little adrift before you lost your God. The effect is that
religion here tends to reflect that homelessness and doubt. Also, religious
doubt becomes so widespread that worldly contests seem like the only reasonable pursuits, and people lose themselves in materialism and competition, entertainment, politics, and the marketplace. On their own these are
never fully satisfying, so alternatives, what I will call graceful-life philosophies, are devised, promulgated, and followed in large numbers. The message of such graceful-life philosophies tends to be: we don't need answers
and we don't need much stuff, we just need to figure out the best way to
live. Cosmopolitan doubt is often harrowing, but it is also experienced as
amusing and empowering—these people feel savvy and free in comparison
to their forerunners. They go to the theater.

Finally, within the mixed, increasingly skeptical community, something
new arises: a committed, ardent belief, where the idea of doubt is written
into the idea of the religion. Here expressions of doubt can feel threatening
very quickly, because the feeling of lost certainty and the pain that accom-
panies it are now very well known. The moral abyss, the friendless world,
seems to be the common state of those outside the community, and they
express this in florid detail: outside the community, people swagger with
the pride of the independent but also bemoan their fate, compete like animals, abuse drugs, commit violence, and generally invite upheaval into
their lives. Those who make a belief commitment reject this and call back
to a period of unquestioned belief—but belief has grown much more self-
conscious and the group often now feels it must consciously police its
membership against doubt. There are communities who simply believe in
something, and communities who believe in "not just plain believing," and
communities who believe in "just plain believing." Doubt is experienced
very differently in these three settings, as we will begin to see in chapter 1.

The basic structure of this book is chronological, but there has been some
thematic bundling. The first four chapters follow the four heroic traditions of
doubt of the ancient world: the Greeks, the Hebrews, the East, and Rome.
These amazing foundational bursts of doubt all fall into the period between
600 BCE and 200 CE. Chapter 5 looks at Jesus and the howl of Christian
doubt: after Jesus at Gethsemane and Augustine in his garden, doubt was
never the same. This period also witnessed the first intermingling of the four
heroic doubting traditions. Chapter 6 watches doubt make a loop around the

Mediterranean in the high Middle Ages, blazing through medieval Islam, Judaism, and Christianity—right through the period commonly thought of as the age of belief. Chapter 7 covers the period that spans the European Renaissance, Reformation, and Inquisition, and we will follow a trail of doubters' trials, many of which end at the stake. Chapter 7 also includes the rise of Zen in Japan, and the meeting, in China, of two great forces of doubt: European science and Asian nontheistic religion. Doubt was becoming an international tradition, aware of its global variety, history, and heroes.

Chapter 8, about the period from 1600 to 1800, includes the scientific revolution with its huge cosmological shift, the scandalous Libertines, the Enlightenment deists and materialists, and some wild doubters of the democratic revolutions. From Newton, Galileo, and Spinoza to Robespierre, Thomas Paine, and Thomas Jefferson is a particularly fascinating journey in the history of doubt—and that's only what went on in the foreground. Untold women and men left a variety of plucky, tortured, pleased, contemplative, or angry records of their doubt in religion, in the afterlife, and in God. Chapter 9 covers the nineteenth century, a period hugely vocal about its doubt; indeed, in this period many people came to believe that atheism was inherently better than religion and that the movement of humanity should be in this direction. The sentiment was a new one and it was widely welcomed: Karl Marx, Sigmund Freud, and Friedrich Nietzsche are only the best known of that opinion. It is worth noting at the outset that the claim that atheism should someday take over the world was born in the nineteenth century because of a specific set of historical circumstances. In all other periods, doubters tended to treat the community of doubt as city-dwellers treat the city: they live there, for better or worse, and they know a lot of country people hate it, but for themselves, they are either stuck there or they love it, but in any case, they do not expect everyone to move there—though it sometimes seems like everybody wants to. I will explain where the idea of an evangelical atheism came from, why it made sense at the time, and how much of the history of doubt it obscures. Finally, chapter 10 will assess the twentieth century and our present times, charting doubt's recent heroes—the doubting poet Georgia Douglas Johnson in the Harlem Renaissance, for instance, and Thomas Edison sounding off against the afterlife to the *New York Times*—and following its silent masses. Doubt twists into new shapes in the horror of the Holocaust. Also, in the twentieth century, Eastern atheism expands remarkably in the West while Western atheism spreads in a great red wave over areas of the East.

From its beginnings to the present, doubt can be identified in seven categories. Two of the earliest were science (materialism and rationalism) and nontheistic transcendence programs (often religions without gods). The next three showed up early as well. There was cosmopolitan relativism as soon as people mixed and as a result began taking their own traditions with a grain of salt and, also, a political need arose for public secularism and tolerance. The moral rejection of injustice, like the doubt of Job and other survivors, appears at the same time, and so do the first graceful-life philosophies, which present themselves as guides for life without belief. Philosophical skepticism, which questions our ability to know the world at all, including our ability to claim God's existence, starts with Socratic questioning but really gets going after the ancient Greeks acquire a multitude of philosophies and that great variety makes some people reject them all. Finally, there is the doubt of the ardent believer. All of these histories are remarkably intertwined.

I wrote this book because as I have studied history I have always noticed doubt, out of the corner of my eye. I came to believe I knew a story that most people did not know. Most scholars, I thought, did know all the elements of the story but did not think of it as a distinct history. Through researching and writing the book, though, I've been surprised to discover that scholars of every period have found doubters and that when these are looked at head on, a strong, cohesive history appears. The names and movements that appear in the first chapters of this book—Socrates, Epicurus, Skepticism, Stoicism, and Diogenes the Cynic; Job and Ecclesiastes, the Jews who fought against the Maccabees; the Buddha, the Carvaka, and Zen; Cicero, Lucretius, and Sextus Empiricus—would be on the lips of doubters ever after. Later, medieval doubters of the Islamic, Jewish, and Christian traditions will be added to these names in various writings about the world's doubters. In fact, throughout the centuries many people have written histories of doubt. Some of doubt's schools of thought, heroes of resistance, and witty anecdotes were remembered, forgotten, and remembered again, across centuries. The great traditions of global doubt cited the existence of one another long before they knew much detail: in their argument against universal belief in God, doubting Christians noted rumors that Confucianism was an atheist religion; doubting Jews cited Aristotle's idea that the world was eternal and thus not created by a Creator; doubters in China embraced the scientific rationalism of the West as soon as they got wind of it—from the Jesuits. The explicit idea that there *is* a history of doubt, one of great antiquity and global expanse, bloomed in and out of common knowledge.

A few things about religion become visible from the history of doubt. One is that there was belief before there was doubt, but only after there was a culture of doubt could there be the kind of active believing that is at the center of modern faiths. Until the Greeks filled libraries with skepticism and secularism, no one ever thought of having a religion where the central active gesture was to believe. Another is that doubt has inspired religion in every age: from Plato, to Augustine, to Descartes, to Pascal, religion has defined itself through doubt's questions. Of course, this extends up to today.

Doubters have been remarkably productive, for the obvious reason that they have a tendency toward investigation and, also, are often drawn to invest their own days with meaning. Many scientists and doctors have been doubters of religious dogma, including the physicist Galileo Galilee, the Jewish theorist and doctor Maimonides, the Muslim philosopher and doctor Abu Bakr al-Razi, and the physicist Marie Curie. Sometimes scientific methodology causes doubt by its example of questions and proof, sometimes doubters are drawn to the sciences, sometimes both. Many ethicists and theorists of democracy, freedom of speech, and equality have also been doubters; in the modern period alone, these include Thomas Jefferson, John Stuart Mill and Harriet Taylor, Frederick Douglass and Susan B. Anthony. Great poets, too, from Lucretius and Ovid to John Keats and Emily Dickinson, have often written because they doubted God and an afterlife and had to work out the question with diligence.

The earliest doubt on historical record was twenty-six hundred years ago, which makes doubt older than most faiths. Faith can be a wonderful thing, but it is not the only wonderful thing. Doubt has been just as vibrant in its prescriptions for a good life, and just as passionate for the truth. By many standards, it has had tremendous success. This is its story.

ONE

Whatever Happened to Zeus and Hera?, 600 BCE–1 CE

Greek Doubt

When we look for doubt among the ancients, in the West we are going to find the most lively cases in the Hellenistic period—the few hundred years between the dominance of Classical Greece and that of Classical Rome.[1] It's not surprising that an in-between period is our main focus: human beings define which are the pinnacle moments of history and which are the in-between moments, and we tend to choose moments of certainty as pinnacles. We praise and envy the certainty, dedication, and meaningfulness of such moments, whether we look at ancient Greece or at a small town in early America. In our modern lives, many of us actively cultivate our differences from these unified communities, in defense of privacy and autonomy. Yet we tend to laud them and long for them, because the ideal members of these societies seem to have been so well nourished by them; intellectually and emotionally, they do not seem bereft. We moderns can't cotton to the constraints and gross inequalities—ideal membership is usually limited, having to do with gender, heredity, and/or wealth—but we marvel at the general ideas of the group, at the rich and jubilant belonging, and at the ideal members' noble and satisfying engagement in civic affairs. Our quickest shorthand for the past is a list of these highly principled moments, their breakdown, and the birth of the next. So the history of doubt looks different than other histories, because it highlights what goes on between periods of certainty: it's like seeing a map upside down—it takes time for the new contours to take shape. The history of being awake to certain contradictions of our condition is the negative image of the history of certainty.

Hence, while usual histories of the ancient world would linger on the certainty of Classical Greece and then rush through its dissolution over the next few hundred years, I will briefly discuss Greek piety and then linger on the budding of doubt at the end of the Classical age and its blooming in the Hellenistic period that followed.

In the heyday of the ancient Greek polis, or city-state, the gods oversaw a very well integrated society. Although every society has some sense of itself as old, as having seen a lot, this was a society with a primary relationship to its religious ideas, and the strength of each of the many poleis had a lot to do with this primary certainty, this lack of doubt. Ideally, you lived for the polis, you worshiped its particular gods, you knew most fellow members by face, and you took part in its governance and defense. It was the central object of identity, politics, and religion. It was an identity that was bigger than the self and bigger than the family. It was often uncomfortable for people to subordinate themselves thusly, but they were extraordinarily well nurtured in doing so.

The polis assuaged confusion and doubt because it was something midway between the world of humanness and the universe at large, and could serve as a shelter. If humanity's central existential difficulty comes from the fact that we have humanness—consciousness, hopes, dreams, loneliness, shame, plans, memory, a sense of fairness, love—and the universe does not, that means that we are constantly trying to wrangle our needs out of a universe that does not tend in such directions. The polis expanded humanness so it seemed longer-lived and larger. The aim of each person's life is to do his or her part in the polis, to serve in a given capacity, to worship the gods of the polis, to fight, to procreate, to keep the thing going.

The Olympian gods were not very remote from humanity. They hadn't created human beings. They were immortal but not eternal. They were often heroic, but they were not particularly honorable in their dealings with one another or with human beings. They were imminent in human life and in the environment: they brought meaningful dreams to sleepers and threw thunderbolts when they were angry. They even lived nearby, on Mount Olympus. They also gave an external cause for human inconsistency or illogic, such as the mystery of why certain people find each other attractive and lovable—as if struck by an arrow. Along with the gods, there were the even more immediate *daemons,* vaguely drawn embodiments of occult power. Sometimes they were doing a god's bidding; at other times they were described as the enacting force of the moment, animating someone to heroism, great speed, or tragic error.

At the height of their cult, the Olympic gods of the Greeks were thought of as very real—not at all the equivalent of parables or half-believed fairy tales. The sun did rise every day, it was indeed the source of all life, it was perfectly consistent in its behavior, and its rising and setting was a vision of spectacular beauty. If we call immense, nonhuman power gods or God, then it is purely descriptive to say that the god Apollo drives his chariot across the sky every day, and perfectly appropriate to express awe at the sight of it. It may be a bit less obvious that Eros is a purely descriptive personification of erotic love, because we don't believe that erotic love exists as a thing outside of human beings. Yet passion can seem to hit us from the outside, and that's how the Greeks saw it.

The great authorities of the culture were Homer and Hesiod, poets who had crafted wonderful praise poems detailing the historical adventures of the gods. In these stories, people were driven in and out of wars, friendships, and adventures because of the whims or ardent desires of gods. Everyone knew these stories, and for centuries upon centuries the lives of ordinary Greeks were interpreted within this engaging and satisfying, if also disturbing, context. As such, ordinary lives generated more evidence for these gods. Life was organized around the gods' rituals and when one participated in a given ritual one experienced the god. Predictions that emerged from dreams, omens, and oracular prophecy came true often enough to feel like evidence. More generally, a trick of light might be interpreted as a fleeting vision, and might subsequently grow more solid, just as rationalists might identify a visionlike image as a trick of light and allow it to grow less strange in memory. For a long time and for most people, it would have been absurd to question the existence of the gods. They were an obvious part of the world; invisible but made apparent by the authority of the poets, the phenomena of the natural world and the heavens, the experience of their worship, and occasional dreams and visions.

Under the gaze of philosophy, this level of belief eroded rather dramatically along three major lines: some people started discussing how the universe actually worked, some people started questioning the reasonableness of the gods' biographies, and some posited a whole other world of meaning that did not rely on the gods in any important way.

THE MECHANISM OF THE UNIVERSE

In the second half of the sixth century BCE, the first Western philosophers were arguing in Ionia. They are the "pre-Socratics"—the philosophers that came before Socrates, Plato, and Aristotle—and what sets this new type of thought apart is that it is an attempt to explain the universe by thinking it through rather than relying on handed-down tradition. Thus the birth of philosophy is, in itself, one of the origins of doubt—when empirical, rational thinking becomes a goal unto itself, that means people have developed a system for checking whether an idea has a foundation outside plain faith. This sort of checking keeps valuing those ideas that have a demonstrable foundation and scuttling the rest. The very behavior gets one into the habit of devaluing beliefs that have no describable, rational foundation.

Thales was the first philosopher in the West. He predicted an eclipse of the sun in 585 BCE, which sets a date for us and indicates his skill as an astronomer. His idea was that the universe was made of one substance, and he thought water the likely candidate. Aristotle is one of our best sources of information on early philosophy (Thales left no writings), and he informs us that Thales also held that "All things are full of gods." Aristotle also explained that Thales believed a magnet had a soul since it can move iron, and Aristotle supposed that this was what Thales had meant when he said that "soul is diffused throughout the whole universe," meaning that the forces that were the gods were very much like the magnetic force.[2] Following Aristotle, some modern scholars have held that Thales' use of the word *soul* was purely naturalistic.[3]

Thales' student Anaximandros was the first philosopher for whom we have any detail, and he explained the world without reference to gods. In his description, human beings were at the center of a profoundly interconnected universe that continues its cycles without the nudging of any gods. It was a tremendous leap into rationalism. But the very movement of the universe, and the precision of its cycles, suggested to Anaximandros that the world was somehow guided. Anaximandros recognized that life on earth seems to undergo constant change—from day to night, life to death, summer to winter—but he saw a constancy behind all this flux, and named that constancy divine. So philosophy overthrew the gods right away. It also spoke of a single God right away, but this philosophical God was very conceptual, as much a matter of physics as metaphysics.

Heraclitus, another of the great pre-Socratic philosophers (he lived from about 535 to 475 BCE), was thinking along similar lines when he said that you can't step in the same river twice. The universe and the beings in it seemed to be part of a single world order that is one unified, but constantly shifting, divinity. God was a force and that force was fire. "This world order, the same for all, no one of gods or men has made, but it always was and is and shall be: an ever living fire, kindling in measures and going out in measures." All life and matter are the same force manifesting itself in a variety of ways—and that's what God is, that's his full description. "God is day and night, winter and summer; war and peace, satiety and hunger; he takes various shapes just as fire does." Was this force a God in the same way Zeus was a God? Could it be Zeus? Heraclitus specified that it was "both unwilling and willing to be called by the name of Zeus." There were reasons, not least social and emotional, for maintaining continuity with the past, but Heraclitus did not offer any logical justification for doing so. Heraclitus laughed dismissively at rites that have people "purify themselves by defiling themselves with other blood"—it was like using mud to wash off mud.[4] He also laughed at the idea that human beings need or get any help from daemons, either in ruining their lives or in getting what they want, writing that "character is for man his daemon."[5] In a similar spirit of jaunty irreverence, the poet Kinesias and his friends hosted an impiety club, meeting for feasts on unlucky days.[6]

Right at the beginning of philosophy, then, the original Greek pantheon was put into deep doubt, in favor of an essentially empirical world. This reinterpreted world was partly understood as a translation of the old gods into more believable terms. As the Greeks had once marveled at the deeds of the gods, now they would marvel at the well-ordered cosmos. Piety could thus continue in much the same way it had in the past. But there was no longer a reason to think of a god as having personality, of being an emotive creature who is in any way interested in us. Religion survived the philosophers, but not intact. As the great historian of Greek religion Walter Burkert has written:

> Only anthropomorphism proved to be a fetter which had to be cast away. . . . In place of the beholding of festivals of gods there is the beholding of the well-ordered cosmos of things that are, still called by the same word, *theoria*. . . . And yet the reciprocity of *charis* [grace] was missing. Who could still say that the divine cares for man, for the individual man? Here a wound was opened in practical religion which would never close again.[7]

Philosophers could conjure proofs of God, but it was God redefined by rationality, without personality, without interest in humanity, and as Burkert says, without grace.

Where Heraclitus saw change everywhere, a constant spark and sputter in the fire of the world, Parmenides of Elea argued the opposite, namely, that change is all a matter of perception—constancy is the rule. Yet, coming from this other direction Parmenides also found doubt and ambivalence. He wrote in the form of a long poem, the first half of which was a description of the universe as unified and unchanging. It did not include God. In the second part of the poem, however, Parmenides offers the universe based on "the opinions of the mortals." Here there are gods: a central female creator god and other gods with more specific portfolios. So, for Parmenides the ultimate reality of the universe is simply the stable fact of being—and this doesn't require God or gods. Humanity, however, needed some kind of theistic religion to explain the universe within the context of its human experience.

All that has survived of Protagoras's *Concerning the Gods* is the first sentence, but it packs a punch. "About the gods I cannot say either that they are or that they are not, nor how they are constituted in shape; for there is much which prevents knowledge, the unclarity of the subject and the shortness of life." It seems Protagoras was indicted for the blasphemy of this book, that he escaped before his trial and drowned while trying to cross the sea to Sicily. We wish we knew more, but what we do know is that an attempt to offer evidence for his theories of the universe threw Protagoras up against the chief obstacles of rational belief in God or gods: we don't know who or what we are looking for and we do not have much time to observe the universe before we die. Protagoras's claim suggests that nothing available to humanity could serve as trustworthy or sufficient proof of the gods' existence; not tradition, nor experience, nor contemplation.

THIS SEEMS UNLIKELY

Pantheon religions have some very attractive traits, but the simple fact that gods are *many* leads to their being identified by somewhat distinct personalities, and that often means they have weaknesses, vices, and bad habits mixed in with the good. The Greek gods were at times lascivious, jealous, scheming, and cruel. Zeus slept around, Hera had a nasty sense of revenge, and few of the gods would scruple against harming human beings. Contemporaneous with the first pre-Socratics, in the sixth century BCE, the poet Xenophanes of Colophon (570–475 BCE) began to criticize the actions of the Olympians—not as a scold, but because he thought that these gods couldn't really exist. Xenophanes complained that "Homer and Hesiod have attributed to the gods everything that is a shame and reproach among men, stealing and committing adultery and deceiving each other."[8] A dedication to rational thought is part of the "this seems unlikely" phenomenon, but it is as often a matter of rationalist history, or linguistics, as it is a matter of natural science.

In the light of knowledge of other cultures, Xenophanes began to feel that the idea of these gods was sort of silly, not just because they acted childishly, but also because they were so very Greek, so much created in the image of Hellenic society. With this critique, Xenophanes began the great tradition of trying to imagine where the idea of gods came from, famously claiming that if oxen and horses and lions could paint, they would depict the gods in their own image.[9] He also noted that Ethiopians describe the gods as black and flat-nosed, while the Thracians picture them with blue eyes and red hair.[10] This cosmopolitan metaphor about oxen gods and redheaded Thracian gods will be adored through the history of doubt. Xenophanes posited that the Olympian gods were nonexistent, but he replaced them with what seemed to him to be a more satisfying deistic conception: one God. This one god was like the one imagined by Anaximandros, but Xenophanes added the notion that the God functions through mind (*nous*); that the universe is guided by mind. This God doesn't look human, and doesn't have a smattering of traits and abilities, but rather "sees as a whole, perceives as a whole, hears as a whole," and, without moving, moves whatever it wants. Xenophanes described this God as dignified, fixed in place, since "it is not fitting" for God to be running from here to there. This God is not gendered and not subject to desires and needs. Because Xenophanes was more a poet, and performer of poetry, than he was a philosopher, he's had a much larger audience than Anaximandros. Religion persisted in its traditional forms, but Xenophanes' ideas were wildly popular.

It has been argued that with these ideas Xenophanes produced the first theology—rational thinking about what God must be like. Also, the rise of this type of critique is often told as a story of the development from a pantheon of anthropomorphic gods to transcendent monotheism. But this is also a story of how gods get put into doubt. Questioning a pantheon leads to monotheism in some cases, but it also leads to rational secularism.

Also working in the middle of the fifth century BCE, just prior to Socrates, Prodicus of Ceos tried to figure out how human beings "learned" the names of the gods. Prodicus was a Sophist philosopher and his method of argument was essentially the linguistic investigations of secular history. He noticed that Homer sometimes substituted the name Hephaestus for the word *fire*. He also noticed that heavenly objects were named the same as the gods, but these, too, were not actually the same as the anthropomorphic gods of the poets. His conclusion was that early human beings worshiped those things that kept them alive, things that gave them light and food, water and warmth. These were the first gods, he guessed, and they were named for their function. The rest of the gods had been individual human beings who gave instruction in farming or production. They came to be revered as gods and goddesses associated with what they had discovered or popularized: Dionysus had provided wine, for example; Demeter brought corn. The fact that Prodicus questioned the origins of the gods does not mean that he questioned the existence of the gods. The Greek gods were always understood as having come into existence at some point, so it's possible that he was trying to puzzle out the truth about them without really questioning them as an absolute reality. His critique, however, seemed sufficient for his contemporaries: ancient accounts of Prodicus take it for granted that he denied the existence of gods, and they later classed him among their famous *atheoseis*.

Democritus of Abdera made a similar hypothesis. People must have invented the gods because they were frightened and excited by what went on in the sky—shooting stars, eclipses, thunder and lightning. At the same time, he continued, people were struck by the precise regularity of the movements of the heavens, and this also gave them cause to admire whatever was controlling these movements. It seemed reasonable to Democritus that such fear and admiration led to anthropomorphized worship: "Some of the men who were able to say something stretched out their hands thither where we Greeks now speak of 'air' and thus they called the whole 'Zeus' and they said: he knows everything, he gives and takes, he is king of everything."[11] That comment that "we Greeks now speak of 'air'" is the heart of his thesis: now that we've got critical thinking, we can see the obvious. He did not seem overly surprised or moved by the conclusion.

Democritus was the founder of the idea of atomism, which suggested that everything in the universe is made of atoms. That is, Democritus suggested that there was some "smallest thing" of which everything else was constructed. The theory can seem less than rationalist since the ancients could not get physical evidence of atoms. It is based on experiment, but of a conceptual rather than an empirical nature. The logic of atomism is that on the human scale of time and space we can see that things grow and decay, a phenomenon that suggests something is being added and taken away, i.e., that solid objects flow. The apple tree is bare in winter, then there is a flower, a fruit, and then the fruit shrivels, disintegrates, and entirely disappears. The fruit morphed into reality and morphed out again, and when you think about it, this is true for everything and everyone in the universe. Thus, everything must be made of some *smallest thing* that teams up with other smallest things to form one object and then another—as sand morphs into dunes and castles. You can get to the same conclusion by cutting any object, a herring for instance, into smaller and smaller pieces, all the way to the smallest piece you can make and then ask yourself, Is this smallest piece still inherently herring? If not, you've invented atomic theory. Democritus essentially guessed how the universe works, in a manner of speaking, because it made sense. It was a stunning insight.

Democritus's atoms fell into an orderly pattern by chance, but he explained that once a pattern is established, the progress of things is not entirely accidental. The orderly pattern allows us to make predictions about the way things will behave and interact. So, not only could Democritus claim that the frightening sky no longer frightened and was no longer so mysterious that it had to be personified, but he could also assert that the beautiful regularity of the universe was neither created nor maintained by the guiding intelligence of a god. Democritus also addressed the emotional-experiential aspect of belief in the divine, best summed up by the question "Well, if there are no gods, why do so many people have religious experiences?" His answer was that when the gods, or suggestions of the gods, show up in dreams and visions, it is because the universe does in fact have some sort of population of phantoms. Democritus described these phantoms as possibly being some effect of atomic behavior, that is, purely natural. There was something that was like the gods, but it wasn't really the gods.

The poet Diagoras of Melos was perhaps the most famous atheist of the fifth century. Although he did not write about atheism, anecdotes about his unbelief suggest it was self-confident, almost teasing, and very public. He revealed the secret rituals of the Eleusinian mystery religion to everyone and "thus made them ordinary," that is, he purposefully demystified a cherished secret rite, apparently to provoke his contemporaries into thought. In another famous story, a friend pointed out an expensive display of votive gifts and said, "You think the gods have no care for man? Why, you can see from all these votive pictures here how many people have escaped the fury of storms at sea by praying to the gods who have brought them safe to harbor." To which Diagoras replied, "Yes, indeed, but where are the pictures of all those who suffered shipwreck and perished in the waves?" A good question. Diagoras was indicted for profaning the mysteries, but escaped. A search was put out for him throughout the Athenian empire, which indicates that the charges were serious, but he was not found.

The reason he was indicted for profaning the mysteries was that nothing broader was on the books. The philosopher Anaxagoras is the earliest historical figure to have been indicted for atheism—in fact, it seems they wrote the law just for him. A meteorite had fallen in 467 BCE and it convinced Anaxagoras that the heavenly bodies, including Helios, the sun, were just glowing lumps of metal. Other people had this information—the meteorite didn't fall in Anaxagoras's backyard—but he was a philosopher and a rationalist and he came to conclusions that were not attractive to everyone. This was the origin of a conflict between religion and science. Here, new information, new empirical data, led to a direct challenge to the way in which the gods were envisioned. This new doubt encouraged a new kind of punishment for doubt. Set up about 438 BCE, the law against Anaxagoras's atheism held that society must "denounce those who do not believe in the divine beings or who teach doctrines about things in the sky."

While Anaxagoras had gone too far for some in talking about the sun as a hot rock, his other ideas doubting the pantheon were too common to be very shocking. At about the same time, Thucydides (460/55–400 BCE) was writing his secular history of the Peloponnesian War and gods did not intervene in the drama. By now, educated people commonly held that traditional belief in the Olympic gods had been fully discredited, and that the most compelling understanding of God was the universe-mind idea of some philosophers. The poet Empedocles wrote that the gods should not be imagined as of human form but rather as "sacred, unspeakably rich thinking," and "swift thoughts which storm through the entire cosmos." Even the believers did not equate God with anything like a personality or even a mystical meaning for humanity—some of Anaxagoras's students equated the universe's mind with Air and others then equated the Air with God. But some of his students continued in the master's way of thinking, referring to the mind of the universe without theistic interpretation at all.

THE OTHER WORLD

Socrates challenged every last conception of life as he knew it, even the idea of having a conception of it. Piety, materialism, hunger for power, and competition were particular targets because of how they distracted people from reality. One must devote oneself to figuring out that one must live for the good, for its own sake. It was a secular morality. Contemporaries did not know what to call a thing like that—he questioned their every faith, their every way of life—so they called it atheism.

Socrates was indicted for atheism, but the wording of the indictment suggests that even his accusers did not think him particularly atheistic, just disruptive and antitraditionalist. From Plato and others we know that Socrates respected the traditional rites of piety. At his famous trial, Socrates responded to his accuser by asking if the idea was that he didn't believe in any gods at all. He went on to say that it had been an oracle that told him to become a philosopher, and he would not have risked the unpleasant social and economic consequences of such a life if he had not believed in the divine origins of the command. There is no reason to doubt his honesty: Socrates wasn't fighting for his life—he would not have been given the hemlock had he merely agreed to stop teaching. Still, there is a reason he was accused of atheism and it has everything to do with his chief claim: that he knew nothing and yet was wiser than most, since at least he knew that he knew nothing. Socrates counts among those great minds who actually cultivated doubt in the name of truth. The Socratic method is an eternal questioning. This is not relativism; there is truth to be found, but human beings may best approach it through doubt rather than conviction.

It was in 399 BCE that Socrates defended himself against the charge, and Plato's description, in the *Phaedo*, of his final day has been a model of a cool philosophical death scene ever since: he comforted his friends and family; sent away anyone weeping too much; joked; and reminded one sad friend that they were all going to die, so there was no reason to be upset for him in particular.

Just at the end he told of an afterlife in which those who have "purified themselves sufficiently by philosophy" live on in ethereal grace, but added, "Of course, no reasonable man ought to insist that the facts are exactly as I have described them." His story, he explained, was "a reasonable contention and a belief worth risking" because it inspires us to be brave. Then he bathed to save the women the trouble of washing his corpse, and drank the hemlock.[12]

In Aristophanes' *The Clouds*, written during Socrates' life, a philosopher called Socrates operates a "Thinkery" and dismisses the gods. His neighbor, an old farmer, asks who makes it rain if there is no Zeus. Socrates answers that the clouds make it rain. If it were Zeus, he could "drizzle in an empty sky, while the clouds were on vacation," but that never happens.[13] When full of water, the clouds rumble like an over-full belly. The farmer asks if it is Zeus that moves the clouds at least and Socrates says, "Not Zeus, idiot. The Convection-principle!" The farmer replies, "Convection. That's a new one. Just think, So Zeus is out and convection-principle's in. Tch, tch." Still, he asks, lightning must be Zeus striking liars? Socrates says no, think of all the un-struck liars you know, and notice that lightning often strikes Zeus's own temples! Then he demonstrates a large model of the universe according to the convection theory. The farmer is convinced. Later, though, he returns to the old gods—and burns Socrates and his fellow atheists alive in their home. Eerie, given how things turned out, but there were other plays at the time where atheism was proposed throughout and punished in the final scene. In a lost drama by Euripides, the protagonist comes to the conclusion that there are no gods, since evil is rewarded and the faithful suffer. He rides his winged horse up to the sky to get a better look, and ends by falling into madness.

We can take Socrates' death, at the hands of a democratic Athenian government, as a signpost of the decline of the great Greek poleis. Athens, which had become a democracy about 510 BCE, had a period of splendor that began with military victory in 490 BCE and ended with military defeat in 404 BCE. Athens had been greedy, had made enemies of too many neighboring peoples, who finally allied themselves against her. That was the Peloponnesian War, and it took decades. Plato was born in 429 BCE and thus grew up under its misery and came of age as Athens breathed its last independent breath. He had been Socrates' student, and the execution of his brilliant teacher, combined with what felt like the mismanagement of Athenian greatness, inclined him to criticize conventional wisdom. He blasted both the traditional Greek poetry that had made characters of the gods and the by now well-established Greek philosophy that had denied the gods.

Greek piety in the period before Plato had two main aspects: ecstasy and sacrifice. You took part in ecstatic rituals of music, dance, and emotional frenzy in order to bring yourself to a divine or near-divine state. You took part in sacrifice in order to enact your humility before the larger wills, the larger hungers, that are denizens of this universe. As historian Michael

Morgan has argued, Plato advanced aspects of both these two traditions: he borrowed heavily from the ecstatic-ritual aspect of Greek religion, replacing the emotional with the cognitive—philosophy was now the route to the divine state—and he also maintained a sense of humility before the larger powers, although he did not see them in their old anthropomorphic terms.[14]

For Plato, the pre-Socratic division between naturalism and spiritualism was not real. As he put it, in a dialog in *The Laws:* "Why, my dear sir, to begin with, this party asserts that gods have no real and natural, but only an artificial being, in virtue of legal conventions, as they call them, and thus there are different gods for different places, conformably to the convention made by each group among themselves when they drew up their legislation." Then these people say that the "really and naturally laudable is one thing and the conventionally laudable quite another," and that there is "absolutely no such thing" as a real and natural right. "These, my friends, are the sayings of wise men, poets and prose writers, which find a way into the minds of youth. Hence our epidemics of youthful irreligiousness—as though there were no gods such as the law enjoins us to believe in. . . ."[15]

The philosophers, Plato tells us, assert that the gods are not real and natural but artificial. They are made up by the law—as Critias had proposed—and are different in each different place—as Xenophanes had noted. Plato is railing because what he sees as the most disruptive thing you can do to a culture has been done: the philosophers have argued that tradition was all a big mistake, that nothing is absolutely true outside the crucible of a particular culture. These ideas had created epidemics of young atheists. That was bad, because people living as if there were no gods were likely to lose the old sacred commitment to living for the community. Plato wanted truth, not just social happiness, and would sweep away everything about the Greek pantheon that he did not feel he could logically support. But that still left him with a complex sense of divinity in the universe and a certainty that human beings need and ought to have a traditional, local religion in which they can believe wholeheartedly.

Like the philosophers before him, Plato had a sense that there was some motive force in the universe. He made a similar inquiry into the question of what moves and animates individual human beings and found himself confronted with the idea of the soul. There had been some mention of souls in the Homeric poems, but the references were vague and did not suggest that the soul much outlasts the body. In the late sixth century Pythagoras of Samos brought the idea of immortal souls into prominence, and later the

Orphic mystery religion made much of it as well, so the idea of humanity as being possessed of immortal souls was in the Greek culture by the fifth century BCE. Plato found it conceptually satisfying because of the relationship between the idea of moving stars and planets in the sky and moving bodies here below. Both move and seem purposeful in a way that separates them from almost everything else: rocks don't move, wind seems random. That suggests that the motive forces of human beings and the motive forces of stars and planets are essentially the same.

We human beings have mind and we are animated. The heavens are animated and much more magnificently than we, so Plato argued that they were possessed of even more splendid mind: "Without intelligence they would never have conformed to such precise computations."[16] As for the question of immortality, just in this period the heavens were beginning to be considered eternal: a recent finding that centuries-old ancient Babylonian astronomical records matched Greek observations of the heavens suggested that the universe was eternal and stable. So by mutual analogy, that which moves purposefully has mind and is everlasting. Plato concluded that some sort of eternal intelligence animates the heavenly objects, and he was comfortable using the old gods' names for them. In his words, a legislator "should strain every nerve, as they say, to plead in support of the old traditional belief of the being of gods," using reason to buttress myth and tradition.[17]

Plato explicitly referred to the soul as a divinity. It's a good reminder that divinity here is that which is primary, self-sufficient, mobile, and alive. The heavenly objects were gods, he concluded, because they are a grander version of our souls, so they must also have a kind of care for the universe, as we do over our affairs, but on a supreme scale. Above these gods, Plato reasoned, there must be a mind that created the whole universe, including these visible gods, and that must be a creator god. Calling it *demiourgos*, he even suggested, albeit vaguely, that there is an even more remote god beyond that. Plato's schema is based on rational attempts to figure out the universe, but what seemed reasonable to his mind was very much shaped by the dominant religious vision of his time and place. The culture's specific claims made certain things—a pantheon, for example—seem more reasonable, more expectable, than they would be to someone outside that culture.

The idea of the immortal soul also made sense in Plato's epistemological theory—his inquiry into how we can know things. Epistemology is still a central issue in philosophy, and we moderns are particularly vexed with the question of how we can come to know anything outside what we already know, that is, how we can climb out of our own culture's basic assumptions, and how we can hope to see beyond our brains' basic formation. Plato's understanding of this issue also led him to believe in the soul: as an essence within us that is possessed of knowledge not gleaned in this life, but rather remembered, somehow, from the past. If one is supposed to be something beyond the fact of one's body, what could that something be other than a kind of mobile remembering? Plato understood the soul as having knowledge about mathematics as well. In this context, a life spent studying and seeking truth was the ultimate religious life. Seeking truth—whether in the realm of math and physics or psychology and metaphysics—was a life of reawakening the soul to its own self-knowing. And this self-knowing was what the soul needed in order to come into harmony with the wider company of higher divinities.

Plato's sense that this was the only religion that would hold up nowadays was thoroughgoing. The only religion that could be really believed by anyone in his time, he said, is based on belief that the stars have intelligence, and that we and they have immortal souls of some sort.[18] The more we learn—and mathematics is the queen of the soul's subjects—the more we will ascend toward self-knowledge and universal truth. This ascension is the drama of Plato's religion.

The process was further conceptualized as the theory of ideals or forms. Plato's tremendous contribution to Western religion was the otherworldliness of thought, and the theory of ideals was a big part of this otherworldliness. Plato took things to be real and true if they were intelligible. The world, then, which is constantly changing, was thus fundamentally unknowable, and could not possibly be real. Rather, it was a flowing variety that only seemed real in human time. The buildings around us feel real enough, but they come down. What is real, then, is the Form that every building shares to some degree. If something is real it is not material, not changeable, and it is intelligible. The great science of the Classical Greek period was geometry, wherein some primary axioms almost magically describe an extraordinary variety of manifestations. Plato thought everyone should study geometry because it was the most developed example of how Forms work.

The theory of ideals suggests that everything on Earth is a specific and flawed copy of an ideal model that actually exists in another reality. The doctrine can seem silly if applied to a chair—it offers a fine metaphor for representation and the problem of language (after all, how can we call each of the variety of chairs by the same name and yet be intelligible?), but it is a stretch to believe that the ideal chair actually exists somewhere. For concepts, like beauty, the theory works much more immediately. The story Plato tells to show this is of a man who falls in love with a boy for his beauty. Erotic and emotional love between a man and a boy was common and idealized in Greek culture, so moderns should not read anything intended to be negative in this pairing—it was intended to suggest perfection. In any case, the man courts the boy and at one point moves to take him. At that moment, the boy turns to the man, and the man is struck motionless by the vision of beauty. In that moment, Plato wrote, the man is sensible to this example of physical beauty as partaking in beauty in general. The boy is so beautiful that the beauty doesn't seem to have much to do with the boy at all, and sexual gratification with this one particular beauty does not seem to bring one closer to beauty itself. After having shown us the man's progression from specific beauty to general beauty, Plato takes us through the further steps that the man takes, from recognition of the true nature of physical beauty, to understanding the beauty of the soul, to beauty of knowledge, and finally—with much struggle—to knowledge of the realm of ultimate beauty, the ideal, otherworldly Form. For a lot of people throughout history, this description of the progress of wisdom has rung true.

It was the first reasoned argument for the existence of another world. Plato famously outlawed poetry, and why? Because the great poets Homer and Hesiod had sung about the unphilosophical and quite immoral gods of Olympus. Plato did not, however, mind a kind of poetry in his philosophy. Although he tried to decipher the world rationally, he also used marvelously engaging stories to illustrate his ideas.

In his "Parable of the Cave," in the *Republic,* Plato described a lesson given by Socrates wherein he imagined humanity as trapped in a cave watching shadows of animal puppets projected onto a wall. Seeing only the shadows, the people made a life's work of discussing the shadows. Socrates then imagines dragging one person up through successive realizations of reality, first taking in the facts of fire and puppets. One's eyes can bear a lake's reflections of animals before one can look up, directly at them. Each of these realizations feels like that first glimpse of fire would feel. Only slowly, with pain, and with the jettisoning of each successive conclusion, did the person eventually see the world and then finally the sun from which all of it comes. Indeed, the sun was so bright that even those who understood it to be there found it very hard to hold in view. Here's the point:

> My opinion is that in the world of knowledge the idea of good appears last of all, and is seen only with an effort; and, when seen, is also inferred to be the universal author of all things beautiful and right, parent of light and of the lord of light in this visible world, and the immediate source of reason and truth in the intellectual . . .

People have insisted for two millennia that Plato meant God when he said the good, but Plato never said anything like that. One can see the temptation, but as the great historian Etienne Gilson put it, "It should be permitted, however, to suggest that if Plato has never said that the Idea of Good is a god, the reason for it might be that he never thought of it as a god."[19] Farther along in the same chapter of the *Republic* as the parable of the cave, Socrates says that until someone has worked hard to understand reality, "he apprehends only a shadow, if anything at all, which is given by opinion and not by science; dreaming and slumbering in this life," he is dead before he ever awakens.

The chief difficulty of interpreting Plato's body of work has been that we do not know the order in which he wrote it. Because there are great differences among the various works, it is very frustrating not to be able to say which was a youthful notion, which a mature claim, and which a parting sigh. Fierce academic battles have been fought over the order of things— computer analysis is now in on it. The *Timaeus* in particular has been hotly claimed as a particularly early work and as a particularly late work, and the heat is generated because the book plays a key role in the great monotheist religions.[20] Some people want it to belong to a comparatively foolish youth and others like it as a culmination. In the *Timaeus,* Plato has the character Timaeus offer Socrates a story, full of rich detail, told as if he'd watched it all from above. The story is of how the universe was created by a fatherly God who happened upon the chaotic stuff of the universe and the Forms, then used the Forms to make the planetary gods, and then human beings, and then helped the rest of the animals to develop from them. It is worth noting both for its fairy-tale tone and because it is one of the earliest descriptions of an idea of natural animal evolution—a crucial notion in the history of doubt. For example, the "race of birds" was created out of "innocent light-minded men," who eventually grew feathers instead of hair: "Land animals came from men who had no use for philosophy and never considered the nature of the heavens . . . their fore-limbs and heads were drawn by natural affinity to the earth. . . ." As for fish, shellfish, and everything that lives in the water: "Their souls were hopelessly steeped in every kind of error" so they were made to breathe water. "These are the principles on which living creatures change and have always changed into each other, the transformation depending on the loss or gain of understanding or folly."[21]

There are parts of the *Timaeus* that sound familiarly religious, describing "the sensible God who is the image of the intellectual, the greatest, best, fairest, most perfect." Much would be made of this in the future, but such lines within the *Timaeus,* and the *Timaeus* in general, were a minor part of all that Plato had to say on the subject. In some of his writings, long taken to be among his late work, Plato softened the notion that the Forms or ideals actually existed as such, but the sense that good had an ultimate version was always at the center of his interpretation of the world. Philosophy was the work of coming closer to it—an intensely pious undertaking capable of generating religious pleasure along with deep knowledge. To the extent that we can call his study theology, it was a naturalist theology, generally much more philosophical than religious in tone. The soul was a descriptive term for that part of us that is beyond the vicissitudes of life, of coming into being and passing away, and can thus reflect upon the realm of truth, reality, permanence, and ideal Forms.

After Plato, one could speak of life after death without recourse to mystery rites or the intervention of anthropomorphic gods; what he'd offered here was a hypothesis, a theory, not a story, not something someone had heard. In the *Timaeus* that is exactly what was shared, a story told to this grown man in his youth, and this book was not particularly popular in its time. Plato's larger religious contribution to his own time was the development of the divine human soul, reaching toward the good, coming to remember itself as immortal and thus becoming so. With the loss of anthropomorphic gods, human beings were profoundly alone, but with the notion of soul, we were also godlike in our immortality, self-sufficiency, and potential wisdom.

The pre-Socratics' rejection of the pantheon was based on the unknowable nature of the universe and the unseemly behavior of the traditional gods, but those arguments only disrupt belief in a particular kind of God. Generally Plato's work was natural science rather than purely religion, because rather than calling in invented gods to be the movers in any given inexplicable phenomenon, he started from the phenomenon and made hypotheses—with simile, with analogy, but without invention. The stars do move. Why? Well, what moves down here on earth tends to have mind, so the stars probably have mind. Since the planets already bore the names of gods, it was a simple step to say that the proper description of the gods is that they are the mindful stars. Natural science and religion were thus brought so close together as to be almost indistinguishable. Individual human beings lost their vibrant, humanlike, intervening gods, but they gained a logical, rational worldview that included distant deities and transcendental possibilities.

The ancient Greek religious rites persisted in the flavor and terminology of even Plato's most rationalist descriptions of the universe—and very much in his prescription for an ideal community. This was not only because these rites were so thoroughly a part of the fabric of Plato's culture. It was also because he assumed that most people were never going to be able to take part in the more philosophical route to truth, beauty, and human comfort; the mass of people needed ritual and ceremony to help them concentrate on the good. In Plato's idealized poleis described in the *Laws,* religion was of paramount social and political importance. For the sake of both individual peace and state unity, atheism was to be a capital offense. It did not, however, take any leap of faith to believe in the state-ordained religion—he does not tout the creator God or the highest God here; the gods Plato speaks of in the *Laws* are the visible gods, the stars and planets. According to Plato, these are manifestly real and thus undeniable. Of course, in another way, the religion Plato prescribes for the polis is less real, more surface-level, than that which he conjures for philosophers, where one searches after one's inner divinity. The larger point is that both are true because no matter how materialist or rationalist your description of the world, if it also includes the possibility of transcendence, all kinds of religiosity become reasonable. If we believe in any possibility of transcendence, then we cannot quite scorn any kind of transcendent work—if ritual, ceremony, and worship function at all in this capacity, for anyone, then they are part of the work of truth and worth their effort.

So, it is the visible gods that the poleis would celebrate, and they would celebrate a great deal. Plato has festivities for one of these gods going on every day. Since these were gods of the same names as the pantheon, the old rites and rituals were to be maintained. Although these rites were full of myth and fantasy, they hid some meaning in them and, most important, they honor the gods and provide occasions for the members of the polis to come together, to know one another, to have a joyous, sometimes wild, experience, bonding them to one another. The cult of the sun would be celebrated as a double cult of Apollo and Helios—religion and natural philosophy. The people would exercise humility, but stretch toward the divine.

The sacrificial part of the Greek religion had to do with submitting to the wild chaotic world beyond one's own will; getting used to the idea that your rational plans will be knocked around by larger forces. The ecstatic-ritual part of ancient Greek religion was a kind of throwing oneself into the chaos, not pitting your rationality against the tempestuous world, but rather leaving your rationality on the shore, letting the waves toss you about, and coming to identify with the waves, with the storm, with the weather. You transcend by letting go of what is human—rationality, pride, and planning—but this is not a cold, dead universe you are submitting to. It is a universe that has had some humanity read into it: an ecstasy, a pleasure, a mind, a divinity. So far these doctrines are all relatively familiar. It's the choice we human beings have. Serious rationalism allows for self-sufficiency, but on a very small scale and with the expectation of many setbacks and the certainty of failure: we cannot win using our logic against the forces of the universe. Yet total submission to—and identification with—a larger force leaves us with no tiny will (or life) to protect and thus renders us free and eternal, although not quite ourselves anymore. What is fascinating is that Plato's solution is both logical and transcendent. Here one does not use logic to conquer chaos. Rather, one uses logic because the logic itself is beauty, is truth. Plato offers the amazing idea that contemplation of *the way things really are* is, in itself, a purifying process that can bring human beings into the only divinity there is.

Plato's mood was one among many. His relative, the Athenian Critias, wrote a play in which religion is an explicitly invented lie, made up in order to keep otherwise brutish human beings honest and law-abiding—the gods had thus been a deliberate deception. Critias had studied with Socrates, too.

Saying there was a God, thriving in deathless life,
Hearing and seeing with his mind, thinking much . . . [This God]
could hear all that mortals speak,
And have the power to see their every act.
. . . With words

Like these he introduced this most alluring doctrine,
Concealing with his lying speech the truth.[22]

The piece goes on to say that it is smart to claim that the gods live in the heavens, since that region is associated with scary things like thunder, lightning, and meteors. Critias was also political, and eventually became one of the Thirty Tyrants imposed on Athens by Sparta. He had a reputation as bloodthirsty, but was liked by Plato, who made Critias a speaker in several of his dialogues. Critias was to become the hero for those who resent it when religious claims are used to justify political power.

MORE MECHANICS AND AN UNMOVED MOVER

Plato had a great student who maintained his doctrines and became his successor as head of the Academy. His name was Speusippus, he was a nephew of Plato's, and his contributions followed along Plato's metaphysics (he hypothesized about the stages one travels in ascending toward the good). He also followed along the Socratic tradition of doubt, arguing that it is impossible to have satisfactory knowledge of anything without knowing all things—and that wasn't going to happen.

Of course, Plato's greatest student was Aristotle, who formed his own school and gave doubt a whole new path. Aristotle took Plato as his main inspiration, but he did not believe that the forms of things existed apart from the examples of things. Aristotle believed that justice and beauty were real, but not that they existed apart from examples of justice and beauty. They are not meaningless because they lack a separate existence, but they do lack it. What was real, then, was what one learned through sensation. With Aristotle's works, as with Plato's, we are unsure of the progression or dates of composition. What's more, none of Aristotle's many actual, finished books survive from the ancient world: all that remains are working drafts of ideas, his notes for his lectures, or maybe lecture notes taken by students.

Aristotle's empirical conception of the universe is important in the history of doubt because it championed rationalism. Throughout his work he called for reason, proofs, and demonstration, and he worked out the beginnings of whole disciplines showing how to do this: from marine biology to logic, political science, ethics, and psychology. His scientific studies, however, were often great jumbles of hearsay examples and thought experiments. Moreover, his philosophy always assumed that the universe made beautiful sense, that it was all going to fit together intelligibly and gracefully. So he often imposed this beautiful sense on things and did not bother to check with nature.[23]

Aristotle himself didn't make much use of the concept of God one way or another. Yet in attempting to find logical causes for all effects, he found himself stymied by the motion of the heavens. Aristotle was so convinced that things needed to be pushed on in order to move that he came up with a naturalistic theory for why a tossed ball keeps moving after it has left the hand that tossed it: the displaced air in front of the ball, he ventured, must come around back and push the ball forward (nature abhors a vacuum, so the air rushes in). A naturalist answer—not right, but rational, i.e., dependent on reasoning and evidence. The ball moved because the air moved, and the air moved because the hand pushed the ball into it, and one can keep going backward this way, tracing the force-origins of any movement, but it begins to beg the question of what started things—what is the ultimate something that is causing all this motion? We moderns don't have a clear idea about where the energy of the universe comes from either. We say it must have come out of the Big Bang with all the matter, but that's not saying much. Aristotle's conclusion was that the world was not made, it has always been here.

For Aristotle, it seemed a logical necessity that behind all the other forces one would eventually find an unmoved mover, the ultimate cause. He also assumed that the whole of philosophy would have an ultimate first truth. Just as with Plato, geometry provided the template for all subjects to someday map out their axioms—certainties from which all else can be deduced. Soon after Aristotle, about 300 BCE, Euclid took this to a new level and reinforced Plato's and Aristotle's idea that through math, the world was magically logical, and might be uncoded and understood.

Aristotle also agreed with Plato that Eros, or sexual desire, plays a catalytic role in our striving toward the good, and the unmoved mover can be understood as causing motion in the same way longing pulls us. Aristotle agreed, too, that the planets moving on their strange paths across the sky, perfectly, eternally, must be moved by something like souls. For him the souls move planets in the same way objects of desire cause movement: by their goodness. Although imaginative, the unmoved mover and the souls in the sky were philosophical ideas. Aristotle certainly treated the old myths as easily dismissible:

Our remote ancestors have handed down remnants to posterity in the form of myths, to the effect that the heavenly bodies are gods and that the divine encompasses the whole of nature. But the rest has been added by way of myth to persuade the vulgar and for the use of the laws and of expediency.[24]

Aristotle thought the unmoved mover had mind; in fact, it was nothing but thought thinking itself. Nothing he had deduced about God suggested that he cared about what human beings were up to, or even really knew that we were here. When Aristotle discussed ethics, a concerned God was occasionally considered as a possibility, but always in the conditional. Yet the atomic theory of Democritus, which required only the patterned unfolding of nature, did not appeal to Aristotle, who saw it as too random, and the universe as too full of fancy shapes, orderly events, and goodness to be thus accounted for. Despite his thoughts on the Olympian gods, Aristotle was convinced that the forms and rites of religion should continue. God may not know or love us, but it is reasonable for us to love God, since the notion of him is so high above us.

Some primary reasons that both Plato and Aristotle had for believing in God were utterly erroneous—simple errors caused by our being stuck to the planet and misled by the sensation that the planet is standing still. If they had been aware that the Earth spins, they would have understood that, by and large, we are making our own light show in the night sky. As it was, the precision of the movements of all the stars seemed astonishing. If we knew how we lined up among the planets, their motion would not seem so strange and willful. Also, had the philosophers been able to leave planet Earth for a jaunt in outer space, they could have seen that, at a distance from gravity and atmosphere, moving things tend to keep moving, without any need for an impelling force. From out there, the motion of the planets would seem natural as well.

In any case, Aristotle's conception of the universe was very mechanical and practically atheistic, and average people, yearning for some cosmic care and interaction, came to populate it with *daemons*. These ghostlike intermediaries between people and gods had been part of Greek religion from as early as Homer, but the term was vague before Plato mentioned them in a story within a story in the *Symposium*. Aristotle wrote that dreams were daemonic but was clear that he rejected supernatural descriptions of them, and he spoke of life itself as daemonic, calling attention to the awesome strangeness of consciousness—but he wasn't implying that it was divine. In the wake of Aristotle's rather materialistic discussion of the universe, however, people came to refer to daemons along the lines of Plato's story in the *Symposium:* active, invisible creatures available for supplication. Now there were evil daemons, too—the germ of our current use of the word. Average people negotiated the world through bribing certain daemons and flattering others. The problem, of course, is that this is no longer an intellectually satisfying world, nor a morally ordered one. The traditional forms of religion were retained, but they had been given new meaning, in which human life was a magical game of fear, supplication, and avoidance in a pointless world.

THE HELLENISTIC AGE AND EUHEMERUS

We conventionally date the Hellenistic period as the three hundred years between Alexander the Great's death in 323 BCE and Queen Cleopatra's death in 31 BCE.[25] The period was so suffused by doubt that its philosophers were the original Cynics and Skeptics. To many, the Hellenistic period is not as familiar as the Hellenic age of Classical Greece that came before it or the period of Ancient Rome that followed just after. Again, in a history of doubt the map of relative value is reversed. This Hellenistic period fits a cosmopolitan model of doubt, where disbelief gets very emotional and painful, savvy and wry. As we've seen, the religion of the Hellenic period was profoundly rooted in each polis, but the poleis were on their way out. During the period in which Plato and Aristotle wrote their philosophy, the individual poleis began to break down, weaken, and take each other over. The great philosophical works that characterize Greek civilization for most moderns were written at the end of the Classical age of Greece, when things were getting very shook up.

When the poleis began to lose their independent identities, it was not long before they were conquered by an outside force. Philip of Macedon conquered the Greek city-states over the course of twenty years; in less time than that, his son Alexander took Egypt, the great Persian Empire and beyond—all of western Asia as far east as modern Pakistan. These vast territories were now under one rule, and Alexander avoided revolt by encouraging his newly won territories to feel like one big, mixed culture. To energize the mixing process, he himself married a Persian princess, despite his preference for men: this new, amalgamated culture was led by a Macedonian but dominated by Greek ideas and predilections. Macedonia was a northern neighbor of Greece, nearby and much under Greek influence; Alexander, for his part, had studied with Aristotle in his youth. The Macedonian army thus helped to dismantle the Greek way of life, but it also spread Greek culture to what felt like the ends of the earth.

In this shifting new world, individuals were much less likely to have the kind of territorial, spiritual, and political home that the polis had once provided. What home they had was likely urban and urbane. Even if Alexander did not found all of the seventy cities he claimed to have founded, he was still a remarkable city builder, shifting the structure of his world from distinct city-states and monolithic empires to a sprawling cosmopolitan network of urban centers. When he died, young, in 323 BCE, no one else proved capable of controlling these vast holdings. They fell into two empires—the Ptolemaic and the Seleucid—under the command of, and named for, two of Alexander's generals. The extraordinary new cities of Alexandria in Egypt and Antioch in Syria served as their capitals, respectively. These were lively, learned places, bustling with people of conflicting traditions and widely divergent cultures of dress, food, religion, and habit. Greek civil servants, soldiers, and business people who settled across the two empires, in Asia and the Near East, brought Greek institutions and influence with them. In rather short order, a common culture was established across these wide-ranging and diverse areas. A common language, Koine, further unified the peoples across both empires.

The Hellenistic period is said to end in 31 BCE, but Rome had been growing in the background for some time, and had started to intervene in Hellenistic affairs as early as 212 BCE. Cleopatra almost managed to reassert the lost power of the Greek-Egyptian dynasty, but things were too far gone, and with the victory of Octavian it was over. Alexander's world was short-lived, yet it was the site of some astonishing cultural innovations, and the cultural life of the Hellenistic world persisted for centuries after it had been otherwise eclipsed by Rome. The innovative spirit that made Hellenistic culture so rich was born of necessity: having been a culture that valued purity, precision, and authenticity, the Greek world was now faced with a status quo of mergers and mixes, influence and collage. The transition was met with distress, but also with a developing taste for cosmopolitan pleasures and virtues.

Somewhat by coincidence, right at the beginning of this period of social and political upheaval, there was a monumental and long-lasting change in the way people understood cosmology. The new Ptolemaic cosmic order that would be established here would persist, with little or no competition, until the Copernican heliocentric revolution of the Renaissance. The dominant idea prior to the Ptolemaic system was very simple: there was earth, heaven, and underworld. The Greeks spoke of the gods living at the navel of the world with humanity tucked up near the mountainous home of the gods; sometimes they described the earth as a disk floating under an overarching heaven. Either way, humanity was located in a central, protected arena. The Ptolemaic cosmology was much more complicated.

Starting with the pre-Socratic philosophers, there were a wide range of possible scenarios, but by the third century BCE, a new consensus had been reached. Earth was still at the center, but now seven planetary (which means *wandering*) realms existed, with the Moon the closest, followed by Mercury, Venus, the Sun, Mars, Jupiter, Saturn, and then the fixed stars. Since the stars were so impressively immutable and life on Earth so busy and changing, it seemed evident that a scale of value existed—and descended toward us. Thus arose the notion of the sublunar realm, where we are, as radically distinct from the superlunar realm, where the gods can be found.

Naturalist and ethical philosophy had made the idea of personified, involved gods appear naïve—especially ones with unseemly biographical details. Instead, there were very nonhuman gods: distant in an emotional sense, and, in the new cosmology, in the physical sense as well. As often happens with distant gods, an underbrush of more quirky and personal spirits began to appear on the ground, spirits one could bribe for favors or blame for mishap. It was beneath the Moon that these daemonic powers held sway. Gods were powerful, but distant and not interested; daemons and other local spirits were on hand, but they were relatively weak, and they themselves had local desires, which might run counter to one's own wishes. The location of the gods in the new Ptolemaic cosmological model added to the general sense that human experience was not guided by any kind of larger meaning. Astronomical speculation had helped to dramatically change the way people felt about the world.

The sociopolitical world had been violently stirred up by the army generals and the spiritual world had been violently stirred up by the cosmologists. That is why this period is Hellenistic rather than Hellenic, Greek-ish rather than Greek. A vast array of varied cultures were bleeding into one another: Persian ceremonial pomp was lavished on once sober Greek politics, and arcane Egyptian cults spread throughout the upper classes of a vast cosmopolitan world. Emperor worship and the mystery religion of Isis were two of the most widely practiced cults.

The rational doubt that characterizes much thought in this period contrasts sharply with the emotional nature of the Mystery Religions, but both were imagined as havens, set against a chaotic background. The Mystery Religions were vast empirewide cults, each based on secret knowledge revealed only to initiates. The "mystery" was the secret knowledge of texts and rituals, but it was also a comment on the atmosphere. The initiation ceremonies were often accompanied by ecstatic celebration: music, dance, darkness, and wine. The wildness here reflected the larger world, which had come to seem like a chaotic mess. Only within the judicious care of the goddess—most of the Mysteries' deities were female—was there any hope of correspondence between one's actions and one's fate. Across the vast territory of the empire, the Egyptian goddess Isis was the most important of them, especially for the intellectual, urban elite. Worldwide now instead of local, Isis left her historical throne on the Nile and took up a lunar throne from which she could see much farther.

All the goddesses of the Mysteries, and several other gods and goddesses as well, were fighting the goddess Tyche, who herself was widely worshiped. Tyche means fortune, chance, and fate. She was worshiped throughout the Hellenistic period, but often with a resignation to her inconsistency that made such worship seem more like respectful terror. For Hellenistic men and women, the world was not constructed as good versus evil; it was order versus chaos.

The sophisticated urbanites who worshiped within the Mystery Religions for generations, over the course of several centuries, were still aware of the Classical Greek gods, of course. They were well aware of the philosophy of the Classical age, too. But these older descriptions of the world as well organized and sensible did not ring true—the world was too much of a muddle. Individuals whose parents or grandparents had been deeply defined by their locality were now "set free," like it or not, into what felt like a world community. What they found was more opportunity, but also more alienation. It was now rather easy to experiment with cultural mores, but it was also easy to lose one's way. In another important change, people went from being citizens, or a citizen's family member, to being subjects. Individuals in the polis had been part of something that explicitly made the whole more important than its parts, and yet that whole was still small enough that the individual could closely identify with it. Without the polis, there was a new kind of freedom and individualism, but all that meaning and sense of community was gone. In response to these shifts, two divergent attitudes arose, so that there are two primary Hellenistic images of humanity.

One was the lonely—even homeless—individual, wandering vast expanses in a vague search for meaning and belonging, and eventually ready to believe in a whole new set of anthropomorphic gods. In the great Hellenistic novel *The Golden Ass,* by Lucius Apuleius (ca. 124–ca. 170), the noble young Lucius gets into trouble in his wanderings and uses his girlfriend's magic ointment in the hope of turning himself into an owl—as she has done for herself—in order to fly out of town and escape. Trying to help, but panicking, she grabs the wrong box of potion for him and Lucius turns into an ass instead of a bird. He spends the novel getting into ridiculous situations and trying to change back into human form. Lucius is an ass in the first place because he is wandering, he is out of place, and he gets in trouble. The novel was hugely popular because its readers were uprooted migrants in a vast, multicultural empire, and, in that situation, too much freedom and too few guidelines sentenced a great many people to difficult lives. Everyone knew someone who had wandered aimlessly and ended up a jackass. The climax of the book occurs when Lucius finally finds the Isis cult and Isis turns him back into a man, declaring that on his own Lucius was lost, even with the help of philosophy, but now: "Let fortune go and fume with fury in another place; let her find some other matter to execute her cruelty. . . ."[26] Without Isis, the lonely wanderer is at the mercy of chaos.

But there was another response to the shift from the Olympian gods to the distant gods of the heavens, along with all the political and cosmological shifting. This other dominant mood in the Hellenistic period was philosophical: a clear-eyed resignation to chaos and uncertainty, and a conviction that reality, even painful reality, is preferable to living under false ideas. What looked like unbearable uncertainty to some was interesting, emancipating, and undeniably true to others. For them, the hard part was understanding how the old model of the world, with its anthropomorphic gods and moral certainty, could have ever been taken seriously by reasonable beings. It was in this climate that a Sicilian doubter named Euhemerus made a big name for himself satirizing the Olympic gods. His famous *Sacred History* was a sort of philosophical novel as well as a travel fantasy— perhaps the first ever. In it the gods had once been great heroes, had been celebrated after death, and eventually had passed into myth and deification. The concept took a lot from Prodicus, who had figured that the inventor of wine was worshiped, eventually, as Bacchus, but Euhemerus made more of a story out of the idea. In *Sacred History* he claimed to have been to the imaginary island of Panchaia and there found evidence that Zeus had been a man, a great king, yes, but he had lived and died in Crete, like any other Cretan. We've got only fragments and summaries, but *Sacred History* was hugely popular and influential in its time and long after. It was one of the first Greek books to be translated into Latin.

The biography of Alexander the Great lent credence to Euhemerus. As a youth Alexander had followed Greek ideals of the ruler: modest, unadorned, sensible. He would run alongside his chariot to keep in shape and encourage his weary foot soldiers. Conquering the Persian Empire changed him, though, and he began to adopt their "oriental" style of leadership, replete with pomp and ceremony and mysterious rites. Then he actually became a god; temples were dedicated to him and he was worshiped all over a vast empire. If rulers were now understood as gods, it was a relatively small leap to imagine that the old gods had once been rulers. That was especially true since Alexander's story so closely paralleled that of Dionysus: the god had waged an identical conquering spree in the other direction, starting in India and heading west. The gods, then, had been heroes, transformed in the local memory as a result of human affection, idolatry, and need. Again the Greek gods were always understood to have come from somewhere, but Euhemerus did seem to be grinning at a mistake rather than lauding an apotheosis.

It is difficult to know how many people agreed with Euhemerism, as it came to be called, but folks clearly enjoyed the idea. Cosmopolitan and enlightened people contrasted themselves favorably to their counterparts of the recent past—men and women who seemed to have lived under a somewhat ridiculous, infantilizing misconception.

So how were the doubters to find meaning? The Hellenistic philosophies are generally considered silver to the Hellenic gold, and judged by the originality and sometimes the sophistication of abstract ideas, there is reason for it. Where the Hellenistic philosophies excelled was the production of what could be called secular religions. They were based on self-help–oriented doctrines often borrowed from the earlier philosophers but interpreted and presented in a way that made more direct sense to a lot of people. I'm calling them graceful-life philosophies to distinguish them from other philosophy. Their goals were practical happiness, and they were not merely theoretical about it: they provided community, mediations, and events. In this they were more like religions, but they did not identify themselves as religions and they had remarkably little use for God or gods.

The Hellenistic graceful-life philosophies had a lot in common. The experience of doubt in a heterogeneous, cosmopolitan world is a bit like being lost in a forest, unendingly beckoned by a thousand possible routes. At every juncture, with every step, one is confronted with alternative paths, so that the second-guessing becomes more infuriating even than the fact of being lost. After a direction is chosen, one is constantly met with another tree in one's path. What do you do if you come from a culture that had a powerful sense of home and local value, and now you are lost in something vast and sprawling, meaningless and strange? The stronger your belief in that half-remembered home, the more likely you are to panic, to grow claustrophobic among the trees and beneath their skyless canopy. Hellenistic men and women felt a desperate desire to get out of the seemingly endless, friendless woods. The graceful-life philosophies of this period were able to achieve an amazing rescue mission for the human being lost in the woods and bone-tired of searching for home.

They did this by noticing that we could stop being lost if we were to just stop trying to get out of the forest. Instead, we could pick some blueberries, sit beneath a tree, and start describing how the sun-dappled forest floor shimmers in the breeze. The initial horror of being lost utterly disappears when you come to believe fully that there is no town out there, beyond the forest, to which you are headed. If there is no release, no going home, then this must be home, this shimmering instant replete with blueberries. Hang a sign that says HOME on a tree and you're done; just try to have a good time. Thus the cosmopolitan doubter looks back on earlier generations with bemused sympathy—they were mistaken—and looks upon believing contemporaries with real pity, as creatures scurrying through the forest, idiotically searching for a way out of the human condition. After all, it isn't so bad if you just settle in and accept a few difficult ideas from the get-go.

THE CYNICS

The word *Cynic* comes from the Greek word *kuon,* which means dog. The Cynics were called *dogs* because they proposed to live like dogs, without shame or convention. Diogenes of Sinope, though not their founder, was their great exemplar. The Cynics felt that the way people lived in civilized society was full of falsehood, emotional discomfort, and pointless striving. Yet honesty, ease, and repose were available to anyone who merely stopped lying, role-playing, and striving. Cynics wanted to live virtuously and calmly, the way the animals do, and so rejected all possessions and social forms and slept outdoors. Diogenes boasted that he performed all his physical functions without shame, like the city's dogs. When he needed shelter he would climb into one of the large earthen storage jugs located in public areas at the time.

Diogenes' rejection of all custom included a rejection of religious observance: he did not take part in any social, political, or sacred rituals.[27] He worshiped no gods. Still, it is possible that he believed in them. What he said on the question was that we should not worship the gods because gods do not need anything from us—in fact, they do not need anything at all. For the most part, though, Diogenes simply ignored the idea of gods and did not in any way include gods as part of his solution to the problem of how one ought to live—or how we ought to understand our predicament. The men and women who followed him gave up everything they had, but it was not in the context of sacrifice or humility. Rather, it was an act of freedom. In walking away from everything that the world both offered and demanded, they made decisions to arrange for their own inner well-being.

The style of their arrangement was to privilege the universe, which does not make value judgments; and does not try to do anything; and does not, for instance, feel shame about its excretions. Diogenes did not complement this with the other half of the religious impulse—to impose humanness upon the universe—so the whole thing can end up sounding less then triumphant, and certainly less than transcendental. After all, these people are the original cynics. But the act of pushing the absence of humanness to its extreme, taking it in, and living it every day, can somehow drive us into timelessness and continuity. We are the stuff of the universe, momentarily sentient, but aside from that little piece of weirdness, we are already home.

It is a startling solution to the human problem. The polis extended humanness into a larger sphere of the universe—it was longer-lived than the individual and had a sense of purpose that arched above his or her tiny, personal will. The Cynics dealt with the same problem by being less consumed with humanness. With the gods gone, the universe seemed like a dead place of violence and chance and we human beings the minuscule representatives of our own emotive fantasy. All that is left of this fantasy is what we maintain in our own civilized, cultured behavior—little creatures holding back the encroachment of meaninglessness with nothing but our body shame and our quest for accomplishment. Diogenes had essentially said, *I give up*, and he found the experience astoundingly liberating.

There is an instructive story that comes down to us: Before a large crowd, Alexander the Great approached Diogenes, who was lying in the street, sunning himself. Standing above him, the young conqueror offered the philosopher anything he wished. It was a sneaky offer, since it was both a reward for Diogenes' wisdom and a teasing effort to tempt him away from it. Diogenes said that perhaps there *was* something he would like and, after a moment, asked Alexander to please stop blocking his sun. What is wonderful about this story is not only the calm that emanates from Diogenes' lack of desires, but also the powerlessness of Alexander the Great in this situation. Here were two men who would normally be seen as decidedly unequal: both had reputation, but one had wealth beyond imagination and the other was without a single possession. But if the poor man truly has no desire for those riches, all we have is two men of reputation, one of whom exerts a tremendous amount of energy and enacts the pomp and ritual of power, while the other one relaxes in the sun. Now the rich one seems ridiculously self-important and the other looks wise and well rested. What is more, given the way of the world, the rich and powerful man seems riding for a fall, whereas the tanning philosopher seems to be in a very stable place. The ragged figure was also surrounded by friends who could only love him for his innate qualities, and who were likely to feel very comfortable with him. The richer figure was more alone, more subject to fear, chance, envy, and exhaustion.

Diogenes did not want anything, so he did not lack anything. Alexander the Great is supposed to have said, "Were I not Alexander, I would be Diogenes," such was Diogenes' independence. Diogenes, by the way, is said to have returned the compliment—saying that were he not Diogenes, he would be Alexander. The exchange brings to mind the two most evident ways of curing oneself of envy: stop wanting what other people have, or go out and get it. The first option may save a lot of time and effort, but, of course, it requires a transformation of one's emotions, or at least a control over them.

Diogenes' advice is that we stop distracting ourselves with accomplishments, accept the meaninglessness of the universe, lie down on a park bench and get some sun while we have the chance. Alexander might have answered that conquering is fun and we might as well do *that* while we have the chance. Still, some control of one's own hungers is clearly necessary for happiness, and for many people what Diogenes had proposed was attractive even in its most extreme form. It asked a lot of a person but seemed to provide great power in exchange, and men and women came from across the empires to observe and imitate his Cynical way of life. The life of a Cynic required and inspired devotion. It was a rejection of meaning and convention, but it had power to sustain and uplift its followers. To be cynical about even the things dogs love is hollow and demoralizing; to be a true Cynic leaves one a few devotions: loyalty, for instance, food, and sleep.

THE STOICS

The Stoics were named for the porch where their founder, Zeno, taught. They shared the Cynics' no-nonsense realism, engaged detachment, and their advice was to concern oneself only with what is within one's control. They were much more enamored of civilization than were the Cynics. Like the Cynics, they did not depend on reference to the gods in their general approach to life. They thought that God was the whole universe. That can sound very religious, but as it was lived by the Stoics, everything being divine was a lot like nothing being divine. The idea served to relax the tension between that which is immortal and omnipotent and that which is worldly, mortal, and weak, and the feeling was rather secular: we are here, this is our situation, there is no hidden other situation. One's task is to become inured to the pain of it.

The central idea of Stoicism was that with the loss of the polis as the center of life, the universe should be conceived of as one giant polis. If men and women acted their parts in the universe as diligently as they had in the polis, they would recreate the sense of belonging to a meaningful and relatively eternal community, larger than themselves in strength and significance. Feeling a part of the community of the universe required a lot more imagination than did belonging to the local and knowable polis, but that is what Stoicism was intended to help with. Adherents of Stoicism had no need to lament their difficult lives, since these lives were merely parts they happened to be playing: it is of little consequence whether one is cast as a queen or as a scullery maid in a play, so long as one does a good job of it. Depending on the play, the part of the servant may well be the better role. In this schema, human beings do not get to know what the play is, let alone the purpose of it, but they are clearly essential to it. The universe, its theater, holds us at its center. Since individual lives were generally held to be preset, Stoics gave a great deal of attention to the notion of fate and looked to the stars for indications of what was to come. Not surprisingly, they were great public servants, and for centuries, institutions of public service encouraged Stoicism within their ranks.

There was a certain amount of ambivalence on the question of the divine. Some Stoics conceived of God as entirely identical to nature. For them, fate or providence was actually materialist determinism; not a prewritten script, but a simple working out of the laws of nature. Other Stoics imagined a more personal God—still very much equated with nature or the universe, but one who cared for individual human beings and arranged their fates for them. Either way, this God was imagined as all good. The only evil in the world occurred when people refused to act their part. The Stoics could never quite account for the fact that the world contains so much suffering and cruelty, but their most common response was to say that these difficulties were all part of a larger and positive grand scheme. Of course, here the grand scheme was a bit mechanical; less the thinking omnipotent being, more the complicated cosmic machine. The whole point was for individuals to be in harmony with the universe; if that could be achieved, there would be no real trouble. Much that seems endemic to religion—prayer, ritual, sin, and miracles—had little to do with Stoic piety. With the Cynics, meaning was generated nowhere; with the Stoics, the universe and its human parts all participated in a general state of meaningfulness, but it was a remote one, without much sympathy for individual hopes or affections.

THE EPICUREANS

Epicurus was a fascinating character in the history of doubt. He said that not only can human beings manage to be virtuous in this chaotic, unsupervised world, they can actually be happy. In fact, there is no reason for them not to be happy. The three chief obstacles to being happy, he explained, are fear of death, fear of pain, and fear of the gods. He dealt with fear of death by arguing that death is an utterly unconscious sleep and nothing more. Death is no problem because when we are alive we are not dead and when we are dead we don't know it. So long as you can possibly worry about it, you've got nothing to worry about.

> Whatsoever causes no annoyance when it is present, causes only a groundless pain in the expectation. Death, therefore, the most awful of evils, is nothing to us, seeing that, when we are, death is not come, and, when

death is come, we are not. It is nothing, then, either to the living or to the dead, for with the living it is not and the dead exist no longer.[28]

He argued that fear of pain is foolhardy since intense pain is usually short-lived and long-term pain tends to be relatively mild, which means it is endurable. Fear of pain is worse than pain itself. Accept the pain, embrace the sting, especially in its absence, and you've vanquished your worst foe: the one in your head. Fear of the gods was more complicated, and part of our interest in this is the simple fact that he rated fear of the gods so high in his list of human problems. It is hard for us to know how much fear people actually lived under in regard to the Greek gods. We have relatively modern examples of torturous fear of God. Nevertheless, Epicurus clearly thought that it was a large fact in his own time and the massive popularity of his doctrines gives that much weight.

Epicurus answered fear of the gods by simply insisting that the gods do exist, sort of, but that they are totally unconcerned with human affairs. Epicurus agreed with Democritus's atomic theory and with his atomic theory of the pantheon: the gods were distinct but rather fluid arrangements of extremely fine atoms. Epicurus elaborated on the consensus argument: because every human group believes in some form of gods, he reasoned, something like them must exist; we must all be getting some sort of valid information through relatively normal sensory means. However, he explained, we have been mistaken in our interpretation of this sensory information. Democritus proposed a naturalist explanation for the gods, but he didn't go much further with it. Epicurus exclaimed with some delight that the gods were essentially images. He believed that these images had some substance and that human beings could witness them, especially in dreams. But despite such appearances, these God-beings were not like traditional notions of God or the gods; they were not in the least like people, and they were not paying any attention to us at all.

Epicurus believed in an infinite number of universes—some very different from our own and some quite similar. What we had understood as the Greek gods, he explained, were really calm and immortal image-beings living in the spaces between cosmic systems. They had but one emotion and it was placid happiness. They did not get angry or frustrated or excited. They did not judge. They had not even made the world—if they had, it would not be so full of suffering. Here Epicurus cited crocodiles as an ugly and terrible danger whose presence did not seem to indicate a benevolent creator.

The one thing that Epicurean gods did provide was the service of example—but this one was a big deal. For an Epicurean, somewhere there are beings that are truly at peace, are happy, and are eternal. The mere idea of this gentle bliss is, itself, a kind of uplifting dream. After all, we human beings know a strange thing: happiness responds to circumstances, but, basically, it is internal. We can experience it when it happens to come upon us; we can induce it with practices or drugs; but we cannot just *be happy*. Whereas certain religions admire a being who has ultra-virtues, Epicureanism admires a being with an ultra-mood—a being that has solved the schism between how it feels and how it wants to feel.

That is one reason for Epicurus to prefer gods to no gods. Another is that, like Democritus, Epicurus said he believed that people really did see gods in dreams and visions. For both, the weight of tradition, going back centuries upon centuries, and the weight of contemporaries' attestations, was too powerful to dismiss. But as far as Epicurus was concerned, what most people believed—that there were gods or a God in charge of the world—was not only wrong, it was a kind of impiety against the truth. Even those who did have rational ideas about the gods when they were thinking clearly, he scolded, often behaved in ways utterly at odds with these ideas. "Not the man who denies the gods worshipped by the multitude, but he who affirms of the gods what the multitude believes about them is truly impious."[29] To his mind, Epicurus was the one being reverent toward the real, while those called pious wasted their affections on the imaginary.

Democritus provided the insight of atomism and materialism; Epicurus proclaimed it was time to use this ability to explain the world rationally to relieve humanity of fear. He actually railed against his much admired Democritus for failing to announce the end of fear of the gods. Now, Epicurus insisted, there is nothing left to fear: we are going to die, but so what? When it is over, it will be over. Pain happens but either does not last long or is bearable, so let it come if it's going to come. And last, there are no ghostly grownups watching our lives and waiting to punish us. Everything is okay. It is all just happening.

What is more, urged Epicurus, life is full of sweetness. We might as well enjoy it; we might as well really make an art of appreciating pleasure. Later his doctrine came to be synonymous with sensualist hedonism, which is why the adjective *epicurean* has that connotation today. But delicate food, drink, and the pleasures of the flesh are not quite what he had in mind. What Epicurus really encouraged was a joyous cultivation of knowledge and friendships. He did write about the delight of food and drink, but he meant learning to experience fully the pleasure of eating even rough bread and water as well as other things; the idea is that you cultivate yourself more than the food.

Epicurus set up "The Garden" as a home for his school, and men and women of various social stations—including courtesans and at least one slave—came and went, studying, talking, and relaxing. Many flourished here. Leontium, an Athenian courtesan and friend of Epicurus, became renowned for her philosophical treatises. Epicurean pleasures did not include politics or much interaction with the world at large. Unlike the Stoics, the Epicureans tended to stay away from public life, seeing it as concerned with false ideals, and likely to trick people into spending their one lifetime running a race no one can win. His doctrines covered a wide range of subjects in science and philosophy, but he geared all of it toward helping people achieve peace of mind (and pleasure) in the full recognition of our peculiar predicament.

Epicurus' theories of astronomy strictly denied that the movement of the heavenly bodies had anything to do with the gods. In fact, he took great pains to demonstrate that human beings had too little information about the heavens to settle on any given explanation of them. He insisted that the Moon's phases might be due to the Moon's rotating, or the interposition of other bodies, or perhaps due to "configurations" of the air. What was important, he stressed, was that "one must not become so in love with" one explanation that one rules out others for no good reason.[30] This was natural science, but it was also an argument against the old gods of Olympus, and perhaps more pointedly, it was an argument against the philosophical religions of Plato and Aristotle. Epicurus believed that those philosophers had done a good deal of damage by convincing men and women that the stars and planets were divine. He wanted to free humanity not only from fear of the wrath of idiosyncratic gods, but also from fear of cosmic necessity and predetermination. As he saw it, some things in life happened randomly and some by necessity, but within this situation human beings had free will. There is no up nor down to the universe and no hierarchy of value outside the human mind. Atoms come together in orderly and disorderly fashions as patterns and chance will have it; nothing purposefully guides them through these various incarnations and nothing is purposefully trying to help us or block our path. The world was not made by the gods and it was not made for us. We may enjoy it in peace.

These are wonderful tenets, but Epicurus understood that such notions are not of much use if one simply understands the ideas. They have to be studied until they are fully integrated and accepted in one's whole being—as with the Cynics, the pursuit of these ideas required devotion. We do not know as much as we'd like about the practices of the Epicureans, but one gets the sense that there *were* practices, that this was a meditative and ritualized life. Epicurus kept enjoining people to work at fully knowing the truth: "Accustom yourself to believing that death is nothing to us, for good and evil imply the capacity for sensation, and death is the privation of all sentience. . . ." This philosophical study should start early in life and end late.

> So we must exercise ourselves in the things which bring happiness, since, if that be present, we have everything, and, if that be absent, all our actions are directed toward attaining it.

And again: "Those things which without ceasing I have declared unto thee, do them, and exercise thyself in them, holding them to be the elements of right life."[31] This wasn't just study, it was exercise; it was something you did.

Rigorous religious practices can have a transformative effect on the human mind and like religions, graceful-life philosophies create meditative, "mystical" experiences—often in the invocation of peace, friendship, study, intellectual play, and joyous calm. These pleasures Epicurus considered immortal, beyond the individual experiencing them at any given time, and thus capable of focusing the individual on the moment. That makes mortality seem much less problematic: in this beautiful moment, one is alive. One of the few documents we have that comes directly from Epicurus is a letter, written to his friend Menoeceus and explaining his philosophy. Having told of the happy finality of death and the glory of prudent, mortal pleasure, the letter ends with these remarkable words:

> Exercise thyself in these and kindred precepts day and night, both by thyself and with him who is like thee; then never, either in waking or in dream, will you be disturbed, but will live as a God among people. For people lose all appearance of mortality by living in the midst of immortal blessings.[32]

Again, coming out of the love and protection of divine care, the great doubter finds himself godlike in happiness, transcendence, and repose.

Epicurus believed there was no real point in praying, both because the gods are not listening and because human beings are entirely capable of making themselves happy on their own. Yet, he also said that the act of prayer was a natural part of human behavior and ought to be indulged. This is a wonderful little idea to which we will return. For the moment, let us at least think of it as a metaphor for human love. Consider the hypothesis that other people do not really exist as we perceive them, that for each of us, the magnitude of our isolation is profound and the possibility of a true bond of love or friendship utterly remote. On some level, of course, it is true: we are all strangers to one another; some of us, for example, are capable of surprising acts we have not yet committed. Ought we not to trust and love anyway? Shouldn't we act as if we could know another person?

Prayer is based on the remote possibility that someone is actually listening; but so is a lot of conversation. If the former seems far-fetched, consider the latter: even if someone is listening to your story, and really hearing, that person will disappear from existence in the blink of a cosmic eye, so why bother to tell this perhaps illusory and possibly un-listening person something he or she is unlikely to truly understand, just before the two of you blip back out of existence? We like to talk to people who answer us, intelligently if possible, but we do talk without needing response or expecting comprehension. Sometimes, the event is the word, the act of speaking. Once we pull that apart a bit, the action of talking becomes more important than the question of whether the talking is working—because we know, going in, that the talking is not working. That said, one might as well pray.

Epicurus adamantly denied the idea that gods were watching, listening to, or even caring about humanity, and yet he did not counsel the sharp rejection of all religious practice. This is strange to modern eyes. Something moderns take for granted about the passionate rejection of religion is missing here: Epicurus does not picture himself in an all-out battle against the institution itself. Without the outside context of a political war between faith and reason, Epicurus does not fear that any single point he might award to the religious will be used against him. Nor is he eager to have his followers shunning prayer or ritual in order to demonstrate publicly their disbelief. Outside the context of a political war between faith and reason, more nuanced arrangements may be safely undertaken.

Epicurus recommended that people take part in the religious conventions of their country. His central purpose in this seems to have been in line with the rest of his advice, i.e., it promoted a trouble-free life. Unlike Zeno and his Stoics, Epicurus was suspicious of society. Still, getting along with people, avoiding confrontation, and making friends in general was much more important to Epicurus than it was to, say, Diogenes and his Cynics. But it was not just for the sake of convention that he advised taking part in public ritual; Epicurus also believed that these rituals were a chance to meditate upon the strange, ethereal beings that existed between the universes. Thinking of them was beautiful and tranquil and brought a calm wisdom.

Epicurus's notion of consciousness was rationalist in a similar way. He had explained that everything was made of atoms and void. The human mind seemed to be other stuff than the rest of the world, but deep down, it was the same. Water and stone, for instance, seem to be made of completely different things, but Democritus and others had made the leap to believe that they were somehow the same. Yet, compared to consciousness, water and stone are obviously the same thing: they are stuff, and consciousness is something else again. Still, Epicurus figured it might well fit into the same schema if, perhaps, mind or "the soul" was made of particularly small atoms. "Keeping in view our perceptions and feelings (for so shall we have the surest grounds for belief), we must recognize generally that the soul is a corporeal thing, composed of fine particles, dispersed all over the frame" and something like wind and heat.

> It is impossible to conceive anything that is incorporeal as self-existent except empty space. And empty space cannot itself either act or be acted upon, but simply allows body to move through it. Hence those who call soul incorporeal speak foolishly. For if it were so, it could neither act nor be acted upon.[33]

The atoms that make up the soul flow more quickly than even those that make up water, clouds, or smoke and thus account for thinking. That part of us that thinks and feels, mused Epicurus, is a part of our physical makeup. Hence, when the body dies, the soul dies, too.

We should note another little fact in passing: Epicurus took care to argue against the idea that the soul lives on in the maggots that swarm out of a corpse. The maggots, he explained, have their own lives; ours is extinguished. The fact that we would not dream of debating this point is a reminder that Epicurus was not answering quite the same question that we are answering today. More specifically, he was denying a different set of hypotheses about where the "self" goes after death. Physics and biology took part in discussions about the fate of human essence—there were maggots in the metaphysics and metaphysics in the maggots. But Epicurus didn't believe any of it, and offered instead a surprisingly modern empirical materialism. For him, human consciousness is an awesome power that requires some special explanation—but not too special. That life ends in death should not be a cause for despair: "The true understanding of the fact that death is nothing to us renders enjoyable the mortality of existence, not by adding infinite time but by taking away the yearning for immortality." It is accepting the finality of death that makes it possible to enjoy the pleasures of the garden. This is a very different garden than the one we got kicked out of in the Eden story. This time you have to eat from the tree of knowledge in order to get in. You build this one yourself, in part in your character, in part in your real environment. It is a lot of work, but you can stay as long as you like.

For Epicurus, living prudently, in deep appreciation of modest pleasures, was not just the route to happiness, it *was* happiness. The key, he said, is coming to know that "the wise person's misfortune is better than the fool's prosperity." Difficult truth is better than wonderful falsehood. Since behavior creates happiness, and accepting reality creates peace of mind, the vicissitudes of chance and worry cannot hold much power. Accept the bad things in the knowledge that they are not really so bad, get over the idea that the gods are watching you, and be happy.

THE SKEPTICS

Despite the deep secularism of Cynicism, and sometimes Stoicism, it was Epicureanism and Skepticism that issued the most powerful critiques of theism. Skepticism began with Pyrrho of Elis, who lived from 365 BCE to about 275 BCE. He wrote nothing, as far as we know, and offered his followers not a system but a manner of living. Pyrrho studied the ideas of the great philosophers and came to the conclusion that a really smart person could convince him that any substance was the primary matter of the universe. What is more, all the philosophers' systems were vulnerable to argument. Pyrrho believed that nothing can be known, because the opposite of every statement could be asserted with plausibility. Also, our senses and minds provide false or merely narrow information. We should attempt to have no opinions. Since we know nothing for certain, we must behave as such, affirming and denying nothing, no matter what the subject. We thus stand aloof from life and thereby attain peace of mind.

Not much of Pyrrho's recommended manner of living is known and, amusingly, what we know conflicts.[34] There is one tradition in which he seems to have been trying to shed human emotion. In that tradition, he was marked by an imperturbability and "oriental indifference" that, at times, had him pass by friends without noticing them. Once, when he was attacked by a wild dog and scurried up a tree, he later apologized to friends for his fear and announced that it was difficult to strip oneself of humanity. In the other tradition, he was not trying to strip himself of humanity but merely to live a moderate life. There is a story of his calming passengers on a storm-tossed ship by pointing out that the pigs on board were quietly munching their food. He recommended such serenity for all. Here Pyrrho seems less the indifferent ascetic and more the caring teacher. Stories of his life support both versions: In 334 BCE he joined the court of Alexander the Great (Aristotle had left the young man, after an eight-year stay, only about a year earlier) and traveled with Alexander to India, where he studied with the philosophers and ascetics of the Indus Valley. In the same court was a philosopher who was a follower of Democritus—a materialist, an atomist, and a believer that sense experience is always flawed by such things as tricks of light and tricks of the human body. When Alexander died, Pyrrho went back to Elis and lived with his midwife sister, and although he taught philosophy to large crowds and regular students, he also helped out at home by cleaning house, washing pigs, and taking fowl to market. Elis named him a high priest. He lived to the age of ninety and, in his honor, the city instated tax exemption for all philosophers. His best student, Timon, was more of a cutup than Pyrrho, writing satires that portrayed anyone who seemed to know anything as an arrogant buffoon. Yet he also wrote beautifully on the Pyrrhonic approach to knowing the world. It was Timon who wrote, "I do not lay it down that honey is sweet, but I admit that it appears to be so." Timon had a wife and kids, and loved food, drink, and a good time as well as philosophy and solitude. Early Pyrrhonism was ascetic, but not exclusively so.

Skepticism became more important in the second century BCE when the philosopher Arcesilaus brought it into the Platonic Academy. This created what we know as the Middle Academy and is the reason that Academic philosophy became a synonym for Skepticism. We do not know much about his opinions on theism either, but his successor, Carneades of Cyrene, left us much more to work with. Carneades was arguably the best philosopher in the five hundred years after Aristotle, and his contribution to Skepticism was immeasurable because he replaced the refusal to believe anything with a sophisticated notion of probability. What he said was that we cannot know anything for certain, but we can carefully determine whether one conclusion is more likely than another. This allowed the Skeptics to bring all their studious doubt to bear on philosophical questions, whereas the old model restricted them to judicious silence.

Carneades did not insist that the gods were a fairy tale, but he unraveled most of the common proofs that they exist. Theists often said, for instance, that there must be gods since so many people had seen them and had described them in a similar manner. As we have seen, even the rationalist Epicurus bothered to offer an empirical explanation for such claims. Carneades dismissed the reports of sightings, both ancient and contemporary, as stories with no basis in reality. Against the claim that the universal belief in gods proved that they existed, Carneades responded that this proved only that people believed in gods—another proof would be needed to show that the gods exist.

Carneades also attacked the theistic position on the basis of the description of the particular gods, especially the God of the Stoics. The Stoic God was a God who was everything good. But in Carneades' opinion, true virtue requires some flaws, some limitations. One cannot be called brave if one has not known fear. There is no meaningful way to be self-disciplined in the absence of temptation. So the description of God as supremely virtuous is inherently problematic. Carneades also attacked the argument by design— the idea that the world was so wonderful that it must have been created by an intelligence—by pointing out problems in the design. We are reminded of Epicurus' feelings about crocodiles. For Carneades, no intelligence seemed to be at work behind agonizing diseases, poisonous snakes, and tidal waves. Furthermore, the Stoics had argued that God's greatest gift to humanity was the gift of reason. Carneades pointedly asked why God had shown such partiality in the distribution of it.

Carneades described this way of arguing as *ataxia,* which comes from a word meaning "heap." People sometimes make arguments by heaping up a collection of ideas and observations that prove their contention. If someone else continually takes one of these ideas off the heap by disproving it, Carneades asks, at what point is the heap no longer a heap at all? It is a clever conception, especially regarding theism. Take three popular reasons to believe in God: (1) How else did we get here? (2) The beauty of the world implies intelligence. (3) People everywhere believe in God. Each can be argued with on its own, and they do not cohere into a single system, but as a heap they are quite formidable. By identifying the metaphor of the argument, Carneades discovered a clear-cut way of examining its validity.

With all the Hellenic and Hellenistic doubters, what is perhaps most striking is how fully invested in the idea of, say, Zeus, they were, even when they doubted to the point of not believing in him. It is what we should expect, but it is somehow still surprising. Greek proofs of a theistic universe hung on attributes of the gods as the Greeks conceived them. They believed in the Olympian pantheon, so one proof was that people had seen these specific characters, in visions and dreams, for hundreds and hundreds of years. They believed that Homer and Hesiod, who had written of the gods so long ago, could not possibly have been entirely lying or entirely mistaken. The Greeks believed in planetary gods because the planets move, so they must be alive, they must have soul. When we look at the way the disbelievers structured their arguments, it was all about these same gods: the dreams and visions were real but were not really images of an Olympian pantheon; Homer and Hesiod were not lying but were faithfully recording an error in human memory—the gods were really just deified heroes; the heavenly bodies are not gods but rather are rocks like the earth. Doubt in the ancient Greek world is rationalism, naturalism, and secular history applied to the Olympic pantheon. The result of that was a world suffused by doubt, within which there were pockets of belief and pockets of real disbelief.

As the philosophers put the gods or God into doubt, according to rationalist narrative and natural science, they sought a philosophical replacement. They were not fighting against the religious impulse; they just reconceived the sacred so that it seemed true. They still thought that a good life could be achieved only through deep and reverent contemplation of reality. The actual human predicament, they cautioned, is very difficult to hold in one's mind; there is a natural forgetfulness that pulls one into the day-to-day world with all its frustration and emotional pain. The philosophers were convinced that thinking about these big ideas, just the pure process of mulling them over, does a person a world of good. A big part of the secret to life was to spend time teasing these ideas out, playing in them, working hard to get to know them. For many, and for Plato and Epicurus certainly, brave thinking about truth is the secret to happiness: concerted and regular contemplation will transform us and let us taste what there is to taste of transcendence.

TWO

Smacking the Temple,
600 BCE–1 CE

Doubt and the Ancient Jews

In the Jewish Hanukkah story, there are two enemies against the pious Jews who triumph in the end: the dominant Hellenistic Empire, and the secular, or "apostate," Jews. Who were these Jewish doubters? In the Hellenistic period, a great numbers of Jews grew secular in their habits, doubting the God and laws of Moses so strongly that they rededicated the Second Temple—*the* Temple, in Jerusalem—to Zeus, and did so in a mood of cosmopolitan universalism, in appreciation of Greek philosophy and culture. This chapter has three parts. The first, "Zeus at the Altar," is a story of the social side of doubt: people no longer wanting to be religious and isolationist. It's a revised Hanukkah story, from the perspective of the history of doubt. The next two parts of the chapter address two sections of the Hebrew Bible, each one a pinnacle of the human expression of doubt. The first of these is the Book of Job, probably written just before the Hellenistic period; the second is Ecclesiastes, probably written right in the thick of the Hellenistic Age. The doubt in these two books feels very different; one is a howl for justice, the other a soulful wink and a shrug. They are responding to two different versions of Judaism. Yet despite important differences, they both have the same central problem: the world is cruel and good people suffer.

Calling something an expression of doubt by no means denies that it is a religious expression, too. Wildly doubting texts can still be religious treatises, and the brilliant ones even shed light on belief. There is pious beauty in these texts. But the fact is, Job and Ecclesiastes are both rather antireligious and antidogmatic. They are remarkable events of doubt and are foundational texts in doubt's history.

We need a little background, both to situate our story and to explain what Jews believed about providence—divine guidance or care—in the centuries at hand. Very early on in their history, the Hebrews had some extremely good fortune on the fields of war and attributed that fortune to their powerful warrior God. There were a lot of "temple peoples": communities based on devotion to a god who was understood as living there. The people would build the god a temple, sacrifice to it, and hope for the best. The Jews were different in that the inner sanctum of the Temple held no statue—this God was invisible. Also, when a temple community started losing wars, they would generally assume that their god was weak or had abandoned them and begin devotions to another god. When the ancient Hebrews started losing wars, however, they built a theology around the idea that *they* had failed *God* and he was punishing them. By doing this they preserved the strength, justice, and presence of their God. They also cultivated a sense that the world of state is an arrangement of moral forces, rather than clever alliances. In its origins, this world justice was strictly for Israel as a whole, but in the eighth and seventh centuries BCE, the prophet Isaiah extended this sense of worth and recompense to individuals as well: prosperity and community respect were proof that you were in favor with God. So now the world was moral at the individual level, too: if you are good, good things will happen to you.

It takes effort to uphold such a doctrine when things start to go very wrong. In 586 BCE Jerusalem was captured by the Babylonians. King Nebuchadnezzar razed the Temple, put the city to the torch, and sent a choice section of the population—thousands of elites, professionals, and skilled workers—into exile to enrich the cities of Babylon. This was a nasty habit of the Babylonians. In the case of the Jews, those left behind soon lost touch with their religion, but those taken to Babylon clung to their Jewish identities. That was unusual; in fact, it may have been the first time that membership in a community of worship was divorced from residence. In Babylon, the Jews were treated well, given support from the state, and allowed to worship as they chose. It was during this period that Judaism began to center on the Laws. They were not even carefully studied before but were thought of as obscure and, in some cases, clearly out of date. But the exiled Jews missed their home and longed for their Temple, and they began to follow their religion in a new way.

That longing is why Psalm 137 says, "By the rivers of Babylon, there we sat, sat and wept, as we thought of Zion," and that is also why the psalm wonders (and both Lord Byron and Bob Marley borrowed), "How can we sing in a strange land?" Psalm 137 swears never to forget Zion, and the strength of that sorrowful commitment created the dedication to the Laws. Only here, in exile, did ordinary Jews begin to keep the Sabbath, to decide upon and live within the dietary restrictions, to practice the rite of circumcision, and to celebrate the various feasts. The absorption of Babylonian astronomical information enabled the Jews to fix these feasts to a calendar and thus come to think of them as regular annual holidays.

When the Persians conquered the Babylonians fifty years later, they set the various captive peoples free, purposefully repatriating the myriad local gods and their worshipers. The Jews were sent home with a mandate to rebuild the Temple and with the gold and silver to do it. The Jews who had stayed behind had intermarried with the local peoples, but the returning, postexilic Jews managed to rebuild the community. Meanwhile, many Babylonian Jews, perhaps even the majority, chose to stay behind, cultivating a great Jewish culture that flourished in Babylon for the next fifteen hundred years. The Babylonian captivity—the phrase has been used throughout history to designate misappropriations of all sorts—is the beginning of the Jewish Diaspora: from the sixth century BCE onward, a majority of Jews have always lived outside Palestine, often shuttling around as the result of one exile or another. The Jews were the first to proclaim a moral, all-powerful God, responsive to the behavior of human beings, and yet they had an incredibly troubled history. That is a paradox, and the Jews came to excel in developing approaches to the problem. Ideas of a just God in a rough world would henceforth be central to the history of doubt.

Both the prophets Jeremiah and Ezekiel witnessed the destruction of the Temple, and the response both came to was to shore up the traditional claims that ours is a just universe. The idea of an afterlife was still absent from Jewish thinking; the doctrine held that the good and the guilty will be recompensed while they yet live. It's a brutal theory for a people to embrace during a disaster, because it suggests that suffering implies guilt, but it also soothes by giving meaning to suffering. The Book of Job seems to have been written in this period, between 600 BCE and 400 BCE, and we shall see that this strict relationship between suffering and guilt was very much on the author's mind. The priest and scribe Ezra was among those who returned after the exile, and he would set the tone for Judaism for the next five hundred years. Like the prophets, he believed in a just God and a morally rational world. As such, he believed the destruction of the Temple and the Babylonian captivity must have been punishment from God for Israel's sins. If things go so badly for the state that its individuals are blanketed by grief, it is difficult to imagine them all personally guilty, to that degree. The explanation was that all the individuals had sinned because no one was following the laws of Moses, from the dietary restrictions—for instance, against eating milk and meat at the same meal—to keeping scrupulously restful on the Sabbath. The exilic Jews invented a way of life while in Babylon and came back to indict themselves, their brethren, and their forebears for not having upheld these laws all along, and for having thereby caused the disaster. Ignorance of the law was no excuse.

As a remedy, Ezra brought together a body of knowledge and created a synagogue system to teach it. It is Ezra who was primarily responsible for determining which of the ancient Hebrew writings would be considered Jewish holy scripture. He conceived the synagogues as local places to study the law and learn how to worship, whereas the Temple was to remain the one place for worship. The synagogues were an organized outpost of the Temple serving to bind together the far-flung community. Diaspora Jews, however, would come to use the synagogue as a place to congregate and eventually to worship as well. So now we have an odd temple people, who somehow survived as a group when separated from their Temple. They had a law code that was modeled to fit with how postexilic people really lived. And they had an organizational system, in the form of Temple, synagogues, and priests. Apparently it was a period of calm and certainty and it went on for over a century.

ZEUS AT THE ALTAR

Then things got very cosmopolitan. As we know, in 332 BCE Alexander the Great conquered this whole area: Greece and the once-indomitable Persian Empire—Egypt and all the way east to India. When Alexander died, Judea had the bad luck to be right in the middle of the Seleucid and the Ptolemaic empires. The Seleucid Empire was like a hand resting flat on Arabia, with its heel in the Persian Gulf and its fingers on the Mediterranean, pointing at Greece. With a slight twist of perspective, the Tigris and the Euphrates may serve as lines of the palm. The Ptolemaic Empire was more compact: a greater Egypt, clustered in a fist around the Nile. Extend the Seleucid pinkie finger down to touch the Ptolemaic fist, and you've got Palestine—a tiny connective stripe between two giants. When she was not a battleground, she was a stomping ground. Broadly, from 332 to 200 BCE, the Jews were ruled by the Ptolemies, from the south, thereafter by the Seleucids, up north.

Although the province of Judah retained its autonomy under its new conquerors, Jews found living in a Greek empire very different from living in the Persian Empire, which had largely stayed out of their affairs. Alexander and his successors viewed Greek culture as the apex of civilization and proselytized it in the lands they ruled. They settled veterans and other Greeks in colonies. They built cities everywhere, bringing Classical architecture to areas that had always been rugged and rural. Greek trade quickly began to generate an economic boom. In no time at all, the Jewish territories—Judah and Samaria—found themselves surrounded by a comparatively wealthy, sophisticated, artistic world. Soon Greek clothing and Greek food became common among the Jews. Especially in the Ptolemaic world, the Jews learned Greek, for business purposes at first, and then more generally, until at last Greek became the common language. Hebrew and Aramaic (the language imposed by the Persians) fell out of favor. Jews often used two sets of names, Greek for travel and business, Hebrew at home. Others simply Hellenized their names and left it at that. There were Jews in Egypt before Alexander the Great came, but now they streamed into Alexandria as soldiers, craftsmen, and professionals. The conquered Egyptians were treated as an underclass to the "Hellenes," which meant everybody else, including the Jews. Jews were among the immigrants who had come to serve the emperor in one capacity or another and were seen as a modernizing force against the Egyptian traditionalism of the local population. Jews were involved in the Greek secular culture around them.

The Greeks seemed to like the Jews.[1] They do not appear to have noticed the Jews' existence at all before Alexander brought them into the Greek sphere, and the initial Greek description of the Jews was that they were philosophers and wise men. One historian from the period guessed that the Jews were to the Syrians what the Brahmans were to the Indians. Some thought the Jews *were* descendants of the philosophers of India. The Bible records the existence of Greek and Jewish business partners in Alexandria at this time. Historians argue about the extent of Greek absorption of Jewish custom and culture, but we know that there were some converts to Israel's God, and there were many more people who were simply attracted to Judaism and wanted to offer sacrifice, take part in annual feasts, and otherwise participate. The opinion of the Jews mattered in the wider syncretic, polytheist world: in part because Jews found animal worship comic in the extreme, Isis no longer appeared cow-headed outside purely Egyptian circles.

The Ptolemaic emperor's attitude toward the Jews of Alexandria was benevolent. The emperor's interest in the Jews was actually at the origin of the great Library of Alexandria. The very first mention we have of the Library of Alexandria is in *The Letter of Aristeas* (ca. 180–145 BCE). Aresteas was a Jewish scholar housed at the library, chronicling the translation of the Septuagint, and he explained that the library and the Septuagint had the same origin. The library was a century old when Aristeas wrote. He tells us that Ptolemy I had sponsored a friend, Demetrius, to create the library. Demetrius (who had ruled Athens for ten years and had been a student of Aristotle) gathered books and scrolls, but he also supervised a massive translation project making works from various cultures of the empire available to anyone who spoke Greek. This process began with the translation of the Jewish holy scriptures into Greek, thus producing the Septuagint, named for its seventy-two translators (they rounded down the number). So the Library of Alexandria was itself a product of the desire for Greeks to get to know the various peoples of the empire, and in the first place, the Jews. And the Jews' enthusiasm for the project was a reflection of Jewish integration into Greek culture.

To celebrate the Greek translation, the Jews of Alexandria hosted an annual feast on the island of Pharos, and each year the text was read aloud to a tremendous congregation of Jews and Jewish sympathizers. Philo, a Jewish Alexandrian philosopher writing in the early first century CE, tells us that many non-Jews participated in this event. Alexandria, with its own Bible and its own holiday commemorating that Bible's translation, began to serve as a second religious center for Judaism. At the same time, the Jews in Alexandria, at stretches accounting for as much 40 percent of her population, were integrating into the wider culture.[2] When the Greek gymnasium in Alexandria was opened up to non-Greeks—though not to Egyptians—a great many Jews accepted the invitation. Jews were the "people of the book," even if that traditionally kept them isolated in a single literature, and they were scholars. With Hellenization, for some Jews, a wider range of literature was coming to be of interest. For the most progressive people of the book, there was an attraction to the great city of the Library.

The Jewish world was still centered in Jerusalem, the heart of these various Jewish communities, site of the Temple, and the one place where interpretation of the Torah was proper. But Alexandria in Egypt was a rival center for Jews, and later, when Antioch in Syria became a great city, it, too, nurtured a lively Jewish culture for both secular and observant Jews. Many smaller places also contained vibrant Jewish populations. We know that sometimes when an emperor wanted to populate an area with a loyal people, he invited Jews: they were offered farmland, vineyards, and cash supplements until the first harvest—along with the transfer of their laws.[3] Jews thus emigrated as a distinct people, with their own law. The great cosmopolitan, syncretic tide was strong, though. Jews began to take part in the urbane pleasures of the larger culture: theater, festivals, poetry, and philosophy of both the Classical schools and the new schools—Epicureans and Skeptics, Stoics and Cynics. Our best sources for this story are some late books of ancient Jewish history, called "Maccabees," that did not make it into the Hebrew Bible. We only have them because they were included in Roman Catholic Bibles, usually at least in part under the title "Apocrypha." From these books it would seem there was a significant population of moderately pro-Greek Jews, and then a prominent but much smaller number of extremely pro-Greek Jews.

Jewish women in particular might have found the Hellenistic world tempting. Ancient Greece had harbored an extreme prejudice against women in public life, but this was easing in the Hellenistic period. Cosmopolitanism often means more equality for women because of the jostling of old roles and hierarchies. Whereas Classical Greek art concentrated on the beauty of the male form, in the art of the Hellenistic Age women began to appear. In frescos and mosaics, we now see images of men and women at table together, relaxing, dressed smartly, enjoying each other's company. As for the Jews, women had some vital roles, but this was a very patriarchal society. The common culture and mores of the Hellenistic world must have been attractive to some Jewish women. In the Jewish experience and in the wider Greek world, the strict kinship ties that bound women to men and men to their fathers were breaking down, being replaced by the cosmopolitan virtues of freedom and the mutual aid of loyal friendship.

The secularist Jewish community began to see the empire and the Greek philosophical tradition as a significant part of their identity. They knew that within the reaches of the Hellenistic empire there were several odd little temple peoples; religion-states that held themselves apart. The ever-changing world seemed headed toward more integration, more trade, more shared knowledge. Some Jews, welcoming those trends, relaxed their observance of traditional law, which now seemed isolating, awkward, or irrelevant. In an ironic twist quite common in the history of doubt, it was often the children of the elite old priestly class who were the most enthusiastic defenders of the new secular culture: they were the ones suitably educated and sufficiently prosperous to take part in the new schools, arts, and other pleasures. Also, the Greeks had expanded the middle class—through egalitarian policy and the generation of government jobs—so a wider group had access to the delights of the culture. Whatever their class, the Jews who enjoyed Hellenistic culture may not have felt any less Jewish, seeing nothing ill in the Greek invitation to civic celebrations, universalist moral philosophy, exercise and education at the Gymnasium, a sense of progress, and a prosperous future for the kids.

By now the Greek mystery religions, especially that of Dionysus, seem to have become important to many Diaspora Jews. By the early first century, popular Jewish legend had it that Ptolemy IV (204 BCE), a great devotee of Dionysus, had issued a decree commanding the Jews of Egypt to become worshipers of Dionysus. The author of Maccabees III explains that many Jews in Alexandria were only too happy to comply. Philo denounced the mystery religions but reproduced their tone in his own religious works; a contemporary Jewish critic warns specifically against Dionysiac revels. But such revels were seductive. Here were secrets and, as one historian has noted, they were "Greek secrets"; they had a powerful cultural cachet and Jews were interested.[4] The Mystery Religions were attractive to Jews for the same reason they were attractive to Greeks: the official worship of both Greeks and Jews had no provisions for an afterlife. The mystery rites allowed individuals to explore the notion. Again, taking part in mystery rites and other Greek diversions did not mean one wasn't a Jew anymore. A man whose parents had neglected to circumcise him, who did not observe ancestral traditions, but instead was an ecstatic worshiper of Dionysus was still legally a Jew.[5] Even if such a man attained citizenship, evidence suggests that he would yet maintain a Jewish identity. When biblical authors wanted to say just what apostate Jews believed, no particular cult came to mind; rather, they were drifting away from religion, forgetting the dietary laws, not observing the Sabbath, and, generally, "changing the practices and estranged from ancestral doctrines," i.e., they were secular and searching Jewish men and women, living in interesting times.[6]

When Alexandria grew overpopulated, the governors stopped offering citizenship to newcomers or their descendants. So there were Jews who were

Alexandrians, and there were newer immigrants who were just Jews and could not aspire to be citizens of Alexandria. It was a frustrating situation for the newcomers, and worse for second-generation newcomers, who began clamoring for equal rights. This meant that when the emperor offered full citizenship to members of all temple peoples in Alexandria who would join in with the general Greek sacrifice, it was not as aggressively coercive as it would have been if there had been no strata of not-very-religious people who wanted citizenship. The emperor was letting the city's most cosmopolitan resident aliens self-select on the basis of universalist spirit and worldly Hellenism. Some Jews were more than happy to show solidarity with the greater community and thereby become officially Alexandrian—ahead of many various newcomers of various nations. And after they became Alexandrian, they were still Jews.

Being a culturally Hellenic Jew became chic. The new vision of Greeks and Jews as natural friends encouraged the production of tales linking their origins. One purported to be an exchange of letters between the twelve-year-old Solomon and his client kings of Egypt and of Tyre; another presented the sons of Abraham as companions of Hercules who, it was proposed, himself married one of their daughters.[7] Since the Jews were an older people, this idea of an early connection suggested that Jews might have taught the Greeks fundamental things—even writing. About 200 BCE the biographer Hermippus supposed that the early philosopher Pythagoras had been a pupil "of Jews and Thracians." Someone, Greek or Jew, invented the idea of a common descent of Jews and Spartans from Abraham. It is clear from Maccabees II that some Jewish circles believed it.

Abraham showed up a lot in such stories and became the preferred hero of the Hellenistic Jews. He was more cosmopolitan than Moses—he had wandered and found a new home, rather than leading people back to their origins. He seemed less exclusionary and far less legalistic.[8] His image reminded people that there was a model for Jews that suggested an extended outward-bound, familial legion, rather than a secluded, inward-looking people of the Law. After all, the reformist Jews noted, Abraham had not followed the law of Moses. He predated Moses. In Genesis 18, he served milk and meat—a boiled calf with butter and milk—to the Lord himself, and both ate heartily under a tree while Sarah made cake. So how important could it be to live one's life according to these arcane injunctions?

The value of the law during the exile was that it kept the Jews united as a separate people, distinct from their conquerors. But isolation was not the goal of Jews living as honored members of these vibrant Hellenistic cities. A conservative Jewish response generated stories that backdated the law—Eve performing postchildbirth purification rites, for example—but a lot of people weren't buying it. Jews gravitated to the Greek world beyond Judaism because it appealed to their sensibilities: progressive Jews moved away from the law of Moses, as Ezra had championed it, because it no longer seemed relevant to their lives. Rigorously lawful Jews did not like any of this and responded with a turn toward isolationist piety. Indeed, this is the origin of a certain kind of religious extremism—some went off into the desert in bands and sharpened their knives. As suggested, the poor tended to be conservative whereas the wealthier segments of the population—those who could afford it—were avid for Greek education and savoir faire. It was a tense situation, but reasonably stable.

That changed when Judah changed empirical hands. Antiochus III had a terrific enthusiasm for Greek culture and he came down from the north and triumphed against Ptolemy at Gaza in 200 BCE. The Jews were suddenly in the empire of the Seleucids—the upper of the two hands. The war that brought this change also brought personal misery for some. But for many others in Judah, the new emperor brought a mood of inclusion and progress.

Antiochus III was enthusiastic, but not very proactive. The Greek-inclined, secularist Jewish community found its great champion in his son and successor, King Antiochus IV (175–164 BCE), who assumed the name Epiphanes (*illustrious* or *revealer*). Pious Jews changed it to "Epimanes" (*cracked* or *mad*), because they despised him, but for a great many Jews, Epiphanes seemed to have the right idea. He encouraged common, secular laws and customs in order to unify the wider community and foster economic growth. Epiphanes' capital city, Antioch in Syria, was enlarged to accommodate Greeks seeking freedom from the growing pressures of Rome, and a very large community of Jews came to live there as well. As Antioch became a major city, it arose as a great Jewish center on the order of Alexandria and Jerusalem. We don't know exactly what secular hellenophile Jews believed or what they practiced, in part because the more pious Jews determined which texts got handed down to us. But what we do know of them directly, combined with the invective of their conservative fellows, suggests that they did not practice many rituals of Judaism. They were uncircumcised and did not keep the Sabbath. They were comfortable with the rituals and spiritualism of the Greeks.

Of course, at this time, Greek culture was more likely to mean Epicureanism than the Olympic pantheon. Even in Judah, Epicureans were very prominent and were well represented in Hellenized cities like Galilee. Antioch and Alexandria were suffused with rationalist Greek philosophy, and we know that Antiochus himself lauded the Epicurean ideals. Antiochus retired the pro-Ptolemaic Jewish high priest—son of the high priest Simon the Just—and gave the post to Simon's more progressive younger son, Jason. Jason was born Joshua but, tellingly, he was happier with the Greek form of his name. Maccabees says Antiochus sold the high-priest position to the highest bidder, and leaves out that he was Simon's son and that he already had a strong following. This following included the large "family" of the Tobiads, who had long countered the religious legalism championed by Ezra.[9] Jerusalem would be reenvisioned as a Greek polis.

The new high priest quickly got to work making the finer things of Greek culture available to Jews. First and foremost, Jason built a gymnasium in Jerusalem, at the foot of the Temple Mount. The gymnasia served as the central nexus of Greek life. Greek culture was exercised and passed along there, and the social connections one made there were crucial to business and prosperity. Attendants had the obvious aim of fitting in, becoming educated, and getting ahead in the Greek world. The gymnasium at the Temple Mount was very popular with Jews and non-Jews alike, and there was nothing really anti-Jewish or antireligious about it, except that, of course, there was. This was most neatly visible in the fact that most gymnasium activities were performed in the nude and there was a Jewish ban on such public nakedness. Walking in the door was already a major statement about one's beliefs. The place was a paean to the male body, to physical beauty, and to human strength. Because the nudity at the gymnasium (the word itself means "to train naked") was coupled with a desire to fit in and to leave old ways behind, some Jewish men actually underwent surgery to reverse their circumcision—the primary sign of the covenant. Many more stopped circumcising their sons. As the Bible tells it, the Temple priests themselves "ceased to show any interest in the services of the altar; scorning the Temple and neglecting the sacrifices, they would hurry to take part in the unlawful exercises on the training-ground."[10] That is, Jewish Temple priests were running off to play naked sports with the Greeks and everyone else. The Jews in both empires who took to Greek culture seem to have felt it was the dawn of a golden age of inspired, secular civilization. That was certainly the gist of the wider discourse.

Jews fought among themselves over all this. Even struggles over doctrine among the most pious groups of Jews came to reflect tension about Jewish interest in Greek culture. The Pharisees, for instance, were following the rationalist Greek impulse when they created an oral law to translate the archaic Mosaic law into the real world of the day. Their rivals, the Sadducees, stuck to the old written law and said that the way of the Pharisees would lead to more respect for "the book of Homer"—implying Greek literature in general—"than for the holy scriptures."[11] This is one of the few times the Talmud mentions any literature other than its own, so Jewish interest in Greek literature must have been considerable.

Whether the more traditional Jews liked it or not, "the book of Homer" *was* respected. We used to think Diaspora Jews were the more progressive populations in this respect, in contrast to a conservative, anti-Greek Judean population. But scholarship over the last four decades has revealed that this idea of a split is very much in error. In Judea as elsewhere, Jewish writing shows an intense interest and involvement in Greek issues and styles.[12] Elias J. Bickerman's brilliant analysis of the Jewish school system created by the Pharisees in the third and second century BCE showed that Greek teaching became so prevalent that it gave rise to the first Jewish school system. There was no precedent for it in Jewish history. Bickerman explains that because the Jewish elite of Hellenistic Judea had so completely embraced the gymnasia, pious, legalist Jews were pushed into offering a conservative Jewish alternative.[13]

For those Jews who embraced the changes, the old ways weren't simply fading away. In many cases, people thought long and hard and decided that some of the old ways, at least, ought to be abolished. There were Jewish reformers and intellectuals who wanted to stop breaking the law when they went to the gymnasium—not by staying home, but by lifting the ban on nudity and generally modernizing the whole religion. Progressive Jewish thinkers at this time produced the first written biblical criticism, asserting that the Law was certainly not as old as Moses, indeed, it was not very old at all. They found the Torah full of allegory and fable and unnecessary restrictions. As Maccabees I puts it:

> In those days lawless men came forth from Israel, and misled many, saying, "Let us go and make a covenant with the Gentiles round about us, for since we separated from them many evils have come upon us." This proposal pleased them, and some of the people eagerly went to the king. He author-

ized them to observe the ordinances of the Gentiles. So they built a gymnasium in Jerusalem, according to Gentile custom, and removed the marks of circumcision, and abandoned the holy covenant.

Jason was acting in concert with his fellows, then, when he began to limit the tremendous number of regular Temple sacrifices, in order to save money for the new Greek-oriented cultural programs. The Temple treasury served as the state bank in a sense—it collected the money and controlled the budget—so the high priest had remarkable power. The devout found it hard to swallow Jason's siphoning off money once designated for Temple sacrifice, but they managed.

The crisis came in 171 BCE, when Antiochus fired Jason and chose a new Jewish high priest, Menelaus, who was even more sympathetic to the cultural mood and financial needs of Antiochus. Jason fled to Sparta, "hoping that for kinship's sake, he might find harbor there"—the "kinship" here refers to the growing legend that Spartans were also descendants of Abraham. Menelaus began channeling some of the money that had once gone to the Temple directly to the king.

Now a new sect of Jews was formed from scribes and their followers. They took the name Hasidim, which means *the pious* or *loyal.* Their idea was to concentrate on the study of the Torah, to observe the Law, and to vigorously reject Greek culture. Yet many Jews saw the new secular developments as supporting the culture that they loved and the king who defended them. Pious and apostate Jew alike had long taken pride in Jewish military heroism in service to the emperor; Maccabees II maintains that the most famous of Seleucid victories, over the Gauls in 275 BCE, was the result of the glory of the Jewish troops.[14] And taking funds away from the traditional Temple expenses did not anger everyone. Indeed, some were happy to strike a blow against an old and constricting system that was literally burning up the community's resources and tying their hands. The Sukkot chapter of the Talmud tells of a woman named Miriam, a relative of the new high priest Menelaus, and, like him, a defender of the new times. Apparently, she marched into the Temple, "struck against the corner of the altar with her sandal, and said to it 'Wolf, wolf, you have squandered the riches of Israel.'"[15] This bright vignette helps illuminate the moment: Savvy, upperclass Jews of the time preferred the advantages of secular Hellenistic society to what they perceived as a narrow, parochial Judaism. Miriam's gesture claimed that the resources of the Jewish people were better spent on the pageant of culture and the needs of modern life than consumed in sacrifices to God.

The progressive, acculturated Jews began to lose hold of things when war broke out again between Seleucid Syria and Ptolemaic Egypt, and a rumor spread that Antiochus had been killed in battle. When the rumor got to Judea, nationalist and Hasidic groups began rioting in celebration. Jason returned from exile—speaking, we imagine, for a group of Jews ill-represented by either devout Hasidim or secularist Menelaus. The rumor was false, though, and Antiochus returned from Egypt to Jerusalem. On the way back, he quelled a revolt inspired by Jason, and when he got to Jerusalem he looted the Temple for gold in punishment. That widened the divide between the secularists and the lovers of the Law.

What happened next, apparently, is that Menelaus petitioned the emperor that the Jews might live under the common law. He proposed a decisive step away from the symbols and practices of separatism. Secular Jews must have already been living this way. What Menelaus did was assert his way of life over the zealot Jew who was being so resistant to the empire. He asked for his people to live within the ordinances of the gentiles, and to make illegal the signs and ways of the old laws—to reject isolationism and ritual purity, the immobilization of business, and the shunning of new pleasures. Circumcision, regular worship, Torah study, the keeping of the Sabbath—it was all separatist and should all be left behind. Antiochus agreed to include the Jews within the law of the gentiles, and the rites of Jewish law were at once made illegal. It seems a strange law for a Jewish man to devise, but as we have seen, these rites had been fashioned by Ezra only about half a millennium before and held different meanings for the Hellenistic Jews than they do for Jews today. Menelaus seems to have been trying to rid Judaism of what he saw as an unnecessary, backward-looking obscurantism.[16]

When the Hasidim went on practicing Judaism after these proscriptions, they became the world's first martyrs. Many went passively to their deaths in defense of their ideals. This history is predominantly told in support of martyrdom, in celebration of the shining remnant of pious Jews. That is not the only way to see it. The story of Chana and her seven sons is a favorite of Jewish martyrdom. Antiochus asked each son to bow to an idol. One after another, from the eldest down to the boy of two years, each hotly refused and went out to die before his mother's eyes. As the little one was being escorted off to be killed, Chana is supposed to have shouted: *Go to Abraham and tell him he bound one son to the altar, I bound seven, and mine were for real!* She went mad days later and fell off her roof to her death.

There were many Jews who would look on such a response as bizarre. Upper and middle-class educated Jews were eager to bring the special qualities of Jewish culture into harmony with the great culture of the Hellenistic world and were happy to distance themselves from the more zealous aspects of Jewish practice. Menelaus asked all Jews for a symbolic show of solidarity with the universalist ideal: a sacrifice on a pagan altar. The Temple was now to be a universal ecumenical center, open to all. An interdenominational god was to stand as a single focus for all, and the god to represent this equality was the old Olympic god Zeus: at this point, almost a secular symbol of Greece. It was not a big deal at all, unless you were a traditionalist Jew who was not part of the Hellenistic culture, in which case, the decree violated the very core of Judaism: to worship no idols; to shun even casual images; to bow to no one but the invisible Jewish God. On December 15, 168 BCE, an altar to Zeus was built upon the Temple altar of sacrifice and sacrifice was made.

To some this was all a wonderful mixing of equally pleasant and sanitized doctrines in an attitude of secular communal devotion; to others it felt like grotesque defamation. In Maccabees I it sounds like Antiochus persecutes the Jews; all other sources, including Maccabees II, speak of the more secular, Greek-influenced Jews as apostates and not, say, victims of forced conversion. Historian Paul Johnson writes that it "was not so much a desecration of the Temple by paganism as a display of militant rationalism."[17] Bickerman tells of Jewish writers borrowing the Homeric style for biblical descriptions, along with Jewish translations of Greek texts that freely translate *Zeus* to *God.*[18] We have a letter from the Samaritan Jews to the emperor from this same period, in which they ask to be thought of as separate from the zealous Jews of Jerusalem, despite their kinship, and request "that the temple without a name be known as that of Zeus Hellenios," that is, they asked him to kindly reconsecrate their place of worship, the sanctuary on Mount Gerizim, to Zeus.[19]

So we know that some Jews avidly welcomed the secular Greek culture while others mildly went along with it. And although we do not hear of it, we can assume that since there was a real likelihood of physical punishment, some Jews went along with the injunction against Jewish ritual because they were afraid. We just do not know how the split went in these early days. Some were secular and sober, boldly walking into the public space of their shared world and insisting that they had a right to do so. Jews who wanted to follow the Law in Jerusalem were obviously infuriated. Beside worldly Jews who saw themselves as important members of an exciting new cultural movement, pious Jews may have felt humiliated, left behind, and frightened. Their taxes supported the reformist program, and now their way of life had been outlawed.

One of the major voices in the legalist, pious Jewish reaction was a priest named Mattathias Hasmon. When the king's commissioners came into the town of Modein for the sacrifices, Mattathias showed up flanked by five militant sons and a number of other followers and put on a display of refusal. As he was finishing, Maccabees I tells us, "A Jew came forward in the sight of all to offer sacrifice upon the altar in Modein, according to the king's command. When Mattathias saw it, he burned with zeal and his heart was stirred. He gave vent to righteous anger." Running to the altar, Mattathias killed the offending Jew and also killed the emperor's officer overseeing the sacrifices. Thus a great revolt of traditionalist Jews began. What had started it was that Mattathias killed a fellow Jew, either because that man felt forced into making sacrifice or because he, like Jason, Menelaus, and Miriam, doubted the relevance of strict Jewish law in the real world of his real life. He may have been just trying to get ahead, a bit of a toady, and not very interested in religion one way or another; or he might have been an enthusiastic man, having fun. If the first case is true, then Mattathias killed the man for failing to make a martyr of himself. If it was not just avoidance of death, though, if it was a variation on deciding to be part of the wider culture and ignoring the proscriptions of one's parents' (or even parents' parents') religion, then this unnamed Jewish man was a martyr for doubt. That he went first suggests he was an actual volunteer. He has been remembered in history, heretofore, only by the words of those who killed him, and therefore remembered as a silence rather than a voice.

Mattathias and his sons fled to the mountains. They were joined by many Hasidim, and soon enough a guerrilla war began in full force. When Mattathias died in 166 BCE, the leadership passed to his son Judah, called Maccabeus, which means "hammer." Judah apparently was one; it is from this bloody nickname that the family name of Maccabee is derived. Judah defeated an expedition sent from Syria to squash his revolt, then occupied Jerusalem, killing sinners—Hellenized Jews—and forcibly circumcising others. Circumcision performed on grown boys and adult men, by force and in anger, is not just a strong hand leading the lost sheep back to the flock. Rather, it's an extraordinary attack on the masculinity and the privacy of those who had chosen to live as Jewish citizens of the wider Greek world. The shift in the status of women must have been enacted in some crushingly brutal way. The Talmud tells us that after the Maccabees took power, Miriam "was punished." The Hammer reconsecrated the Temple in 165 BCE, and Hanukkah celebrates the event. As we learn from Maccabees I, it took another year before the progressive Jews were starved out of the stronghold of Gezer, and the citadel of Jerusalem, and chased into exile.[20] Note that in 162 BCE, Judah Maccabee strode a battlefield after a victory and was appalled to find pagan amulets on the corpses of his own fallen soldiers.[21] Even there, a little doubting.

The emperor didn't give up, but to everyone's surprise the Maccabees were able to keep winning against impressive forces and were thus able to set up the Hasmonean dynasty. They couldn't have done it without Rome's support, so they had traded one master for another, but, for the time being, this new one was less interested in the Jews' internal affairs. And that was the end of that particular experiment in interdenominational temple worship.

As I've said, Hanukkah is the celebration of the revolt that reclaimed the Temple, and it is generally remembered as marking a clash between powerful pagan oppressors and determined Jewish victims. But the revolt's first victim was a secular Jew at the hands of a zealot Jew, the further struggle entailed murder and forcible circumcision, and it ran the cosmopolitan Jewish way of life right out of town. When secular Jews celebrate this one, they might want to do it with some care—let one candle burn for the other side. If Judah and his hammer make one kind of hero, devoted to one kind of wisdom, Miriam and her little sandal make another kind of hero, storming in defense of a wider field of wisdom and the virtue of an open mind. And let us not stop there. Chana and her sons were martyrs, but so was the Jewish man Mattathias killed, and just as the pious Jews were persecuted, so, too, were the men forcibly circumcised, and the boys torn from their gymnasium. Again, we cannot say much about the shift in education, mores, and general choices for girls and women, but we can surmise that the public life of Jewish women was suddenly and forcibly curtailed.

In the Hellenistic Diaspora, Jewish-Greek sophistication proceeded merrily onward. We learn much from their opponents: Philo denounced those "who express their displeasure with the statutes made by their forefathers, and incessantly censure the law." He also took it for granted that sons of wealthy Jewish merchants would attend gymnasium.[22] As we've noted, Philo also tells us that many non-Jews took part in the annual Jewish festival on the island of Pharos to celebrate the translation of the Septuagint.[23] In the same period, the Seers added: "Cursed be the man who rears a pig and cursed be those who instruct their sons in Greek wisdom," which certainly suggests that some Jews were raising pigs among their animals and were raising children among the Greeks.[24]

Despite the separatist rigors of the pious Jews of Jerusalem, Greek culture had a powerful impact on Jewish thought. To quote Bickerman again, "The fear of death . . . was the impelling force behind the Greek mystical current in the early Hellenistic world. The achievement of the Pharisees was to channel this current into the mainstream of Jewish tradition."[25] In the next centuries, the branch of Judaism most concerned with this Greek idea of life after death would create a new religion. Their two great ideas—a single, caring, ethical God, and a life after death—were joined together in Christianity, and in this form took over much of the world for a period of two millennia and counting. But both Greek and Jewish culture had traditions of secularism that also grew, and they, too, would learn more together. Theirs has been a story of sophistication, cosmopolitan delights, progressive politics, tolerant intellectualism, and comfort with different but urbane neighbors. The Jewish-Greek secular mix has been as fruitful as the religious.

JOB

The earliest evidence of doubt in the Hebrew Bible is in the psalms, all, it seems, written by the fifth century. The first psalm mentions avoiding "the counsel of the ungodly," indeed, in this very short psalm "the ungodly" are mentioned three more times: he who delights in the Law of the Lord is prosperous, "the ungodly are not so," "therefore the ungodly will not stand in the judgment," and furthermore "the way of the ungodly shall perish." Psalm 14 says that the fool, in his heart, says there is no God.

We do not know the date of the Book of Job either, but we can get closer.[26] Many scholars believe it to have been written between 600 and 400 BCE; some think a little later, i.e., in the early part of the Hellenistic period. Whatever the date of the biblical book, the folk story of Job as a good man who suffered, kept his faith, and was rewarded had been around for a long time. We are by no means certain that the folk story was of Jewish origin, or that Job himself, if he existed, was a Jew. Someone, somewhere, wrote a poetic masterpiece that took that Job story, a story of faith, and reimagined it as a philosophical question—as a moment of truth. This, then, came to be included in the Hebrew Bible. The biblical Job is a story of faith, but a faith that is pushed past its limits, into fury, revolt, and doubt.

The biblical Job was a "perfect and upright man."[27] He was pious, generous, and kind, and as a result, he was extremely prosperous. He had a great household: a good wife, seven sons and three daughters, many serving hands, and so much land and cattle that he was "the greatest of all the men of the east." He fed the hungry, looked after the fatherless, welcomed the stranger, and was honored for it. He was so righteous that every day he made extra burnt offerings to God just in case his sons ever "cursed God in their hearts." The drama begins when, gazing upon Job, God bragged to the devil of the good man's piety. The devil made light of it, telling God that it was Job's prosperity that kept him pious. God insisted that Job would stay righteous even without all his blessings, and the two embarked upon an awful experiment.

In one of the most brutal and crafted scenes in the Bible, God sends a messenger to Job reporting that the Sabeans have stolen his oxen and ass and have slit the throats of his servants in the field. "While he was yet speaking," another messenger shows up and says that Job's sheep, and those tending them, have been burned to death by fire from the sky. Another messenger arrives, again while the last is still speaking, this time announcing that the Chaldeans have stolen the camels and killed the remainder of his working hands. Finally, and once again before the last messenger has finished speaking, another messenger appears. This one tells Job that a fierce wind has blown out of the mountains, knocking down the house in which his entire brood of children, all adults, were feasting. All are dead. Job's response to this is to rip his clothes, shave his head, and fall on his knees, saying, "The Lord gave and the Lord hath taken away; blessed be the name of the Lord." Job bore his sorrow and trusted God.

When God then asks the devil if he is convinced, Satan shakes it off, saying, "Skin for skin, yea, all that a man hath will he give for his life"; that is, Satan promises that if he hurts Job right down to his flesh, the righteous man will curse God to his face. God tells Satan to go ahead and do as he pleases, cautioning only that Job must survive the ordeal. Thus invited, Satan covers Job in "sore boils" from head to foot. The devil, by the way, does not show up much in the Hebrew Bible, and even in this story he disappears at this point. In any case, Job's response to his sudden skin disease is to take a broken piece of pottery—to scratch himself—and go sit in the ash pile. His wife howls at him: "Dost thou still retain thine integrity? Curse

God and die."[28] But Job stays righteous, answering that having taken the good he now must not refuse the evil. As in the older folk story, the biblical Job passes the test. He has the stomach to bear calamity and not reject the feeling of submission and trust he has for his God.

The philosophical end of the matter was not so easily borne by grit and determination. As the story continues, three of Job's prosperous, pious friends come to counsel him and help him through his grief. When they see him they can hardly recognize the man they knew. They rip their clothes at the sight of him and put dust on their own heads in sympathy and sadness. Job's sorrow is so great, his loss so inestimable, that for seven days and seven nights they sit with him in silence.

Finally Job speaks to them and, in beautiful verse, wishes that he had never been born or that he would die now. "Why died I not from the womb? Why did I not give up the ghost when I came out of the belly?"[29] As his friends try to answer the lament, they upset Job so keenly that it produces a new kind of doubt. It is here that the biblical Job pulls away from the folktale. In the folktale, Job suffers, but waits faithfully for decades upon decades until God shows up and gives him riches, new loved ones, and long life. The biblical Job does not wait in silence for long, and it is because here there is an attempt at rational conversation and it leads quickly to a paradox. The friends attempt to help Job make sense of what has happened to him. The first friend argues that, despite the calamity, the world is still a good and just place, worth living for, and that God is great and wise. In fact, he suggests, Job must have deserved this punishment, since it was happening, so it must be all for the best. "Behold," offers this first friend, "happy is the man whom God correcteth."

It is not the worst take on one's own suffering, especially if you have a sin or weakness for which you wish to repent. It is a bit unkind when offered from one friend to another, in a time of need. But worse, Job's suffering was too extreme to fit the model. He had not been overtly sinful, and there's no way ambient misdeeds or vague bad thoughts could possibly account for the horror his life has become. Job knows his own innocence and, given his level of pain, he cannot tolerate the suggestion that he brought this upon himself—it doesn't make emotional or intellectual sense. Job's other two friends, and then a "young man" (scholars think the young man's part was added later), each deliver a speech attempting to rationalize Job's misfortune. None questions that God has the power to create or prevent all these circumstances. All insist that God is good and just and deals honorably. It is in reaction to this defense of providence that Job's critique of God begins to escalate.

After each of the visitors' speeches, Job lets fly a torrent of accusations and challenges. God is described as wildly powerful but inanely capricious. He is not only unjust, he is uncaring: "Behold, I cry out of wrong but I am not heard: I cry aloud, but there is no judgment. He hath fenced up my way that I cannot pass, and he hath set darkness in my paths."[30] Not only is God ignoring Job and failing to deliver justice, he's an actively nasty, aggressive power: "He hath destroyed me on every side, and I am gone: and mine hope hath he removed like a tree. He hath also kindled his wrath against me, and he counteth me unto him as one of his enemies. . . . My flesh is clothed with worms and clods of dust; my skin is broken, and become loathsome." God is deliberately and gratuitously vicious. When the reach of Job's indictment begins to frighten his friends, they redouble their insistence that the world, and with it Job's suffering, is fair.

Job says, "Miserable comforters are ye all," and we can see what he means.[31] He begins to catalog the great darkness of his suffering in contrast to the certain light of his innocence. The effect is a visceral reminder of the experience of injustice and wrongful humiliation. To one friend Job explains, "He teareth me in his wrath, who hateth me: he gnasheth upon me with his teeth; mine enemy sharpeneth his eyes upon me." Further, "God hath delivered me to the ungodly, and turned me over into the hands of the wicked. . . . He poureth out my gall upon the ground. . . . My face is foul with weeping, and on my eyelids is the shadow of death; not for any injustice in mine hands: also my prayer is pure." To another friend Job delivers a yet sadder lament, hollering, "He hath put my brethren far from me, and mine acquaintance are verily estranged from me. My kinsfolk have failed, and my familiar friends have forgotten me. . . . My breath is strange to my wife. . . . Yea, young children despised me; I arose, and they spake against me." Job says that God has put his feet in stocks and let him waste away "like a rotten thing."

When his guests insist upon God's justice, Job is pushed into considering God's dealings not only with himself but with all humankind. He draws a poignant picture of the fortunes of the corrupt.

Wherefore do the wicked live, become old, yea, are mighty in power? Their seed is established in their sight with them, and their offspring before their eyes. Their houses are safe from fear, neither is the rod of God upon them.

Their bull gendereth, and faileth not; their cow calveth, and casteth not her calf. They send forth their little ones like a flock and their children dance.[32]

And afterward, whether or not a man or woman has had a chance to enjoy life, Job mutters, it ends the same for everyone: "One dieth in his full strength being wholly at ease and quiet. His breasts are full of milk, his bones are moistened with marrow. And another dieth in the bitterness of his soul, and never eateth with pleasure. They shall lie down alike in the dust and the worms shall cover them."

The poetry of it is arresting. It plays us through his fall over and over, from good health to festering agony, from the top of social blessings to the worst of curses, and each time it pounds us with the question, How could it be just for a good man to suffer such agony? Prior to God's bet: "When the Almighty was yet with me, when my children were about me. . . . The young men saw me, and hid themselves: and the aged arose, and stood up. The princes refrained from talking, and laid their hand on their mouth. . . . Unto me men gave ear, and waited and kept silence at my counsel. . . . And they waited for me as the rain."[33] They treated him thus because not only did he do no evil, and not only did he help those who asked him, Job actually sought out misery that he might give succor:

I delivered the poor who cried out, the fatherless and the one who had no helper. The blessing of a perishing man came upon me, and I caused the widow's heart to sing for joy. I put on righteousness, and it clothed me; my justice was like a robe and a turban. I was eyes to the blind, and I was feet to the lame. I was a father to the poor, and I searched out the case that I did not know. I broke the fangs of the wicked, and plucked the victim from his teeth.

Now consider his present days: "Now, they that are younger than I have me in derision, whose fathers I would have disdained to have set with the dogs of my flock. . . . They were children of fools, yea, children of base men: they were viler than the earth. And now I am their song, yea, I am their byword. They abhor me, they flee far from me, and spare not to spit in my face."

How could this be fair? Job never refers to the specific texts of Judaism, but for any Jew reading the text soon after it was written, the key question would have been, Could Isaiah have been right about providence, given this level of suffering, and this level of innocence and goodness? In God's world, Job accuses, the vicious get away with everything:

> Some remove landmarks; They seize flocks violently and feed on them; they drive away the donkey of the fatherless; they take the widow's ox as a pledge. They push the needy off the road; . . . some snatch the fatherless from the breast, and take a pledge from the poor. They cause the poor to go naked, without clothing; and they take away the sheaves from the hungry. . . . The dying groan in the city, and the souls of the wounded cry out. . . .[34]

Yet God does not charge them with wrong. It gets worse. The murderer "rises with the light" and "kills the poor and needy," and gets away with it. "The eye of the adulterer waits for the twilight, saying, 'No eye will see me'; and he disguises his face," and gets away with it. "In the dark they break into houses, which they marked for themselves in the daytime"; for such premeditated evil, "their portion should be cursed in the earth." Why should the cruel prosper? "The womb should forget him, the worm should feed sweetly on him; he should be remembered no more, and wickedness should be broken like a tree. For he preys on the barren who do not bear, and does no good for the widow."[35] Justice here is not merely absent, it is perverse. God allows a man who "does no good for the widow" to prosper, but he aggressively destroys Job's wealth and, with it, Job's capacity to continue serving the needy of his community. Job never formulates the idea that human beings have justice and that God has none, but he lays out the argument: Job looks after widows and orphans, God steals their support and coddles their enemies.

Job finds God responsible for great cruelty and the indictment is most effective in its details: "When I say, 'My bed will comfort me, my couch will ease my complaint,' then you scare me with dreams and terrify me with visions, so that my soul chooses strangling, and death rather than my body. I loathe my life."[36] The Job author here lets Job defend his brazen doubt, saying: why hold back since we all die anyway? Job moans that it is wanton cruelty that suffering people, brutalized or in mourning, should have to relive their horrors in dreams. In fact, why should any of us, in seeking rest, be met with nightmares? Job is scared by making such accusations, but encourages himself with a reminder of mortality—angry at it but keenly aware of the refuge from God it provides: "Oh, remember that my life is a breath! . . . As the cloud disappears and vanishes away, so he who goes down to the grave does not come up. He shall never return to his house, neither shall his place know him anymore. Therefore I will not restrain my mouth; I will speak in the anguish of my spirit."[37]

Over and over Job asks why God is absent and why wisdom is hidden and wishes he could just have his day in court. And God, strangely enough, shows up and responds. He does not, however, offer a court; this isn't wish fulfillment, it's philosophy. There is no court. What God does is to show up as he is said to show up: in natural wonder and paradox. With much sound and fury, God appears and begins to list all the things he has done and the secrets that he knows. "Where were you when I laid the foundations of the earth?" he roars, and what follows is one of the best tirades ever written. Some key phrases include:

> Have you walked in the depths of the ocean? Have the gates of death been opened to you? Where does light come from? And where darkness?
>
> Hast thou entered into the treasures of the snow? Hast thou seen the treasures of the hail? Hath the rain a father? Who hath begotten the drops of dew?
>
> Canst thou bind the sweet influences of the Pleiades or loosen the bands of Orion?
>
> Out of whose womb came the ice?
>
> Who has put wisdom in the inward parts? Or who hath given understanding to the heart? Who can number the clouds in wisdom? Or who can stay the bottles of heaven, when the dust groweth into hardness, and the clods cleave fast together?
>
> Wilt thou hunt the prey for the lion? or fill the appetite of the young lions, when they crouch in their dens, and abide in the covert to lie in wait?
>
> Gavest thou the goodly wings upon the peacocks?
>
> Hast thou given the horse strength? Hast thou clothed his neck with thunder? Canst thou make him afraid as a grasshopper? The glory of his nostrils is terrible.[38]

God here raises all the pertinent questions: the origin of consciousness and wisdom, the nature of death, the majesty of the stars, the wild animals, the complex wonders of nature, the magic of mechanics, the hugeness of the planet. He's even got the sheer awesome display of the horse, not only as a gorgeous, striding power, but also as a little, mortal creature, timorous as a trapped grasshopper, yet breathtaking in its glorious terror. That glory and quivering are embodied in the same animal seen from two perspectives is a reminder of the paradox of scale; and a reminder that this God is presented as solving that problem by inhabiting all realms, from the infinitesimal to the inconceivably immense.

God also asks Job whether he gave "wings and feathers unto the ostrich." The question is a kind of joke since ostriches cannot fly, but it's a serious joke. Life isn't only good, it is wonder-full. It feels so inventive that people look to an imaginative capability that transcends that of humankind. When God describes the ostrich he does not exalt her—he teases, instead, that her strangeness is beyond the understanding of human beings. She "leaveth her eggs in the earth and warmeth them in the dust, and forgetteth that the foot may crush them, or that the wild beast may break them. She is hardened against her young ones, as though they were not hers: her labor is in vain without fear; because God hath deprived her of wisdom, neither hath he imparted to her understanding. What time she lifteth up herself on high, she scorneth the horse and his rider." The image of the ostrich is inexplicably foolish, yet so magisterial in her own mount that she scorns the horse and rider. And this is matched by the brave strangeness of a great variety of creatures. The eagle seeks her prey and "her young ones also suck up blood: and where the slain *are,* there *is* she." This is how God accounts for himself. He does not say, Here is the proof of justice or of my existence; he simply cites the weird glory of the natural world. The Job author says that life may make us reject divine justice, but nothing is going to help us explain the bloodthirsty nestling or the seven-foot, flightless bird, wandering off from her eggs with her head in the sky, oblivious.

It is interesting that God includes in this list of accomplishments the creation of some of the nasty creatures mentioned by Epicurus to argue against the idea that the gods made the world. The Hebrew God is saying, *You couldn't have a clue how I made the crocodile* (here called the leviathan), *let alone why.* The crocodile is dangerous and not particularly beautiful; its relationship to humanity is largely antagonistic and often fatal. God's point is that it is intolerable hubris for Job, or his well-meaning friends, to attempt to explain God's actions.

God concludes with a stunning portrait of a crocodile, in one chilling description after another—"He esteemeth iron as straw, *and* brass as rotten wood. The arrow cannot make him flee. . . . Darts are counted as stubble:

he laugheth at the shaking of a spear." Again the descriptions are about imaginative excess: when the crocodile plunges into the sea he makes it bubble like a boiling cauldron.[39]

At this point, Job throws himself down before God and admits the fault of his presumption: "Therefore have I uttered that I understood not; things too wonderful for me, which I knew not . . . I have heard of thee by the hearing of the ear; but now mine eye seeth thee: wherefore I abhor myself, and repent in dust and ashes."[40]

God's response is first to punish Job's three friends for getting everything wrong, and then to return all of Job's estranged friends and living relations to him along with his livestock (twofold), and again Job is given seven sons and three daughters. Indeed, the new daughters are the most beautiful in the land, and it is specified that Job gives them inheritance along with their brothers. Job lives another 140 years, and is thus able to see his sons' sons down four generations before eventually dying, old and full of days.

Many people think of the Book of Job as a story of faith and divine justice, and it is not surprising. Throughout history, Church and rabbinic readings have managed the problem of the Book of Job by reverting back to the folktale: they'd excise or play down the middle section of the poem—the rebellion—and thus make it a simple tale of faith and patience. Others have dealt much more frankly with the text but still found justice in allegory or interpretation. Most of those willingly admit that to locate an interpretation of divine justice in this poem, they had to set aside some quandary, like the bet with Satan in the beginning, or the actual death of Job's first set of children. But in the text God made a bet with Satan, slaughtered a good number of innocent people and animals, and allowed Satan to torture Job with a skin disease. Job could handle that, but when his friends said this all fell within a rational system of divine justice, he was infuriated. The clash that follows is described by modern philosopher Martin Buber as a clash between Job's emotional belief in God and the more legalistic "religion" of his friends: "there now came and sought him on his ash heap *religion*."[41] So, "[i]nstead of the 'cruel' (30:21) and living God, to whom he clings, religion offers him a reasonable and rational God, a deity whom he, Job, does not perceive either in his own existence or in the world, and who obviously is not to be found anywhere save only in the very domain of religion." Unlike his friends, "Job knows of justice only as a human activity, willed by God but opposed by His acts. The truth of being just and the reality caused by the unjust acts of God are irreconcilable."

Buber describes Job's searching for justice in the world and only finding it within himself. Consider Job before his friends arrive. It seems his God was among the most shadowy versions of imposing humanness on the universe. Job seems to have fully accepted the idea that sometimes you're a winner and sometimes you're a loser and you've got to learn to take it on the chin. Sometimes a wind blows out of the mountains and knocks your house down, crushes the ones you love, leaves a hole in the sky where there was a beacon. The friends insist that justice is orderly, that the universe works according to decisions and justice. Job chokes on the notion because, as Buber explains, he recognizes these to be traits of his own much more than they are traits of the mind of the universe, of God. Buber called Job "one of the special events in world literature, in which we witness the first clothing of a human quest in form of speech."[42]

When God does show up in the end, he does not even address Job's questions about justice. Leaving behind divine justice, he instead touts all the remaining mysteries and paradoxes: Where did humanity come from? Where did the earth come from? The animals and plants and geography seem too beautiful and complex to have simply occurred on their own. How do the heavenly bodies move so perfectly in their orbits? What is consciousness? What is emotion? How can we understand the varieties of scale from the infinitesimal to the universal? Or the varieties of time from lives that last an instant to the never-changing stars? What is death? God's argument is of the heap variety: its individual tenets are magnificent but not conclusive, yet the aggregate effect of the whole extraordinary list is persuasive. It is enough for Job. Modern science, of course, has seriously thinned the pile, but some tough existential questions still remain.

God's list in the Book of Job provides an interesting parallel to the night sky; today we've got explanations for the heavens that do not rely on God and, in any case, we are not confronted with the night sky in the same way we used to be, because so many of us live in well-lit areas that block out the show. We have an explanation for most of the natural wonders in Job; we have also succeeded in chasing away most of these wonders. Our poets now talk about bridges, skyscrapers, and planes along with hawks and wild horses, so that we feel the opposite of how Job did (small and powerless). Also, some things God mentions are inventions, like unicorns, and these just aren't as impressive as they once were.

The Job author had God appear and cite the mysteries of the world and barely mention justice—he does not, for instance, rehearse for Job the intricacies of communal justice versus individual justice, or challenge him with stupefying moral quandaries. Since God has overtly changed the subject and argues his outraged supremacy on other grounds entirely, it is reasonable to see the Job author as saying that God is the sum of all the world's secrets and powers but unconcerned with justice. It may be tempting to interpret God's speech as meaning that God's justice is above our understanding, but he does not say that. He is outraged that Job is questioning him, but he does not offer comfort and say, *Foolish child, it all works out.* He just says, *How dare you?* We must remember that he has killed a number of people beloved by Job, on a vain bet, and never speaks to it. He lets the devil toy with Job's body, on a vain bet, and never speaks to it. There is no afterlife for God to make good on people's innocence or guilt. And he's told us it is wrong to assume that those who suffer are guilty. That does not leave a lot of room for justice. God makes his peace with Job in gifts, but the implication is that Job goes off to live with his new family under a new, far less orderly, regime.

We may well ask if it is possible to envision divine mind without divine justice. Certainly it is possible to have mind without true justice—that goes for most of us. But to have as little sense of justice as the universe exhibits and still have mind? If one arrives at the conclusion that there is no divine justice, and yet still believes there is a God, then what kind of God is that? Either this God is powerless to put his care into action, or he does not care at all but still has mind, so he is a sort of sociopath. If the main reason for persisting in believing in God is that he made the world and all the creatures in it, it will be hard to argue that he does not have the power to make it a less actively dangerous and chaotic world. If he really does not care, but has mind, it cannot be the kind of mind we usually mean.

Job was never given a reason to repudiate his conclusions about justice, but faced with the remaining questions, he threw himself beneath them and was willingly trampled into submission. The pose is worth considering for a moment. With dignity, standing tall despite outrageous adversity, Job delivered his rage on injustice. Then he was reminded of the remaining questions and he essentially collapsed as an independent agent. Job tells God that having seen him with his own eyes, "I abhor myself and repent in dust and ashes."

Why abhor oneself? What had changed since Job's speeches of confident self-possession? God had not yet cured Job's illness, restored his possessions, or given him a new family, so it is not gratitude. And yet Job began to grovel. What is this groveling about? The Job story gets going when Job starts to take as his business a question he had always been humble about.

He was not an arrogant man, but he went through a horrendous experience of seemingly random pain, violence, and loss—and to his mind the experience gave him special knowledge about justice. Yet, hearing God's list, or just thinking about the existential questions it poses, Job guessed that despite the lack of providence, there was a God. He was drawn with more nobility when he was covered in skin disease and thinking for himself. Apparently, according to the Job author, facing the great questions is empowering, while submitting to a higher power is a relief, and perhaps necessary, but not especially dignified. Perhaps not Job, but the Job author, and the rest of us, have been elevated by the rebellion and the compassion of its questions. Jack Miles's wonderful literary reading of the Hebrew Bible as a biography of God offers the insight that after the Book of Job, God never speaks again.[43] God may seem to silence Job, but Job silences God. It is lovely that Job silencing God is part of the text (though likely an accidental order of the books), because it reflects a real change in the real world after the Book of Job came into it.

There is something grand about a story that tries to reconcile human beings to loss, to letting go of the things that the universe has allowed us to amass and keep for a while—including the idea that after we lose everything, there is a good chance we'll get it all back someday. Could the Job author have been satisfied with this as a parable of divine justice? It is not a parable of divine justice. It is a parable of resignation to a world-making force that has no justice as we understand justice. God comes off sounding like a metaphor for the universe: violent and chaotic yet bountiful and marvelous. The Job story is a story of doubt. God's list brings Job back into the fold, but the fight has transformed that fold. With Job, that paradigm of a just God was carried to an extreme that immediately identified the problem with the idea: the world is not just. If justice exists, the Book of Job concludes, it does so in a way inconceivable to humanity. Job asked deep questions and they have lingered for millennia.

A final word: Despite Job's tremendous contribution to our history, the figure of greatest doubt in the book is certainly Job's wife. It is she who issued the astonishing advice "Curse God and die." Women are often cast in the sneaky role in ancient texts, and those who wanted to sneer at women have made much of this depiction. But it is also true that part of the reason these stories have persisted, while countless other tales have disappeared, is that women liked them, and even liked their roles. Job's wife is a tough bird. She can see a guy who is out of luck when she looks at him, and she suggests he cash in his last two chips: allegiance to God and life. What she planned for herself is singularly unclear. We do not know, at this point, if she had any faith at all. Unlike Job and his friends, she fades in and out of the story like Satan and God. The tendency is to not notice that this woman declares a complete rebellion from God and to see her instead as an aspect of Job's own mind, the part of him that doubted. But in the actual text an actual woman—pious and good, helpmeet to a man of great right-eousness and benevolence—this actual pious woman proposes that it is time to dismiss God and be done.

The Book of Job is a section of a Bible sacred to millions of people throughout history, and so it has claimed the attention of great religious minds. The glory of the poetry itself, the subtlety of the questions, and the paradox of the text make it rich with human feeling and psychological com-plexity. Many religious people have offered interpretations of it. To see the history of doubt we have to stop rushing in and providing someone's moment of certainty, of belief. Just as the Maccabees and Hasidim stand in front of the cosmopolitan reform movement, blocking it from our eyes, so the way religious people have answered Job's author throughout history has hidden the story from our eyes. But, of course, whatever else, Job doubts.

ECCLESIASTES

Ecclesiastes is so out of sync with the rest of the Bible that scholars and theo-logians have long marveled that it got in at all. The book was too beautiful, too smart, and too pious, in its own odd way, to leave out. It is an amazing study on doubt, both in its philosophical viewpoint and in its advice on how to best live in a world without divine justice, without an afterlife, and without any overarching meaning. The author of Ecclesiastes was Koheleth (Hebrew *Qohelet*); Ecclesiastes is the Greek translation of that name.[44] We do not know much about him. Writing in the Hellenistic period, he seems at times to be referencing such Hellenistic notions as Tyche-Chance, and he often sounds like Epicurus. But as he never calls these Hellenistic phenom-ena by name, we cannot be sure of the extent of Koheleth's real familiarity with any of it. The book was written in the third century BCE (250–225), and later scribes softened the text a bit, even giving it an epilogue that takes back some of the conclusions announced in the main text. The main text is pretty strange stuff for a holy book.

Koheleth doubted every aspect of religion, from the very ideal of righteousness, to the by now traditional idea of divine justice for individuals: "Be not righteous over much; neither make thyself over wise: why shouldest thou destroy thyself? Be not over much wicked, neither be thou foolish: why shouldest thou die before thy time?"[45] Koheleth saw the relationship between merit and recompense as chaotic. One of the greatest doubting lines ever written was his, and it has flashed a glint of light in myriad lives, for millennia: "Under the sun, the race is not to the swift, nor the battle to the strong, nor bread to the wise, nor riches to the learned, nor favor to the skillful: but time and chance happeneth to them all." There is such a thing as the swift here, and the strong, and the wise. Koheleth is not rejecting the idea of value. Train, and excel, and rejoice in your labors, but don't assume the battle goes to the strong, or riches to the learned, or favor to the skillful. It just doesn't. Time and chance happens to them all.

Who is this person doling such keen advice across two thousand years? Apparently, he was a teacher, well schooled in traditional doctrines, but now with his own philosophy. He'd begun life as many do, out for a little mirth and pleasure, but he soon decided it was a vain pursuit. Good times did not seem sufficient or real: "I said of laughter, *It is* mad: and of mirth, What doeth it?"[46] So Koheleth set himself to study and accomplishment, building many houses for himself; planting vineyards, gardens, and orchards, and digging ponds of water, "to water therewith the wood." He built himself a large family of servants, herds of oxen, flocks of sheep, and his wealth in silver and gold equaled that of kings. He was wise, learned, industrious, rich, and beloved, and knew how to have a good time: "I got me men singers and women singers, and the delights of the sons of men, as musical instruments, and that of all sorts. . . . And whatsoever mine eyes desired I kept not from them . . ." And then something went wrong.

At some point, Koheleth was struck by the impression that nothing he had done really meant anything. It was a deep experience of meaninglessness, extending from a dismissal of material accomplishments to a dismissal of even wisdom and truth. "Then I looked on all the works that my hands had wrought, and on the labour that I had laboured to do: and, behold, all was vanity and vexation of spirit, and there was no profit under the sun." Koheleth began to see even wisdom as a kind of useless strife and competition: "Then said I in my heart, as it happeneth to the fool, so it happeneth even to me; and why was I then more wise?" So "this also is vanity."

In the end, "there is no remembrance of the wise more than of the fool forever; seeing that which now is in the days to come shall all be forgotten. And how dieth the wise man? as the fool. Therefore I hated life; . . . for all is vanity and vexation of spirit."[47] The real crisis was that wisdom turned out to be a sham—it cannot be obtained. Here is the birth of Jewish skepticism: "When I applied mine heart to know wisdom, and to see the business that is done upon the earth. . . . Then I beheld that a man cannot find out the work that is done under the sun: because though a man labour to seek it out, yet he shall not find it; yea further; though a wise man think to know it, yet shall he not be able to find it." And further along: "I said, I will be wise; but it was far from me. That which is far off, and exceeding deep, who can find it out?"[48] These mournful lines would be remembered throughout the history of doubt.

He began to hate everything, even life itself, because without an overall meaning everything he had created was as nothing. The idea of leaving it all for the next generation didn't help at all: "Yea, I hated all my labour which I had taken under the sun: because I should leave it unto the man that shall be after me. And who knoweth whether he shall be a wise man or a fool? Yet shall he have rule over all my labour wherein I have laboured, and wherein I have shewed myself wise under the sun. This is also vanity."[49] One might labor a whole life to amass a fortune only to leave it to an idler or a fool. All this pointlessness is so much worse, he argued, because struggling to accomplish anything is frustrating—"vexing" as most translations have it—and even tormenting. The one who works hard lives in a state of stress, "his days are sorrows, and his travail grief; yea, his heart taketh not rest in the night. This is also vanity."[50]

Yet Ecclesiastes would not be so beloved if Koheleth had offered nothing but his nihilism. The book works in its punchy, practical way because Koheleth's horror at the meaninglessness of life in the big picture is matched by his pleasure at the goings-on of daily life, and he gives splendid advice on how to focus on the mundane and thereby become happy. The tension is never reconciled because remembering death is his favorite technique for learning to be in the moment. Thus, his advice on taking time to laugh with loved ones is broken with grumblings about how empty it all is, and his grumblings are broken with sighs of contentment and joy.

The advice begins with some startling verses about accepting the moments of life in which one finds oneself, even the painful ones, and even the final one. His two-by-two list of these life-moments flattens them so that they all seem the same: manageable and sweet, like steps in a dance; only a fool would embrace some and protest others. It feels good to read:

> To every thing there is a season. . . . A time to be born, and a time to die; a time to plant, and a time to pluck up that which is planted; a time to kill, and a time to heal; a time to break down, and a time to build up; a time to weep, and a time to laugh; a time to mourn, and a time to dance; a time to cast away stones, and a time to gather stones together; a time to embrace, and a time to refrain from embracing; a time to get, and a time to lose; a time to keep, and a time to cast away; a time to rend, and a time to sew; a time to keep silence, and a time to speak; a time to love, and a time to hate; a time of war, and a time of peace. What profit hath he that worketh?[51]

There is nothing for a person to do but "to rejoice, and to do good in his life."[52]

It is a comforting speech, but as I've suggested, the mood doesn't hold. Koheleth turns directly back to the brute facts of death. He's resigned and even bemused, but he's also, well, vexed, and he stays vexed. His point this time is that if it is frustratingly unfair that the universe shows no difference in the fate of the good and the bad, the wise and the foolish—if the race is not to the swift nor the battle to the strong—it is almost comic when that lack of differentiation is seen to extend all the way down to the animals. Yet for him there is no other conclusion: men and women are animals and they die like animals, every time. The Job author gave no sign of having heard of the idea of an afterlife; Koheleth has heard of it, but does not believe it.

> For that which befalleth the sons of men befalleth beasts . . . as the one dieth, so dieth the other; yea, they have all one breath; so that a man hath no preeminence above a beast: for all is vanity. All go unto one place; all are of the dust, and all turn to dust again. Who knoweth the spirit of man that goeth upward, and the spirit of the beast that goeth downward to the earth? Wherefore I perceive that there is nothing better, than that a man should rejoice in his own works; for that is his portion: for who shall bring him to see what shall be after him?[53]

Koheleth brushed aside the dream of an afterlife with a simple appeal to reason—*Who knows this?*—and the conclusion that human beings have nothing above the beasts in this regard; "all is vanity." Then immediately, with a kind of poetic magic, he confronted the tragedy with a call to rejoice in one's own works. He used to hate life because of the awful paradox of it all, but has found happiness through a degree of reconciliation. He argues that there is no meaning to life, there is no life after death, and although this state of things is absurd and empty, life is good and one's own work is worthy of joy.

Having gained fame in wisdom and enterprise, he rejected both as vain pursuits. Now he reports that despite having abandoned meaningless struggle, he finds that life, in itself, is blighted by injustice. He now "considered all the oppressions that are done under the sun" and was struck by "the tears of such as were oppressed," that "they had no comforter," and that "on the side of their oppressors there was power."[54] The knowledge of this, of the vicious awfulness that some people have to live with, is unbearable. From this he concluded that the dead are better off than the living. Worse, he tells us, "Wherefore I praised the dead which are already dead more than the living which are yet alive. Yea, better is he than both they, which hath not yet been, who hath not seen the evil work that is done under the sun. Again, I considered all travail, and every right work, that for this a man is envied of his neighbour. This is also vanity and vexation of spirit."[55] The first part of this embodies the density of poetry: what level of horror could make a person who merely hears of it wish he were dead—and then take back the comment and wish instead that he'd not yet been born? The phrase is an effective shorthand for getting us to revisit the vilest horrors of which we have each, individually, heard. He brings into the discussion the emotion thereby produced in each reader's breast. The latter notion, about envy, is a nice little psychological insight: success is as often born of envy as of love, and it will bring envy as often as it brings love. The urge to do very well is born of wrong-thinking and doesn't make us happy.

I quote Ecclesiastes at length, rather than in bites, because it is important to see how the book creates its concentrated trance, how it multiplies its meanings with odd juxtapositions. A sentence about the misery of some lives is raised to an unbearable pitch, and then tumbles into a reminder that successful people will always be surprised at the level of envy they meet, just there where they'd hoped for affection. These are two very different ideas, but they both wake us to doubt. The collapsed distance between them is a poetic device that helps jostle us, and resettle into a more useful paradigm. The collapsed distance between them is also a hypnotic device that helps the information come in under the mind's defenses; it gives its information in a way that is so kaleidoscopic that it is hard to even begin to argue with it. Here Koheleth begins with a rant against a single person frantically climbing for success:

> There is one alone, and there is not a second; yea, he hath neither child nor brother: yet is there no end of all his labour; neither is his eye satisfied with riches; neither saith he, For whom do I labour, and bereave my soul of good? This is also vanity, yea, it is a sore travail. Two are better than one; because they have a good reward for their labour. For if they fall, the one will lift up his fellow: but woe to him that is alone when he falleth; for he hath not another to help him up. Again, if two lie together, then they have heat: but how can one be warm alone?[56]

So the rant against working constantly for yourself alone concludes in another paean to daily living. There is vexation and vanity, but there is also the sweet life: get yourself someone to share your days and warm your nights, and lift you when you falleth. The crowning suggestion is:

> Live joyfully with the wife whom thou lovest all the days of the life of thy vanity, which he hath given thee under the sun, all the days of thy vanity: for that is thy portion in this life, and in thy labour which thou takest under the sun. Whatsoever thy hand findeth to do, do it with thy might; for there is no work, nor device, nor knowledge, nor wisdom, in the grave, whither thou goest.[57]

Again, his happy advice is mixed with talk of death and misery because the real facts of death and misery are being used to awaken us into a sensibility in which we could take his advice to be carefree. Koheleth does not see transcendence anywhere but suggests instead that we may best live in this life by finding love and doing our work with devotion.

We've seen some incredible doubt in providence, justice, and righteousness, and we're closing in on the question of Koheleth's idea of God. First, though, we need to note his take on history. Koheleth's famous statement that there is "nothing new under the sun" is framed by similar claims of eternal sameness:

> Vanity of vanities; all is vanity. . . . What profit hath a man of all his labour which he taketh under the sun? One generation passeth away, and another

generation cometh: but the earth abideth for ever. The sun also ariseth, and the sun goeth down, and hasteth to his place where he arose. The wind . . . whirleth about continually, and the wind returneth again according to his circuits. All the rivers run into the sea; yet the sea is not full; unto the place from whence the rivers come, thither they return again.[58]

Everything is locked into cycles of repetition, and your efforts to do anything are like waiting for the rivers to finally fill the oceans: there's progress all the time, but you're not getting anywhere.

The thing that hath been, it is that which shall be . . . and there is no new thing under the sun. Is there any thing whereof it may be said, See, this is new? it hath been already of old time. . . . There is no remembrance of former things; neither shall there be any remembrance of things that are to come with those that shall come after.[59]

It is an extreme vision of stasis. History does not quite exist in it: generations simply replace one another, forgetting everything. There was no fall from grace and there is no progress toward a better day. "Say not thou, What is the cause that the former days were better than these? for thou dost not inquire wisely concerning this."[60] To understand the strange piety of Ecclesiastes, one must consider this idea of sameness, along with its insistence on the finality of death and the pointlessness of accomplishment. Taken together, these three ideas do not leave a great deal for God to do.

Many have suggested that Koheleth did not believe.[61] In the text he left us, Koheleth believed in God, but a pared-down God. The Lord exists, but not much. His most significant action is that he condones worldly pleasure. He doesn't mind our good time. Thus, there is nothing better for us to do than relax. Here Koheleth is directly responding to the idea of history repeating endlessly, with no memory: "Behold that which I have seen: it is good and comely for one to eat and to drink, and to enjoy the good of all his labour that he taketh under the sun all the days of his life, which God giveth him: for it is his portion. Every man also to whom God hath given riches and wealth, and hath given him power to eat thereof, and to take his portion, and to rejoice in his labour; this is the gift of God."[62]

Our proper and possible relationship with God was very limited. Koheleth tells us not to break vows to God, "for an unfaithful and foolish promise displeaseth Him"; in fact, it is better "that thou shouldest not vow."

Indeed, Koheleth enjoins humankind to leave God alone in general, to the benefit of the both: "Let not thine heart be hasty to utter any thing before God: for God is in heaven, and thou upon earth: therefore let thy words be few."[63] It's a short sentence, but it's the most Koheleth directly says about God and religion. It suggests a way of living that is radically distanced from God. After all, how else do human beings commune with God? His injunctions sound pious (do not bother God, do not risk saying something you cannot live up to, do not assume you know what to say), but they function as a command to live in a secular way—to not engage in an act that at the very least reminds one of the possibility of the divine. He seems to think that is more sensible. Interestingly, he does tag on the notion that although there is a lot of malarkey out there, there is some truth in it: "For in the multitude of dreams and many words there are also divers vanities: but fear thou God."[64] It's funny, when one thinks of how much Koheleth's philosophy resembles Epicureanism, that this is the active instruction he offers.

In keen contradistinction to the Book of Job, Ecclesiastes does not argue with God, and certainly doesn't hold the injustice of the world against him. The only way to real peace was to concentrate on the fact that there is no real justice: Death comes to everyone the same way, "to the righteous, and to the wicked; to the good and to the clean, and to the unclean," to the sinner, and alike to "he that sweareth, as he that feareth an oath." It is not fair. "This is an evil among all things that are done under the sun, that there is one event unto all: yea, also the heart of . . . men is full of evil, and madness is in their heart while they live, and after that they go to the dead." He sighs that "a living dog is better than a dead lion."[65]

This is billed as good news. There is no reason to struggle against the great, and the crazy, and the evil: we ought simply to be glad we are alive. Many kings and lions disintegrate now in their dusty tombs, and meanwhile, although we may be but dogs, we may drink cold, clear water under the blue sky. Being alive transcends all envy when we remember death. Good news indeed, but we also should not miss what a sharp and wide blade this is, and what it is cutting. Doubt may clear the path for carefree living, and may increase pleasure in the life one has—but while scything away much nonsense, you have to be careful not to get caught in the blade. After he rejects divine justice and purpose, it takes real work for Koheleth to avoid falling into nihilistic misery. He offers a gentle nudge toward a new response. There are "just men" whose lives go as if they had done "the work of the wicked," while "there be wicked men" who prosper: "Then I commended mirth, because a man hath no better thing under the sun, than to eat, and to drink, and to be merry."[66]

God here reads like another fact of nature, something one should not anger, but that is otherwise benign. He was a force within a world, and the world was chaotic. That's the bulk of what Koheleth said about God, but we can read a bit more from what he said about the world. It is worth wondering how much Koheleth could believe in a historical divinity once he had given up on change. If nothing ever changes, then God has no plan. What is more, if there is nothing new, then neither are we new; if nothing is new, individuals are the same. The distance and disinterest of God is profound here, and it is best articulated in the fact that we individual human beings are going to be forgotten. Any effort to stand out is entirely futile. Koheleth makes this clear with a tidy vignette: "There was a little city . . . and there came a great king against it, and besieged it, and built great bulwarks against it: Now there was found in it a poor wise man, and he by his wisdom delivered the city; yet no man remembered that same poor man."[67] This forgetfulness is built into both the form and the content of Koheleth's speech as he repeats and repeats "All is vanity."

Of course, he has to say all this so strenuously because there is another level at which things certainly are remembered. The man Koheleth cited as having been forgotten was remembered by Koheleth; even if the memory has grown figurative, the merit persists. Koheleth himself is engaged in a profoundly disciplined, artistic, and philosophical project—the writing of this text—but he is trying to shake off any particular ideas about recompense. He has a deep morality, too; it is Koheleth who says, "Cast thy bread upon the running waters: for after a long time thou shalt find it again."[68] For the one striving furiously to accomplish with his or her eye on rewards, for the envious who sit by and pine, Koheleth brings the news that it is all an illusion. And yet the human experience is one in which we feel different and we yearn to do something memorable, something that speaks out of our own hearts. So go ahead and speak. But remember: nothing lasts. Seen from a bit of distance, the sun, wind, and rivers—and the generations of humanity—all seem locked into laws of repetition, full of gorgeous and inconsequential variation. The most fabulous of us is like a delightfully curious little eddy in the burbling of a stream one summer morning, years ago.

That little ripple came and went unseen. Koheleth doesn't mention God in a watching capacity; this God is not interested in what we are up to—he certainly doesn't want regular reports. Our great need to be seen and understood will not be met by God, for according to Koheleth there is nothing to see; our separate identities are as interchangeable as asteroids floating around in space. We mislead ourselves about everything.

The last mention of God in Ecclesiastes is as follows: "As thou knowest not what is the way of the spirit, nor how the bones do grow in the womb of her that is with child: even so thou knowest not the works of God who maketh all."[69] We are back to the unanswered questions; the mystery of the world. How bones generate may seem less of a mystery to us moderns, since we know about the double helix and the gene, but what we know is limited: we do not have much understanding of how the whole thing actually transpires. Yet even if we knew the technical bits, it is still an example of staggering complexity and beauty. And that is Koheleth's point. He advocated resolving oneself to seeing what is true in our human situation, and that meant doubting or dismissing many hopeful beliefs. Yet this realism did not blind him to wonder: we see, with beauty in mind, a world that may not have been designed with beauty in mind. Indeed, we see with design a world that may not have been designed. That is the meat of the mystery of the world: it is not a question of how it all works; rather, it is a wonder that it all is and that it strikes us as so splendid.

Religion works through practices and so does doubt. Along with his larger philosophical message, Koheleth offers ideas on how to save your own life. Accepting the abyss at the edge of life will strip away any reason for envy or longing inside us and leave us only with the awareness that for the moment we are here. Accepting the abyss is relaxing whereas fighting it is exhausting. So the first step toward a good life is to not become entranced by extraordinary success in wisdom, business, or anything else. Control your desires to match your resources, not the other way around. Work hard, but forget worldly recompense; forget the afterlife; forget being watched or judged by God; and forget being remembered. Along with his prescriptions for love, friendship, and work, he advises contemplation of somber truths. Think dark thoughts. He says the day of one's death should be thought of as better than the day of one's birth; that it is "better to go to the house of mourning, than to go to the house of feasting: for that is the end of all men; and the living will lay it to his heart. Sorrow is better than laughter: for by the sadness of the countenance the heart is made better."[70] Let yourself know that you are small and mortal and therefore free to go about your little concerns. "The heart of the wise is in the house of mourning; but the heart of fools is in the house of mirth. It is better to hear the rebuke of the wise, than for a man to hear the song of fools."[71] His suggestion is that we try to stay awake. We remember death so that we remember life, but also so that we will be prepared when disaster comes: "But if a man live many years, and rejoice in them all; yet let him remember the days of darkness; for they shall be many," and in dark times, the concerns of good times will appear as vanity.[72] It helps to remember this information during normal times; disaster is normal, too, just spaced more widely.

Sounding much like Epicurus, Koheleth said not to fear death, "For the living know that they shall die: but the dead know not any thing, neither have they any more a reward; for the memory of them is forgotten. Also their love, and their hatred, and their envy, is now perished; neither have they any more a portion for ever in any thing that is done under the sun."[73] Koheleth names three things that die along with living beings: their love, their hatred, and their envy. Of these, two are better off dead—death is a release from them. Koheleth smirks here that our impulse to live longer is connected with proving things to people, and we may rest assured that our interest in these people is going to disappear from the universe.

Koheleth's God was more than hidden; he was uninvolved and almost entirely irrelevant. Koheleth's injunctions followed suit: Don't pray much; we all die pretty randomly; it does not matter if you are good or bad; beasts and human beings take part in the same death; we all are dust and return to dust; no one is remembered for very long; there is no plan; and nothing ever changes. Love your spouse. Get some work to do, do it with all your might; enjoy the simple pleasures of food, drink, and love. Everything else is vanity. Koheleth was a premier figure in the history of doubt. He was rationalist, while maintaining a bright sense of paradox and mystery; he suggested behaviors and meditations to learn to bear a seemingly harsh reality; and he had fabulous distribution: his book became part of one of the most famous cultural works of humankind. It is a sublime contribution to the history of doubt, and a wildly surprising message to find nestled between Proverbs and the Song of Solomon.

There was some relationship between Hellenistic society and the rise of Jewish doubt, but it is difficult to say what that relationship was. Job and Ecclesiastes clearly speak to a crisis in Judaism, and a Golden Age of Jewish Doubt. Was this crisis a result of the interaction between the Hebrew and Hellenistic worlds? James Crenshaw, one of the foremost scholars of wisdom literature, expressed it thusly: "There may indeed be some truth in the claim that the confrontation between Hebraism and Hellenism produced a compromise position, best exemplified by [Koheleth]. However the Jewish tradition alone had its share of ambiguities, and these disparities between religious conviction and actual reality found expression in [Koheleth]'s realism."[74] Historians differ on whether Koheleth knew of Epicurus, but all agree that their doctrines bear a family resemblance worth pondering. Either from shared thought or similar circumstances, Koheleth and the Hellenistic philosophers who were his contemporaries championed similar ideas of doubt at the same cosmopolitan moment in the ancient world. Hellenistic Jews found themselves part of a new world, and some clearly embraced it. It feels like the pious and anti-Greek Hellenistic Jews kept Judaism going—because we are inheritors of their version of it. Granted, without them, Judaism might have disappeared; but it might not have— there were Jews who maintained their identity in a more secular way, outside the Hasmonean state. In any case, in studying history it is not necessary to choose sides. To understand oneself, however, and one's culture, it is best to get a good, open-minded look at the losers in some of these historic victories, and note this greatest early drama of Jewish doubt. As for Job and Ecclesiastes, they are canonical texts in the history of doubt.

THREE

What the Buddha Saw, 600 BCE–1 CE

Ancient Doubt in Asia

O ur subject is the history of those who have doubted the existence of God. Asia does not have a history of much discussion of this subject, because for the East the existence of God was rarely the central question. Nonetheless, it is central to our story that a world of individuals, across half the earth and several millennia, did not live under the gaze and regulation of God or gods. We need to find out how they managed that. A great many of them believed in other unprovable, world-correcting contentions, such as karma. Although that fact helps to answer our first question about how they lived without God, it raises others: Did anyone doubt karma? And if so, was doubting karma similar to the experience of doubting God? We will see that there were those who doubted karma, and in many ways it was marvelously similar to doubting God in the West. The Carvaka of the late seventh century BCE in India were the earliest example of radical doubt in the human record. Remarkably, here in its cradle, doubt was already as joyous, mournful, sly, and calm as it would ever be, and as brash.

We begin our story with a quick study of Hinduism, both for what it will tell us about religions that do not understand the world as remedied by gods, and because the features of the Hindu world are precisely what the Carvaka would attack, along with the idea of gods. Soon after the Carvaka first issued their dismissal of the tenets of Hinduism, two extraordinary god-denying religions also came into being. These were Jainism and

Buddhism, and although they were likely inspired by the Carvaka, these two religions take the idea of a world without gods in brave new directions. We will look at each in turn.

HINDUISM

Hinduism is the earliest of the major religions on the planet. It has no founding figure or central institution, but it coheres around a collection of ideas and texts. These texts stretch far back in human history: the most ancient, the Vedas, was already put into a final form by 1200 BCE. They were brought together under the name Hinduism when British scholars in the nineteenth century turned their attention to the religions of the subcontinent.[1] The word *Hindu* is just a reference to the Indus River, as is the word *India*. In a post-Enlightenment classifying mood, they drew borders around concepts that had always flowed rather freely. The Hindus then told their own story—a defense was often called for—but they had accepted the notion of having a single religion, as such.

The Vedas were long understood by modern historians to have been written by Aryas, nomadic pastoralists whose language forms the basis of Greek, Latin, French, Spanish, Italian, German, English, Persian, Sanskrit, and most of the languages of northern India (they never got to the south). It was said that they moved into India between 2000 and 1200 BCE and wrote the Vedas. The battles recorded in the Vedas would then be the battles waged by the Aryas in conquering northern India. Further, this would suggest that the caste system in India originated from the light-complexioned Aryas' desire to keep the darker, indigenous population in check. Historians now contest this interpretation, arguing that the idea of Aryan origins was invented by nineteenth-century Europeans seeking a more Caucasian source for the great Vedic works. Instead, the Arya language is said to have diffused into India without conquest. The Vedas were then written by a more native population detailing its own struggles and its own concern with controlling a less privileged population. Either way, the Brahmans were the highest class, and the Shudras were the laborers.

In this early period, the religion of the Vedas was crowded with personalities, godly and human, but it was relatively simple in practice. The Brahmans were not merely the rich, they were the priests. The culture supported them, and in a style far above that of laborers, because they officiated over rituals and sacrifices as well as taught religion, and because they were increasingly honored as superior beings. The religion they oversaw was mostly about soma and sacrifice. Soma was the name of a plant that one crushed to make an intoxicating drink. The drink was also called soma, as was the god who personified the effects of the drug. Sacrifice was important because the Rig Veda told of gods who won battles and they had to receive sacrifice in order to do this.

The gods in the early Vedic period were a family of humanlike characters, and the problems they mediated were generally worldly problems such as hunger and power. These gods were seen as being above us in the way pets may be seen as below us, and human beings tried to charm the gods for favors. It was a utilitarian bargain with almost entirely material aims. Yet the great philosophical quests that would emerge out of the Vedic tradition were hinted at, here and there, even in this ancient text. The Vedas make reference to the oneness of the real world, and that this real world is not immediately apparent to human beings. The idea was already emerging that worldly existence goes through almost unfathomably long cycles of generation, destruction, and regeneration. People, things, worlds, and the whole universe would disappear, but existence would not. It would just keep cycling. Indeed, as time went on, the gods that were identified with these world-changing processes became central and edged other gods out of favor: Vishnu was the preserver, Shiva was the destroyer, and Brahma was the creator. They became the Trimurti, the Vedic trinity. From there things seems to have been heading toward worship of Brahma alone, either as the one god of a monotheism or as a naturalist version of the mechanics of the universe.

And then came the Upanishads. Between 900 and 600 BCE a number of texts were written that came to be seen as supplementary to the Vedas. The Upanishads were among them and are generally considered the culmination of the Vedas. They introduce a way of understanding life and the universe that has stood as an alternative to belief in God for millions of people across thousands of years. It is here that we first see the notion of transmigration, the idea that people are born and die, and are born again and die again, over and over. The process is called *samsara*. It is in keeping with the Vedas' sense of natural, cyclical change, but it is something new, in part because of *karma*. According to the idea of karma, when people die they are reborn into an embryo and start again—and everything about the new infant's life, from its caste to its daily luck, is appropriate to how the reincarnated soul had behaved in past lives.

The evidence for this is interesting. First of all, the continuity of life suggests more continuity. The general reported experience is that every time we wake up, we find ourselves the same—we cannot remember everything we have done, but we feel it has always been us, ourselves, acting. The fact of our own continued presence so far, rationally suggests more of the same. Many people say that they cannot believe they are going to disappear entirely from this world. Second, sometimes people seem to have firsthand knowledge of things that they have never experienced. For millennia, the West casually believed in the inheritance of acquired characteristics: we believed that a carpenter's son would be born more apt for carpentry than he would have been if the father had taken up dentistry instead. In fact, a half-century before Darwin, Jean-Baptiste Lamarck suggested this as a mode of evolution: animals who stretched their necks to reach the highest leaves would have offspring with slightly longer necks until, at last, the process produced the giraffe. By the early twentieth century it had been generally accepted that genetic information is not influenced by behavior, but we can see why the feeling persisted so long. It speaks to the mystery of personality, and even more to the mystery of the way human culture perpetuates itself. Its transmission can be so subtle and ubiquitous that each generation seems oddly possessed of a wealth of untaught knowledge and habits. Karma has that going for it, too: it explains and supports the continuity of the community, and it legitimizes all sorts of untutored authority.

The third great suggestion of karma is the unfairness of things. The idea of karma is not just that we are endlessly reborn, but that we keep getting born into situations that we earned in a past life: we climb up and down the philological ladder of animals, human beings, and gods. Animals move up the rungs, life by life, rather automatically, but once an *atman,* a soul, reaches the human stage, its progress responds to its behavior. The idea is generally phrased in terms of good karma: one tries to store it up in order to move forward rather than stay the same. But if you are very bad in one life, you will die and become a less enlightened, more miserable creature with an even shorter life span—the beetle is a favorite horrid fate for a human being. It is all very gradual, over almost endless human lifetimes, but if you are good, your incarnations are ever more sophisticated, happier, and longer-lived. As a social doctrine karma justified the differences between the social strata as innate and deserved. You are who you are because you deserve it. One moves to a higher station in the next life by loyally fulfilling the obligations of one's present incarnation.

Gods in this schema are beings that live about a thousand years in extraordinary states of peace and bliss. But even so, gods are subject to rebirth and if they behave very badly over the course of their long lives, they can be demoted down to human beings again—sometimes they can even be brought down to beasts of burden. It is good to note that a culture concluded that endless life on earth was the true reality of the human condition, and furthermore that the prospect of these endless lives was awful. Apparently, the idea of eons of repetitive effort, strain, suffering, and death gets unpleasant. Each of these lives ends in a variety of decay, disease, mutilation, and death. Over and over. People wanted out. Their primary spiritual desire was to attain release, *moksa,* from the ridiculous treadmill of samsara.

For along with the notions of samsara and karma, the Upanishads also introduce the idea that every once in a while, individuals can be released from the endless cycling. How did the Upanishads suggest you get yourself out of the world? The answer was renunciation: you leave the world by separating yourself from the world as much as you can, by turning inward. The Upanishads suggested that the adept leave home, enter the forest, and live a very pared-down life: sleeping on the ground, eating little, practicing controlled breathing, and meditating. By living this way over long periods, one seeks to change one's very being, to become increasingly and profoundly disengaged from the human. Other people, outside the Vedic tradition, were involved in similar psychophysical practices at this time, but the logic of renunciation and asceticism was first articulated in Upanishad texts.

What is all this renunciation about? Here it may suffice to say that certain rigorous practices of solitude, stillness, and silence seem to bring human beings to a different kind of reality—a different basic idea about themselves and the universe. By myriad attestations, the people who practice these techniques do seem wiser, calmer, and even happier. Even before we start talking about what a rigorous program of meditation can do for a person's mind, we can simply note that a little quiet can markedly change our thoughts. The concerns of daily life do not mean much in the grand scheme of things, and yet most of us spend all our waking time thinking about them; solitude, stillness, and silence get us thinking differently. But it is not easy. The classic Hindu metaphor for the restlessness of the human mind is a monkey, but one drunk on wine and stung by a hornet. Even five minutes of meditation—of keeping one's mind focused on a word or an image, for example—is as difficult as calming down that monkey for a similar spate. It takes a great deal of practice. Even just a bit of quiet, let alone specific concentration, can be difficult without training. Many people manage to hear music or talk almost all the time. The mind leaps about. All this suggests that rigorous silent concentration would entail a different way of being, and this could have a considerable effect on the mind. It seems to affect the chemical balance of the body and the function of the brain. In extreme stillness, one enters a third state, something different from sleep or wakefulness, where a different part of the mind is available. In other terms, the lack of industry or fidgeting lets one feel like a rock among the breakers.

The Hindu notion of meditation is essentially that when we can manage silence and stillness, we get a glimpse of our real self, our atman, the "I" that directs all the rest of the sensation. If we meditate longer, this true self emerges more fully. To be at peace, one must clear away everything that is not the true inner self. This "everything" includes one's own body in particular, because it is the source of so much useless, distracting, and redundant desire. This is not just an idea to be believed; one does not become wise, tolerant, and quiet by force of will. It is a transformative process: the meditation and renunciation have an effect on the adherent such that he or she really becomes a different sort of person. This quieter, wiser, more tolerant self seems to the adherent to be a truer self. At last, one would slip out of the cycle of birth and rebirth in an exquisite and final release.

What we have seen so far is the general background of various Eastern religions: The gods are usually an assumed part of the universe, but they did not create or maintain the world. Rather, the universe works according to almost mechanical principles of organization and maintenance, and life within it cycles along on its own logic of appropriate rebirth. Justice is automatic, death is an illusion. This principle takes care of a lot of the same needs as does the idea of God. Responding to the problem that we are human and the universe is not, samsara, karma, and moksa breach the gap by adding justice to the universe but insisting on deep patience from humanity. Implementing the idea involves training oneself out of humanness while at the same time coming to know, and thereby dissolving into, the wide unhuman universe. Over much of history in the East many people, perhaps most, were practicing members of several religions. There were religious tensions and oppositional relationships in the East, but often the situation was analogous to trying to get into good physical shape—you can do a variety of exercises toward the same goal. Even when the idea of an absolute God would begin to dominate one of these religions, that God was presented as an exit, moksa, release. The rest of the time, the way out of these traps was dominated by practices: ways of thinking, acting, and being that help you find release.

Hinduism does not suggest that everyone drop his or her work and go live in the forest. In this period, only male Brahmans were invited, for one thing. But more fundamentally, no one is supposed to proceed toward enlightenment until he or she wants to. The idea is that people will want to on their own because, eventually, nothing else will satisfy. First of all, the desire for physical pleasure, once indulged, does not really bring happiness. To refuse one's desire may be more repression than transcendence, but many who do indulge get bored. When this happens, people often move on to seek achievement. As a goal, achievement is a step up from hedonism, because it is communally directed—to be successful is to be valued and lauded by at least some part of the community—but the project is still rather ego-oriented. Still, Hinduism suggests that we indulge this desire so long as it satisfies. Once we grow weary of physical pleasures and tired by accolades and honors, Hinduism recommends that the next stage of life is to leave all this behind and seek internal happiness and, eventually, release.

To obtain that happiness, four major techniques, four yogas, were devised. One of the most fascinating aspects of the yogas is that they reflect the belief that people come in great varieties, and thus need various paths, yet are all capable of reaching the same truth. The *jnana* yoga is the path through knowledge. For people who live through the intellect, the best path to truth is to think about the real nature of one's self, to learn that a more real self lies somehow behind the noisy surface-self, and to spend time coming to know this deeply. There are thinking exercises to help achieve this. Westerners can recognize this notion in the very Latin roots of the word *personality,* for here, too, the daily self was seen as a kind of fake: actors take on persona when they put on a mask and sound, *sonare,* their parts through, *per,* them. This suggests a conviction that there is something behind the personality. What is more, by the logic of modern biology, all of our body's cells are replaced in any seven-year cycle, which brings up the very serious question of what part of us is actually consistent over time. It is also open to all of us to notice that every person sees the universe through his or her own eyes and must therefore be seeing a very skewed view. Hindu philosophy suggests that if we want to know reality better, we need to practice seeing outside ourselves, perhaps starting by thinking of ourselves in the third person, from a bird's-eye view. The effect of seeing oneself in this way can be quite jarring, and that is the point. This jnana yoga is supposed to be the shortest path to truth, but also the steepest. It's best for people whose minds tend in that direction anyway.

Then there is *bhakti* yoga, the yoga for those more invested in emotion than thought. Bhakti yoga is about love. Becoming one with everything might be seen as a journey in two directions: forgetting the self and embracing the whole. The bhakti concentrates on the embracing. It is a common Hindu idea that those on the bhakti path do not want to become one with all truth and reality, because then there would be nothing outside themselves to love! In the words of a Hindu classic: they want to taste sugar, they don't want to be sugar. Bhakti explicitly advises the novice to choose a god image to worship. The novice chooses this god image based on what kind of love he or she wants to express in worship: love for a child, for a parent, for a friend, or for a lover. Parent and friend may seem most usual to those in the Judeo-Christian tradition, and at Christmastime Christians worship God as an infant. They dote on his tender vulnerability. As for the lover, the Song of Solomon in the Hebrew Bible and much mystical literature of the Middle Ages has an unabashedly erotic component. Bhakti is easily the most theistic of the yogas, but it would be wrong to see it as being fundamentally different in its path or its goal. The point is still release from the false self.

The first two yogas may be thought of as primarily renunciatory: one leaves the world, enters the forest, stops doing mundane labor, and distances oneself from mundane relationships. The third, *karma* yoga, is the path to truth through work—through staying in the world—and karma yoga can be performed through either of the first two yogas: jnana or bhakti, knowledge or devoted love. The secret of getting to enlightenment while working is to do the work for its own sake, with no thought of its results or benefit. One can work for the sake of the work and thereby move one's attention away from the planning, greedy, false self or one can work for the sake of a god image, offering up one's labor in adoration, and therefore move one's attention away from the planning, greedy, false self. Either way, it takes years of intense concentration, but one eventually learns to let go, and see other people, and the world, in a stunningly new way.

The final yoga is the "royal road to reintegration," *raja* (royal) yoga, and it is understood as the most empirical of the four. Practitioners of raja yoga essentially experiment on themselves, trying to induce the state that will help them to come to truth, to see reality. Solitude and subsistence living were common approaches toward enlightenment; this was more. Raja is a varied, curious, and intense approach to finding altered states and shaking free the true self. The basic rules are to abstain from lying, injury, stealing, sensuality, and greed, and to aim for cleanliness, contentment, self-control, studiousness, and contemplation—but that is just to get the most common difficulties of life out of the way. From here, one can begin the training of postures, fasting, and controlled breathing that have helped others in their quests. Some of these techniques are purposefully excruciating. Raja is an aggressive approach to pulling the mind out of its ordinary, repetitive somersaults of thought. Buddhism was to borrow a great deal from this yoga, although with a dramatic new twist.

All the yogas offered time-tested instructions for creating a permanent rupture in our illusions and obtaining fantastic states of well-being. Is it true? Many cultures and a vast number of people have practiced and acclaimed it. With the human arts, the working proof is the clatter of recognition and release that the person reports. Apparently, these astonishing discoveries can go on and on, over years of effort and discovery. Some people seem to reach a state of being that is noticeably different and massively impressive to others. To contemplate how much a behavior may transform a person's perceptions, over time, we may consider feral children, hypnotism, psychotherapy, cures for autism, advertising, and the effects of wearing eyeglasses with lenses that invert the scene. A regular program of practices can change something, such that one thinks and sees in a different way.

Most people of the early Vedic period might feed a "forest dweller" and think of it as a karma-enhancing experience, but they did not generally go off into the forest themselves. After all, progress toward moksa was imagined as taking thousands upon thousands of lifetimes, and for one who had reached the Brahma level the fast track was a collection of extremely difficult practices: some simply very hard to do and some terribly painful, involving isolation as well as physical strain. For average people, the early religion of the Vedas was about Brahman priests making sacrifice and performing rites that helped to hold the cosmos together, feeding the mechanism of the world.

THE CARVAKA

In the distant past, before even the Carvaka, we have evidence of some serious doubting minds. When the ancient king Agatasatru asked a wandering ascetic, Purana Kassapa, what were the benefits of the life of a recluse,

Purana Kassapa's answer denied that there were any. There was no justice, the guilty were not punished, and no good behavior helps: "In generosity, in self-mastery, in control of the senses, in speaking truth, there is neither merit, nor increase of merit."[2] Another figure from that shadowy time, Ajita Kesakambali, wrote that "There is neither fruit nor result of good and evil deeds. . . . It is a doctrine of fools this talk of gifts. . . . Fools and wise alike, on the dissolution of the body, are cut off, annihilated and after death they are not." These early doubters may have paved the road for what came next.

An extraordinary materialist doctrine came into bloom in India in the seventh century BCE. It was called Lokâyata and its adherents were the Carvaka.[3] We do not know very much about them, as no copy has survived of their central text, the Brihaspati Sûtra, which dates from 600 BCE. In fact, the Lokâyata texts would appear to have been systematically destroyed by the Brahman class, defending their dogma. In a pleasing twist, much of what has been preserved of the Carvaka was also arranged by their enemies: Lokâyata texts are embedded in works that argue against them. Commentaries and arguments against the Carvaka appeared until the sixteenth century CE, and one gets the sense that even then they were arguing against a living doctrine with devotees among the current population. Many of these works contained extended quotations because authors cherished Lokâyata as the best materialist to quote with disdain. Materialism is intuitive when you first start thinking about it, but it becomes extremely counterintuitive when you take it a long way in certain directions, and the Carvaka did.

The Carvaka believed that there is no afterlife whatsoever, and they thought it was pretty funny that anyone believed otherwise. The idea was that we are our bodies, these bodies think and feel, and after a while they wear out and die. Since the thinking and feeling part of us was always just an effect of the body itself, there can be nothing to live on after death. In the ancient play *The Rise of the Moon Intellect,* a personified Passion gave a little speech describing and promoting the Carvaka position. Smiling, he began by attacking the Brahma, and he was pretty aggressive right out of the gate: "Uncivilized ignorant fools, who imagine that spirit is something different from body, and reaps the reward of actions in a future state; we might as well expect to find excellent fruit drip from trees growing in the air."[4] The Brahma, he says, are uncivilized fools for pretending that there is such a thing as spirit, separate from the body, that can go off after the death of the body, benefiting from the things that spirit and body did when they were together. There is no separation, insists Passion; spirit cannot hang from nowhere any more than a ripe mango can hang in the air.

The Carvaka found claims of an independent spirit—one that could exist without a body—to be more than just wrong, they were dishonest. Not surprisingly, the Carvaka resented the righteousness of believers and their scolding of unbelievers, especially because according to Lokâyata, the truth is so plainly evident. That position could be summed up as such: the truth is what is evident. In the words of one text, "Only the perceived exists; the unperceivable does not exist by reason of its never having been perceived. . . . How can the ever-unperceived, like things such as the horns of a hare, be an existent?"[5] Passion says of the Brahma, "Assuming the existence of what is the mere creature of their own imaginations, they deceive the people. They falsely affirm the existence of that which does not exist and by their frequent disputations endeavor to bring reproach upon the unbelievers who maintain the word of truth. Who has seen the soul existing in a state separate from the body? Does not life result from the ultimate configuration of matter?"[6] Excellent questions. The Carvaka believed that the sense perceptions were the only source of knowledge. The world was what it purported to be.

The whole universe, they posited, was constructed of earth, water, fire, and air. There was no spirit or life force; consciousness was only a modification of the four elements in the body. The Carvaka themselves admitted it was strange that consciousness and life could come out of earth, water, energy, and air, but it seemed to be the truth. The description of the Carvaka in Madhava Acarya's *Sarvadarsanasamgraha* says that "In this school the four elements are the original principles; from these alone, when transformed into the body, intelligence is produced, just as the inebriating power is developed from the mixing of certain ingredients; and when these are destroyed, intelligence at once perishes also."[7] They were very explicit in their claim that, weird as it is, matter sometimes simply has intelligence, as in the case of a human being; and no intelligence can exist without matter, just hovering around somewhere. For them, the reality was that we simply are our bodies, and they loved pointing out that when someone says *I am fat* or *I am tall* they are being utterly literal; there is no *I* distinct from its body.[8] Consciousness does not imply an in-dwelling spirit.

The Carvaka thought that the ascetic's approach to life was a waste. This, after all, is the only life we have, so we ought to enjoy it as much as possible. To stick with *The Rise of the Moon Intellect* for a moment, after

Passion introduced the idea of the Carvaka, he explained that it had been taught by materialists ever since the teacher of the Brihaspati Sûtra. Then he introduced two new characters, Materialist and Pupil, who took the stage as Passion receded. Materialist began by attempting to explain the world to his student. The key was this: the chief realities of existence are pleasure and pain, and the point to life is to avoid pain and to seek pleasure. Upon being told of the behavior of the ascetics, Pupil incredulously asks why anyone would renounce sensual pleasures and submit to physical pain? His teacher replies that it is ridiculous to do so, for how can fasting and exposure compare with "the ravishing embraces of women with large eyes, whose prominent breasts are compressed with one's arms?"[9] Another good question. Pupil asked, "Do these pilgrims indeed torture themselves in order to remove the happiness which is mingled with this miserable existence?" And Materialist said it was true, they don't want pleasure because it is mixed with pain, and again lamented, "But what prudent man will throw away unpeeled rice which encloses excellent grain because it is covered with the husk?"[10] From the sidelines, Passion applauded this answer.

Materialist also explained that the only reason people behave properly is because they fear punishment. "The three Vedas are a cheat," he claimed, because they pretend there is a higher system of justice in the world. They are also a cheat because they prescribe all sorts of inefficacious ritual. Materialist insists that even those who perform these rites do not believe them: after all, animals slain in sacrifices were said to ascend to heaven, but if people really believed that, surely they would sacrifice their parents and thus give them an express ride to paradise! But no one does, so they must not really believe it. Neither was Materialist impressed with the idea that funeral oblations actually served to feed the deceased. He wanted to know just how that worked, since the flame of an extinguished lamp is never revived by pouring on more oil. The scene closes with Passion offering "Materialist, you are my beloved friend!" to which Materialist replies, "May thou be victorious. Materialist salutes thee."[11] It's a nice alliance.

The Carvaka proclaimed that there were no gods, and that there was no heaven; the only hell there is, they insisted somewhat gleefully, is here below, caused by normal pain and frustration. But these were all side issues. It was samsara, karma, and moksa that they most devotedly denied. Lokâyata did not offer any replacement for these things, that is, its adherents did not dedicate themselves to human society or some particular vision of a purposeful existence. For them, no gods, no karma, and no supernatural world of any kind meant that there was no justice in the universe, and therefore no strict reason to maintain justice in our little human existence. No morality could have any meaning because the whole system had no purpose; virtue and vice, the Carvaka explained, were merely social conventions. We ought to keep up a degree of kindness because it generally functions to our own advantage—it works—and that simple functional issue should be our guide in life. Again, the only goal for a human life was pleasure; we should seek sustenance and love and sensual delight of all variety. We should not worry about the meaningless whole.

And what of the wonder of the world? How did we get the snakes and the sky, the flowers, the earth, and ourselves? According to the Carvaka, the answer, again, is evident: it just came to pass—the same way things come to pass every day. The world followed its nature and thus became itself and no one had to help it. They knew that some would find this hard to believe but were themselves convinced. In the Sarva-darsana-samgraha, a fourteenth-century description of the Lokâyata—which rejected the doctrine but quoted it at length—the point was put succinctly: "An opponent will say, if you thus do not allow any unseen force, the various phenomena of the world become destitute of any cause. But we cannot accept this objection as valid since these phenomena can all be produced spontaneously from the inherent nature of things."[12] They were not minimizing the majesty of the natural world, they just thought it began and developed by its own internal logic. "Who paints the peacocks, or who makes the cuckoos sing? There exists here no cause excepting nature."[13]

The Carvaka saw themselves as debunking Hinduism by arguing against the idea of gods, karma, and rebirth. In their words, "Others should not here postulate [the existence] of merit and demerit from happiness and misery. A person is happy or miserable through [the laws] of nature; there is no other cause." For them, "there is no world other than this," and all sorts of religious promises to the contrary "are invented by stupid impostors of other schools of thought."

Two materialist philosophical schools were founded in the same early Classical period that saw the origins of Lokâyata. These were the schools of Logic (Nyaya) and Atomism (Vaisesika), and both specialized in questions of epistemology, studying the method and grounds of knowledge, especially having to do with its limits and validity. These schools were analytical and secular: they called for limiting claims about humanity and the universe to those things that could be logically demonstrated. Logic and Atomism both claimed that all we could know was what our senses told us, plus the inferences that we could make from that sensory information: we can know that a fire is hot by sticking our hand into it, and after a bit of similar research we can conclude that all fire is hot. We cannot go ahead and draw conclusions about who made the fire hot, or if there has always been fire.

The Carvaka outdid even the doubt of Logic and Atomism. The Carvaka did not believe in the validity of inference. This is mentioned in most surviving discussions of them, because it helped make their position seem particularly untenable. It is also the central concern of the Tattvopaplavasimha, a seventh-century text that is the only extant treatise considered an authentic text of the Carvaka. No inference means that no dependent ideas are real knowledge. The problem is that we think by using dependent ideas: we know water is wet because all the water we have ever encountered has been wet. According to the Carvaka, all these experiences with individual cases only serve to prove that in those cases, water was wet. We do not know anything about water or, in fact, about any generality. It is a sophisticated skeptical notion. Even after seeing ten thousand swans and all of them white, we must not assume that swans are white. Along the same lines, the Carvaka did not believe in cause and effect. Events that are linked in a before-and-after way are simply that: events that human beings have often perceived as following one another in time. No more conclusions should be drawn because no link can be proved. So when the Carvaka said that nothing can be known except the information of the senses, they really meant *nothing*. This was tricky stuff, and most polemics merely describe the basic ideas of the Carvaka epistemology before returning to discuss their more obvious attacks on religious doctrine.

The Sarva-darsana-samgraha cited the Carvaka as saying that the rituals of the Brahma are useless, and the Vedas are "tainted by the three faults of untruth, self-contradiction, and tautology"—a sharp critique. This text also contains a long section of verse detailing the Carvaka position. A fraction of it reads:

The fire is hot, the water cold, refreshing cool the breeze of morn;
By who came this variety? from their own nature was it born.
And this has been also said by Bṛhaspati—
There is no heaven, no final liberation, nor any soul in another world
Nor do the actions of the four castes, orders, etc., produce any real effect

> *The Agnihotra, the three Vedas, the ascetic's three staves, and smearing*
>   *oneself with ashes,*
> *Were made by Nature as the livelihood of those lacking knowledge and*
>   *manliness.*
> *If a beast slain in the Jyotistoma rite will itself go to heaven,*
> *Why then does not the sacrificer forthwith offer his own father?*
> *While life remains let a man live happily, let him feed on ghee even though*
>   *he runs in debt;*
> *When once the body becomes ashes, how can it ever return again?*
> *If he who departs from the body goes to another world,*
> *How is it that he comes not back again, restless for love of his kindred?*
> *Hence it is only as a means of livelihood that Brahmins have established*
>   *here all these ceremonies for the dead—there is no other fruit anywhere.*[14]

These were the main points of the Carvaka position, a dynamic and early episode in the history of doubt.

The same period that saw the birth of the Carvaka saw the origins of Samkhya, the oldest of the six orthodox schools of Hindu philosophy. Samkhya upheld an interpretation of karma, but it was otherwise naturalistic and atheistic: the world was made of stuff and of souls that cycle through life until they finally end up free. In the words of the scholar Ninian Smart, eventually "The emergence of the cosmos out of chaos at the beginning of each cycle [of the universe] is explained through an evolutionary theory. . . . The whole process is explained without reference to a personal Creator."[15] Samkhya understood the world to have come to be of its own accord; just as milk flows for the young animal and is not sentient, the primordial substance of the world goes along, following its nature. Although the Samkhya texts usually simply ignore theism, they argue against God at times, holding, for example, that if God were free he would have no desires to make him create and if he were not free he would be unequal to the creation he is purported to have made—therefore he cannot exist.[16]

The sixth century BCE was a period of innovation in Hinduism. It saw the birth of the Samkhya philosophy as well as a great range of religious movements, the most prominent of which were Buddhism and Jainism. Lokâyata began just at the start of this tumultuous and productive century. Although Buddhism and Jainism took Hindu philosophy in a different direction, modern scholars suggest that Lokâyata may have influenced them. If all the Carvaka had done was reject karma, harsh ascetics, and sacrifices, its effect would have been enormous—but, of course, they put even larger matters into question. Jainism and Buddhism were not devoted to the material world as was Lokâyata, but they had learned to question the fanciful metaphysics of the Vedic priests.

JAINISM AND BUDDHISM

The great leaders of Jainism and Buddhism both lived in the sixth century BCE. Both movements began, in a sense, as Hindu reforms that explicitly rejected much of the Vedas—especially their supernaturalism. Both seem to have been influenced by the materialism of the Carvaka, and the dates would fit such a hypothesis, although we are not really sure.

Jainism supposedly existed deep into prehistory. Mahavira, the great founding figure who created the Jainism that we know, was, it seems, the twenty-fourth Tirthankara, or Jain leader. Jains had apparently been a small sect of ascetic monks before Mahavira. It is said that during his periods of extreme isolation he pulled out his hair at the root and went about naked at all times, whatever the weather. He had an experience of enlightenment and soon became the leader of a small group of monks who followed his example of nakedness, hairlessness, and other aggressive acts of physical asceticism. Mahavira taught that the gods and goddesses, sacrifices and rituals of Hinduism were all nonexistent and/or irrelevant. What the religion of the Vedas had right was samsara and moksa—and that was about it. Jains understand the universe to have no end; it merely cycles in and out of existence over the course of eons. According to them, we happen to be living at the end of a cycle, in a period of decay, and that explains our troubles. Through most of the cycle of millions of years, *jiva*, spirit, and *ajiva*, worldly stuff, are pristinely separate and the universe is a paragon of harmony and peace, but in periods of decay jiva and ajiva start collapsing into each other and that is the reason for struggle and pain.

Jains believe in karma, but as an actual substance—they spoke of vapor and in modern times often speak of a fine dust of atoms—that gets on your jiva in the course of living in the world. Any interaction might cause a bit of this karma to adhere to your spirit, but when you do harm to any other living thing, the karmic burden is very high. And it is karma that keeps the wheel turning, causes the endless births and deaths to keep repeating. If you want release, the thing to do is to minimize the amount of this stuff that you accrue over a lifetime. Jain monks and nuns—and men and women are both seen as capable of spiritual release—thus live very carefully in all their actions. They are vegetarian, first of all, and also known to place a cloth over their water as they drink in order to be sure they do not accidentally ingest a living creature. Some Jain monks and nuns sweep the path before them as they walk so as to minimize the chance of crushing a bug. The most popular image in Jainism is of a monk who has stood still so long that vines are twining up his legs.

Jainism is generally understood as an atheist religion. The Hindu gods were rejected and were not replaced by any supernatural force. Its great leaders, especially Mahavira, were adored, and their images adorn the Jain temples, but for the most part these great leaders have been remembered as such, and not transformed into supernatural saints or gods. In fact, the temples were and are more for the laity than for the monks and nuns. Lay Jains keep up a number of rules and prohibitions—they embrace vegetarianism, for instance—but are on a less rigorous program than the monks and nuns, whom they hope to emulate in future lives. The most important Jain prayer, to be said once a day, is a statement of respect for those who have made headway in their travel down the path of release. Jainism did not spread worldwide like Buddhism did, but it did manage to remain a living religion in its own birthplace, India, across these many centuries. There are about two million Jains in India today. The big godless religion to begin in India at this time, however, was Buddhism.

The Buddha was born Siddhartha Gautama, about 566 BCE. He was a prince at a time and place in which princes were plentiful, and therefore more like what we would understand to be a nobleman, but still, he was a prince. When he was born, the story goes, his father brought in soothsayers to report on his newborn son. They said that the boy would have one of two extraordinary futures: if he stayed in the world, he would unify India and become her universal king; if he forsook the world, he would become a world redeemer. The father preferred that his son meet the kingly destiny, and he spared no effort in keeping Siddhartha attached to this world. As some versions have it, the young man had forty thousand dancing girls and three palaces. More crucially, his father kept Siddhartha in the confines of royal palaces, hiding pain, poverty, illness, and even old age from the prince so that he would not question the beauty of the world. Some say that each night the dead leaves were cleared from the garden so that the young man would have no melancholy. He was trained in all the fine arts of living, involved himself in their high standards, and enjoyed them to the fullest. Soon he had a beloved and beautiful wife and an infant son. He was happy.

Alas, one day Siddhartha slipped away from the palace and met up with all the difficult realities of life—death, sickness, sorrow, and decay—and he was horrified. He also saw a forest dweller and was surprised to find that this skeletal, unkempt, silent creature was not most pitied among men but was rather considered to have achieved freedom and true happiness. Nothing was ever the same for the prince. Once Siddhartha knew that all bodies will age and suffer, and that suffering is everywhere, the things of the world and the pleasures of the flesh appeared empty to him. Meanwhile, he kept thinking about the reputed freedom of the forest dweller. When he was twenty-nine he left his family and went off into the wilderness to seek enlightenment.

The first thing he did was to find two of the foremost Hindu masters and study raja yoga and Hindu philosophy with them. When he felt he had learned all they had to offer (we don't know how long this took), he found a group of ascetics and joined their practices of physical trials. He fasted and strained himself; did exercises like clenching his teeth so intensely, for so long, that he dripped with sweat. At one point he lived on a grain of rice a day. He was respected among his fellow ascetics as the most intensely devoted and the most advanced. It all worked rather well, too: he reported going through epiphany upon epiphany, knowing ever more refined states of consciousness, experiencing ecstasies and states of blissful clarity. But he still felt that he had not come to his true self. One day someone offered him a beautiful meal—one story speaks of an array of delicacies, another says it was a woman with a bowl of milk—and Siddhartha, to his own surprise, takes it and gulps it down. Once the good food is inside him, he is struck by one of his great revelations: starving oneself is no better than feasting. Neither one nor the other—and note that he had taken each to its most refined excess—would lead to enlightenment and happiness. He did not know what to do, so he went off on his own to meditate. One evening, sensing he was ready, he sat down under a Bodhi tree and vowed not to get up until he had reached enlightenment.

Buddha means awakened one, and apparently, that is how the Buddha came into being. Under the Bodhi tree Siddhartha woke up and now he could see the truth. The first formulation of that truth was that "all is *dukkha*," suffering. Pleasure always entails some sorrow when it ends, pales, or has ill effects, but more than that, "all is dukkha" means that all is out-of-joint: human existence is dislocated. There is a rupture between our feelings and desires and the vast, unresponsive universe. We don't fit. Yet in his awakening, the Buddha found a way to solve this rift and thereby generate unspeakable bliss, all by what amounted to a shift in perspective. He was thirty-five. He spent the next forty odd years teaching, wandering, and ministering. Apparently, people responded to him as if he were another sort of being. Historian Huston Smith has observed that in the human record only Jesus and the Buddha seem to have constantly evoked the question "What are you?" rather than "Who?"[17] He founded an order of monks. Texts of his teachings—carefully memorized by his disciples and later recorded—would fill a small room. From the second to the seventh centuries, Buddhism was probably the predominant religious community in the world. There are now more than 300 million Buddhists worldwide.

The Buddha denied a central Hindu notion, that of the atman. The religion of the Vedas held that human beings who had reached the stage of the Brahman could come to know their true self, their atman, and thereby reach bliss and release. The Buddha said, Look all you want, you are never going to find your atman. Why? Because there is no atman. There is no such thing as a self. Meditation had brought the Buddha to enlightenment, so he was obviously still very keen on it, but in a revolutionary way. Instead of using its techniques to find our atman, our truest self, we are to use these techniques, the Buddha instructed, to come to understand Anatman, the doctrine of no-self—the Sanskrit negation "an-" attached to the great Hindu ideal, "atman."

Making yourself less important is a common religious goal. There is a logic to this, in that we know ourselves to be infinitesimal specks in the universe, and yet each of us feels extremely important to ourselves. This experience of feeling important but not being treated as important by the universe is the source of much woe. Religions thus advise their followers to feel what they logically know to be true: that our self is not the center of the universe. It is something else again to argue that the self does not exist. As we have seen, materialism is obvious at first and then starts to get counterintuitive when taken in certain directions (you can say you will believe only what you can prove, but you cannot really prove any cause and effect). Buddhism seems to go in an opposite progression: we think we have a self, we assume it, it's the thing reading this right now, mulling it all over. But spend some time seriously looking for that self and one's internal experience starts to fragment into disparate flickers of thought and sensation. All that noise had been seen as the stuff blocking you from getting to your atman. The Buddha said all that noise and all the more subtle sensations deep inside, all that is all there is to you, and the reason you could not find a central, coordinating self in there is because there is none. The idea that the self does not exist becomes increasingly intuitive. We are each a composite of sensations and actions.

There are feelings and doings, but there is no distinct feeler or doer. Modern psychotherapy offers another take on the Buddha's claim that the self does not exist. As psychiatrist and theorist Wilhelm Reich explained it, the personality is created around moments of pain so uncomfortable that they are blocked off.[18] A person's character or persona is thus a chronicle of where he or she has been alienated from his or her self. Ways of coping with psychic pains, which are themselves assiduously avoided, create our individuality—that's the real reason why the ways of coping are so (boldly) defended. We hang on to our cocooned pain because the thought of looking directly at it seems unimaginably worse than what we suffer living around these cocoons.

For the Buddha, once you see that you have no self, there is no self to defend and therefore no reason to avoid one's pain. Now there may be embarrassment, but no one to be embarrassed. Clinging to our personalities is thus our undoing. The Buddha's program is complete concentration on the particulate sensations that one gathers from the world and feels inside oneself, and the point is to see that none of this is really tied up into the narratives and conceptual structures that we usually assume. Take a bad feeling, for example, and concentrate on every aspect of its physical sensation: it reveals itself to be a constantly shifting physical phenomenon. It exists, but is a sensation, a curiosity, not a package of meaning or a spur to action. Do this exercise often enough, combined with other practices of intense concentration, and when you are confronted with formerly painful sensations, the curious-yet-calm response will replace the response of anger or sorrow. That's just a small part of the transformation the Buddha promised, for if you can really come to believe that there is no self, how could your ego get bruised in the first place? The Buddha said that happiness was available to us if only we could simply set down our unfounded conviction that the self exists and must be protected. It does not exist, there is nothing to protect, and we are living in a very different world than we had ever dreamed.

Having come to truly understand Anatman, the Buddha had been freed—from everything. He had managed this transformative realization through entirely experimental means. No one had to wait to be reborn as a Brahman in order to achieve moksa. All the ideas of the religion of the Vedas—castes, samsara, karma—were, for the Buddha, beside the point. I'll discuss what he thought of those things, but for now it is enough to note that he gave delightful and nuanced answers about being agnostic toward all that. He advised accepting that such things are immune to investigation and essentially forgetting about them. The real excitement was in this actual, natural, real world. The Buddha said that we are tiny creatures, convinced of a sense of me-against-the-world, and possessed of a comically small vantage point from which to see the social world of human beings and the universe as a whole. He suggested that we go through a mind-training program to realize that we are at one with the universe and can therefore partake in its eternity, peaceful mindlessness, and grandeur—rather than envying it. He liked to quip that what he was arguing was obvious; the problem was only in taking it in. We know we are fists of nerve amid a placid field; the Buddha said just let go and all will be well. The only proof was in the doing.

When the Buddha came to believe that neither self-indulgence nor self-abnegation could work, he championed the idea of The Middle Way, between the two. This may sound easy, but the Buddha said it was as narrow as a razor's edge—one had to be vigilant against the seductions of either tendency. Yet the Buddha said that once you do manage to get rid of your sense of self, the truth of the universe is yours. You are no longer living from a single vantage point. Since you are not a separate self, your compassion is limitless; you are all compassion, all empathy, because not being *you* entails being everything else: since we are not at all separate, we must be all. A related and equally important concept is that everything in the world we know is constantly coming into being or disappearing, and it is all basically made of the same stuff. There are no true nouns, then, only verbs. As a billion various ocean waves are all in fact water, wave-ing, so the water is just the universe ocean-ing—holding in the form of water. The part of the universe that *is* you, is really just "you-ing" right now. There is no reason to fear anything, or to take pride only in those things particular to you, because there is no you, there is just a momentarily you-ing universe from which you could not be separated any more than an ocean wave can be separated from the water. The terrible separation of death and the alienation of individuals within communities and within the vast universe—these ruptures don't exist in reality. We are tricked by the default time frame of our minds, so we do not see the flow of a oneness. With a lot of work, we can reconfigure our default settings so that we see things as they really are: flowing, timeless, interconnected. It leaves one bemused, gentle, and unflappable.

The Buddha had a wonderful talent for metaphors and kept them lively and undogmatic. He described in detail how a goldsmith purifies gold and how afterward, he can stop purifying it and form it into whatever he wants. It was the same for monks. First they get rid of "gross impurities" like "bad conduct of body, speech, and mind." Then they get rid of sensual thoughts, thoughts of ill will, and violent thoughts. "When he has abandoned these, there are still some subtle impurities that cling to him, namely, thoughts about his relatives, his home country and his reputation," and an earnest monk has to eliminate these, too. "When he has abandoned these, there still remain thoughts about higher mental states experienced in meditation." At this stage the monk is still working hard to remain pure and is not yet truly calm and at peace. "But there comes a time when his mind becomes inwardly steadied, composed, unified, and concentrated. That concentration is then calm and refined; it has attained to full tranquility and achieved mental unification; it is not maintained by strenuous suppression of the defilements." Then the Buddha said that in this state, like the gold that can be anything, the monk so trained can realize any given mental state "by direct knowledge." The person who has reached this level of reality can achieve anything, even with such wild wishes as these:

> May I wield the various kinds of spiritual power: having been one, may I become many; having been many, may I become one; may I appear and vanish . . . dive in and out of the earth as if it were water; walk on water without sinking as if it were earth; travel through the sky like a bird while seated cross-legged; touch and stroke with my hand the moon and sun, so powerful and mighty. . . .
>
> May I understand the minds of other beings, of other persons, having encompassed them with my own mind. May I understand a mind with lust as a mind with lust; a mind without lust as a mind without lust; a mind with hatred as a mind with hatred . . . a mind with delusion as a mind with delusion . . . a distracted mind as distracted . . . an exalted mind as exalted . . .[19]

The Buddha insisted that the way to enlightenment was not at all supernatural, or even spiritual, but he could be very fanciful about the world as it was experienced by one who is enlightened. Yet the goal of all this is not really to do with flying, but with having a mind that will allow us to partake of the flight in the world, and even more, to be able clearly to see and hear the human beings around us.

This situation of ours is bliss, said the Buddha, and we are already there. If you can root out the sense of self entirely (and the best way to do that is to go looking for the self), you find you are a collection of thoughts amid the universe, with nothing to do but be delighted with that surprising truth, and with the whole range of experience, without preference, without hurry, without dread. Every moment is a marvel of being. One gets to this state by practicing it. One extracts the self by acting the way one would act if one had already extracted the self. Pay alert attention to every moment; so that washing out your rice bowl is as good a moment for wonder as setting foot in a foreign land. The world is in constant flux—it all changes all the time, mountains melt, universes collapse—all we have to do is learn to accept it.

The Buddha promised that he had held nothing back, that he had offered everything he knew about reaching enlightenment. He emphasized again and again that this was not a revealed doctrine—it had no magical component, there was no one to whom one should pray—he had figured it out through experimentation and concentration. Knowledge about the human condition and practice of yoga meditation were the routes. The body should be made reasonably comfortable and not overindulged. An eightfold path should be followed: right views, right intent, right speech, right conduct, right livelihood, right effort, right mindfulness, and right concentration. All these together mean that you should saturate your deepest assumptions with the truth about the self and about impermanence; get it into your very marrow. With the right ideas logically clear in your mind, dedicate yourself to them, and do nothing that betrays them. If you choose to be a butcher, you are working against convincing yourself that you are one with everything and the embodiment of empathy. You should try not to kill in general: monks are often vegetarians, and sometimes even try to eat only foods that fall off plants naturally. If you lie, you are scared of revealing the truth of your nature to yourself and to others, and you thus fortify the very wall around your ego that you have been trying to tear down. You should associate with the wise, work according to the principles described, and put this labor of seeking truth above all else. Devote every moment to it.

As for the great metaphysical questions, the Buddha would not answer them one way or another. He explicitly refused to say whether the world is eternal or not or both or neither; whether the world is finite in space or infinite (or both or neither); whether an enlightened being exists after death (or not, or both, or neither); and whether the soul is identical with the body or different from it. He said that worrying about these things would be like a man pierced by an arrow asking questions about the family origins of the man that made the weapon. It was not just the waste of precious time that he was trying to highlight, but the wrongness of the question in this context. He said that to ask where the soul goes after death is like extinguishing a campfire and then asking whether the fire went east or west when it left. "The question is not put rightly." Was there a God? Were there gods? The Buddha said these are questions "which do not edify."

So what about karma, the great nontheistic religious belief of the East? The Buddha suggested that we imagine a line of candles, and using the first to light the second, and the second to light the third, we progress down the line illuminating each one. He then asked if the flame on the last candle is the same flame as the first. Of course, in some sense it is, and in some sense it isn't. It is to this degree, the Buddha suggested, that we are reborn: it is a matter of causality, of considerable influence, but not the passing along of any kind of enduring substance. But he gave different versions of his doubt to different seekers, explaining that each needed something different in order to come to understand. Sometimes as he advised people in right behavior, he spoke at length about some after-death experience (he was vague) corresponding to one's behavior here and now. Confronted with such contradictions, he explained, people learned in different ways and there was no way to say anything on the matter that was decidedly and specifically true. It was the same with nirvana. Etymologically, the word *nirvana* breaks down to mean *to extinguish* or *to blow out*—to extinguish the boundaries of the separate self. The result was pure bliss, but the Buddha would not say much more. Was it eternal life or true annihilation? Nirvana, he said, was "incomprehensible, indescribable, inconceivable, unutterable." We cannot speak of what we will know once we have rid ourselves of every aspect of the only consciousness we have ever experienced.

The Buddha created no priest class. Although this was based in his notion that all "should be lamps unto themselves," it also fit nicely into the Indian religious world at the time because it did not directly threaten the livelihood of the Brahman. Still, the Buddha was a threat. He made it clear that sacrifices and rituals couldn't be of any interest to someone seeking enlightenment. He also encouraged a shift in the use of people's resources such that what had once gone to Brahman sacrifice now went to support the monks: they would walk through town with their bowls once a day and the peasants would come out to fill the bowls with food, and would feel honored in so doing. This behavior has persisted through the ages. In all such things the Buddha gave careful and nuanced guidance. When asked if people should give alms to only his own followers, the Buddha replied that he would never caution against any gesture of generosity, for the sake of both the givers and the receivers: "Even," he said, "if one throws away the rinsings from a pot or cup into a village pool or pond, wishing that the living beings there may feed on them—even this would be a source of merit, not to speak of giving a gift to human beings. However, I do declare that offerings made to the virtuous bring rich fruit and not so much those made to the immoral."[20]

We have already seen that the Buddha rejected the caste system. In his forty years of preaching, he ignored social status, giving everyone his teaching equally as suited each one's need, and encouraging everyone to seek nirvana. The path he proposed was that each person seek truth individually, but that they become part of the *sangha,* the Buddhist community, and that they make use of the *dharma,* the Buddha's teaching. The dharma was a relatively solitary road to follow, but the sangha was designed as a cooperative society in which all helped all, in whatever ways they could, to reach moksa. But again, this began as a training program for individuals, a therapy, and one that could work only as the result of one's own discipline and labor. When the Buddha lay dying, after a long life, a beloved disciple is supposed to have asked him about a successor and the Buddha replied, "You should live as islands unto yourselves, being your own refuge, seeking no other refuge, with the Dharma as an island, with the Dharma as your refuge, seeking no other refuge." No God, no karma, no dependence on community for meaning—the only focus for one seeking enlightenment was this strict yet joyful internal exercise.

At the heart of it, as with so many practices of self-mastery and secular happiness, was the injunction to remember death. The Buddha loved to regale his listeners with various brief things and announce that such is the brevity of the human life. At one point he mapped out a life in numbers so that it would appear as limited as it is. He said that one who lives a long time lives a hundred years, and a hundred years is three hundred seasons, a hundred winters, a hundred summers, and a hundred rains. He broke that down into months, then how many winter, summer, and rain months, how many fortnights, how many fortnights of each season; 36,000 days or 12,000 days of each season; 72,000 meals—including times of "taking mother's milk" and times of fasting:

> Thus, O monks, I have reckoned the life of a centenarian: the limit of his lifespan, the number of seasons, of years, months and fortnights, of days and nights, of his meals and foodless times. Whatever should be done by a compassionate teacher who, out of compassion, seeks the welfare of his disciples, that I have done for you. These are the roots of trees, O monks, these are the empty huts. Meditate, monks, do not be negligent, lest you regret it later. This is our instruction to you.[21]

His gift was a message that seemed harsh but was, he assured his followers, the simplest road to the greatest happiness. The dharma, he said, was made for most people: most people could not reach enlightenment on their own but could indeed manage to do it with the dharma.[22]

Scholars and practitioners widely agree that the Buddha taught the existence of no personal god, and that only by the vaguest interpretation of the idea of the godhead as the bliss of the true and the real, can we suggest that nirvana was something like a Buddhist godhead. He specifically said it was a sin against right living for anyone to claim to have supernatural powers.[23] Once Buddhism was out of the Buddha's hands, the ideas of prayer and worship, a universal mind, magic, gods, and, of course, karma began to creep into many of the Buddhist sects that arose across the centuries. But not all. In his teachings and in certain groups of his vast following, Siddhartha Gautama created a way of living that actively addressed the seeming rupture between the human world and the nonhuman universe, and he did so while profoundly doubting God or gods, karma, or any other universal justice. One gives up many aspects of what we have been defining as humanness in following him, but it is not like the Cynics who emulated dogs, i.e., who became a part of nature. The Buddha invited us to use our human consciousness to realize that we are not a part of nature, we are all of nature. It was a transcendent secularism, an empirical guide out of the limitations of the human mind as it is generally configured. The Carvaka stand as champions of wry materialist skepticism about knowledge. Buddhism is a nontheistic graceful-life philosophy and a nontheistic transcendent program. Through them both, countless people across the millennia lived by doubt.

FURTHER DOUBTS AND
DOUBTS THAT LINGER

Early in its history, Buddhism split into two branches: Theravada and Mahayana. Theravada is sometimes thought of as the orthodoxy and Mahayana as the reform, but that is not accurate, since both emerged from the original Buddhist teaching. It can seem like Theravada is older because it was dominant early on but has since been quite overtaken by Mahayana, which itself split into several major and myriad minor subsets. Thus Mahayana reminds us of Protestant Christianity, but the metaphor should not be taken too far. Theravada Buddhists see practice of the Buddha's dharma as the way to reach enlightenment—there is no supernatural help. They believe that reality, the truth, is the bliss they are seeking, so they value nothing above wisdom. Nirvana is understood as the result of penetrating contemplation of the nature of the universe. They see the Buddha as a historical figure, a man who made important discoveries and who has since served as the supreme teacher and inspirer. Over the centuries, many other people following the Buddha's program had also "awakened," and they, too, were called Buddha, so that the Buddha who had been Siddhartha had to be designated as such, and in Theravada these other Buddhas are also thought of as great teachers.

The main idea of Mahayana is that anyone enlightened enough to enter nirvana would refuse to do so, out of compassion for his or her fellow human beings. The classic story tells of three men, dying of thirst in the desert, who come upon a high wall. The first makes the effort, reaches the top, shouts out *Oasis!*, and jumps over. So does the second. The third reaches the top, sees the oasis, and stays put in order to direct other thirsty people toward the water. Someone who was about to step into nirvana, but who could not bear to go in without all of us, is called a bodhisattva—after the Bodhi tree underneath which the Buddha had come to enlightenment. Eventually, and in various places, the bodhisattvas were worshiped for this generosity: the best loved of all bodhisattvas is the Chinese goddess of mercy, Kwan Yin. She is often pictured carrying tears of compassion or pearls of illumination in a little vase. Like all bodhisattvas, Kwan Yin is praised for having refused Buddhahood until she knew that everyone had finished their painful cycles of death and rebirth, but Kwan Yin is understood as the most compassionate of all. She is so merciful that she will not rebuke someone even if he or she well deserves a lesson, and even people so corrupt that they would meet only penance in any other Buddhist system are renewed through her kindness if they call out to her in sincerity. Not all bodhisattvas made it that easy, but a central point of Mahayana was the claim that seeking enlightenment did not require leaving one's life and joining a monastery. One could stay in one's life, work toward enlightenment through some degree of stillness and silence meditation, and seek to commune with a bodhisattva through whose grace alone one might be saved.

Both Buddhisms saw themselves as rafts intended to ferry people over to enlightenment, and the Mahayana saw themselves as the big raft—that's what *Mahayana* means. They tend to call their rival the little raft, the *Hinayana,* but the term is mildly derogatory. Theravada, *the way of the elders,* is what they call themselves. That term, raft, is important because the Buddha said the dharma was like a raft: once you get to the shore of enlightenment you can ditch the raft; there's no reason to fetishize the thing if you don't need it anymore. As I have said, the Mahayana split into many subgroups, but what they tend to have in common is an openness to a wide range of interpretation. Some of these have been accused of fetishizing the raft a bit, that is, of making the rituals and images of Buddhist practice an end unto themselves. There is something both looser and more egalitarian here. In contrast to Theravada, the Mahayana sects give later texts more importance, and the spiritual abilities of women are treated with considerably more respect. In many of the Mahayana sects, human efforts to reach nirvana are supported by divine powers, but not necessarily. Mahayana Buddhism is supposed to be for everyone, so it can be interpreted at whatever level works for each person.

The chief philosophical idea that allowed the great variation of Mahayana Buddhism is generally called *shunyata,* and its philosopher was Nagarjuna, a Brahman priest living from 150 to 250 CE. Nagarjuna wanted to know how the Buddha's system really worked, how awakening really came to pass. He didn't actually say anything brand-new, but drawing on the sutras, especially the Prajnaparamita, he argued that everything our human minds come up with is equally wrong. Anything that can be framed as a duality is equally wrong on either side of the duality: so truth is neither being nor nonbeing. Nagarjuna concluded that there was ordinary knowledge and then there was transcendent, intuitive knowledge, or *prajna.* Just as the Buddha had suggested: really knowing that ordinary knowledge is useless for seeing the truth is the condition that can allow the rising of prajna. The mind naturally understands everything in divisive categories, defining night by differentiating it from day. The objective is to free the mind from its attachment to concepts. One has to somehow jostle the grip of the thinking mind until it lets go and you can begin to see the boundary-less, unified being. It is in this state that an awakening can happen. Alan Watts has described Nagarjuna's doctrine:

> There comes a moment when this consciousness of the inescapable trap in which we are at once the trapper and the trapped reaches a breaking point. One might almost say that it "matures" or "ripens," and suddenly there is what the Lankavatara Sutra calls a "turning about in the deepest seat of consciousness." In this moment all sense of constraint drops away, and the cocoon which the silkworm spun around himself opens to let him go forth winged as a moth. The peculiar anxiety which Kierkegaard has rightly seen to lie at the very roots of the ordinary man's soul is no longer there.[24]

Nagarjuna's vision is compelling because it spells out the transformative aspect of the experience: you are not going to realize something, you're going to be knocked out of this whole "realizing" realm and into the total reality. We can't even believe that there is no self because that is just a dyad with the belief that there is a self, and the thing about reality is that it isn't the opposite of anything of which we can conceive. The world isn't one, and it isn't not-one either; it isn't this and it isn't that. Apparently, if you press your mind against this for a long while, some difficult thing pops. The Buddha's notion of *right thinking* meant getting the ideas logically straight in your mind, but this new doctrine is not very respectful of right ideas because the content of the doctrines is no longer crucial—they just have to work to rattle you out of normality of concepts. That does seem to contradict the Buddha's very clear instruction, but on another level, it is in perfect alignment with his message. One result of Nagarjuna's doctrine was to allow the development of spectacularly diverse cosmologies and a lot of Buddha worshiping where there had once been a program of personal inner training.

It is not uncommon to see the atheist versions of Buddhism described as truer to the original Buddha's message and considered more sophisticated than the others. Yet the more popular, metaphysically fanciful versions are also regarded as true in their own way. Since there is no philosophically justifiable conception of reality, the more fanciful is at least in the right spirit of things—it's at least different from the everyday rational mind. Of course, some people who think of the Buddha as a savior are not doing so to jostle their rational mind's hold on reality; they are believing in a savior. In Mahayana there is ritual and petitionary prayer. Also, in most Mahayana sects, karma and samsara are back: in fact, the whole concept of becoming a Buddha is seen as possible only within the superstructure of karma—it takes a lot of lives for a person to reach this level of enlightenment. Mahayana also teaches that there are thousands of Buddhas, each presiding over a distinct Buddha-realm or Buddha-verse. Again, all of this appears to have moved away from the original teaching, but the Mahayana say that Buddha's compassion was his chief message and that these doctrines are true to that compassion in a way that makes Theravada seem self-centered and cold. Mahayana values compassion above wisdom.

The most popular Buddhism today is the Pure Land sect of Mahayana, which is also the most metaphysically whimsical of the religion's incarnations. Pure Land centers its attention on one of the Buddha-verses—specifically, the western paradise of "Pure Land," located at an inconceivably great distance from the earth and presided over by Buddha Amitabha. What makes it so special is a surprising vow Buddha Amitabha once made: that anyone who trusted in him purely would be given enlightenment and welcomed into Pure Land! That makes it a lot like the bhakti yoga of Hinduism: intense love and selflessness can bring one to enlightenment. Obviously, this is exactly what the Buddha who had been Siddhartha said could *not* happen, but for the masses this promise of devotion and reward was attractive. Add to it the idea that one could spend eternity with one's loved ones in the bliss of Pure Land, and the attraction becomes even stronger. Yet Pure Land also appeals to many people who do not take its promises literally—who see Pure Land as a state of being, not a real place—but who are more inclined to the romantic, emotional, artistic, and fantastic aspects of life than to the strictly philosophical. They enjoy the descriptions detailing world structures and systems: Buddha-verses of every imaginable kind where, for example, all sustenance including enlightenment itself is in the very perfume of the flowers, and one sniffs to bliss. The great mass of Buddhist temple art in the world is Mahayana, not Theravada, and the huge portion of it that depicts Buddha-verses is essentially all Pure Land.

Hinduism underwent a profound change as a result of its absorption of Buddhism, becoming much more inward-looking. And at about the time that the Mahayana sect arose in Buddhism, Hinduism also came into a kind of theistic version of itself. The Mahabharata is one of the classics of post-Buddhist Hindu epics (we think it was written somewhere between 200 BCE and 200 CE). It was one of the major sites of this revival of theism, especially the much beloved section called the Bhagavad-Gita. In it, a noble young warrior, Prince Arjuna, is about to go into a terrible battle where both sides are peopled with friends and family members, and he freezes. He gets out of the chariot and declares that he cannot do his worldly duty and fight this battle because he is one with these people, and to kill them is against everything he believes and wants to believe about the nature of the universe. But Lord Krishna, a prince and a counselor to princes, persuades Arjuna that since all is one, and this reality is illusory, he best serves truth if he goes through the motion of his role, knowing that it is all unreal and trusting reality to powers beyond himself. Krishna then majestically transforms into the god Vishnu. Arjuna pledges himself to Vishnu and then goes out into the fray.

Robert Thurman has quipped that, after the Buddha, it took eight hundred years to get Arjuna back in the chariot—meaning that Buddhist-influenced Hinduism survived for almost a millennium in a basically non-magical form wherein one's behavior determines one's inner life in the same way it influences one's physical fitness. There must have always been forces who wanted to attend to worldly goals and still find inner peace. Eventually a device emerged, but Thurman's joke is funny because of its suggestion that the power of Buddhism had so prolonged the process. It is a useful insight because it reminds us that when people are responsible for their own salvation, a great deal of their time and effort is required—time and effort that might otherwise be devoted to the wealth and defense of the state. Also, absenting oneself from the battle of life is very difficult for some people, and meditation and yoga are very difficult for everyone, so there are a lot of reasons that people might have wanted Arjuna to exalt in his worldly duty without betraying any basic principle.

The sangha that Buddha had devised contained a large element of political theory, setting monks at the center of a community whose aim was bliss for everyone. Monasteries spread throughout India, especially after King Asoka (ca. 272–232 BCE) championed the religion with zeal and secured the practice of Buddhism among a great portion of humanity. Over the course of several centuries, however, Hinduism reabsorbed so much of Buddhism that there was little explicitly Buddhist practice outside the great monasteries.

Around the time that Buddhism was born in India, a period of political warfare and philosophical speculation in China produced both Taoism and

Confucianism. Although the first was interested in metaphysical questions and a kind of transcendence and the other was much more mundane in its concerns, both had atheist origins. The doctrine of Confucius had its own statecraft. In a way Confucianism was all statecraft: it was about strict attention to a chivalry of social details, such that the state would click along like a perfect machine. As was true for chivalry, the ornate social ritual of Confucianism was a revival of behaviors that had once been the unselfconscious way of life of a great power: now the golden age was long over, and Confucius wanted to reinstate the behaviors and self-consciously recommit society to its principles. Just as Buddha had redirected sacrifice to the gods into support for the monks, Confucius redirected sacrifice to ancestors into sacrifice to living family members. Both are better for the state in that they take resources that would have been burned and feed them instead into the living community. In the Confucian system, each familial relationship had a clear higher and lower partner, based on age and gender—the higher was supposed to look after the lower, in exchange for deference and respect. At the core of the system was the notion of *ren,* "human-heartedness": generosity and compassion must animate the chivalry and the submission. Is Confucianism a religion? It is generally considered so. There is nothing supernatural about it, but it is concerned with morality, has revered texts, and is deeply based in practices as well as theory. But Confucius left God out of it. When asked direct questions on spiritual matters, he tended to be agnostic and dismissive. When asked about humanity's supposed duty to ancestors, he said, "We don't know yet how to serve men, how can we know about serving the spirits?" When asked about death, he offered another pragmatic question: "We don't know yet about life, how can we know about death?"

The Confucianism that was eventually established as the official state religion of China, in the Han period (206 BCE–220 CE), was a theistic, later version. Nevertheless, some important thinkers were championed for their naturalist and even ardently secular vision of the universe. Hsun Tzu (third century BCE) was a celebrated voice of doubt in God and spirits in the Han period, and is understood as having founded naturalistic Confucianism in purposeful contradistinction to the idealist Confucianism of the period. Hsun Tzu wrote such delightful claims of naturalism as: "When people pray for rain, it rains. Why? I say: There is no need to ask why. It is the same when it rains when no one prays for it." We do not influence the natural world through supernatural powers. Neither, he argued, does the world send us supernatural messages:

When stars fall or a sacred tree groans, the people of the whole state are afraid. We ask "Why is it?" I answer: there is no [special] reason. . . . These are rare events. We may marvel at them but we should not fear them. For there is no age which has not experienced eclipses of the sun and moon, unseasonable rain or wind, or strange stars seen in groups . . . but when human ominous signs come, then we should really be afraid. Using poor ploughs . . . spoiling a crop by inadequate hoeing and weeding . . . these are what I mean by ominous human signs.[25]

There were other doubters in this same period: Huan T'an wrote that "Life is like the flame of a lamp, going out when the fuel is exhausted," and Han Fei Tzu wrote, "If the ruler believes in date-selecting, worships gods and demons, puts faith in divination, and likes luxurious feasts, then ruin is probable." Clearly, Confucianism had developed a theism and a trend of divination. Yet historians generally understand the entrance of Buddhism into China as a phenomenon linked significantly to the fundamentally secular nature of Confucianism, to its interest in the world as we commonly know it. Because China already had a unifying political ideal in Confucianism, when Buddhism entered China its political dimension was left to the side.

China's two great—and essentially opposing—native doctrines were Confucianism and Taoism. As Buddhism changed when confronted with Confucianism, Buddhism was also much influenced through its encounter with Taoism. Whereas Confucius prescribed the revival of defunct codes of behavior, the Taoists advised a rejection of all codes of behavior. They were a lot like the Buddhists in the sense that they were seeking happiness, they believed in the oneness of true reality, and they worked with psychophysical practices. Taoists found Mahayana Buddhism particularly harmonious with their worldview. Kumarajiva, the great translator of Sanskrit Buddhist texts into Chinese (born in 344 CE), translated the Mahayana text known as the Prajnaparamita. It became a favorite among the Chinese in part because its ideals resonate beautifully with Taoism. Indeed, it has been said that the Chinese preference for Mahayana owes fundamentally to their love of the Prajnaparamita sutras and their relationship to Taoism.

Still, there were significant differences between Taoism and Buddhism. For one thing, although meditation was already important to Taoism before Buddhism arrived in China, the meditative practices and techniques of Buddhism were much more fully defined. Also, the Taoists had a more ornate theory: they believed in a life force coursing through everything, and to obtain more of it—or in some versions, to expend less of it—they experimented with all sorts of foods, yogas, and sexual practices. Whereas most psychophysical programs tried to forget the body, the Taoists proposed that we use the body to help us achieve the internal states we are after. Therefore, the Taoist inner arts have a great deal of precisely choreographed motion, such as Tai Chi. Buddhism in China was very much influenced by Taoist ideas. In the Middle Ages, this influence would come to its most momentous fruition in the development of Zen Buddhism, one of the greatest doctrines of doubt ever imagined.

Early Taoists argued that there was nothing about the world that demanded a creator God for explanation. Our internal organs do not need to be controlled by some central figure; rather, they follow their nature as they function alone and as they interact. Taoism was as atheistic a system as was Buddhism or Confucianism when they were all in their beginnings, but it did not stay that way. Like Buddhism and Confucianism, Taoism was born under the notion that human beings can transcend human misery through practices, but all three religious programs were very demanding, perhaps too demanding for the majority of people who had been excited by their doctrines and promises. Into that space was born a "religious" Taoism, which allowed people to benefit by association with practitioners, or by the grace of a distant enlightened person. This was a bit of magic. It would not be wrong to see this as social magic, or rather the real effect that human beings can feel from their community and its emblematic leaders, but for some the result was the growth of a kind of saint worship. In general, the beginning of the new millennium in Asia was marked by increasing superstition and magical thinking. The ancient age of fierce secularism and widespread interest in nontheist doctrines was over.

In such a moment it is particularly delightful to find the great naturalist thinker Wang Ch'ung. He lived from 27 CE to 97 CE and was an extraordinarily independent thinker. Legend has it that he was so poor in his youth that he could not buy books and so, thanks to his prodigious memory, he educated himself while standing in bookstores. By his time, both Taoism and Confucianism had grown ornate with spirits and superstitions. Wang Ch'ung was not attached to either of those schools, or any other, but learned from the best they had to offer and brazenly critiqued what he saw as their excesses. His great work, *Discourses Weighed in the Balance* (Lun-Hêng), was a compendium of arguments against magical thinking.[26] His arguments were often loose and meandering; he did not have a system, but rather met each idea under critique armed with his canny sense of the unlikely. Consider, for example, his insistence that the heavens above are a place of spontaneous rather than intentional motion.

> Why must we assume that heaven acts spontaneously? Because it has neither mouth nor eyes. Intentional activity is associated with a mouth and with eyes: the mouth wishes to eat, and the eyes to see. These desires manifested outside come from inside. That the mouth and the eyes are craving for something which is considered an advantage, it is due to those desires. Now, when the mouth and the eye are not activated by desire, there is nothing for them to seek. Why should there be activity then? . . .
>
> When the heavens are changing, they do not desire to produce things thereby; things are produced of their own accord. That is spontaneity. Releasing matter and energy, the heavens do not desire to create things, but things are created of themselves. That is spontaneous action without intention or desire.[27]

Wang Ch'ung dedicated the book's eighty-five chapters to questions on the nature of the world and to more specific legends or practices. In all, the point was to champion a profound naturalism. In a chapter entitled "The Indifferent Heavens," he addresses the famous story that Po Ch'i had been made to commit suicide by the heavens. As the story goes, Po Ch'i was being punished because when the Chao army of several hundred thousand men had recently surrendered to him, explained Po Ch'i himself, "I deceived them, and caused them to be buried alive. Therefore I deserve to die." Wang Ch'ung insisted this had nothing to do with the heavens and used the story to argue the absurdity of such ideas of justice:

> Po Ch'i was well aware of his former crime, and acquiesced in the punishment consequent upon it. . . . If heaven really had punished the guilty, what offence against heaven had the soldiers of Chao committed—those who surrendered? If, instead, there had been wounding and killing on the battlefield by the random blows of weapons, many of the four hundred thousand would certainly have survived. Why were these also buried in spite of their goodness and innocence? . . . We see from this that Po Ch'i was mistaken in what he concluded.[28]

It was not just the question of justice that bothered Wang Ch'ung, but the whole sense of this world as intentionally created. As he put it, "If the heavens had produced creatures on purpose, they ought to have taught them to love each other, and not to prey upon and destroy one another." He recognized that people had arguments for why the heavens had arranged things this way, but he insisted that if things had been arranged, they would have been arranged better. He mused that some believe that a clay dragon will help bring rain—derived, he tells us, from an idea in the I Ching that dragons and clouds attract each other. To counter the notion, Wang Ch'ung wrote that the Duke of She in Ch'u was very partial to dragons and had them painted on his walls, panels, plates, and dishes. Wang Ch'ung thus winkingly concluded that there must have been a continual rainfall in the state of the Duke of She.[29]

He tells us that when we see a god in dreams, that does not prove the gods are real; after all, other things seen in dreams are often patently unreal. He admits that we also have "direct dreams" in which "we dream of so-and-so, and on the next day see him." But these direct dreams, he insists, "are semblances." When we question the dreamed-of person, he will reply that he has not appeared to us in our dreams. Since the persons we saw in our dreams know they did not appear, what we saw was merely their likenesses. "Since so-and-so [was a] likeness, we know that God, as perceived by Chien Tse, was solely [Chien Tse's] representation of God."[30]

Confronted with the new spiritualist forms of Confucianism, Wang Ch'ung quoted Confucius on the belief that milfoil (an herb) and tortoises could be read for predictions.

"That is not correct," said Confucius, "for their names are essential. The milfoil's name means old, and the tortoise's, aged. In order to elucidate doubtful things, one must ask the old and the aged." According to this reply, milfoil is not spiritual, and the tortoise is not divine. From the fact that importance is attached to their names, it does not follow that they really possess such qualities.[31]

As Wang Ch'ung quoted Confucius to demonstrate the great thinker's naturalism, he also tried to rein in the magical thinking growing within Taoism. He relates a story of a congregation of Taoists who vied with one another "exhibiting strange tricks and all kinds of magic." The prince among them then "attained to Tao and rose to heaven with his whole household. . . . All who have a fad for Taoism and would learn the art of immortality believe in this story, but it is not true."[32] Man is a creature, Wang Ch'ung insisted, and even when man's status is princely or royal, his nature cannot be different from that of other creatures. "There is no creature but dies. How could man become an immortal?" Further, how could man fly? Some creatures are equipped to run, some to fly. Their bodies are differently organized according to the nature with which they are endowed. "Now man is a swift runner by nature, therefore he does not grow feathers or plumes. From the time he is full-grown up till his old age he never gets them by any miracle. If amongst the believers in Taoism and the students of the art of immortality some became feathered and winged, then we might see them eventually fly and rise up." Wang Ch'ung did not have a systematic set of beliefs; he did not replace one positive philosophy with another. Rather, he put all contentions to the test of his own sense of how the world works and of what is likely. The functioning of the workaday world was his model for the world in general. Winged things fly. If a man tells you he can fly, check him for wings.

Consider his claims about eternal life:

> There is a belief that by the doctrine of Lao Tsu one can transcend into another existence. Through quietism and absence of desire one nourishes the vital force, and cherishes the spirit. The length of life is based on the animal spirits. As long as they are unimpaired, life goes on, and there is no death. Lao Tsu acted upon this principle. Having done so for more than a hundred years, he is said to have passed into another existence and became a true Taoist sage.
>
> Who can be more quiet and have less desires than birds and animals? But birds and animals likewise age and die. However, we will not speak of birds and animals, the passions of which are similar to the human. But what are the passions of plants and shrubs, that cause them to die in the autumn after being born in spring? They are dispassionate, yet their lives do not extend further than one year. Men are full of passions and desires, yet they can become a hundred years old. Thus the dispassionate die prematurely, and the passionate live long. Hence Lao Tsu's theory to prolong life and enter a new existence by means of quietism and absence of desires is wrong.[33]

This is a beautiful reminder of the cunning of doubt. Wang Ch'ung was one of the great minds of evidentiary rationalism.

* * *

When men and women take on a quest for inner transformation, they become engaged in grappling with doubt. That is true whether they struggle against doubt or strive for complete doubt; in both cases the spiritual quest is conceived as an aggressive confrontation with one's ambivalence. These Eastern religions all assert that our deepest assumptions about life, even about ourselves and the way we think, are misleading in the extreme. What's more, these erroneous assumptions are held to be the cause of our suffering. Thus, adherents of these religions are entranced by the act of doubting, for the sake of shaking themselves free. The Buddha suggested we doubt a great many things; later followers went further, doubting even the conviction that we doubt the supernatural. Many of these have swung all the way over into belief on this very basis. It's a curious state of affairs, but one can see their point.

No matter how we define it, doubt was vibrantly alive in the ancient East. The Eastern religions have not been concerned with God, and often not even with gods. And yet they have been engaged in the same paradox, the same struggles against distress, that we see in the Western fixation on the idea of God. It seems worth considering the religious struggle to free oneself from samsara as a flipped image of the religious struggle over the truth about God. We are human and the universe is not. We are out of joint here and will, at times, go to supreme lengths to close that gap, to make sense of meaning and death, creation and destruction; the living worlds of beauty, love, passion, sadness, fear, and horror and the seemingly unthinking and unfeeling expanse of galaxies, of sand on the beach, of swarming atoms and their field of empty space. Without God, there was still a language to discuss the awful weirdness of being carnivorous and hungry creatures capable of intense compassion. In spite of the development of much accompanying theism in the last two millennia, many of the Eastern religions solved the rift by advising that we change ourselves, inwardly, such that the noisy human experience is progressively silenced and we come to fit better in this universe. Of course, with the ideas of karma, samsara, and moksa, many of these religions were dealing with a universe that had some mechanistic relationship to us—we were not utterly accidental and unregulated creatures.

Within this general paradigm, the Carvaka, the Jains, and some traditions of Buddhism came to doubt the supernaturalism of such solutions. They did not base the weight of their attack on the belief in God, because that was not the major point with which they were confronted. Instead they laughed at or respectfully dismissed a whole range of ideas that seemed designed to make people feel better but that actually made them miss the true and real aspects of the beauty of life. For the Carvaka, that meant feasting on ghee and other treats and spending time with delightful and attractive people. For the Buddha, that meant following a testable enlightenment program that would lead to true happiness, based on reality: we need only learn to notice that we are in paradise. Here again we see struggle, a struggle to believe the reading of reality that the Buddha reported (which seems intellectually self-apparent but feels all wrong). The Buddha made it clear that it would take a lot of work to learn to respond to the world through these insights.

We now turn our eyes to the West to have a look at the marvelous pagan doubters of ancient Rome. In the chapter after that, we'll start to see that the traditions of doubt in the ancient world—Greek, Hebrew, Eastern, and Roman—will all begin to come together and coalesce, and doubt will mature in a variety of incarnations as we plunge into the Middle Ages and beyond.

FOUR

When in Rome in Doubt, 50 BCE–200 CE

Empire of Reason

We have seen the philosophical doubt of the ancient Greeks, the ancient Jews, and the ancient East. All of that took place between the sixth century and the first days of the Common Era: in 600 BCE the Carvaka had their materialist text, the Brihaspati Sûtra; in 536 BCE the Jews were headed back to Jerusalem from the Babylonian exile; and about 500 BCE the Greek philosopher Heraclitus was explaining the world in rationalist terms. The Buddha died in 480, Confucius died in 479, Socrates was born in 470, and by 458 Ezra was working out the postexilic interpretation of the laws of Moses. Plato's *Symposium* came in 387 BCE; Alexander the Great invaded India in 327 BCE; Epicurus died in 271 BCE; and the Greek Hebrew Bible, the Septuagint, was composed about 255 BCE. In 168 BCE an altar to Zeus was built at the Temple in Jerusalem. At the turn of the millennium there was a turn toward theism in the Far East; in the Middle East, Jesus was born and in his name the whole tremendous Western culture of doubt would change profoundly. But not right away.

In the centuries around the turn of the millennium, the Romans created some of the most exquisite doubt of all time. It is worth noting that in this vibrant Roman world, early Christianity was taking hold in the poorer sectors of the Middle East, but at this point educated Romans took little notice. It was not until very late in the Roman period that the intellectual Roman world addressed the Christian idea. The second-century pagan physician and philosopher Galen spoke of Jews and Christians as philosophers and compared the cosmology offered by Moses with those offered by Plato and Epicurus (of the three, he favored Plato's). In writing about the Christians' philosophy, Galen criticized their dependence on "faith" and dismissed them as lacking the epistemological evidence for what they claimed.[1] Educated Romans lived in a rationalist, doubting world. But we must begin at the beginning.

The early Romans were pantheists and their major gods, Jupiter, Mars, Juno, Minerva, and Venus, were of Italian or Etruscan origin. Before Greek influence transformed Roman religion, it was a utilitarian affair, used to help ensure the family's need for food and procreation and the state's need for military power and domestic tranquillity. In the family, religion was overseen by the paterfamilias; in the state, it was overseen by the magistrate. Although the absence of a separate priestly class did not persist throughout the Roman Republic and Empire, it helped establish the importance of religious ritual for individuals, such that support of the state was linked with participation in ritual but not with private beliefs.

We used to say that ancient Rome was a society of almost no religiosity. Historians explained that the relationship between religion and the state in ancient Rome was so explicit, and so clearly in the service of the state, that the whole thing was more bureaucratic than pious.[2] One visited temples and made sacrifices in order to please the gods so that they would keep safe the Roman Republic and, later, keep the Roman Empire powerful and unified. Most people seem to have believed that the complicated communities of human institutions were extended into a less visible realm of big gods and local, average gods who were attentive—in varying degrees—to human beings. In this official Roman religion, there was no religious education for the young or old, no cosmology or doctrine, no message about one's own soul, and no ethical code: it was a system of sacrificing to the gods for the sake of the state. Religion was about reading the world for messages from the gods: omens, auguries, and prophesies read from entrails and other natural things. Divination was much less personal in ancient Rome than it had been in Greece. So it was not a very metaphysical or emotional religion, but it would seem that many people found it satisfying: they took part in the glory and persistence of a great state, and auguries helped them manage their communal anxieties and ambitions.

Later historians looked more closely and found evidence of religious passion and interiority in rites that had once seemed to us more cold and public.[3] There are some subtle indications that the man at the emperor's altar was weeping—engaged in managing his inner life, not only the fortunes of the state. Still, this merely adjusts our persistent overall impression that ancient Roman religion was not centered on the internal experience of the individual. Some have further modified this vision of Roman areligiosity by pointing out that alongside the more secular gestures of the official religion, Romans took part in a number of imported cults that offered more passionate or personal religious experience. By the period of the empire, the Hellenistic mystery religions, from Isis to Mithraism, had grown very popular throughout the Roman world—enough so that Isis was felt to be a threat and was banned from the city of Rome. The other source of an inward religion was Judaism and, as we'll see, it was very tempting, too. It eventually took over.

The Roman religion was stable and varied. Rome did not care much what else her subjects believed or to whom else they sacrificed, so long as they also believed in Rome and sacrificed to her gods and, later, to her emperors. The whole structure of the Roman gods was a result of the Romans conquering the Greeks militarily and the Greeks conquering the Romans intellectually. The Greeks claimed their culture represented the height of human accomplishment and the Romans believed it, coming to identify their local gods with gods of the Greeks and thereby taking on the glories of the Greek past. Later, as the Roman Empire grew, the Romans followed their own experience and encouraged the identification of the gods of the empire with the local gods of newly conquered or incorporated areas. There were many full identifications, and lots of double temples.

The apotheosis of Alexander the Great, and other Hellenistic kings after him, was the model for the deification of the emperors, first at death and then, later, while they still ruled.[4] There's something in this that feels very religious, very full of belief. But that seems to be the opposite of what was happening. As the master classical scholar Arnaldo Momigliano has argued, it was not that the state was so religiously valued, but rather that the words *god* and *godlike* had lost meaning to such a degree that they could be applied to anyone in affection or admiration.[5] Cicero called Plato a god and, more wonderfully, Lucretius used the term for his idol, the essentially atheistic Epicurus. In Momigliano's words, "people were finding it easy to call exceptionally powerful men gods because they were losing faith in the existence, or at least in the effectiveness, of their traditional gods."[6] The deification of emperors was meant to justify the extraordinary power of the

Roman leaders and help consolidate the far-flung provinces. Emperors during their lifetimes were more godlike in proportion to their absence: communities who had the emperor in their midst tended to treat him as emperor, while in the provinces, the imperial cult's sacrifices, temples, games, speeches, banquets, and statues all helped give the sense of a present ruler.

In other important ways, the empire's subjects were well served by the imperial cult—the celebrations were lively communal events at which poets and musicians excited the crowd, athletes competed for prizes, local rulers made speeches in honor of the great ruler, and children and teenagers took part in performances. But it is reasonable to distinguish such social and political ceremony from religiosity. The people were celebrating themselves, their community, the empire, the universe, and the fact that none of it had come crashing down. We do not all know how it feels to have our city besieged, but when it happens, we learn that we ache to take part somehow in the alchemy of mourning and reviving and to conjure again the illusion of mundane safety. Some reveled for the glory of peace, others for the militarism that they saw as keeping war at a distance from their homes. They were also having a good time. More traditional religion was not even doing that for them anymore. As Momigliano puts it, "The imperial cult was primarily a sign of indifference or doubt or anxiety about the gods."[7]

As for the intellectuals, many scoffed at the idea that a job title could hoist one into the sphere of the immortals. Lots of authors showed a bit of a sneer or a smirk in their discussion of imperial deification, and it is likely that many cultivated people felt this way. In this secular world, where the masses showed religious reverence for the emperor and went to statist celebrations, and intellectuals did not even do that, spiritual people flocked to the Mystery Religions and to Judaism. The Mystery Religions that had emerged in the Hellenistic world had spread to the reaches of the empire. They addressed the darkness of the universe and of our own hearts, and the fate of the individual after death. Judaism was yet a step further away from the statist Roman religion because it had those spiritual features but, unlike the mysteries, could not tolerate any other gods, including emperors. The Romans put up with this, for a long time, for three reasons: first, they respected the antiquity of the Jews; second, the Jews had proved, militarily and otherwise, loyal and reliable; and third, the Jews were shockingly willing to die rather than to tolerate even minor infringements on their law. In the next chapter, we will see the result of mixing Judaism and the Mystery Religions (and then adding Greek philosophy and Eastern asceticism), but for now these two spiritual ideas, and the Christianity that derived from them, must be seen as they were by most Romans: as exotic or bizarre minorities. Their secrecy, separatism, and passion remind us of the public, vaguely bureaucratic nature of mainstream Roman religion.

In the several decades before and after the turn of the millennium, Julius and Augustus Caesar championed a religious revival to go along with their new, more authoritarian regimes. How much this revival affected the average person's beliefs is difficult to tell. Augustus also gave the people other things to think about. The Plebeian and freed population of Rome vastly outnumbered the Equestrian and Patrician classes, and often enough a large number of them could not find work and could not afford adequate food. In response, to keep the average Roman from revolting against the rich, Augustus legislated food prices so that the poor could manage their basic needs, and also began a system of free food—in the form of grain distribution. At the same time, he offered free entertainment in the form of chariot races, bloody gladiatorial combat, lavish spectacles in amphitheaters, and the Circus Maximus. The poet Juvenal (55–127 CE) coined the pertinent phrase in his fourth satire, writing that "the people," who once demanded and received "commands, consulships, legions, and all else," had grown passive and "now concerns itself no more, and longs eagerly for just two things—bread and circuses!" These circuses, and even the sharing of bread, may have served basic religious needs, providing an emotional outlet and a sense of community. It is a question suited as well for our own age and our own public spectacles.

A few dates will be helpful before we go on: Octavian (who would become Augustus) defeated Antony and Cleopatra in 31 BCE and, as Augustus, established the Principate in 27 BCE. Jesus' ministry seems best dated as lasting from 27 CE to 30 CE, but the movement he started did not show up on the radar of any intellectuals for the following two centuries. The Pax Romana, or Roman Peace, a period of great stability and cosmopolitanism, stretched from Augustus's reign to 180 CE. We speak of Imperial Rome as beginning in 27 BCE and extending to 284 CE. In 330 Constantine moved the capital of the empire from Rome to Byzantium (renamed Constantinople), and began turning the empire toward Christianity. The years between 300 and 600 used to be called the Dark Ages, though they are nowadays called Late Antiquity, both of which give some valuable clues as to the period's nature. For now, we begin in the Late Republic, just as it falls.

The Roman world invented some stylish new doubt and brought many older traditions of doubt to a culmination. The six greatest disputants of that doubt were Cicero, Lucretius, Pliny the Elder, the Emperor Marcus Aurelius, Sextus Empiricus, and Lucian of Samosata. These men of the Roman Empire robustly schooled themselves in the history of Greek thought and then turned to explain it all to their peers: in brief and in Latin. In so doing, they offered the most shapely theories of doubt yet arranged. This is both old and new doubt, for these authors were often more straightforward in their versions of Greek doubt and they reimagined these ideas according to Roman concerns. The first was Marcus Tullius Cicero. Few texts in the history of doubt rival the importance of *The Nature of the Gods*.

CICERO

Cicero began writing *De Natura Deorum* about 45 BCE. By then he was weary of politics: he had been an important figure on the political stage, serving as praetor in 66 BCE and consul in 63 BCE. He had also been exiled, only to be recalled by Pompey and hailed as a hero. In 56 BCE he was made an augur, an official diviner, which was a lauded religious and political position. Cicero would write a book arguing that divination was hokum, but he set a great value on attaining this prestigious post. He had been against Caesar in defense of the Republic, but when Caesar won, Cicero made peace with the situation and lived with honor in Rome. When he retired from public life, he began writing a series of philosophical works with the express purpose of giving Latin-speaking Rome the opportunity to know the Greek philosophers, and to give them a philosopher of their own. Where he found Latin unsuited for philosophy, he coined new words and phrases and thereby left the language capable of greater subtlety than before.

The conceit of *The Nature of the Gods* was that many years earlier Cicero's friend Cotta, a great orator and priest, had invited the young Cicero to his home. When Cicero arrived he found himself in the company of three famous men—one an Epicurean, one a Stoic, and one, Cotta himself, a Skeptic from the Academy—engaged in a heated conversation about the gods. The Epicurean and the Stoic have some very definite ideas about the matter; Cotta, the Skeptic, claims to know nothing for sure, but also claims to be expert at seeing falsehood. Cicero—much younger than the great men—merely listens and offers a few words about his own opinion on the last page of the text. Readers have always understood that the Cotta character represents the critical eye of the adult, authorial Cicero.

The Epicurean, a man named Velleius, had already been talking when Cicero showed up, so he is asked to give a précis of his opinion. Book one of the text is made up of that speech and Cotta's answer to it. Book two is the Stoic Balbus's turn, and the third and final book is Cotta's answer to him. It would be well to note that when he was a boy, Cicero's family had moved to a town outside Rome so that his education could be of the highest affordable order. There he studied with several brilliant Greek teachers in succession: the first was an Epicurean, to whom he was devoted, but in 88 CE he found a new master, Philo, the leading exponent of the Academy's Skepticism, and Cicero experienced a sharp conversion. Although he remained fond of his old teacher, he was never again very tolerant of Epicureanism. Later in Cicero's youth, a famed Stoic and friend of his father came to live with the family. Cicero learned a great deal from him but did not reject Skepticism. Thus, in a sense, the story recapitulates Cicero's personal philosophical journey.

Cicero frames *The Nature of the Gods* by saying that most philosophers have affirmed the existence of the gods and that this assertion is "plausible and one to which we are all naturally inclined." But he immediately offers the reminder of a history of doubt: "Protagoras however professed himself in doubt on the matter and Diaogoras of Melos and Theodorus of Cyrene did not believe them to exist at all."[8] Then he explains that even among those who assert that gods exist, disagreements abound as to their form and character. For him, the key to the debate is whether the gods do nothing, care for nothing, and made nothing, or whether they made everything and continue to rule over everything, on into eternity. Before giving the floor to the proponents of these two opinions—the Epicurean and the Stoic— Cicero tells his readers that society might be in serious trouble without religion, or even without an understanding of the gods as caring and interested. Cicero takes this seriously, worrying that without the belief in concerned gods, piety would become mere convention and eventually disappear along with religion and sanctity. "And when these are gone, there is anarchy and complete confusion in our way of life. Indeed I do not know whether, if our reverence for the gods were lost, we should not also see the end of good faith, of human brotherhood, and even of justice itself." Many who would doubt religion would also worry about the social consequences of losing it.

There are a few instances in the book in which Cicero has one character or another—usually Cotta—nudge one of the other speakers to be more honest, since this is, after all, "a private conversation," in a private home. Cicero thus communicates that for the health of the state, these things should not, perhaps, be broadcast, but that in a quest for the truth, the wise must speak their true minds, if only to one another. He asks everyone to "come into court, weigh up the evidence, and return their verdict" on the matter of the gods, because there is too much at stake to be cavalier. "Surely even those who believe that they have attained certainty in these matters must feel some doubts when they see how widely wise men have differed about so crucial a question."[9] This plea for serious investigation would be paraphrased or repeated by many future doubters.

Velleius's Epicurean speech starts off the conversation, and it is deeply irreverent, especially because Cicero manages to tease the Epicurean for his own certainties at the same time that he uses him as a mouthpiece for teasing yet other people's certainties.

> Then up spoke Velleius, with all the confidence of men of his school (whose only anxiety is lest they may seem to be in doubt on any point), as if he had himself just returned to earth from some council of the gods held in one of those abodes of theirs "between the worlds" which Epicurus talks about. "Listen," said he. "From me you will get no mere figments of the imagination, such as the god whom Plato describes in his *Timaeus* as the creator and artificer of the world, or the fortune-telling old witch whom the Stoics call Providence, or some theory of the universe being itself endowed with mind and senses, a sort of spherical, incandescent and revolving god. All such marvels and monstrosities as these are not philosophy but merely dreams."

It is a remarkably forthright beginning by all accounts; it is fun to see Cicero refer to the God shown in the *Timaeus* as a very particular instance in Plato's work, and a silly one at that. Velleius then proceeds to give a little synopsis of the Greek philosophers' opinions on the gods. It is almost a comical performance, because Cicero is exaggerating what he saw as an Epicurean tendency to casually dismiss great minds (of other opinions). From its origins in the Garden, the Epicurean school had spread across the Hellenistic world and through the Roman Empire, and along that journey it had nurtured some dogmatic incarnations. The philosophy had become a way of living for countless people across several centuries, and followers could be loyal to the conclusions of the master without always being able to reconstruct the argument. So Velleius's tour of the Greek philosophers offered Latin readers a quick lineup of philosophy so far, and allowed Velleius to swashbuckle his way through history. He laughs at Anaxagoras's idea of the mind of the universe, saying that it makes no sense to imagine a mind without a brain; he scoffs at Pythagoras, who said the whole world was God—that, Velleius argues, would mean God was part of us and since we are so often unhappy, that would mean God was unhappy, and that seems to him unlikely. Even if there could be conscious intelligence separate from a body, all the spinning the universe does would make a person sick, "so why would it not be unpleasant to a god?"

But he saves the most space for Plato, Aristotle, and the founder of Stoicism, Zeno. He dismissed Plato for inconsistency and his cosmological imagination. "How could your friend Plato in his mind's eye comprehend so vast a piece of architecture as the building of a universe, and how God labored to create it? How did he think God went about it? What tools did he use? What levers? What machines? . . . It would be tedious to say more, for it is all the stuff of dreams rather than the search for truth."[10] Velleius then says that he has the same problem with the Stoic Providence that he has with Plato's God:

> I ask you both, why did these creators of the world suddenly wake up, after apparently having been asleep from time immemorial. . . . I ask you, Balbus, why that "Providence" of yours remained quiescent through that mighty lapse of time? Was she work-shy? . . . Why should God in any case wish to decorate the universe with lights and signs, like some Minister of Public Works? . . . I suppose he had previously always lived in darkness, like a pauper in a hovel?[11]

Valleius thought Aristotle had been "very confused" because he considered God to be bodiless: "And how can this universal god have motion if he is bodiless?" Velleius accuses the Stoics of reconciling the tales of Hesiod and Homer with their own views and, "In this way the most ancient of our poets, who never dreamt of such a thing, are converted into Stoics unawares!" Mythic dogma had been reframed as allegory for a more secular age, but now this allegory was itself becoming dogma.

Velleius concluded by saying, "So far I have been dealing in a general way not so much with the opinions of philosophers as with the fantasies of lunatics"—as absurd, he complains, as the gods described by Homer. Velleius then turns to explain the happy, totally unconcerned, immortal gods described by Epicurus: ethereal creatures that are just a bit more than images; shaped like human beings, but inactive; floating in the space between the worlds. Valleius agrees with Epicurus that gods exist because "an idea that by its nature commands universal agreement must be true," and since we all feel the gods are happy and immortal, it is reasonable to assume that this is true, too. Also, the gods show up in dreams occasionally. The end of Velleius's speech starts with naturalism and wraps up in an attack on the idea of a caring God:

> Our master has taught us that the world was made by a natural process, without any need of a creator: and that this process, which you say can only be effected by divine wisdom, in fact comes about so easily that nature has created, is creating, and will create, worlds without end. But as you cannot see how nature can do this without the intervention of mind, you follow the example of our tragic playwrights, and take refuge in a divine intervention to unravel the intricacies of your plot. You would have no need of such divine handiwork, if you would only consider the infinite immensity of boundless space in all directions. . . . In this immensity of breadth and length and height there swarms the infinite power of atoms beyond number, and although they move in a vacuum, they cohere amongst themselves, and then are held together by a mutual attraction. Thus are created all the shapes and forms of nature, which you imagine can only be created by some divine blacksmith with his anvil and his bellows! So you have smuggled into our minds the idea of some eternal overlord, whom we must fear by day and night. Who would not fear a god who foresees everything, ponders everything, notices everything? A god who makes everything his own concern, a curious god, a universal busy body? . . . Epicurus has saved us from all such fears and set us free . . . [12]

A deus ex deus ex machina! Velleius also criticizes the Stoic notion of Providence as foolish and dangerous, in that it makes people dependent on soothsayers, seers, and "every charlatan who will read for us the riddles or our dreams." Again, Epicurus is most praised for having delivered people from such dependency and its attendant fear.

Cotta's refutation is harsh, but it never contests the Epicurean's dismissal of the various ideas of God. Instead, it attacks the one vision of God that

Epicurus did defend. Indeed, after bestowing a few compliments, Cotta goes right to the heart of the matter, asking why "universal agreement" of belief in God had been either posited or taken seriously. Diagoras and Theodorus had openly denied the existence of the gods, reminds Cotta. Protagoras had doubted and for it he was banished and his works burned in public. "I suspect," Cotta explains, "that his example made others more reluctant to express such sentiments, when they saw that even agnosticism could not escape such penalties." It is a remarkable comment, suggesting that there has been much doubt that never dared speak its name. Cotta moves on and demands to know how Epicurus or Velleius came to know all the opinions of humankind on the gods:

> It is difficult, you will say, to deny that they exist. I would agree, if we were arguing the matter in a public assembly, but in a private discussion of this kind it is perfectly easy to do so. Now I myself hold a religious office, and believe that public religious worship and ritual ought to be reverently observed: so that I could wish to be certainly persuaded on this first question, that the gods exist, as a matter of fact and not of faith, I confess that many doubts arise to perplex me about this, so that at times I wonder whether they exist at all.[13]

Cicero has Cotta say that despite all this, he does believe that gods exist, and that he was only challenging anyone's ability to prove it. But other than offering these few quick disclaimers, Cotta makes a powerful argument for deep doubt.

Cotta has no patience at all for the specifics of the Epicurean gods. He has lots of detailed arguments, such as: why, if we are to believe our dreams of gods had some substantive origin, should we move on to assign the visiting image-beings any kind of divinity? "Suppose our minds are visited by such images, and that at most some apparition of this sort does come our way. Why should it be blessed? Why should it be immortal?" And why are these image-gods in human shape? Cotta jokes, "What a fortunate clash of atoms, from which suddenly men are born in the images of the gods!" What's more, he argues, it would be egocentric to so favor our own shape as to imagine the gods thusly. "No doubt an ant feels the same way," quips Cotta. "And which human shape are we speaking of? How many human beings are handsome?" Cotta digs in at the point where the gods are supposed to look like us. "Are we to imagine that . . . some are pug-nosed, some flap-eared, and some beetle-browed or big-headed, like many of ourselves? Or with them is everything made perfect?" And if they are all perfect, they are all the same; "then among the gods there can be no telling who's who or what's what!"

Cotta also finds a lot of fun with the idea of an unmoving god who is featured like a human being, and baits Velleius along these lines: "So your God will have a tongue, but he will not speak. . . . And the organs of procreation, which nature has added to our bodies, will be useless to a God." Epicurus thought the gods looked like us because we are so beautiful, but Cotta argues that the beauty of our bodies has everything to do with their function—our bodily organs are beautiful because of their complex and precise workings. For a god to have them would be foolish.

Cotta gives other reasons to doubt the Epicurean gods—he finds their idleness childish, for example—but his most powerful argument is that the whole idea of these gods is unfounded. Yes, many people believe in gods, says Cotta, but they all believe differently and according to the conventions of their local culture. "Are you not ashamed as a scientist, as an observer and investigator of nature, to seek your criterion of truth from minds steeped in conventional beliefs?" Since universal agreement and dreams were the reasons offered by Epicurus for belief in gods, Cotta had pretty well finished off Velleius at this point, but there is still a tone of frustration here. It is the misuse of logic that really burns Cotta. "The whole theory, Velleius, is ridiculous. . . . Is there no end to the old wives' tales which you permit yourself? . . . I do not believe these gods of yours exist at all."

Cotta's disbelief in these gods is not offered in favor of anyone else's conception of the gods. Cotta mocks Velleius for treating the great minds of ancient Greece with so little grace, but he does not do so in defense of any of their gods. Quite the contrary. He scolds that Velleius "called some of the most famous men fools, dreamers and lunatics. But if none of these could discover the truth about the nature of the gods, we may well wonder whether they exist at all." [14]

At the very end of his rebuttal, Cotta's argument takes a little turn. After a lengthy argument against Epicurean gods, Cotta wraps up his response with the accusation that by offering people the idea of uninvolved, totally uninterested gods, Epicurus became one of the important destroyers of religion. Along with him, Diagoras, Theodorus, and Protagoras are accused of demolishing religion, reverence, and worship by their denial of or doubt in the gods. Cotta's diatribe at this juncture is a little history of doubt. It is a delightful miniature of my central task and well displays the tone of Cicero's book. The idea of religion was defended, and the denial of the gods was seen as a destruction of religion, and to be avoided. Yet, the text manages to rehearse every argument *against* God and to dismantle every argument *for* God. This text picks up after Cicero has mentioned the work of Diagoras, Theodorus, and Protagoras.

> Then there are those who have argued that all our beliefs about the gods have been fabricated by wise men for reasons of state, so that men whom reason could not persuade to be good citizens might be persuaded by religion. Have not these also totally destroyed the foundations of belief? Or Prodicus of Chios, who ascribed divinity to everything which benefits mankind: what room did he leave for religion?[15]

Cicero then remarks that some say "brave and famous and powerful men" were deified "and that these are the gods whom we have now become accustomed to worship." People who say such things, he muses, must have no religious feeling. "This line of thought has been especially developed by Euhemerus," explains Cicero, "and our own Ennius has been his foremost disciple and interpreter." Ennius (239–169 BCE) was considered the founder of Latin poetry (he introduced Greek forms and changed Latin to do it), most famed for his epic history of Rome, the *Annales,* but he also brought Euhemerus to the attention of the Roman Republic, and, eventually, the Empire. Cicero asks if when Euhemerus describes these deified heroes "and where they lie buried," did he seem to have strengthened religion "or to have utterly undermined and destroyed it?" The same goes for the "holy and solemn shrine of Eleusis," or of Samothrace, "for when these are examined by the light of reason, they seem to be a recognition of the powers of Nature rather than the power of God."

Cicero added that Democritus, "one of the truly great men, from whose springs of thought Epicurus watered his own little garden," had erred on the issue of the gods. "Who can understand what he means by these images?" Why worship them? "Then comes Epicurus and uproots religion entirely from the minds of men by taking away all grace and favor from the gods." As Cicero had it, and it may have been a message from Cicero to sympathetic readers: "It is obviously true, as our mutual friend Posidonius argued in the fifth book of his work on the nature of the gods, that Epicurus did not believe the gods existed at all and that what he said about them was said merely to avoid the odium of atheism." He continues with a final attack on the Epicurean gods on grounds that would apply to the specifics of many other visions of God:

> A god endowed with human limbs but with no use for them! A god transparent and insubstantial, giving no sign of grace or favor to anyone, inactive and indifferent! In the first place such a being could not even exist, and Epicurus knew this, so that he merely paid lip-service to gods whom he had in fact destroyed. And finally, if this is all that a god is, a being untouched by care or love of human kind, then I wave him good-bye.[16]

This good-bye ends the first book, and it is surely Cicero's. We recall that his first important teacher was an Epicurean, that his second was a Skeptic from the Academy, and that after his conversion, when he passed from one teacher to the next, he never went back. This philosophic conversion was perhaps the issue at the center of this text: If you want truth, you have to avoid making up anything. If you want to attend to people's needs, and take seriously the knowledge they glean from their emotions and dreams, don't give them something that looks like the comforts of old but is in fact an almost entirely useless, unsubstantial nothing.

Now it is the Stoic Balbus's turn to speak, and Cicero treats him very differently than he had treated Velleius, praising him and giving his position a much longer and more subtle hearing. The gist of Balbus's argument comes down to two major notions, which he circles again and again. First, the rotation of the heavens seems too beautifully coordinated to be without divine control. Second, if there is no God, then human beings are the most reasonable, wisest, most powerful creatures in the world, and that seems arrogant and childish. I've remarked on this before, but it is worth remembering that most moderns live in densely populated areas where the stars are obscured by light and pollution. Even when we are in the country and confronted by the awesome spectacle of the firmament, we know that we are the ones that are moving—the whole question of how the light show works no longer packs a wallop. It's because we thought we were at the center of things, with everything moving around us, that we were so amazed at the concert of motion above us—once we see the Earth as out there somewhere, spinning around, the precise pageantry of the firmament is less staggering. For most of human history, however, the moving night sky was an excellent argument for theism. Listen to Balbus's confidence: "What could be more clear and obvious, when we look up to the sky and contemplate the heavens, than that there is some divinity of superior intelligence, by which they are controlled?"[17]

He has four reasons to believe in God: (1) foreknowledge of future events—he believes divination works and that it could work only if a God had preordained events; (2) the blessings of nature—climate, abundance of food, and so on—and how perfectly they fit our needs; (3) awesome natural spectacles such as thunderbolts, cloudbursts, blizzards, hailstorms, and floods; and (4) the regularity and motion of the heavenly bodies. This last he says is "perhaps most important," for if you go into a house or gymnasium or anyplace else and everything is beautifully arranged and in perfect order, you assume that someone is in charge—these things just don't happen by themselves.[18] "Their constant and eternal motion, wonderful and mysterious in its regularity, declares the indwelling power of a divine intelligence. If any man cannot feel the power of God when he looks upon the stars then I doubt whether he is capable of any feeling at all."[19]

Balbus thinks it is the extreme of hubris to imagine that we are the smartest thing going. Moderns might agree, insofar as, if other life exists in this vast universe, it may well be smarter than we are. But the only extraterrestrial Balbus is thinking about is God, so if there is no God, we are the smartest things in the world, and that seems like puerile narcissism. Balbus says, "Only an arrogant fool would imagine that there was nothing in the whole world greater than himself. Therefore there must be something greater than Man. And that something must be God."[20] This argument is mingled with the idea of design in the universe, because Balbus is impressed enough with humanity to insist that someone better than we must have made us. "If you see some great and beautiful building, would you infer, because the architect is not immediately visible, that it must have been built by mice and weasels?"[21]

Balbus cannot exclaim enough about the absurdity of a small part of the universe, humanity, being in some way superior to the entirety of the universe. Therefore, since the universe gave birth to intelligence, it must itself be a living intelligence. The Stoics spoke of God in quite naturalist terms: "That which we call Nature is therefore the power which permeates and preserves the whole universe, and this power is not devoid of sense and reason." For Balbus, not only is the universe God, but the heavenly bodies are also divine. There are two kinds of warmth, he says: the kind from fire and the kind from living bodies. The sun never goes out and it sustains rather than destroys life, so it must be alive and so must all the stars be.

For Balbus, the stories that had been told about the gods—arising from the poets, from philosophy, and from the coincidence of certain earthly events with certain celestial periods—are all "wild errors." Of Homer's stories: "These tales are frivolous absurdities and both those who tell them and those who listen to them are a pack of fools."[22] Yet, since a divine power permeates everything, Balbus concludes that we might as well say that it permeates the earth under the name Ceres and the sea under the name Neptune, and so on, and worship these gods each in the way custom has established. Stoic allegory aside, he says that in truth the universe itself is God, and he laughs at the Epicurean belief that our world came into being by chance. "Is it not a wonder that anyone can bring himself to believe that a number of solid and separate particles by their chance collisions and moved only by the force of their own weight could bring into being so marvelous and beautiful a world?"[23]

He suggests that we'd have as much luck if we poured a sack of golden letters on the ground and hoped they'd spell out the *Annales* of Ennius. Balbus continues: "If these chance collisions of atoms can make a world, why can they not build a porch, or a temple, or a house, or a city?"[24] As we'll see, Cicero will have Cotta help Balbus with this apparent problem by pointing out that we are talking about nature, not a poem and not a city. The secular explanation is that life, and other matter in other ways, is an accident that has fallen into a replicating pattern. By contrast, when we make predictions of a highly complex end point whose arrangement has nothing to do with the intrinsic properties of its parts, the odds of getting there by random action drops to the infinitesimal.

Balbus's argument that there must be intelligence to the universe is based on his awe of the heavens, and, to share his awe, he also catalogs the wonders of the earth: "Think also of the beauty of the sea . . . and the shellfish clinging to the rocks!" He argues that the world must have been made for us since only we appreciate it. Just as Athens and Sparta were made for the Athenians and Spartans, so the universe is for us. We are the ones who have measured out the course of the stars, so the stars must be for us. He observes, "What use is the pig, except to be eaten? Indeed Chrysippus says that the life in a pig is merely the salt which prevents it going bad."[25] The line highlights the comic hubris of Balbus's idea that the world is made for us.

Balbus concludes even more about the existence of God from the existence of humanity, and cites Zeno as having said that "Nothing . . . which is devoid of life and intelligence can give birth to any living creature which has intelligence. But the universe does give birth to living creatures which partake of intelligence in their degree. The universe is therefore itself a living intelligence."[26] Balbus finishes up by mentioning that injustice in the world is no proof against the gods, because the world is unjust only in its minutiae and "the gods . . . have no interest in trifles. For great men all things always turn out well . . ." Then he reminds Cotta that, as a priest, he ought not to argue against the gods.

Cotta turns to Balbus with a bit of a smile at this little reminder, and Velleius, catching it, expresses his hope that Balbus will be as well skewered as he himself had been. Cotta's response assures the company that he has much to say against the Stoic notion of God, but first he takes a last shot at Epicurus, accusing him again of having been afraid to deny the gods' existence: "it seemed to me that he had his tongue in his cheek."[27] The Stoics at least seemed devoted to truth. Cotta then turns back to Balbus and says that he needs no encouragement regarding his public duties: "Nobody, be he learned or unlearned, shall ever argue me out of the views which I have received from my forebears about the worship of the gods."[28] He says he has never been contemptuous of Roman religion and "I have even persuaded myself that Romulus by his reading of omens and Numa by the institution of religious ceremonies together laid the foundations of the Roman state."[29] There is certainly enough evidence here to see that Cicero may have been informing his readers that his belief in God was social and political rather than spiritual or intellectual. Regarding the actual existence of the gods, Cotta turns to Balbus and says, "I am persuaded of this belief by its traditional authority: but you have given me no reason why I should believe it."

Cotta mocks Balbus's contention that the stars and planets are divine, reminding him that "Velleius and many others would not even consider [them] to have any life at all!"[30] And he laughs even louder that Balbus could give any weight to the idea of universal agreement, saying, "You are content that such matters would be decided by the judgments of fools, you, a Stoic, a member of a sect which regards all folly as a kind of mental disorder!"[31] Cotta teases Balbus about the idea that the gods had appeared to men, and Balbus breaks in, dramatically citing some famous and revered examples. "You palm me off with hearsay, Balbus," answered Cotta, "when what I want from you is reasoned argument."

Cotta dispatches Balbus's four proofs just as briskly. He doesn't believe in divination, and he acknowledges that there are both pleasures and terrors that impress people so mightily that they believe them to be of divine origin, but that does not make it so. Cotta then takes on the claim that since the universe is superior to a person and a person can think, the universe must be able to think. He proposes a parallel: nothing on earth is superior to the city of Rome; therefore, either the city of Rome can think or an ant is superior to the city of Rome. Cotta plays with this, proposing that "A being which can read is superior to a being which cannot. But nothing is superior to the universe as a whole. Therefore the universe as a whole can read."[32] And since for Balbus the universe has intelligence because it gave birth to creatures with intelligence, Cotta insists that the universe is also a lute player. He concludes that there is no reason to imagine that the universe is God, and if the universe is not God, neither are the stars. "You are right to wonder at them," he assures Balbus, but that they are amazing does not mean that they can't be natural phenomena.[33] Listing some other marvels, he says, "We must seek an intelligible cause for all these phenomena. The moment you fail to find one, you run off to a god like a suppliant to an altar."

As for the craftsmanship that went into the universe, "I would agree," says Cotta, "if only the universe were a house and not, as I shall show, a work of nature." Cotta is impressed with Balbus's description of the harmony and interrelation of nature, but insists that this does not mean a divine spirit is needed. "Nature persists and coheres by its own power without any help from gods. There is indeed inherent in it a kind of harmony or 'sympathy' as the Greeks call it. But the greater it is in its own right, the less need it be regarded as the work of some divine power." Cotta then asserts that there are no immortal beings, citing the great Skeptic Carneades for his argument that every living thing is subject to change, suffering, and destruction. He even says that "there is in fact no immortal body, no individual atom which cannot be split and pulled apart. Every living thing is therefore in its nature vulnerable. . . . There are countless other cogent arguments to prove that every conscious being must perish in the end."[34]

In a beautiful argument, Cotta then revives Carneades' idea about the moral qualities of God, writing that "a being who is not and cannot be touched by anything of evil has no need to choose between good and bad. What then of reason and intelligence? We use these faculties so as to proceed from the known to the unknown. But nothing can be unknown to God."[35] This problem is also true for all the other virtues, he argues, noting that justice is a product of human communities, that temperance entails temptation, and that courage occurs in situations of pain, toil, and danger.

What could God know of any of these? At this point Cotta says that he cannot feel contempt for the ignorant masses when such rubbish is being said by the Stoic philosophers. Returning to the notion that the universe is God, Cotta asks, "Why then add a number of other gods? And what a crowd of them!" After listing the names of star gods, he turns to the idea of Ceres, the corn god, and Bacchus, the vine god, and abruptly asks, "Do you think that there is really anybody so mad as to believe that the food which he eats is a god? As for the human beings who are said to have become gods, can you give me some rational explanation of how this could happen in the past but not in the present?"[36]

The last important argument that Cotta puts forth addresses the issue of whether the gods care for human beings. Balbus had argued that they did, since they gave us reason, the greatest possible gift. Cotta isn't so sure, because reason is so double-edged: "It is only a few, and those rarely, who use it for good, while many use it constantly for evil." He insists that the gods could have endowed us with reason in a form that excluded vice and crime and asks how the gods could have made such an awful mistake. As for punishing the bad and rewarding the good, Cotta doesn't see any evidence of it. "Diogenes the Cynic used to say that Harpalus, who was regarded as one of the most successful robbers of his time, was a living witness against the gods, because he enjoyed so long a run of good luck." Cotta then tells a number of enchanting stories of sacrilegious temple robbers, including one who apparently said, "It would be silly to pray to the gods for their gifts and then to ignore them when they are on offer." Such behavior, we are informed, led to the thief's prosperity, not his downfall. "We may consecrate shrines to Reason, Faith and Virtue, but we know that it is only in ourselves that they are to be found. From the gods we ask the fulfillment of our hopes of safety, wealth and success. Therefore the prosperity and good fortune of the wicked, as Diogenes so often said, absolutely disprove the power of the gods."[37] Reason, faith, and virtue come from within ourselves—so the only thing remaining that argues for the gods' existence is divine justice, which is why it is so important to notice that there is none. Cotta also reminds us of "Diagoras the Atheist," who dismissed the votive pictures testifying to divinely saved sailors by asking where the pictures were of those who had prayed and yet perished.

Most poignantly, Cotta wants to know why Balbus's all-powerful Providence allows beauty to be destroyed. He speaks of ravished towns and wonders:

> Could not a god have come to the rescue and saved those great and splendid cities? . . . [Y]ou say the divine power of a god can create, set in motion, or transform anything at will. You do not offer this belief as a superstition and an old wives' tale, but as a reasoned proposition of physical science. You argue that matter, which comprises and composes all things, is malleable and changeable throughout the universe.[38]

Divine Providence was supposed to be able to "accomplish anything it pleases" and yet it lets people die. "It does not even care about nations. Nations? It does not even care about whole peoples and races, and so we need not be surprised if it shows contempt for mankind altogether." That would be bad enough, laments Cotta, but the Stoics also argue, "in the same breath," that the gods send dreams to people and that it is proper for people to make vows to the gods.

> But vows are made by individuals, so the divine mind does give heed to the concerns of the individual? So, you see, this divine mind cannot be quite so busy as you thought. But even suppose it is at full stretch, turning the sky, protecting the earth, calming the sea, why does it permit all the other gods to idle about doing nothing? Why not appoint some of those who are unemployed to be Commissioners of Human Affairs? There seem to be any number of them available.[39]

That's the last real statement of the book. After this—and Cotta's final assertion that he hasn't been denying the existence of the gods but merely reminding his listeners how difficult a question it is—Cicero suddenly comes back into the picture to sum up. In a brief, rushed conclusion he compliments Cotta on his masterful analysis but says that, overall, he agrees with Balbus and the Stoics. And that is that.

Cicero's book provided a ready source for anyone interested in the arguments for doubt, and it made the ideas available for those who read only Latin. What did Cicero believe? He seems to have been religious in his earlier philosophical works, *The Republic* and *The Laws*. In the introduction to *The Nature of the Gods,* Cicero tells us that he was writing philosophy for three reasons. First, he had time on his hands because political changes had shoved him out of government. Second, he thought it would be good for the nation to learn philosophy and many could not read Greek. Last, he confessed, he "was moved to these studies by my own sickness of mind and heart, crushed and shaken as I was by the great misfortune which I had to bear." This great misfortune was the loss of his daughter Tullia in 45 BCE. He was inconsolable, and his second marriage broke up in the strain. For a while he planned to have Tullia deified, so he could build a little temple to her memory on his estate, but he did not do it. Instead, Cicero wrote this masterpiece of doubt, though he never said the one was a replacement for the other.

Some scholars have wondered at this—should not he have turned toward religion? Yet, clearly, he didn't. This is another possible response to loss. This response places the quest for demonstrable truth above the vicissitudes of the mad and thieving universe. The search for provable truth can be an unalienable comfort. What Cicero seems to have concluded is that we cannot know if the gods exist—but it seems unlikely.[40]

It has been suggested that Cicero grew more skeptical because his contemporaries were growing more credulous.[41] His friend Terentius Varro had just written *Divine Antiquities,* a treatise on religion that cataloged the vast array of Roman gods and called for a revival of their worship. Varro noted that ways of understanding the gods changed across time, but the specific reality of their details interested him less than the mandate that the gods be worshiped for the sake of the state. Varro's book helped inspire the conservative revival of Roman religion by Julius Caesar and then Augustus. Cicero celebrated Varro's work in 46 BCE, and in his own work of this period he also took care to emphasize that Rome must actively preserve its religious traditions. Now it was 44 BCE and Varro's work was inviting a restoration of obsolete cults. A public sacrifice was decreed to celebrate Caesar's birthday, and Caesar's statue was increasingly exhibited among the gods. The tone of the times, among both Cicero's friends and his enemies, was increasingly sanctimonious. That may have helped drive Cicero in the other direction. Yet his skepticism was not only a distaste for Caesar's swagger. Cicero's beloved Tullia was dead, and we may presume her father asked himself some pointed questions about the nature of mortality. Seen thusly, Cotta's words on immortality sound sadder and sterner. Consider his final remarks to Balbus on his God: "It follows from this theory of yours that this divine Providence is either unaware of its own powers or is indifferent to human life. Or else it is unable to judge what is best. 'Providence is not concerned with individuals,' you say. I can well believe it."[42] These words are among the book's final statements.

Cicero's next book was *On Divination,* and it, too, was supposed to be a record of a private conversation, this time between Cicero and his brother.

The brother spoke first, in favor of the divinatory arts. Cicero demolishes that argument, hammering on the disproportionately small number of correct or useful prophecies and the inherent improbability that bird entrails or the position of far-off stars could actually contain information about the details of a given person's life on a given day. It was one of the last things Cicero wrote. He got on the wrong side of Augustus and was put to death in 43 BCE. He is remembered, among other things, as the unsurpassed master of Latin prose. He will remain a key voice into the future of doubt.

LUCRETIUS AND THE EPICUREAN POEM

Titus Lucretius was a contemporary of Cicero and the great poet of Epicureanism. His *On the Nature of Things* is a book-length poem detailing a complex and subtle philosophical system—perhaps the only successful venture of this nature in Europe's history. The ideas were faithfully reproduced from Epicurus, yet the poem is a highly original work representing the concerns of its own historical moment. Lucretius felt the decline of the Roman Republic: the military was taking over, there was a rash of scandals, and one heard of corruption everywhere. In the circumstances, Lucretius believed it would be best to follow Epicurus out of public life and into the garden of friendship.

In the poetry of Lucretius, Epicurus is the savior of humanity. This is heroic poetry of doubt and disbelief. It celebrates Epicurus as the great champion of rational thought and as the conqueror of religion. The text also hints at the pride people felt in coming to these conclusions. A little of the verse quickly demonstrates the gratitude of centuries of followers:

When before our eyes man's life lay groveling, prostrate,
Crushed to the dust under the burden of Religion
(Which thrust its head from heaven, its horrible face
Glowering over mankind born to die),
One man, a Greek, was the first mortal who dared
Oppose his eyes, the first to stand firm in defiance.

Not the fables of gods, nor lightning, nor the menacing
Rumble of heaven could daunt him, but all the more
They whetted his keen mind with longing to be
First to smash open the tight-barred gates of Nature.

His vigor of mind prevailed, and he strode far
Beyond the fiery battlements of the world,
Raiding the fields of the unmeasured All.

Our victor returns with knowledge of what can arise,
What cannot, what law grants each thing its own
Deep-driven boundary stone and finite scope.
Religion now lies trampled beneath our feet,
And we are made gods by the victory.[43]

Traditions regarding the lives of Greek gods were reduced to fables as Epicurus went "raiding the fields of the unmeasured All." The last two lines of the quoted passage are particularly strong.

For Lucretius, as for Epicurus, the finality of death and the absence of the gods did not seem depressing; indeed, they seemed to add to the sweetness of life. Consider his calm musings on life after death:

Besides, if the soul by nature stands immortal
And slips into the body right at birth,
Why can't we recall as well the times gone by,
Preserving traces of our former lives?
But if the spirit's power is so altered
That all its hold upon past action fails,
Well, that, I think, strays not too far from Death.

That is to say, if you cannot remember your past lives, it is pretty much the same as not having had them, and if you live after death but with no memory, it is pretty much the same as having died. As for reincarnation, Lucretius notes that "if souls never died and could swap bodies, all creatures would become a welter of cross-traits: . . . the hawk would flee the swooping of the dove; men would be foolish and fierce creatures wise."[44] No, he concludes, the soul dies with the body.

On the matter of the origin of consciousness, Lucretius explains that this, too, arises from the laws of nature: "Now of necessity you must admit that sensate things must even so consist of insensate atoms. . . . From the insensate, animal life is born." The fact that we are made up of insensate parts teaches us that the insensate can produce (or add up to) the sensate.

"So nature turns all food to living flesh / And from that food gives birth to animal senses / In much the same way as she make dry tinder / Explode in flames and turn all into fire."[45]

This naturalism, he insisted, was good news, and he was as eager as Epicurus had been to free humanity from worry over death.

> *Death, then, is nothing to us, no concern,*
> *Once we grant that the soul will also die.*
> *Just as we felt no pain in ages past*
> *When the Carthaginians swarmed to the attack*
>
> *So too, when we no longer are, when our*
> *Union of body and soul is put asunder,*
> *Hardly shall anything then, when we are not,*
> *Happen to us at all and stir the senses,*
> *Not if earth were embroiled with the sea and the sea with heaven!*
>
> *Now if you happen to see someone resent*
> *That after death he'll be put down to stink*
> *Or be picked apart by beasts or burnt on the pyre,*
> *You'll know that he doesn't ring true, that something hidden*
> *Rankles his heart—no matter how often he says*
> *He trusts that there's no feeling after death.*[46]

The problem is, "He posits, unknowing, a bit of himself left over." He imagines his corpse, feels sorry for it, and "Can't see that when he dies there'll be no other / Him living to moan that he's bereft of him."

Our feelings indicate that our selves are real and lasting, and that if we are not included in a game of pinochle a hundred years after our own death, we'll feel a little left out. Lucretius suggests that if we can understand that we will not, in any sense, persist after death, we will realize that after death we will be free from all possibility of pain, anxiety, humiliation, and other nastiness. For instance, we will be entirely spared the feeling of being posthumously picked apart by beasts. You will not have an opportunity to miss your own pretty face, Lucretius reminds us, saying that when a man dies, "there'll be no other him living to moan that he's bereft of him." Also, whatever section of time you have is the only time there is, as far as you are concerned, so the urge for more time does not make any sense.

Survive this generation and the next—
Nevertheless eternal death awaits,
Nor will the man who died with the sun today
Be nonexistent for less time than he
Who fell last month—or centuries ago.[47]

A bit grim, certainly, but the point was to stop trying to grasp on to life, to enjoy it, and to stop worrying about death.

I mentioned earlier that Epicurus was "a god" to Lucretius. What Lucretius says is that Epicurus is "a god, the founder of a way of life called 'wisdom' now." If Prodicus of Ceos and Euhemerus were right about the gods originating from normal people who brought great things to humanity, then, Lucretius giddily suggested, Epicurus had earned the title, too:

Bring the old inventions of those other gods!
For Ceres, they say, first brought us grain, and Bacchus
Gave mortal men to drink of the juice of the vine—
And yet life can go on without these things,

Without a clear heart, though, no one lives well.
All the more does he merit from us the name of god
Whose way of life, by now spread worldwide, brings
Sweet soothing solace for the minds of men.[48]

Solace is the gift that seems most godlike to Lucretius—far above a glass of red and a dinner roll. There are other things to eat and drink, and life is sustained by more than food. Lucretius's idea of the origins of gods was that people saw wondrous things in the sky, and the changing seasons, and could not understand them. "Their refuge, then: assign to the gods all things." But it did not stop there. "Unhappy human race—to grant such feats to gods, and then to add vindictiveness! What wailing did they bring forth for themselves, what wounds for us, what tears for our descendants!"[49]

Lucretius believed in the Epicurean gods, but given that they were absent from the world of humanity, his world was godless. He teased people who paid lip service to the conventional gods from a rationalist standpoint. His words on the matter have a nice dismissive shrug to them and delightfully bring his subject before our eyes: "If a man insists on calling the sea

'Neptune' or the grain 'Ceres,' and would sooner abuse the name of 'Bacchus' than to call wine what wine's called, we'll give way, let him tell us and tell us the world is the 'Goddess Mother' so long as in truth he still keeps his mind clean of the taint of vile religion."[50] Lucretius perfected the irreligious sneer.

He set out to describe how religion worked because he thought mystery and fear kept driving otherwise intelligent people back to their old beliefs. He warns: "Don't you scurry like a brainless fool" to check where the thunder and lightning of a storm come from in order to understand the minds of the gods.[51] Thunder and lightning, he assures his reader, are natural effects of the clouds and the sky in a storm. His details are fun. (Some thunder comes from clouds crashing into each other, some from overfull clouds popping: "No cause for wonder. Blow up an animal bladder until it bursts: it gives its great big pop.") But the point is always to assure his reader that there is a naturalist explanation for everything.[52]

He explains the physical and emotional world with such passion—for both the exquisite and the grotesque—that his poem stands as a sublime answer to God's litany in the Book of Job. His curiosity and wit celebrate everything: why fruits are "faithful to their trees," instead of all things bearing all things; why "Nature cannot fashion giants to ford the sea in a few strides"; why stone walls "weep plenty of big drops"; why equal-sized balls of wool and lead have different weights; why children get dizzy when they spin around. His descriptions of the movements and behavior of atoms are so detailed that the text feels at times overly technical, but any search for a mechanical explanation might carry him off into his own surprising, imaginative musings:

Don't suppose atoms link in every way.
You would meet freaks and monsters wherever you turned:
Races of half-beast men would spring up, tall
Branches might sometimes sprout from a living torso,
And land-dwelling members link with the life of the sea
And Nature, mothering anything anywhere,
Would feed Chimeras snorting stench and flame.
None of this happens, we know, for everything
Is made of certain seeds, by certain parents,
And in their growing they preserve their kinds.
Of course they must; a fixed law makes it so.[53]

He goes on to explain the "fixed law" in terms of atoms. He begs his reader to "Hear the truth of reason! A new fact fights to clear its way, to accost you and show you a new aspect of the world." And then he proceeds to versify Epicurus' wondrous cosmology:

> *In no way now can it seem plausible*
> *That while space yawns in every direction, endless,*
> *And numberless seeds in seas unfathomable*
> *Fly this way and that, driven on in ceaseless motion,*
> *Our world and sky should be unique creations,*
> *And all those seeds out there accomplish nothing!*
>
> *When after all our world is made by nature*
> *Of her own, by chance, by the rush and collision of atoms,*
> *Jumbled any which way, in the dark, to no result,*
> *But at last tossed into combinations which*
> *Became the origin of mighty things,*
> *Of the earth and the sea and the sky and all that live.*[54]

Since our world of atoms happened to create amazing things, Lucretius assures us that natural law and accident must have created endless pockets of wonder. These other worlds are populated by "different kinds of men and animals." Neither the "sky and earth" nor the "sun, moon and sea" are unique; all are in abundance throughout the universe. This leads Lucretius to claim that gods simply couldn't be running all of this—it is too much work. Instead, "nature is free," and unconstrained; "rid of all gods, she works her will herself." It is a big moment in the history of doubt:

> *Who can wheel all the starry spheres, and blow*
> *Over all land the fruitful warmth from above*
> *Be ready in all places at all times,*
> *Gather black clouds and shake the quiet sky*
> *With terrible thunder, to hurl down bolts which often*
> *Rattle his own shrines, to rage in the desert, retreating*
> *For target drill, so that his shafts can pass*
> *The guilty by and slay the innocent?*[55]

The stanza develops the idea that a world constantly managed by an intelligent force is much less efficient than a universe that generates itself according to regular principles. The stanza also takes a jab at the silliness of an intelligent God raining lightning bolts on his own shrines, and laughs at the sheer work involved in the idea of divine justice. Many have said that this world is so complex that it must have been conceived by an intelligent, powerful force. Lucretius says the opposite—it is so vast and majestic that it must be self-propelled, because it is ridiculous to imagine a force trying to enact all of this when one has an alternative mechanistic theory.

For its delightful inquiries—into such questions as whether women and other female animals enjoy sex (he thinks yes)—its poetic style, its imaginative descriptions, and its intellectual rigor, Lucretius was read and studied by the general populace as well as the leading figures of Roman literature for centuries. He had a tremendous impact on Virgil, for example, and Ovid loved his materialism and mockery of the traditional gods. Ovid's own best statement on the subject was, "It is convenient that there be gods, and as it is convenient, let us believe there are." That does seem to have been the mood. Lucretius, by the way, died without putting his poem into a published form; Cicero did that for him. Cicero outlived Lucretius only by a little, both men dying in the middle of the first century BCE. We now turn our attention to three later figures of the early empire.

PLINY THE ELDER

Pliny the Elder lived from 23 to 79 CE. Nero's reign created much anxiety and real persecution, mostly among the privileged classes. Stoicism provided the philosophical platform for the opposition, who were attempting to revive and empower the senate. Pliny the Elder, however, was a Stoic who was reasonably comfortable under the emperors, serving the state by day and writing his largely inoffensive, and often Imperially flattering, treatises by night. *Natural History* is the only work of his to have survived down to us, and it is a treat. The book is a compendium of information about the world, some of it merely interesting to the modern reader, such as the section on "The papyrus plant and the invention of paper," and some of it historically enlightening and entertaining, such as the section called "Man occupies a small fraction of the earth, itself a mere dot in the universe" and the one on "Alexander the Great's famous dog" (the dog fought lions and elephants!). But in his section on "The revival of people pronounced dead," he showed more caution, writing that "life is full of such predictions but they should not be collected, since more often than not they are false."

In his section on "The search for God," Pliny opines that it is a sign of human weakness to try to find out the shape and form of God.[56] He doesn't think that the world "is properly held to be a deity," but his list of its characteristics (everlasting, boundless, an entity without a beginning and one that will never end) does not include consciousness or intelligence. Indeed, when he speaks of the more personified notion of God, he displays more skepticism.

> Whoever God is—provided he does exist—and in whatever region he is, God is the complete embodiment of sense, sight, hearing, soul, mind and of himself. To believe in . . . an infinite number of deities corresponding to men's vices, as well as their virtues . . . plumbs an even greater depth of foolishness.[57]

Pliny introduces a few new reasons not to believe in the specific gods of the pantheon. For one thing, he thinks if everyone were correct about his or her gods, there would be more gods than people. He dismisses the details of these gods because they seem unlikely: "Some nations have animals—even repulsive creatures—as gods," but worse, "to believe that some gods are always old and grey-haired, while others are young men and boys," or to believe in gods who are "lame, born from eggs, or who live and die on alternate days—such beliefs are little short of the fantasies of children." He also dismisses the pantheon for the old reason of the gods' adulterous and treacherous behavior. It is at this juncture that he asserts: "God is man helping man: this is the way to everlasting glory."[58] This assessment allows him to flatter the emperors, for when Roman leaders aid their people, "The apotheosis of such men is the oldest method of rewarding them for their good deeds." He was not saying that those who do civic good actually become gods, but he was supporting the practice of treating them that way.

As for the Stoic idea of God, "It is ridiculous," Pliny wrote, "to think that a supreme being—whatever it is—cares about human affairs. Don't we believe that it would be defiled by so gloomy and complex a responsibility?" Anyway, why should God care about or even judge us when "some men have no respect for the gods, while the regard shown by others is shameful." Pliny also gave witness that the Hellenistic worship of chance, Tyche, was still vibrant in the first century CE. "Throughout the whole world, in all places and at all times, Fortune alone is invoked, alone commended, alone accused and subjected to reproaches . . . to her is credited all that is received . . . and we are so subject to chance that Chance herself takes the place of God."

What Pliny says regarding "the power of the gods" demonstrates his wit and sarcasm as well as any other passage in the book and offers us a new kind of irreverence. "The chief consolation for Nature's shortcomings in regard to man is that not even God can do all things. For he cannot, even if he should so wish, commit suicide, which is the greatest advantage he has given man among all the great drawbacks of life."[59] Pliny goes on to say that "God cannot give mortals the gift of everlasting life, or recall the dead, or cause a man who has lived to have not lived, or someone who has held office not to have held it." Furthermore, "He has no power with respect to the past, except to forget it." Also, God "cannot make twice ten other than twenty." Here's the kicker: "These facts show without a shadow of doubt the power of Nature and prove that this is what we mean by 'God.'"[60]

Pliny is certain that there is no life after death. The period after your last day is precisely the same as the period before your first day: "neither body nor mind has any more sensation after death than it had before birth."[61] It is just "wishful thinking" that imagines an afterlife, an idea that we never extend to the animal kingdom, even to those animals that live longer than we do, "as if man's method of breathing differs in some way from that of other animals." Pliny poses numerous questions about the soul, asking what it is made of, what is its power of thought, how does it hear or touch, and what use it gets of these senses. He asks where the soul resides and how great is the crowd of souls from so many ages past. Then he dismisses all these questions as "characteristic of childish gibberish and of mortal men greedy for an everlasting life." At last he dismisses the whole argument, exclaiming, "A plague on this mad idea that life is renewed by death!" Throughout the text, Pliny peppers his jaunty, upbeat discussion of the world with little remarks about how death is one of the greatest favors bestowed upon humanity. Rather than imagine an afterlife: "How much easier and much surer a foundation it is for each person to trust in himself, and for us to gain our pattern of future freedom from care from our experience of it before birth!"[62]

Pliny the Elder was a naturalist, and in this work he was a compiler of observations and interpretations of the known world. In this kaleidoscope of truths and notions, he manages to communicate an ebullient and curious personality and to let us in on a calm, self-assured, and exacting mood of doubt. Pliny believed that sometimes it rained blood and a hundred other things that would seem magical by our standards; he did not have enough evidence about the natural world to be certain that blood showers were anything other than a rare but natural phenomenon. On the other hand, he understood life after death as supernatural, born of wishful thinking, and he did not believe in the supernatural.

MARCUS AURELIUS

A hundred years after Pliny, the rationalist mood that he embodied would still reign over Classical Rome. The emperor Marcus Aurelius (121–180 CE) is often described as a philosophical agnostic and a practical atheist. He is seen as the quintessential figure of his age in this respect, for though he was more scholarly than most, he represented the period's relative indifference to religion. Marcus Aurelius admired Stoicism, of the variety that did not give much credence to the notion of Providence, but he could never choose between the Stoic idea of a somehow-ordered universe and the Epicureans' idea about atoms and chance.

Aurelius stands out as a man struggling to internalize the truths of philosophy; his *Meditations* read like a sage counseling himself through some dark night or ethical confusion. That he was emperor, and perhaps as close to a philosopher-king as the West would ever know, has long fueled interest in his *Meditations,* but it needn't have. The book is a marvel of insight and advice. It is not particularly original in its ideas—it is mostly a mixture of Stoicism and Epicureanism—but the voice here is new and warm, and the advice, on all sorts of subjects, is good. It feels good to read it. The book does impart information on the nature of the physical world, but it is mostly what it says it is, a call for meditation, a guide for thought. The Emperor knew what so many passionate doubters have known: it takes as much repetitive reading, ritual, and practice to live well as a doubter as it does to live well as a believer. The book reiterates some ideas over and over in different forms, to help the reader and, we suspect, the author, actually learn the precepts that he or she has come to recognize as true. Here is the heart of it: "Whether the universe is a concourse of atoms, or nature is a system, let this first be established, that I am a part of the whole which is governed by nature; next, I am in a manner intimately related to the parts which are of the same kind with myself." We are all one. The result of this realization would be not only our own inner calm, but also an attitude of patience and generosity toward other people, even fools. "Men exist for the sake of one another," counseled the emperor, "Teach them then or bear with them."

Marcus Aurelius lived at a moment and in a position in which he did not feel pressed to come to a decision on the existence of God or gods, so it appears that he did not. His thoughts include casual references to such gods or God—which do not feel at all forced—but when he really gets down to the subject, he is in a good deal of doubt. His basic position is as follows:

> The periodic movements of the universe are the same, up and down from age to age. And either the universal intelligence puts itself in motion for every separate effect, and if this is so, be thou content with that which is the result of its activity; or it puts itself in motion once, and everything else comes by way of sequence . . . ; or indivisible elements are the origin of all things. In a word, if there is a God, all is well; and if chance rules, do not thou also be governed by it.[63]

Either all is planned or we must learn the rules for living in a world of chance.

These rules are familiar to us by now: remember death so that you are vividly aware that you are alive and so that you will take the right things seriously; remember that you have nothing to fear from death since you won't be around for it; don't lust for money or praise because they do not bring happiness, but try to control desires; and don't lust for posthumous fame since you won't be around for it anyway, even if it were worth having. What then should we do? Where other philosophers of the good life counseled devotion to knowledge, as with Plato, for example; or to family and work, as in Ecclesiastes; or to friends, as with Epicurus, Marcus Aurelius joined those who counseled devotion to the community, to the great multitude of one's fellow human beings.

Many times throughout the *Meditations,* Aurelius mentions that perhaps the universe has God and meaning, and perhaps it does not. After soothing himself with thoughts of the interconnectedness of the universe under the sure hand of Nature, he at once pushes himself a little further: "But if a man should even drop the term Nature (as an efficient power), and should speak of these things as natural," even then it would be "ridiculous to affirm" that everything changes and at the same time "to be surprised or vexed as if something were happening contrary to nature" when things fell apart.[64] Mood is his chief subject here, but through that conversation we hear a lot of doubt about the existence of God.

As in almost all graceful-life philosophies, the central piece of advice is not to forget the big picture, but rather to remember it constantly, especially when you feel lost or unloved, abused by chance or by your associates. The emperor puts this in rationalist yet exuberant terms: "Everything harmonizes with me, which is harmonious to thee, O Universe. Nothing for me is too early nor too late, which is in due time for thee. Everything is fruit to me which thy seasons bring, O Nature: from thee are all things, in thee are all things, to thee all things return."[65] How shall we bear loss if we cannot be sure that everything has a meaning? The emperor reminds us, things change and we must attune ourselves to expect that. "Observe constantly that all things take place by change, and accustom thyself to consider that the nature of the Universe loves nothing so much as to change the things which are and to make new things like them." And elsewhere: "Loss is nothing else than change. But the universal nature delights in change . . . and will . . . to time without end."[66] He even leaps to the notion that death will not end some stable "you" that began at your birth, but will only further change a "you" that has never stopped changing: "Do not imagine that the solid and the airy part belong to thee from the time of generation. For all this received its accretion only yesterday and the day before, as one may say, from the food and the air which is inspired. This, then, which has received the accretion, changes, not that which thy mother brought forth."[67] We must stop trying to defend the stability and coherence of a self and a world that are always changing.

Aurelius approaches the question of souls and life after death carefully: he is not concerned to prove any given system, or to disprove them all, nor does he even seem to be interested in the religious consequence of his conclusions. Yet the idea does not sound likely to him and he pokes at it rather amusedly. "If souls continue to exist, how does the air contain them from eternity?" He tells us that some people say souls shrink like bodies after death and are received "into the seminal intelligence of the universe," and thus make room for new souls. "And this is the answer which a man might give on the hypothesis of souls continuing to exist. But we must not only think of the number of bodies which are thus buried, but also of the number of animals which are daily eaten by us and the other animals. For what a number is consumed, and thus in a manner buried in the bodies of those who feed on them!"⁶⁸ If souls took up space, we would need an awful lot of room by now, given the number of creatures that die on any day, even just for the tables of Rome. The emperor didn't believe in life after death and by the time he wrote these meditations he had spent a long time teaching himself to be at peace with annihilating death. He leaned on Epicurus here, but his own phrasings are very satisfying and show what this emperor made of the consequences of doubt. His themes—again—are time, reminders of death, and the solace of contemplation:

> Soon will the earth cover us all: then the earth, too, will change, and the things also which result from change will continue to change for ever, and these again for ever. For if a man reflects on the changes and transformations which follow one another like wave after wave and their rapidity, he will despise everything which is perishable.⁶⁹

Aurelius was not saying we should actually hate the changeable world, but when our arrangements are cradling us in some self-satisfied bliss, we must not be anxious to keep it, nor too terribly saddened when it all changes.

> Acquire the contemplative way of seeing how all things change into one another, and constantly attend to it, and exercise thyself about this part of philosophy. For nothing is so much adapted to produce magnanimity. Such a man has put off the body, and as he sees that he must, no one knows how soon, go away from among men and leave everything here, he gives himself up entirely to just doing in all his actions, and in everything else that happens he resigns himself to the universal nature.⁷⁰

> Constantly contemplate the whole of time and the whole of substance, and consider that all individual things as to substance are a grain of a fig, and as to time, the turning of a gimlet.⁷¹

And my personal favorite:

> Severally on the occasion of everything that thou doest, pause and ask thyself if death is a dreadful thing because it deprives thee of this.⁷²

These excerpts remind us again that we need to be reminded again and again. He speaks of the nature of reality, but also about the necessity of regular meditation on the situation. One must sit still and reflect on the fact of change; one must "acquire the contemplative way of seeing"; one must "exercise" oneself on this matter; one must "constantly contemplate the whole of time"; one must think "severally on the occasion of everything" you do. As a result you will be happy, calm, and generous. In fact, nothing will make you a good person faster, says Aurelius. "Do not act as if thou wert going to live ten thousand years. Death hangs over thee. While thou livest, while it is in thy power, be good."[73]

His take on prayer is that even if there are gods, we ought to ask them only for the maturity and the fortitude not to need anything. This way, whether there are gods or not, we have not wasted our time:

> Either the gods have no power or they have power. If, then, they have no power, why dost thou pray to them? But if they have power, why dost thou not pray for them to give thee the faculty of not fearing any of the things which thou fearest, or of not desiring any of the things which thou desirest, or not being pained at anything, rather than pray that any of these things should not happen or happen? . . . Begin, then, to pray for such things, and thou wilt see. One man prays thus: How shall I be able to lie with that woman? Do thou pray thus: How shall I not desire to lie with her? Another prays thus: How shall I be released from this? Another prays: How shall I not desire to be released? Another thus: How shall I not lose my little son? Thou thus: How shall I not be afraid to lose him? In fine, turn thy prayers this way, and see what comes.[74]

By the end of those examples Aurelius has offered us a rather heartbreaking prayer, but it helps. Elsewhere he returns to the idea of worrying over a sick boy and says that all we know is that the child is sick, not what will happen, and we must not project onto the situation any positive or negative fantasies. The boy is sick. That's all you know. Don't worry.

More than once he bids his reader to "Look down from above on the countless herds of men and their countless solemnities, and the infinitely varied voyagings in storms and calms, and the differences among those who are born, who live together, and die." From this vantage, "raised up above the earth," he tells us to observe humanity, and to consider how many people have lived before you, how many will live after you are gone, "how many know not even thy name, and how many will soon forget it." What is more, those who are now praising you may very soon curse you and "neither a posthumous name is of any value, nor reputation, nor anything else." From high above, "thou wouldst see the same things, sameness of form and shortness of duration."

All these philosophers who have told us to forget fame are remembered these millennia later, and were famous in their times. The problem of fame is theirs in a special way. Emperor Aurelius counseled against striving for renown but was wrestling with the meaning of his own:

> He who has a vehement desire for posthumous fame does not consider that every one of those who remember him will himself also die very soon; then again also they who have succeeded them, until the whole remembrance shall have been extinguished as it is transmitted through men who foolishly admire and perish. But suppose that those who will remember are even immortal, and that the remembrance will be immortal, what then is this to thee? And I say not what is it to the dead, but what is it to the living?[75]

As I have said, the *Meditations* are contemplative songs, prayers to the self, things one comes to know slowly. Although they sometimes refer to gods or God, Marcus Aurelius's work is about how to live as a human being in a universe that is not human and that does not bend toward human desire. To live well in the world as it presents itself, we need, not to assign possible traits to the universe, but to internalize the traits we do see. That means accustoming ourselves to believing that we are each a little nothing in a great expanse of space and time, and are therefore free of worry. Aurelius, however, had the doubled task of remembering that even if you are emperor, you are still a little nothing and therefore free of worry. To drive this point home to himself and to his readers, he lists famous people from the near past who he feels are forgotten by the present generation. It is poignant for us to read these lists, for indeed most are lost to history, although a few have come down to us with full biographies.

> Augustus' court, wife, daughter, descendants, ancestors, sister, Agrippa, kinsmen, intimates, friends, Areius, Maecenas, physicians and sacrificing priests—the whole court is dead. Then turn to the rest, not considering the death of a single man, but of a whole race, as of the Pompeii; and that which is inscribed on the tombs—The last of his race. Then consider what trouble those before them have had that they might leave a successor; and

then, that of necessity some one must be the last. Again here consider the death of a whole race.[76]

Letting this shock of forgetfulness do its work, Marcus Aurelius then tells us to what we should devote ourselves, given such circumstances: "Thoughts just, and acts social, and words which never lie, and a disposition which gladly accepts all that happens as necessary, as usual, as flowing from a principle and source of the same kind."[77] The text is full of such gems of calm, but it also contains passages in which he seems to encounter anew the problem of life in a possibly godless universe, and he thrashes around trying to galvanize his inner forces, to go and do whatever it is that can actually and honestly be done on any given day in such a world:

> The universal cause is like a winter torrent: it carries everything along with it. But how worthless are all these poor people who are engaged in matters political, and, as they suppose, are playing the philosopher! All drivellers. Well then, man: do what nature now requires. Set thyself in motion, if it is in thy power, and do not look about thee to see if any one will observe it; nor yet expect Plato's Republic: but be content if the smallest thing goes on well, and consider such an event to be no small matter.

He adds that if Alexander and others have "acted like tragedy heroes, no one has condemned me to imitate them. Simple and modest is the work of philosophy. Draw me not aside to indolence and pride."[78] In their own ways, Plato and Alexander had conquered much of the known world about five hundred years earlier; Marcus Aurelius suggested asking for considerably less.

The emperor did not suppose that his own path was without fault. He says that there is no man so fortunate that when he is dying there are not some who are glad of it. "Suppose that he was a good and wise man, will there not be at last some one to say to himself, Let us at last breathe freely being relieved from this schoolmaster? It is true that he was harsh to none of us, but I perceived that he tacitly condemns us. This is what is said of a good man." There is no perfect way to be, we can only do our best; and some people will dislike us no matter what. He suggests we remember such depressing truths when we are scared of death, since it will remind us that this world is not worth clinging to.

Aurelius constructed a worldview that attended to religious needs without religion. He made peace with death, found an ambivalence he could live with on the question of meaning, and learned to pray for the one thing for which prayer is a self-fulfilling activity, the prayer to remember one's own strength. He did not argue that the world was mechanistic and therefore free of wonder. He was awestruck at the world. With delighted reverence he marvels at human generation, "that a man deposits seed in a womb and goes away, and then another cause takes it, and labors on it and makes a child. What a thing from such a material!"[79] He similarly marvels at the unseen forces of the universe: "I observe then the things which are produced in such a hidden way, and see the power just as we see the power which carries things downwards and upwards, not with the eyes, but still no less plainly." Gravity and reproduction are always the showstoppers, and such wonders were effective reminders of the interconnected grand scheme of things, that giant truth, which, could we only remember it, would set us free to live in joy and die in peace. It's just a matter of keeping things in mind:

> Thou canst remove out of the way many useless things among those which disturb thee, for they lie entirely in thy opinion; and thou wilt then gain for thyself ample space by comprehending the whole universe in thy mind, and by contemplating the eternity of time, and observing the rapid change of every several thing, how short is the time from birth to dissolution, and the illimitable time before birth as well as the equally boundless time after dissolution.[80]

The emperor says that we may gain for ourselves "ample space" by coming to know the big picture of the universe. There's something in that advice that resonates with all graceful-life philosophies and living religions. A wise heart must be made: we need to master a certain amount of pain, anxiety, and fear before we have the space to be generous, and that space must be defended by study and meditation on reality.

SEXTUS EMPIRICUS

After drinking deeply of the emperor's good counsel, it is a decided contrast to turn to our next great Roman doubter, for he did not have anyone's heart on his mind. Sextus Empiricus is the best exemplar of the Skeptics of

Roman times. He lived from the mid second century through about the first quarter of the third. Skeptics of this period went about their arguments by dividing any notion into two possible, oppositional ideas, and then positing a dependent notion for each side until they found something contradictory or absurd, at which point they would dismiss the original proposition. In so doing, they sought not to isolate the truth, but rather to prove that certainty, on any issue, made bad sense. The Skeptics of this period were robustly against the Epicureans, Stoics, and Neoplatonists, all of whom they referred to as "the dogmatists" for believing that they knew truth. It was the philosophy of "no," and it reigned for centuries at the Academy. On the question of the gods, it is quite something to read Sextus's endless juxtaposition of conditionals. The relativism is a lofty aim, but taking a subject and arguing against all opinions on it is a very strange way to proceed and makes for weird reading.

There are three surviving works by Sextus, each in several volumes. They are all classic works of doubt—doubt of everything—but for the question of the gods, two essays are important. The first is in book three of *The Outline of Pyrrhonism*—Pyrrho being the founder of Skepticism. Chapter three of this text is "On God," and Sextus begins with a tellingly pat avowal of official belief: "[W]e conform to the ordinary view, in that we affirm undogmatically the existence of gods, reverence gods, and affirm that they are possessed of foreknowledge. But in reply to the rashness of the dogmatists, we have this to say." And then he launches into a dismissal of being able even to conceive of the idea of deity without making a logical mess of it. "But granted that God can be conceived, it is necessary . . . to suspend judgment on the question of his existence or non-existence," primarily because "if the impression of him proceeded from himself," everyone who believed in God would have basically the same idea of him. But they do not. In any case, he argues, it is impossible to prove the existence of something that does not make itself apparent. He picks up Epicurus's idea that a powerful god who knows all doesn't make sense in a world so full of evil as our own.

> If [God] has the power of forethought for all things, but not the will, he will be considered malicious. And if he has neither the will nor the power, he is both malicious and weak. But to say this about God is impiety. Therefore God has no forethought for the things in the world.

But if he takes no thought for anything, and no work or product of his exists, a person will not be able to say where we get the idea that God exists, seeing that he neither appears of himself nor is apprehended by means of any of his products. For these reasons, then, it cannot be apprehended whether God exists.

Sextus concludes that all those who assert the existence of God are "guilty of impiety." If they say God is involved with us, then God is responsible for evil, and if they say he ignores us, "they will necessarily be saying that God is either malicious or weak," and that is "manifest impiety."

Sextus's other meditation on the existence of God is found in his *Against the Dogmatists,* which takes the form of five books: *Against the Physicists, Against the Ethicists, Against the Logicians,* etc. In these, Sextus does just what the titles suggest; he argues against all the ethicists, for example, setting up their views and knocking them down, without ever posing an alternative. There is a beautiful tedium to it. The argument has an almost liturgical singsong quality, such that in the deficit of knowledge Sextus posed, this song and its debunking scythe could perhaps soothe its own wound, its own harsh claim that we can know nothing.

But Sextus did not remain completely agnostic on the question of the gods. He began his *Against the Physicists* with a section called "Concerning the Gods," offering a rare assessment about the general state of doubt. On the question of the existence of God, he says, some assert his existence, some assert his nonexistence, and some "say that he is 'no more' existent than non-existent." He also reports: "That he exists is the contention of most of the dogmatists and is the general preconception of ordinary men. That he does not exist is the contention of those who are nicknamed 'atheists' such as Euhemerus . . . , and Diagoras of Melos, and Prodicus of Ceos, and Theodorus, and multitudinous others."[81] Ordinary people, he says, believed in God, but the group who did not was still worthy of the word *multitudinous.*

He then summarizes the arguments of each philosopher: how Euhemerus said gods were men of power, deified in memory after their deaths; how Critias believed "the lawgivers of ancient times invented God as a kind of overseer of the right and wrong actions of men" especially to prevent secret wrongdoing; and how Prodicus of Ceos contributed the idea that the ancients equated the sun and rivers and other beneficial things with gods. As for Epicurus, Sextus comments that, according to some, Epicurus allowed the existence of God in his popular works, "but not where the real nature of things is at issue."[82] We are told of Democritus and Epicurus and their images of giant personages that show up in dreams. Sextus asks why Epicurus' dream images of giant people didn't give rise to the belief in giant people, rather than gods. Also, saying that great heroes became gods after death doesn't explain where the idea of gods came from. And if people were going to believe that all beneficial things were gods, like the rivers and the sun, then why didn't they think people were gods, "especially philosophers" because they benefit our lives (just as Lucretius said), "and most of the irrational animals[,] for they help to perform work for us," and our domestic furniture and whatever else "of an even humbler character."[83] He was having a good time. Since this view, he concludes, is "extremely ludicrous," the whole idea of these origins for the gods must not be sound. That's all he offers as argument against the atheists. Note that he argues with their positive notions of where the gods came from, but he never gets near arguing with their claim about the absence of God.

Sextus then coolly offers some propositions in favor of the claim that gods are real. It's worth considering one before we turn to his argument on the nonexistence of the gods. In the singsong of his Skepticism he argues that if the universe is powered by something, either that something must be eternal or it must have leapt into being. Since nothing would have been there to cause a leap into being, the universe must be eternal; and since human beings are intelligent and they were made by the universe, the universe must be intelligent; what is eternal and intelligent is divine; therefore the universe is divine. Therefore the gods exist.

Then Sextus turns to refute the argument for God. Again, there's something both amusing and annoying about the way he argues: If the gods exist, they are living beings. But if they are living beings, they have sensation. If human beings had more senses than gods, they would be superior to gods, and anyway, "to prune away from God this or any other of his senses is an altogether unconvincing procedure," so the gods have taste, which means they can taste bitter, which means some things displease them, which means things can harm them. "But if this is so, he is perishable. Consequently if gods exist they are perishable. Therefore gods do not exist."[84] God would also be able to smell and touch and hear. "But if this is so there must be certain things which are vexations to God; and if there are certain things which are vexatious to God, God is subject to change for the worse, hence also to destruction. Therefore God is perishable. But this is in violation of what was the common conception of him. Therefore the Divinity does not exist." Sextus says that sensation itself is a kind of alteration and change, and if God is receptive to change "he will at all events be receptive of change for the worse. And if this is so, he is also perishable. Therefore it is also absurd to claim that he exists."[85]

Following the example of Carneades, Sextus also argues that most of humanity's best qualities have to do with enduring pain or avoiding temptations, so without pain or temptation, God cannot really be said to be virtuous. It is the person "who holds out under the knife and the cautery" who shows endurance, not the one who is drinking honeyed wine.[86] Moreover, "if the Deity is all-virtuous, he also possesses courage . . . and if this is so there must exist something which to God is fearful. . . . Hence if divinity exists it is perishable. But it is not perishable, therefore it does not exist."[87] The divinity, he insists, would also have to possess greatness of soul, "to rise above events," but if he does, he must be subject to consternation and therefore perishable. God must deliberate, because it is a virtue, and therefore he cannot be omniscient; God must know pain in order to know its opposite, pleasure, and therefore he must not be eternal and perfect; God must possess wisdom and temperance, but both imply struggle and temptation; so for Sextus, in answer to all these, the gods must not exist. In a particularly interesting variation, "If nothing is non-evident to God . . . he does not possess art . . . since art appertains to things which are non-evident and not immediately apprehended." And if he does not even possess the art of living, he won't possess virtue. "But if God does not possess virtue, he is non-existent."

Through all of this, Sextus has a few convictions, and first among them is that we cannot imagine any truly noncorporeal being able to do, think, or feel anything. This suffuses his discussion of the voice of God, for he says that God either speaks or cannot, and the latter "is in conflict with the common notions of him," which means God speaks, which means he has lungs and a windpipe and tongue and mouth. "But this is absurd," he insists, "and comes close to the story-telling of Epicurus. Therefore we must say that God does not exist." So Sextus, too, scolds Epicurus for keeping the bare image of gods. Since you need a body to do anything, God must be corporeal, but if so, it must be either a compound or a simple body. If it is a compound it comes apart and is therefore perishable; if it is simple it is just a thing, like fire or water. If it is one of these, "it is inanimate and irrational, which is absurd. If therefore, God is neither a compound nor a simple body, and there is no further alternative, one must declare that God is nothing."[88]

Many arguments for God suggest that our frail virtues must exist in perfect form somewhere. Sextus turns this on its head, stretching Carneades' argument to its full power and entering a new idea into the annals of doubt. God's reputed virtues, he explains, were fully realized versions of human virtues, and that did not make sense unless God had our weaknesses. Wisdom and courage are aspects of human struggle. They do not exist in pure form. Skepticism thrived in the ancient world from the fourth century BCE to the third century CE, and Sextus is our best and almost our only surviving source for it.

LUCIAN OF SAMOSATA

Lucian of Samosata (ca. 120 CE–ca. 190 CE) was a Greek satirist born at Samosata on the Euphrates in northern Syria. He was familiar with all the schools of philosophy, and made fun of them all, especially the Cynics. In *Hermotimus,* one of his longer dialogues, his big question is how a person is supposed to choose between the philosophies, when it would take more than a lifetime to properly learn them all. In *Timon* a poor man who has been rich scolds Zeus for his indifference to the injustice of man. Zeus admits that the strident disputes in Athens had kept him away from the place lately, and when the man becomes rich again, he praises Zeus and forgets his criticisms. In several works, Lucian shows himself aware of Christianity, satirizing the Christians in his *Passing of Peregrinus,* a story of a sage who poses as a leader to the Christians:

> These deluded creatures, you see, have persuaded themselves that they are immortal and will live forever, which explains the contempt of death and willing self-sacrifice so common among them. It was impressed on them too by their lawgiver that from the moment they are converted, deny the gods of Greece, worship the crucified sage, and live after his laws, they are all brothers. They take his instructions completely on faith, with the result that they despise all worldly goods and hold them in common ownership.

So, he explains, anyone "who knows the world" can get rich tricking "these simple souls." It was no worse than anyone else got: Lucian tells stories of a Stoic philosopher losing his cool and throwing a tantrum about wages—among other indignities. Lucian was trying to make his audience laugh, rather than start a revolution, and his jokes had a way of lasting. He treated the Olympian gods as obvious fictions. It is worth noting that Lucian's *True History* made him the founder of science fiction: his characters went to the moon, comets, and other sites of outer space and met inventive extraterrestrials.[89] Lucian told of lamp-people—most "were obviously pretty dim"—and described a world where men married men and carried the babies in their thighs, and a world where Tree-men reproduced by cutting off and planting a testicle. We also learn of vines that grow grapes of water rather than wine (strong winds blowing their harvest make for our hailstorms); and that "bald men are considered very handsome on the moon."[90]

Lucian was long thought to be the author of another important work in the history of doubt: *Lucius, or The Ass*. These days the author is often called Pseudo-Lucian, because many think it wrongly attributed to him. It was another take on the story Apuleius told (perhaps from a common source) about a man changed into an ass. Apuleius's version had offered detailed wandering in a soulless world. The Lucian version was shorter and simpler, and even more doubting. Nobody answered prayers. Life here simply unfolded in a series of ribald accidents. Future doubters would often praise Lucian with this text in mind.

Cicero, Lucretius, Pliny the Elder, the Emperor Marcus Aurelius, Sextus Empiricus, and Lucian of Samosata each offered a different model of the mature doubter: Cicero with his lost daughter and his lost Republic and calmly "calling everyone to court" on the matter of the gods; Lucretius with his secular savior and his beautiful psalm to a world in which nature tumbles along with no God; Pliny with his cool compendium of mundane and fabulous information and his hearty doubt in the immortal; the emperor with his book of warm advice and resignation; Sextus with his verbal assault on every certainty; and Lucian just laughing. Ancient secular philosophy is one of the greatest trees in the orchard of the history of doubt, and it produces some of its most mature fruit here. This doubt was fresh and lively for centuries, well suited for peace and for guiding countless generations of men and women through the strange turns of life on earth. The cellars that held the preserved fruit of ancient doubt were not well defended against time and violence, but whatever pots of jam survived the dark millennium would be even sweeter when opened in that strange new world. For the change in climate that shut down doubt's living branches here, we turn next to a group always up for a little religious fervor: the monotheistic, pious Jews of Palestine.

FIVE

Christian Doubt, Zen, Elisha, and Hypatia, 1–800 CE

Late-Classical Mix

In the early Middle Ages, something curious happened to the ideas of faith and philosophy. For the first time, belief itself became the central religious duty. A new kind of doubt appears here, in response. This new doubt doubts the other side of the equation—us. This one is believer's doubt, and it hurts more. Whereas before there was not much reason to try to believe (it was more a question of what to say in public), now the religion is set up around the idea that belief is difficult and that we must work toward it. Doubt will never be the same. We hear the cries of wretched doubt in this period, and we are reminded that the Book of Job is the only place we have heard such wailing before. We will have to look at how belief changed within Judaism just before Jesus, and then turn to see how Jesus, and then Saint Paul and Saint Augustine, so altered the nature of belief in religion that doubt was never the same again. Then we will take a look at two very important figures in the story of doubt: Rabbi Elisha ben Abuyah, the most notorious doubter in the whole Jewish tradition, and Hypatia, a woman renowned as the last secular philosopher of the ancient world. Last, we will turn to the Far East for a look at Zen Buddhism, which arose and flourished during this period, and created a dedication to doubt all it's own.

THE JEWS AT THE TIME OF JESUS

Reflect now on how Judaism's certain and passionate commitment to God must have looked, and felt, in the midst of pagan Rome and her philosophers of resignation. While the Romans kept their gods for the sake of the state, the Jewish religion held that the people's ability to follow God's law was more important than the state: the state was there to serve that primary task. Judaism, at this point, offered one God, reigning over an orderly, just universe; and an afterlife. It had not always been so. As we observed earlier, Judaism began like many other "temple religions" with priests who carried out a series of washing and eating rituals and sacrificed to their God in a temple—though the God Jews worshiped was invisible. The disaster of the Babylonian captivity changed the religion forever. Through the invention of the synagogue system and the common keeping of the laws, the returned exiled Jews had recreated Judaism so that it was not necessarily attached to the Temple proper. They had managed their sense of having been punished by a just God by developing and obeying what they took to be his commandments. They could survive without the Temple as their local source of identity, and they could come to a new understanding of themselves as God's chosen people, for now they behaved as a nation of priests. Within that idea, they became interested in seeing the whole world brought together under their one, transcendent God.

At about this same period, the Hebrew people had come in contact with Persian Zoroastrianism and become influenced by that religion's vision of the world as divided between the truth and the lie—good and evil—the forces of light and the forces of darkness. The Jews were beginning to think of their God as all-powerful and all-good, and that raised the problem of where bad things come from. Many peoples see that as a problem wrongly put—there is disjuncture and sorrow and heartache, but there is not a *force* that is *evil*. So it was distinctive when the Jews took on this Zoroastrian idea. They did not yet believe in an afterlife, but good and evil were forces in the living human world. By the period of the Second Temple, the age of the prophets was understood to be over and new inspiration took the form of apocalyptic literature in which good and evil finally go to war and Israel finally converts all the pagans and enters its age of triumph and happiness. The kingdom of God was coming. Jews served God's law, had an explanation for sorrow, felt that they had recourse in the universe, and had something to look forward to.

The dynasty of the Maccabees then converted the gentile populations living in Palestine to Judaism. The large territory of Idumea, in the south, was converted as a whole and annexed to Judea, and the Idumeans became an integral part of the Jewish nation.[1] The Jews were evangelical in this period. The Maccabee period ended because the governing Jews fought among themselves: the powerful Queen Alexandra managed to expand the Jewish territory and to keep Rome at bay, but when she died her sons battled each other for her throne and eventually Rome intervened to decide the succession. Not surprisingly, having resolved the civil war, the Romans stayed. These Roman governors taxed the Jews, and now and again provoked an uprising by breaking the Jewish laws—as when Pompey had imperial guards march through Jerusalem even though their shields were decorated with animal images, or when Caligula tried to put a golden statue of himself, as a god, in the Temple. In such encounters, Jews were martyred and Rome backed down with stunned respect and some disgust. All this while, after the Maccabees' victory and through the rise of the Roman Empire, some Jews were convinced that God was soon going to send another great warrior who would chase out the Romans, convert all the Jews in the area, and bring on the next, and perhaps last, great phase of Jewish history. Historians widely agree that when the Jews of this period spoke of waiting for a *Messiah,* an anointed one, they were waiting for a king, a worldly leader who would give them back their own powerful independent state.

Along very different lines, the Jews had also been growing increasingly convinced that some kind of afterlife was in store for the pious Jew. The idea of an afterlife had arisen and grown strong after the Maccabean period began in 168 BCE. It was a result of the outside influence of the Mystery Religions, as well as the internal logic of Judaism, sparked especially by the prophet Isaiah. At the same time, the notion of "believing" arose as a criterion for being in God's good grace. At its very beginnings this *believing* was contrasted with Greek rationalism. There is evidence of this new challenge to believe in the text known as the Mishnah. The Hebrew word *mishnah* means "study," and the title refers to the first postbiblical codification of Jewish oral law. During and after the Babylonian captivity, the biblical laws had been intensely studied, and all sorts of decisions had been made about their meaning. The laws were updated by prominent leaders and supplemented by traditions of popular observance. All this had been known as the Oral Torah. It was finally collected in writing, as the Mishnah, about 200 CE. Along with the Gemara—later commentaries on the Mishnah itself—it forms the Talmud. In the tractate Sanhedrin of the Mishnah, there is a remarkable statement about the afterlife and, also, the only mention in the whole Talmud of a Greek philosophy or philosopher:

> All Israel has a share in the world to come, as Isaiah said: And all of your people who are righteous will merit eternity and inherit the land. And these are the people who do not merit the world to come: The ones who say that there is no resurrection of the dead, and those who deny the Torah is from the heavens, and Epicureans.

Modern Jews use "apikoros" as a generic term for atheist, but even if the author was speaking broadly of unbelievers here, he was singling out followers of Epicurus to do it. This passage, then, makes very clear that at the beginning of the Common Era, Jews were developing a notion of belief—belief in life after death, in the idea that their text was received from God, and in God himself—as a major human responsibility for the reward of eternity. Note that the afterlife was seen as a given for the whole group. The individual's sole responsibility was to not step out of the group by rejecting the doctrine. We are reminded that the Hebrew notion of divine justice also started out referring only to the fortunes of the group and later became meaningful for individuals. This excerpt also makes it clear that some Jews were Epicureans or otherwise engaged in Greek doubt. This one line from the Mishnah tantalizingly suggests that Jewish doubt, although severely frowned upon by those who kept the records, did exist. There were Jews who doubted life after death, there were Jews who doubted that the Torah came from the heavens, and there were Jews who were Epicureans. It is an amazing little passage for its annunciation of the afterlife, for its requiring only belief for that afterlife, and for its evidence of unbelief. So now along with ethical monotheism, the Jewish religion offered an afterlife. Such bounty did not go unenvied.

Jews made up 10 percent of the Roman Empire and 40 percent of the great city of Alexandria. There was a lot of integration and mutual assimilation between Jews and the other populations of the empire. There were Romans who added the Jewish God and Jewish rituals to their own cycle of worship. We have explicit evidence of a systematic attempt to propagate Judaism in the city of Rome itself as early as 139 BCE. Many non-Jews were interested. They were attracted by the religion's antiquity, by its intellectualism and philosophy, by its transcendent God, by its already ancient texts, by its sense of righteousness and loyalty, by the welfare system it had developed through its synagogues (the only one like it), and by its jubilant feasts and festivals. Most Jewish families in the Diaspora sent a regular contribution to the Temple and made a pilgrimage to Jerusalem for at least one of the three major yearly festivals. They reported a mood of relaxation and rejoiced to be among like-minded people, at their own temple, celebrating their mutual support. That seems to have carried over to sympathetic non-Jews who joined them at these events, for many seem to have made a point of joining the Jews year after year, assisting at the sacrifices and honoring the rules. There was a whole class of semiconverts who had done everything short of circumcision; these were sometimes called God-fearers.

In the colder religious world of the Roman Empire, Judaism could be very attractive. In the Septuagint, Jews had a venerated Bible that had been translated into the common Greek of the Hellenistic world. Although the West would lose its Greek literary fluency in favor of Latin during the Roman period, across centuries many in the Eastern empire could read the Hebrew Bible because there was a Greek version. In the early second century, the poet Juvenal recorded that Roman families "degenerated" into Judaism when fathers allowed themselves to take on some of its customs, and the sons then became Jews in every respect.[2] In the great urban centers of Syria there were also numerous converts to Judaism. Most remarkably, the Adiabene royal house in Mesopotamia converted to the Jewish faith.[3] Their Queen Helena was particularly active in Jerusalem, and the dynasty became a permanent factor in the Jewish social world.

In 70 CE, the more pious and disaffected Jews in Jerusalem again became thoroughly convinced that God was leading them to their long-awaited golden period, and they revolted against Rome. So many Jews got swept up in this that their forces did well against the Roman troops, at first. When the Romans finally diverted enough legions to beat them, the Jewish state ended, not to appear again until the twentieth century. Almost a thousand Jews slaughtered themselves and their families at Masada rather than surrender to the Romans, and the Romans, outraged by this obstinacy, destroyed the Temple. In 135 CE there was a final revolt, and after it the Romans forbade Jews to enter Jerusalem. All Jews were in diaspora now. Many newly converted Jews, such as certain members of the Adiabene royal family, participated in the revolts against Rome. After these revolts, however, the Jewish religion lost a great deal of its appeal for pagans because the

Jews had been traitorous to the Empire. Judaism entered its Rabbinic period: the synagogue replaced the Temple, and Torah study replaced sacrifice. The religion grew insular. Yet, in the last decades of the Jewish state, two Jewish men, Jesus and Paul, carried the Jewish God and their version of Judaism to a much wider world. What had happened between the Romans and the Greeks happened again between the Romans and the Jews: the Romans crushed the Jewish state and then willingly converted the great Roman Empire to a sect of the Jewish God.

It was not as unlikely as it sounds. The fight had started because this group believed terrifically in a terrifically powerful God, and that was appealing, especially in a place where many people had come to think of religion as a dry matter of fulfilling social and political rituals. Weary paganism was tinder to Judaism's flame, but the rites of circumcision and the legalist, separatist mood had always kept these two elements too far apart for any real ignition. Paul of Tarsus lit the blaze. He broke away from the singularly Jewish worship of Jesus, led by Jesus' brother James, by preaching to the gentiles and by declaring that circumcision and other Jewish rites and laws had been supplanted by the death of Jesus. Everything about Jesus' life and ministry would suggest that he did not have this in mind: he had been a practicing Jew his entire life, honoring all the common and active commandments, prayers, fasts, and rituals, and he spent his short ministry preaching to the Jews. Furthermore, several members of the singularly Jewish Jesus community had known Jesus personally and well, whereas Paul had never met him. But Paul offered something much more practicable and enticing to the vast gentile world. The Romans destroyed James's group in 70 CE, because they were Jews in Jerusalem, and that made it all the easier for Paul's alternative group to get a foothold. It was in this situation that Paul developed and evangelized some of the amazing new ideas that were brewing in the Judaism that raised him.

THE WORRIED GOD

One of the fascinating things about the new Jesus religion is that its central figure was, several times in his short and deeply convicted life, quite wracked with doubt. These moments of doubt—and the weight of the new religious idea that had given rise to them—permanently changed the history of doubt. Forever after, we have had an image of agonizing doubt as part of our model of a religious life. This was not framed as doubt in the existence of God. It was doubt in the ability of the human being to inhabit his or her side of the new equation. It could be very hard to bear.

It is not uncommon for religions to have stories of doubt in their periods of origin, when God or gods first introduced themselves. When Abraham and Sarah met the Hebrew God for the first time, he promised them a child, and Sarah doubted. The beginnings of Christianity had its conversion doubt, too. But Sarah was well convinced by the time she felt the first kick. By the time of early Christianity, the level of the human relationship with God had been stepped up a notch and confirmation was no longer so easily imagined. The part about divine justice for all individuals had not been around in Sarah's time, and had since given Job many agonized days and nights. Now there was the matter of an afterlife as well. It was a lot to believe, and there was more. The God who talked to doubting Sarah all those years ago did not claim to be the ultimate truth toward whom one struggled one's whole life, as toward the sun—that was a Neoplatonic idea that suffused all later talk of a transcendent God. Sarah was not asked to see God as the living intelligence of the universe, nor to accept the brotherhood of humanity—these were Stoic ideas that became enmeshed with the Jewish God.

By the time of Jesus, the Jewish God was firmly connected to the idea of an afterlife, but no precise details had been worked out. The Jews have generally avoided the logical messiness of detailed theology. We've seen what was required to enter the Jewish afterlife: you should not reject the community's belief in an afterlife. The doctrine was passive and expectant, and there were few specifics about splendor in heaven. In Judaism, the afterlife never became central, and neither did the idea of belief. Another quote from the Mishnah tells this: "Better that they [the Jews] abandon Me, but follow My laws."[4] The Greeks and Romans had said that belief was not as important as practicing the rites of the local gods; the Jews here said that belief was not as important as following the laws of their God, notwithstanding location; the Christians had neither rites and location nor the Law to bind them. They focused instead on belief. Doubt was thus an accepted aspect of Greek and Jewish life, but not the center of it. With Christianity, managing one's doubt, that is, husbanding one's faith, became the central drama. Search the Hebrew Bible for the word *belief* and it shows up rarely in this meaning—belief in God. Daniel got out of the lion's den because he believed in his God, and here and there people are said to not believe what a given prophet has said, but that's it. Then, when Jesus appears, the word *believe* blooms like a patch of poppies in a great green field. Suddenly, it is the heart of the matter.

Jesus is a difficult historical figure, but not impossibly difficult. We have no indication that he wrote anything, and our earliest descriptions of his life and work, the first three Gospels, were written about half a century after he died. These make an interesting historical source because they are synoptic (they tell the same story and can be compared piece by piece), but they vary in the order and meaning given to the various events. It seems that after Jesus was gone, a community of believers told stories about him, and as time went on these stories became quite anecdotal, divorced from any definite setting: "once a woman came to Jesus and said . . . and he said . . ." When people finally got down to writing about Jesus, they no longer knew much detail; they did not know the real order of these stories or how much time had passed between them. Because of the structure of the synoptic Gospels, historians believe they were constructed out of these pericopes, or "cut out" minidramas.[5] John's Gospel, the fourth, does not fit with these, but given our knowledge of the cultures involved, we see it as superior on some matters. Later Christians wrote all sorts of things about Jesus' life, and this material holds much less sway with historians. So we do have some historically decent material to work with, and we have developed linguistic and culturally based techniques for figuring out what likely happened and what likely was written in later or misconstrued. We may thus proceed, if with caution.

Jesus seems to have called for reform in the practice of Judaism—his angry scene with the money changers is suggestive—but many historians have argued that a reformer would have left more evidence of having been one.[6] It is reasonable to think of such religious reform in terms of doubt. Historians have also argued that Jesus' doubting of establishment values had so much in common with the Cynic way of life that he was quite possibly influenced by them. Religious historians Burton Mack and John Dominic Crossan have both championed this view; wrote Crossan, "Maybe Jesus is what peasant Jewish Cynicism looked like."[7] This vision of Jesus as a part of the tradition of graceful-life philosophies is fascinating, but there is much that is difficult to know. What we can know better is what the earliest Gospels report that Jesus said about doubt and the way that they describe the doubts Jesus had in his own mission.

Jesus, who would come to be understood as one with the God of this new religion, had a moment at which he doubted his ability to do what was asked of him and another moment when he doubted God's loyalty. The first moment was in Gethsemane. The scene is after the Passover meal: Jesus has taken a few of his disciples and asked them to keep watch while he goes off a little way to pray. Jesus, who is feeling "sorrowful and troubled," turns to Peter and the two sons of Zebedee and tells them, "My soul is overwhelmed with sorrow to the point of death. Stay here and keep watch with me." Then Jesus walked a little farther off, "he fell with his face to the ground and prayed, 'My Father, if it is possible, may this cup be taken from me. Yet not as I will, but as you will.'"[8] He asks three times if he really had to allow himself to be brutally sacrificed, and before each new plea, he goes back to check on the boys. Each time, they've all fallen asleep and no one is watching. Jesus rebukes them at first, but by the last time he simply tells them to get their rest. Then Judas comes and kisses him, and he is delivered into the hands of the authorities. He is calmly resolved in their hands, but we see him doubt one more time: on the cross, suffering, after he has been up there for many hours, he calls out, "My God, my God, why have you forsaken me?" In two out of three of the synoptic Gospels (Mark and Matthew), these are the last words Jesus says before he dies. It sounds like he was expecting something that did not seem to be happening.

Jesus' final question is the first line of Psalm 22, the second line of which is "Why are you so far from saving me, so far from the words of my groaning?" Psalm 22 is written in the person of a pious yet suffering man, or nation, frustrated by waiting for God's redemption, but it closes with a certainty that God will indeed come through. Thus, many people have interpreted this final cry as a gesture of faith—Jesus was referring to the whole psalm and therefore never really doubted God. It has been posited that someone added these words to Jesus' story since they provided another connection between Jesus and the Hebrew Bible. Another interpretation is that such scenes of doubt sat so uncomfortably with the idea of Jesus as God that they had to be true and well known, or they would never have been left in the text.

The Christian world that we are about to enter had at its very center two images of a man, who was a God, in an agony of doubt. The stakes had been raised to such a degree that doubt would now be part of religion in a way it had not been before; for many people, their very God had wailed of it. For Jews, the encounter with the Mystery Religions, with Platonism, and with Stoicism melded with the developing logic of their own idea of theocracy.

Now along with his old persona of a warrior god of one nation, the Jewish God had taken on qualities of the ultimate truth of Plato; the distant, universal, logical God of the Stoics; the caretaking genies and daemons; and the provider of an afterlife, like the gods of the Mysteries. For Jesus the stakes had been raised because he believed that something very big was about to happen to all of creation, or rather, was already happening: John the Baptist, a prophet who was Jesus' teacher, had announced that the kingdom of God was actually coming, and very soon. Jesus proclaimed that it had already begun; all you need do, to see it, is to believe it. Even with the caveat that unbelievers are not able to see the change (at least at first), people are going to have expectations; either something happens, or it doesn't. Furthermore, Jesus was aware that if he went into Jerusalem and preached, in the way he planned to, he would likely be tortured and killed, and he seems to have been upset about it. For those who followed him, his anxieties became their own, for he sent them out to tell the world that an imminent change was already unfolding, and he told them to let themselves be abused and persecuted in his name.[9] Suddenly the question of how much one believed became the central religious issue and one that was going to be tested in the most dramatic ways: the world was either going to change or not, and the believer was either going to withstand torture and submit to martyrdom or not. There were two more major factors contributing to the heightened focus on belief: Jesus' magic is one. The other mostly comes along after Jesus, and that is the struggle to have enough belief to conquer one's sexual and material lusts. We will begin with Jesus and, later, the heresies. And Augustine will explain the problem of doubt and the taming of our inner beasts.

Jesus worked miracles and his apostles doubted them and, indeed, his whole mission, repeatedly. Just as the word *belief* becomes important in the Bible only when Jesus arrives, the word *doubt* is hardly ever mentioned in the Hebrew Bible—and is almost always tucked in an innocuous phrase— "no doubt" this or "no doubt" that—but is prevalent in the Christian New Testament. In the whole Bible, "doubt" mostly comes up in direct reference to the claims and behaviors of Jesus. He was always doing things that some people did not believe. We will take a few examples. First, in Matthew 14 the disciples are keeping watch in a boat on a lake. Jesus walks out to them on the water and they "cry out in fear."

> But Jesus immediately said to them: "Take courage! It is I. Don't be afraid."
> "Lord if it's you," Peter replied, "tell me to come to you on the water."

"Come," He said. Then Peter got down out of the boat, walked on the water and came toward Jesus. But when he saw the wind, he was afraid and beginning to sink, cried out, "Lord save me!" Immediately, Jesus reached out His hand and caught him. "You of little faith," He said, "why did you doubt?"[10]

Jesus spoke of doubt as a force that could erase the support beneath one's feet. Belief is very powerful in this world, which means doubt is very powerful here, too. Consider the Gospel of Mark's version of a story about a boy afflicted with terrible fits, because, as his father explains, he is possessed by an evil spirit:

> Jesus asked the boy's father, "How long has he been like this?" "From child-hood," he answered. "It has often thrown him into fire or water to kill him. But if you can do anything, take pity on us and help us." "'If you can'?" said Jesus. "Everything is possible for him who believes." Immediately the boy's father exclaimed, "I do believe; help me overcome my unbelief!"[11]

In Matthew 17 this same story is told, but here, the unbelief mentioned above is missing, and there is another. The disciples had tried to cure the boy and now they ask Jesus why they could not. "So Jesus said to them, 'Because of your unbelief; for assuredly, I say to you, if you have faith as a mustard seed, you will say to this mountain, "Move from here to there," and it will move; and nothing will be impossible for you.'"[12]

Consider how belief is set up here not as a matter of belonging to a group but in terms of winning the battle against one's unbelief. Another clue to the nature of belief here is that Jesus' experience of preaching on his home turf was so disappointing. We think of it as a lesson in how things are if your audience *knew you when,* because when he began to preach they dismissed him, saying, "Is this not the carpenter, the Son of Mary, and brother of James, Joses, Judas, and Simon? And are not His sisters here with us?" But it is also about the function of belief in Jesus' teaching—what in Mark is beautifully called his "mighty work": "He could do no mighty work there, except that He laid His hands on a few sick people and healed them. And He marveled because of their unbelief. Then He went about the villages in a circuit, teaching."[13]

The unbelief shut him down and he was unabashed about it, that is, he took it as public knowledge that for his explanations to work the people had to bring some belief to the equation. The Buddha, too, was never embarrassed when an audience had a closed mind to his teaching—he knew he needed them to work hard so there was no point in haranguing anyone too resistant.

Then there is the disciple who has been known these last millennia only by his doubting. John 20 tells of Thomas, one of the twelve apostles, remembered ever after as Doubting Thomas:

> Now Thomas (called Didymus), one of the Twelve, was not with the disciples when Jesus came. So the other disciples told him, "We have seen the Lord!" But he said to them, "Unless I see the nail marks in his hands and put my finger where the nails were, and put my hand into his side, I will not believe it." A week later his disciples were in the house again, and Thomas was with them. Though the doors were locked, Jesus came and stood among them and said, "Peace be with you!" Then he said to Thomas, "Put your finger here; see my hands. Reach out your hand and put it into my side. Stop doubting and believe." Thomas said to him, "My Lord and my God!" Then Jesus told him, "Because you have seen me, you have believed; blessed are those who have not seen and yet have believed."[14]

Yet in Luke 24, other disciples also need some hands-on proof that he had risen again:

> Now as they said these things, Jesus Himself stood in the midst of them, and said to them, "Peace to you." But they were terrified and frightened, and supposed they had seen a spirit. And He said to them, "Why are you troubled? And why do doubts arise in your hearts? Behold My hands and My feet, that it is I Myself. Handle Me and see, for a spirit does not have flesh and bones as you see I have." When He had said this, He showed them His hands and His feet. But while they still did not believe for joy, and marveled, He said to them, "Have you any food here?" So they gave Him a piece of a broiled fish and some honeycomb. And He took it and ate in their presence.[15]

The odd note about Jesus' needing a snack was significant. Ghosts, it was said, could not eat, so the meal was a proof offered to the reader—lest he or she have any doubt.

Jesus' miracles have been interpreted by myriad scholars and theologians. I will here merely note that he was a powerful teacher who created events in which belief and doubt were purposefully thrown into confusion. Miracles were a common feature of the wandering Jewish preacher of the time. Still, Jesus' miracles were delivered with more authority than usual, and people were often genuinely surprised by them—so much so that they asked him to leave or gave some other telling hint of actual amazement. The miracles functioned to show people that Jesus was someone worth listening to, but they were also a way of starting a conversation about belief and doubt. What this conversation meant to Jesus is difficult to say, but it is clear that he saw some connection between belief in his miracles and belief in the arrival of the kingdom of God. Following him, in the coming centuries, the descendants of the ancient Greeks and the inhabitants of the Late Roman world would once again worship a god with a face and other human features; a miracle worker. But this time, the questions of swallowing the mythology and believing the promises of justice and an afterlife were right out in the open—front and center. This time, when the religion was born, the culture was already in possession of a large, written tradition of doubt.

Jesus himself was a Jew speaking to Jews and not promoting much of a mythology—and, of course, his conversation about belief and doubt was not about Greek philosophical objections to faith. But his ideas and his image came to the real attention of the Roman Empire after he was long gone. He came to Rome in a story from the East, told in common Greek and already incorporating major tenets of religions that were familiar throughout the empire. The ubiquitous image of Isis holding her divine son Horus was transformed into Mary and the infant Jesus. How did it come to be that the far-flung cities that had once sold, bought, and borrowed the works of Cicero and Lucretius could now lose every sign of such doubt and apparently take on universal belief in an anthropomorphic God? The philosophers had even rejected the idea that a God could move or change in any way. The philosophers and the Jews had both rejected the idea of any God with a biography: a face and a mother, a handshake and a style of speech. And then here was God, a man. It is a stunning shift. Jesus presented a leap of belief: in the invitation to believe that the kingdom of God had come, and in the miracles, but also in the predictions and claims to be able to forgive sins. That leap was made even more explicit after Paul reimagined the religion in terms of the magic of resurrection and life everlasting.

Jews were very wary of magicians and messiahs after the rise of Christianity and no longer encouraged them. Although the afterlife did remain a factor in Jewish ideas, for the most part this new emphasis on belief was a gentle shift in Judaism, whereas Christianity took it and ran.

PAUL: FOLLOWING ABRAHAM ALL MORNING

Through Paul, God was so connected with the afterlife that the story of Jesus' miracles grew increasingly symbolic for this one great miracle: he was going to save humanity from death. Not from hell, mind you. Before they were offered eternal life through Jesus, people throughout the empire were not worried about being judged and damned; they were afraid of death, of rotting in the ground. In Paul's hands, Jesus' death and resurrection became the center of the new religion.

Paul was convinced that human beings could not earn their way to heaven; they could get there only through the strength of their faith. In an attempt to fortify his interpretation that the Jewish law no longer needed to be followed, Paul looked back to the Hebrew Bible. In Romans 4:11, Paul said that Abraham's blessing from God was not "justified by works" but because he had faith in God. The Hellenistic Jews had looked back to Abraham as a model of Jewish piety without Mosaic law. Paul stretched this even further, taking as his model Abraham in the period after he had met God but before he was circumcised. It was a few hours. God first told Abraham that he would be mightily blessed in the future, and later that day Abraham had himself and all the males in his household circumcised. Paul wrote triumphantly of that blessing, "Under what circumstances was it credited? Was it after he was circumcised, or before? It was not after, but before!"[16] Thus Abraham "is the father of all who believe but have not been circumcised," as well as those "who not only are circumcised but who also walk in the footsteps of the faith that our father Abraham had before he was circumcised." His language here emphasized that faith is what is most important; it is needed even by the circumcised. Keep in mind that "the footsteps of the faith that our father Abraham had before he was circumcised" wouldn't get you down your garden path.

When God said to do it, Abraham cut himself and his kin, with no proof yet of God's power but only the claim of it. "Against all hope, Abraham in hope believed," says Paul, and that was enough. This is stirring because it marks the moment when the world shifts into the new concentration on faith. We have seen it announced in the Mishnah and variously discussed by Jesus, but until Paul, there was still the law. Paul says, "It was not through law that Abraham and his offspring received the promise that he would be heir of the world, but through the righteousness that comes by faith." And further: "Therefore, the promise comes by faith, so that it may be by grace and may be guaranteed to all Abraham's offspring—not only to those who are of the law but also to those who are of the faith of Abraham. He is the father of us all."[17]

Now the law has been deemed unnecessary. Why? Paul explains: "We know that the law is spiritual; but I am unspiritual, sold as a slave to sin. I do not understand what I do. For what I want to do I do not do, but what I hate I do. . . . I know that nothing good lives in me, that is, in my sinful nature. For I have the desire to do what is good, but I cannot carry it out. . . . What a wretched man I am! Who will rescue me from this body of death? Thanks be to God—through Jesus Christ our Lord!"[18] So the law is too hard, but he also argues that not only is it too hard, it is a hindrance:

What then shall we say? That the Gentiles, who did not pursue righteous-ness, have obtained it, a righteousness that is by faith; but Israel, who pur-sued a law of righteousness, has not attained it. Why not? Because they pursued it not by faith but as if it were by works. They stumbled over the "stumbling stone."[19]

Belief is everything. Jesus had made belief and doubt matters of philosophi-cal and religious importance by challenging people to believe by fiat—to have faith. Paul said that this faith could function instead of the Jewish law, instead of Torah, instead of circumcision. Belief had never been so charged with redemptive power. "Christ is the end of the law so that there may be righteousness for everyone who believes," he says.[20] Naturally, there was new anxiety over the quality of one's belief.

To speak of Paul in the history of doubt, we must glance at two state-ments of his that did a lot to stifle doubt in the coming years. In the first, Paul shut down the questioning of divine justice, one of the most obvious forms of doubt in the world. How can the world be called just when there are inno-cents suffering unthinkable deprivation and horror every hour of every day on earth? Romans 9:14–15 says, "What then shall we say? Is God unjust? Not at all! For he says to Moses, 'I will have mercy on whom I have mercy, and I will have compassion on whom I have compassion.' It does not, therefore, depend on man's desire or effort, but on God's mercy." And Paul swept up with a little humbling: "But who are you, O man, to talk back to God? 'Shall what is formed say to him who formed it, Why did you make me like this?' Does not the potter have the right to make out of the same lump of clay some pottery for noble purposes and some for common use?" Divine justice, according to Paul, is not ours to ponder. Second, Romans 13 extols obedience to authority: "Everyone must submit himself to the governing authorities, for there is no authority except that which God has established. The authorities that exist have been established by God. Consequently, he who rebels against the authority is rebelling against what God has instituted." Paul said that is why we have to pay our taxes—because governors are God's servants and deserve obedience. Faith was going to protect the temporal authorities, and the temporal authorities, in turn, were going to protect faith. Belief was going to become part of the structure of the state in much the same way that ritual had.

How did such talk fall upon the ears of the people who had been Cynics, Stoics, Epicureans, and Skeptics? How did the average intelligent thinker in the Christian world come to think of philosophy as a thing that stopped short at faith in a mythological god, i.e., a detailed and anthropomorphic figure? Well, at first such talk didn't fall on their ears. This was a Jewish conversation in the beginning (but even then it took place in Greek, in the Hellenistic world, quoting from a Greek Bible). At that time, the meaning of Jesus was understood in terms of the Jewish categories of the Law and the Prophets and within the Jewish Messianic tradition. When Paul began preaching to the gentiles and the Greek-speaking eastern Roman Empire did hear of the new religion, he already had a common language with them and there was a central text in that language. Moreover, his message was Hellenized to suit them: in Acts 17, when Paul goes to Athens and delivers a lecture to Epicurean and Stoic philosophers, he puts his religion into terms designed to convince an educated philosophical audience.[21]

After Paul, many early Christian texts were composed with the intention of convincing wise and careful rulers, like Marcus Aurelius (to whom some such texts were dedicated), to end the persecution of the Christians. They were being persecuted because, having broken away from the Jews, they no longer had special dispensation from worshiping the Roman Imperial gods. They were an emotional cult that had arisen among the poor, and, to many citizens of the empire, the salient fact about the Christians was that they scorned the state symbols and regularly broke the law by dishonoring them. So the early Church Fathers' "apologies," their defenses of Christianity, tended to address the problem in terms that would mean something to the philosophical men whose opinion meant nothing less than life or death.[22] In the second century, Justin's *Apology* defended Christianity by suggesting that Jesus should be seen in the tradition of Socrates since both had died for a more powerful vision of God. Justin pointed out that the Christians were being persecuted as atheists because they denied the state gods, and that, of course, had been Socrates' crime as well. This made sense to the Greeks, who had always understood the Jews as a "philosophical race" and were open to understanding this offshoot, with its parallels to Stoicism and Platonism, as a kind of philosophy.

Another link between Stoicism and Christianity came from the Christian adoption of some tenets of Epictetus (ca. 50–ca. 138 CE), a man who started out as a slave in Rome and ended up a Stoic philosopher in Greece. He lived about four hundred years after the Stoic school of Zeno was established in Athens. Epictetus's work concentrated almost exclusively on how to live: the Stoic teacher was to encourage his students to live the philosophic life according to virtue, reason, and nature. The point of it all was to be happy, to flourish. Imperturbability and freedom from passion were the route. On these themes, his is the best Stoic writing to survive down to us. Far in the future, Thomas Jefferson would admire Epictetus as creating an exemplary ethos for the secular citizen. Epictetus's Stoicism was also outstanding in its insistence on the doctrine of the brotherhood of man, and it was this aspect of it that was incorporated into Christianity.

That's what happened to philosophy. Judaism, in its Christian form, became the dominant philosophy, and it did so by taking on large aspects of the old philosophies. Because they were persecuted, Christian thinkers found that they had to justify the Jewish God in terms of Greek philosophy. They did not much mention the old Jewish story, which had, at its base, a mythic, anthropomorphic warrior God who had opinions, traded favors, and loved to hear of his fame. Although the Jews had grown far beyond that, they had not really theorized the experience of their universal God. Plato had. Platonism could not give you the whole experience of Judaism—the imminent yet transcendent god, the beauty and antiquity of the Bible, the songs and the psalms, the nurturing ethics of the creator—but it could express the central tenet in solidly philosophical terms. The secret was in the mix.

If Christianity swallowed philosophy, we may mourn philosophy, but we should remember also that in matters of the mind, eventually you are what you eat. By calling upon philosophical precedent in their life-and-death struggle, the early Church Fathers let philosophy into their concerns. One early figure, Tertullian, argued against it, famously asking, "What has Athens to do with Jerusalem?" But by now the answer was: a lot. Tertullian's opinion would echo in the medieval West, but most of the early Church Fathers supported the notion of Christianity as a philosophy, or at least as a continuation of philosophy, and considered that it might well make use of the wisdom that had come before. They delighted in saying that the Greek philosophical past was their own early history in much the same way they claimed the Hebrew Bible as their own. Beyond the threat of persecution, the early Church Fathers had been educated in philosophy, often before they came to be Christians, and they felt they had to speak to it. Then, too, every aspect of Greek education and pedagogy was attached to the great philosophical and pagan texts. How would one even think of teaching rhetoric or grammar without them? So they kept them. The early Church Fathers were not the first to take such a position: Philo (ca. 25 BCE–ca. 50 CE), the Jewish philosopher who lived in Alexandria and wrote the history of his people in Hellenized terms, had also advocated the study of the seven liberal arts in order to study philosophy, and the study of philosophy in order to study theology.

Meanwhile, something very strange happened in the history of religion. We have seen that Plato and Aristotle provided some philosophical arguments for the idea of a single, unfathomably great God. Yet they also spoke of many other gods. Also, Plato's most emotive and creative vision of God was offered only in the *Timeaus;* elsewhere, ideas of religious ascent and the Forms suggested other kinds of "other" worlds. Aristotle's first cause was very remote and did not even know us. Five hundred years later, a man named Plotinus changed everything.[23] We do not have many details about his life. We know that he was born in 205 and that as a young man he wanted to study philosophy in India. We also know that he thought Christianity was an offensive, mythic little cult. For a while he studied in Alexandria with the same teacher as had Origen, and later he joined the Roman army hoping it would take him to India, where he still hoped to find a teacher. The army was routed, though, so Plotinus ran to Antioch, continued to study, and later started what became a prestigious school of philosophy in Rome.

What he taught came to be known as Neoplatonism. It described Plato and Aristotle using only those parts that added up to a God. Highlighting Aristotle's rational arguments for the first cause, it had the best time with Plato's descriptions of communing with the other world and of working to find one's deepest self. This was mostly Aristotle's God: he had no personality, did not know of us, had not created the world, took no interest in it, and would never judge it or anyone. One could not even say that he had being. Yet Plotinus took Plato's idea of "emanation" and said that, in a sense, God emanated the world into being and was the world. Plotinus's descriptions of communing with this God and finding the true self were full of trance states and ecstatic meditations, and surely drew on whatever he knew of Indian philosophy and religion. He did mention that he managed to reach such ecstatic communion only a few times, but it was enough. Neoplatonism became a tremendous force and remained so for centuries upon centuries.

Plotinus wrote it up in a book called the *Enneads*. Even today, religions call it philosophy and philosophy calls it religion, so it does not usually take center stage in histories. Yet Plotinus's creation of Neoplatonism was one of most important events in the history of both philosophy and religion. It was the single most powerful conduit by which monotheism drew upon philosophy. Neoplatonism drew on the authority of the great philosophers and it was emotionally satisfying. In creating it, Plotinus invented a mixed, reshaped version of Plato and Aristotle that would change forever the way everyone would see those philosophers. We have seen that Plato and Aristotle had religious beliefs and contentions, but we have also seen that these were conflicted, and more searching than believing. It was Plotinus who made Plato and Aristotle seem religious.

So far, Christianity was a mix of the Greek culture and the Jewish tradition, with their two universalist ideas. It had a cosmology that, although based ultimately in faith, was complex and philosophically Greek in tone, and it had the passion and commitment of the Jews. Built right into the very nature of Christianity, then, were the doubts of both these traditions: the Greek questions about truth, science, and reality, and the Jewish worry about the unjust world. These would not come to light for a while, but it is good to notice the idea here, as we watch these two traditions jockey and jerk into alignment, merging in a way that felt right to the inheritors, that converted the people into the religion, and that kept individuals from a martyr's death.

EASTERN INFLUENCES,
GNOSTICISM, AND THE HERESIES

Christianity found an increasingly large following in the Roman Empire. Members' willingness to accept martyrdom, borrowed from their parent religion, Judaism, drew attention and vigor to the new cult. We do not know if Emperor Constantine converted Rome to Christianity because the Christians seemed an unstoppable force, or if Christianity only became an unstoppable force because Constantine—whose mother had been a Christian but who was raised pagan at pagan courts—had a genuine conversion. In any case, in 313, having seen a cross in the sun (and having previously worshiped Apollo), Constantine made Christianity a lawful religion in Rome and set out to strengthen it through the construction of churches, the calling of councils and synods, and the elaboration of church networks. He himself continued to worship Apollo as well as Jesus and did not accept baptism until he was on his deathbed. But publicly he and his followers promoted the notion of "one God, one Empire." Everyone was supposed to believe and practice in much the same way, but they never came close. Over the coming centuries the intellectual and religious life of the Roman Empire would become increasingly dedicated to ascertaining just what the rules and beliefs of Christianity ought to be, and converting the population to them. The ancient world had few doctrinal disputes, yet had room to doubt the gods in general. In early Christendom the energy that would have gone to doubt was channeled to "heresy," i.e., doctrinal disputes.

In 270, the year that Constantine was born, a Christian Syrian farmer named Anthony did what preachers in that area of the world had long done: he gave up the world. This meant celibacy and a wandering, lonesome life, and the name such people earned, *monachos*—which became *monk*—simply meant solitary one. Anthony wandered into the desert and stayed for decades, emerging about 310 as the first "man of the desert," *erémétikos,* which became *hermit,* a Christian monk. Christians in the West found this a fascinating option of total commitment.

Meanwhile, in the villages of Persia, east of the Roman Empire, a new religion was on the rise and it brought about a mighty fusion between the great traditions we have so far been considering. This new religion was Manichaeism, started by Mani, who lived between 216 and 277. Mani knew Zoroastrianism, Buddhism, and Christianity and he thought they were all magnificent, but also that they were flawed because each was confined to particular languages and locations and because each had long ago bastard-ized the true teaching of its founder. He reenvisioned them as one and saw himself as the final successor in a long line of prophets, beginning with Adam and including Buddha, Zoroaster, and Jesus. He made his text canon-ical in his lifetime, to ensure that his teachings did not degenerate, and the religion spread dramatically through Persia, through the Roman Empire (and as far west as Spain), and into India and China. It would be stomped out by Christians in the West, but not before it profoundly altered the young religion by adding a tremendous dose of Eastern asceticism. Meanwhile, in the East, Manichaeism remained vibrant for a millennium and a half.

At first, Manichaeism was understood in the West as a Christian reli-gion. It was a mix, as Mani had wanted it to be. Like Zoroastrianism, it understood the moral world as a battle between good and evil. Like Buddhism, it was devoted to strict religious exercises, silence, stillness, and solitude, coaxing inner transformation. Manichaeism was only gradually cut off from the fold, but once it was, its dualism and its emphasis on Mani rather than Jesus made it particularly abhorrent to mainstream Christians. By that time, though, Manichaeism had profoundly infused Christianity with Eastern asceticism. This is when Christianity learned the immense power of a fevered ascetic movement that called on people to dedicate themselves entirely to the spirit. There were glimmers of an ascetic move-ment in the life of Jesus—but forty days in the desert does not a hermit make. Jesus ate heartily and fed others when they were hungry. He told people to give up their goods and their families, but not in a purposeful attempt to strain the body and thereby enter a new consciousness or puri-fied state. That came to Christianity from a good deal farther East.

Manichaeism was a type of Gnosticism—a dualistic religion that offered salvation through special knowledge, *gnosis,* of spiritual truth—and it is to Gnosticism that we now turn. Gnosticism grew up within Judaism, the mystery religion Orphism, and Plotinus's Neoplatonism, and it had a par-ticularly rich career in Christianity. Gnostics in any of these groups tended to see themselves as the elite of that group. In fact, they were often the com-munity leaders. But like many mystics before and after them, they believed that not everyone could handle the secret truth of the world: in this case, that the creator God was not good. With Gnosticism, the whole question of doubt gets spun on its head. The Gnostic idea was that human beings have within them the spark of something absolutely transcendent, something completely alien to this world. This spark, our consciousness, is a spark off the fire of an unimaginable God. Our humanity is the same stuff as this entirely distant, otherworldly God. This God did not make the world. Instead, the world was made by a creator God, a much less extraordinary figure. In many interpretations, the creator God was downright evil. Human beings have been worshiping this creator God by mistake, explained the Gnostics, but they should not do so. Gnostics called this creator God "Saklas," the Blind One; or "Samael," God of the Blind; or "the Demiurge," the Lesser Power. As Gnostics saw it, human beings are of more value than the creator God since we contain a spark of what is true, good, and transcendent.

Whereas religions generally looked at the cosmos and reported that such a wonderful world must have been made by an amazing intelligence, the Greek and Roman philosophers had wondered if there could be a God since the world was such a cruel series of ruptures and distress. The Gnostics took this idea in another direction: they saw the world as a limiting, nasty, frustrating cage and assumed a cruel God had made it. They cursed this God and felt superior to him. His limitations or villainy explained evil in the world, and his lack of the transcendent spark made a new kind of sense of our alienation from the world. Humanity has humanness—meaning, compassion, and love—and the universe does not, but just outside the universe, somehow, lives the true God and this true God has humanness, too. In fact, that is where we got ours. Gnostics were believers. They belong to the history of doubt because over the centuries, within the history of Judaism and Christianity, Gnostics doubted all of the personal characteristics of that God. Doubting God's benevolence is as fundamental a matter as doubting God as thinking, creating, all-powerful, or eternal.

For the Gnostics, all the crowing about the magnificent order of the cosmos was suddenly cast as wrongheaded: order and natural law were not to be celebrated, they were to be derided. Why marvel at the economy or grace of a law that effectively keeps you trapped on the surface of the planet, destined to die and rot in the ground or go up in smoke? Why marvel that God made beaches, wheat, and honeycombs if, on the important questions, any fairly decent human being would have done a better job; would, for instance, neither invent torture nor allow it to be invented? That would go for any kind of torture, and meanwhile, look how many kinds there are. They were very clear on the point that human beings owe no allegiance to the creator God. Our sense of ethics, pity, and care makes us far superior to the universe in which we are trapped.

It is a celebration of the human. Our only mission is to come to know who we are, to realize that we belong elsewhere, and to try to find our way back to our home outside this world. We find our way back by cultivating our alienation here below. People were to wean themselves from life—not to reconcile themselves to it, but rather actively to seek alienation from it. Most religions suggest that there is something to be learned from the observation that, for the great material universe, all our striving, our vanity, our longing, is meaningless. The Gnostic paradigm says No, there is no lesson there for us. Whatever we have inside us, whatever is most like us and least like the rest of the known universe, that is the actual reality, and all we have of truth.

Gnosticism was in the mainstream of early Christianity: about 140 CE, one of the most prominent and influential early Gnostic teachers, Valentinus, seems to have been under consideration for election as the Bishop of Rome. By the end of his life some twenty years later, he had been forced from the public eye and branded a heretic. There was growing hostility to Gnosticism's secret knowledge and its continuous creation of new scripture. By 180 CE, Irenaeus, Bishop of Lyon, was attacking Gnosticism as heresy. By the end of the fourth century Gnosticism was eradicated, its remaining teachers were murdered or driven into exile, and its sacred books were destroyed. Until recent finds, all we knew of it was the polemical denunciations and fragments preserved in the Christian documents on heresies. But thrilling new finds have given us the other side of the story. The newest finds include the Nag Hammadi collection, discovered in 1945 by an Arab peasant digging in his field. In the large earthenware jar he uncovered were a library of texts that Gnostic monks had buried sometime around the year 390 to save them from the hands of the orthodox Church. Elaine Pagels, one of Gnosticism's most eminent scholars, has written that "to know oneself, at the deepest level, is simultaneously to know God: this is the secret of *gnosis*. . . . Self-knowledge is knowledge of God; the self and the divine are identical."[24] There is something very individualist in a doctrine that allows each person to access truth.

It was an odd choice, dogma over mysticism, because the mystic vision is less susceptible to doubt. It makes few universal claims about details—meaning there is little to be contradicted; it encourages self-altering practices that try to bring the adherent to God through experience, not reason; and last, it never claims that philosophy or texts are proof of anything, so that whole approach to questioning religion is invalidated. So the mystic position would have been easy to defend but, as I noted, it afforded each believer a lot of interpretive power. Gnosticism recreated the problem of doubt by imagining a way to find the world woefully beneath humane standards, reject the Creator God in those terms, and yet preserve belief in God. It cursed the Judeo-Christian God for death and disease, heartache and loss, drought, fire, flood, and famine, all that has gone wrong with history all these many years. Furthermore, it left room for men and women to do their own thinking, to work out their own relationship to their inner self and the strangely hostile world in which it finds itself.

Along with Epicureanism and Stoicism, Manichaeism and the larger world of Gnosticism were the major competitors of mainstream Christianity. Then there were the heresies. These took over huge swaths of Christendom, sometimes for many centuries, because the people who followed them believed that the orthodox Church was wrong and that following it would lead to damnation. These heresies were about several major things. Here are a few: some people thought that sacraments performed by sinful priests were worthless and thus saw the orthodox Church as horribly corrupt because a lot of priests were caught sinning; some people did not believe that Jesus could have been all God and all man at the same time, and they worked out some other way to understand him; and some people saw Jesus, who was not around from the time of creation, as a somewhat junior partner to God. Those who followed Nestorius—the Nestorian Christians— were horrified at the idea that God had ever suffered and claimed that God's human son and servant, not God himself, had died on the cross. When the Nestorian teachings were condemned at the Council of Ephesus, Nestorius reported that the foolish crowds of Constantinople mocked him by dancing around bonfires, chanting that they had won the point: "God has been crucified. God is dead."[25] The Nestorian issue began over what to call the Virgin Mary: even today Nestorians venerate Mary, but they do not speak of her as the mother of God.

To note another important example, Pelagianism suggested that human beings were quite capable of deserving heaven by leading reasonably good lives—there was no need for chastity, poverty, or extensive fasting, nor was there a need for infant baptism, since children were considered to be born without sin. There was also no need for God's grace; people could earn heaven themselves through "works." Variously interpreting Paul and Jesus, supporters of acts and supporters of grace as the determining factor in salvation began a tug-of-war that would never leave Christendom.

By the time we get to the thinker who most shaped Christianity for the six centuries of the early Middle Ages, orthodox Christianity had already had to reconcile itself not only with the ancient philosophers but also with diverse variations of the Christian vision. Furthermore, although Paul had insisted on faith over law and works, various factors—most notably the influence of the Manichaean religion—had placed faith in the context of strict physical challenges, beginning with chastity, poverty, and fasting and eventually including self-flagellation and the wearing of hair shirts. Christianity inaugurated a harrowing new form of doubt: doubting one's ability to believe *enough,* and to enact that belief in dramatically painful processes. The story of doubt would now include all those who struggled to meet these challenges and who, at least for a while, found that they could not do it. It was doubt's time for dark nights of the soul.

AUGUSTINE (354–430)

Christian writers leapt away from the Classical world on only one issue, the definite existence of one transcendent God. In every other feature they did not leap but crawled cautiously, continuing to respect the questions of the past and their forms of expression as the highest achievements of humanity, and seeking only to reframe them in this new condition of unimpeachable faith in God.

The most famous scene in Augustine's marvelous *Confessions* is his conversion, and it comes late, in the eighth of thirteen chapters. He spends the chapters up to that point wrestling with doubt and temptation. At the beginning of the book, Augustine struggles with whether to be a Christian, as his mother, Monica, was and heartily wished him to be, or to continue to be a Manichaean. He had lived in a Manichaean community for nine years and was struggling now, among other things, with his discomfort with the Manichaean solution to the problem of evil, with some of their more fabulous astronomical claims, and with his knowledge that his dear mother wanted him to be a Christian. Augustine also had a lot of trouble with lust before and after becoming a Christian. He enjoyed worldly pleasures, from poetry prizes and important career appointments, to sex and food.

In an early chapter called "My consort sent home," Augustine had to part with the woman with whom he "had slept for many years," and with whom he'd had a son. She was sent away so that a proper marriage could be made for him, and he tells us that this was his mother's idea, in the hope that the married state would lead him to Christianity. Augustine did not take it well. "My heart which was deeply attached was cut and wounded, and left a trail of blood." But his parentally arranged marriage had to wait two years for the girl to come of age (he liked the look of her and was willing to wait), and in the meantime he could not resist his passions and descended further into sin than before. In commenting on this, Augustine offered his thoughts on Hellenistic graceful-life philosophy, and it is worth hearing the passage in his words:

> Nothing kept me from an even deeper whirlpool of erotic indulgence except fear of death and of your coming judgment which, through the various opinions I had held, never left my breast. With my friends Alypius and Nebridius I discussed the ultimate nature of good and evil. To my mind Epicurus would have been awarded the palm of victory, had I not believed that after death the life of the soul remains with the consequences of our acts, a belief which Epicurus rejected; and I asked: If we were immortal and lived in unending bodily pleasure, with no fear of losing it, why should we not be happy? What else should we be seeking for?[26]

So the only thing keeping him from embracing Epicurean happiness was his belief in a life after death, and that one received reward or punishment there.

It was in this state of physical indulgence and emotional anguish that he read the works of Plotinus and became devoted to Neoplatonism. Many people were now reading Plato only in terms of Plotinus's interpretation, and many Neoplatonists did not even do that; they just read Plotinus. Before being won over to Neoplatonism, Augustine had always thought, like Cotta in Cicero's study of the gods, that there could be no noncorporeal mind. Neoplatonism convinced him that such a thing was conceivable and that seeking it out is the path to wisdom. He later hypothesized that if God had given him the Holy Books before he had read "the Platonist books," he would have been seduced by their sweetness and hence never would have learned the "solid foundation of piety."[27] But it was not enough on its own. It was at this juncture that "with avid intensity," Augustine "seized the sacred writings of your Spirit and especially the apostle Paul." For Augustine, Plato had solved the problem of an intelligence without a body, so what had always seemed contradictory in Paul's vision now made perfect sense. Augustine said that what Paul had that was missing from Plato, was that Plato wasn't offering tenderness and attention. Plato's God was too far off.

> None of this is in the Platonist books. Those pages do not contain the face of this devotion, tears of confession, your sacrifice, a troubled spirit, a contrite and humble spirit, the salvation of your people, the espoused city, the guarantee of your Holy Spirit, the cup of our redemption. In the Platonic books no one sings: "Surely my soul will be submissive to our God? From him is my salvation; he is also my God and my savior who upholds me; I shall not be moved any more."
> No one there hears him who calls "Come to me, you who labor.". . .
> It is one thing from a wooded summit to catch a glimpse of the homeland of peace and not to find a way up to it, but vainly to attempt the journey along an impracticable route surrounded by the ambushes and assaults of fugitive deserters with their chief, "the lion and the dragon." It is another thing to hold on to the way that leads there, defended by the protection of the heavenly emperor.[28]

Augustine's description of what it is like to try to come to true insight by Plato's route was alarmingly torturous—even the sentence that houses it clangs around breathlessly. The philosophical religion of Neoplatonism offered him tiny glimpses o f "the homeland of peace" but was too hard. Under the protection of the "heavenly emperor," the journey suddenly seemed possible.

Augustine's intellectual and emotional acceptance of Christianity did not quite add up to a conversion experience in his own appraisal. He did not feel he was a Christian until he could give up all sex, all food beyond his barest needs, and all worldly enterprise, including his job as a teacher. The day came in a chapter called "The birth pangs of conversion," and the pangs hurt. Our famous image of Augustine's conversion begins with his learning of the *Life of Anthony* from a friend. After hearing the story, Augustine hollered at himself:

> Many years of my life had passed by—about twelve—since in my nineteenth year I had read Cicero's *Hortensius,* and had been stirred to a zeal for wisdom. But although I came to despise earthly success, I put off giving

time to the quest for wisdom. For "it is not the discovery but the mere search for wisdom which should be preferred even to the discovery of treasures . . ." But I was an unhappy young man, wretched as at the beginning of my adolescence when I prayed you for chastity and said: "Grant me chastity and continence, but not yet." I was afraid you might hear my prayer quickly, and that you might too rapidly heal me of the disease of lust which I preferred to satisfy rather than suppress.[29]

Along with the great joke at the end, note the terms in which Augustine credited Cicero with having awakened in him the search for meaning—not the finding, but the searching. Augustine was reeling from Anthony's story, and wished he himself had made such progress. He wailed to his friend, "What is wrong with us? What is this that you have heard? Uneducated people are rising up and taking heaven by force while we, with all our high culture and without any heart—see where we roll in the mud of flesh and blood." Alone with his friend, Augustine was suddenly weeping, agonized by his inability to choose the holy life: He had recognized that worldly contests brought only strife and anxiety, and yet he still sought after such victories and so many other pleasures. Would he ever fully convert? He ran out to the garden to be alone, but his friend, who was having similar problems, followed him and looked on as Augustine thrashed around. As Augustine described it:

I was deeply disturbed in spirit, angry with indignation and distress that I was not entering into my pact and covenant with you, my God, when all my bones were crying out that I should enter into it and were exalting it to heaven with praises. . . .

Finally in the agony of hesitation I made many physical gestures of the kind men make when they want to achieve something and lack the strength, either because they lack the actual limbs or because their limbs are fettered with chains or weak with sickness or in some way hindered. If I tore my hair, if I struck my forehead, if I intertwined my fingers and clasped my knee, I did that because to do so was my will. . . . Yet I was not doing what with an incomparably greater longing I yearned to do, and could have done the moment I so resolved.[30]

This is one of the most beautifully written scenes of Christian doubt: it was an "agony of hesitation," and he went through a series of gestures and facial expressions that reminded him of nothing less than the look of amputees, shackled men, or men faint from disease when they struggle to do something beyond their strength. He tore his hair, struck his forehead, and folded himself over his knees. In this ecstasy of externalized pain he realizes that his will could move his body in an instant, but that his will could not command itself. We are now around the other side of doubt for the first time, hearing not from someone whose doubt is all about getting to the bottom of what's real, but rather from someone whose doubt is all about actively trying to commit oneself to belief and, momentarily at least, failing.

He called it a sickness and a torture; he berated himself for his continuing connection to doubt. "I was twisting and turning in my chain until it would break completely: I was now only a little bit held by it, but I was still held." God was pressuring him, "wielding the double whip of fear and shame," trying to keep him from succumbing to that last-holding chain. In his mind he cried out, "Let it be now. Let it be now."[31] But he could not actually make the decision. "The nearer approached the moment of time when I would become different, the greater the horror of it struck me." His old passions and "empty-headed" desires pulled him back. "They tugged at the garment of my flesh and whispered: 'Are you getting rid of us?' And 'from this moment this and that are forbidden for ever and ever.'. . . What filth, what disgraceful things they were suggesting!" He was already mostly past these desires, he said—they were not confronting him face-to-face but whispering behind his back. But still "the overwhelming force of habit was saying to me: 'Do you think you can live without them?'"[32]

Then "Lady Continence" appeared to him and reminded him of how many people, young and old, male and female, have lived chaste lives, and told him that none of them could have managed it without God's help. That is, there was a leap to be made here. Lady Continence asked, "Why are you relying on yourself, only to find yourself unreliable? Cast yourself upon him, do not be afraid. He will not withdraw himself so that you fall. Make the leap without anxiety; he will catch you and heal you." But Augustine was still listening to the mutterings of impure desires while Lady Continence lectured him to ignore them. "This debate in my heart was a struggle of myself against myself." The friend stood waiting in the silence.

Now Augustine, choking with tears, again retreated from his friend, threw himself under a certain fig tree, and yelled out to God, "How long, how long is it to be? Tomorrow, tomorrow. Why not an end to my impure life in this very hour?" And then he heard "from the nearby house" a little boy or girl chanting over and over: "Pick up and read, pick up and read."[33] Augustine thought about it and could not remember a game in which that chant was used. He stopped crying, stood up, went over to the book of the apostle Paul that he had left by his friend's side, and read a line in which Paul counsels to "make no provision for the flesh in its lusts." The line is from Romans 13, a few lines after the instruction to respect authority and pay one's taxes that we saw earlier. Luckily, Augustine looked at the passage more pertinent to his problem. It was the sign he needed: "I neither wished nor needed to read further. At once, with the last words of this sentence, it was as if a light of relief from all anxiety flooded into my heart. All the shadows of doubt were dispelled."[34]

For Augustine, doubt had reigned for years, and now it was over. Following his lead, other Christians would see this wrangling with doubt, even to smacking oneself on the head and screaming, as an integral part of religious experience. And what was the hope? That all shadows of doubt would disappear.

In the very next paragraph of Augustine's *Confessions,* two interesting things happen: first, Augustine shows the book to his friend, who notices a further passage on the page that reads "receive the person who is weak in faith," and the friend takes that to mean himself and he converts, too; and second, the two young men run off to tell Monica, Augustine's mother, of the conversion. It is more than she had ever hoped, for Augustine "did not now seek a wife and had no ambition of success in this world." This, he declares, is a far greater joy than the grandchildren she had been looking forward to. There is a family drama in this doubt story about which we cannot hazard much, but we can notice that some problems in accepting one's parent's idea of the world are being acted out in this scene of doubt and conversion. With Augustine—because he passed through a period of being Manichaean, with its Far Eastern influences; and a period of Neoplatonism; and also knew the philosophers and sympathized with Epicurus—we have finally come to a moment when all the major traditions we have been following—the Greek, Hebrew, Far Eastern, and Roman—are represented in this one harrowed soul. Meanwhile, the whole conversion scene is understood as questionable autobiography, since so much of it was lifted straight from the *Enneads* of Plotinus. Augustine may have simply had a similar experience and borrowed phrases and structure to tell the story, but the connection reminds us of the importance of Neoplatonism here. Just after his conversion, Augustine and his mother, Monica, have an ecstatic experience together, which has had great meaning for Christianity. It too was written in language borrowed directly from Plotinus: "Our minds were lifted up by an ardent affection towards eternal being itself." Then, "Step by step we climbed beyond all corporate objects" and beyond the sun, moon, and stars.

After the conversion scenes, the rest of Augustine's book is about wrestling with the intellectual problems presented by the idea of God. In this struggle, Augustine shadows Cicero's concerns, but, unlike Cicero, he finds answers that satisfy him. These answers are viable because they do not try to give proof of the existence of God; rather, they are predicated on the existence of God. It is as if Cicero asked the question "How can it be?" in a tone that made it clear the question meant "Clearly, it cannot be." Augustine asks the same question but is already devoted to figuring out how it can be. He would either come to an answer or declare that he did not know how it worked, but still, so "it is." Velleius, Cicero's Epicurean character, had laughed at the idea that God created the world, pointing out the enormity of the task by inviting his listener to imagine the construction site. He had argued that no God made the world, it made itself. Augustine was not asking these questions to debate the existence of God. He was certain that the scripture was true, but he did not understand *how* it could be true. Indeed, he devoted the final three chapters of this thirteen-chapter book to an analysis of Genesis, all in an attempt to discover why God did not need a construction site.

Speaking directly to God, Augustine asks Velleius's questions. "How did you make heaven and earth, and what machine did you use for so vast an operation?" God could not have made these things in air or on land since neither had been made yet. "Nor did you make the universe within the framework of the universe. There was nowhere for it to be made before it was brought into existence. Nor did you have any tool in your hand to make heaven and earth. . . . Therefore you spoke and they were made, and by your word you made them. But how did you speak?" God had called out to Augustine, through words, in the garden; there was the child's rhyme and the apostle's book. Drawing on the description of Jesus as "the Word" and on the one line in Genesis that suggests that God spoke the world into existence, Augustine answers Cicero's question to his own satisfaction: creation was managed through words.

Cicero had also asked what God was doing before he made heaven and earth. Augustine presents the question and then mentions a snide answer that he has heard around, to wit: "He [God] was preparing hells for people who inquire into profundities." Augustine saw the humor, but says the question deserves a better answer. "I would have preferred him to answer 'I am ignorant of what I do not know' rather than reply so as to ridicule someone who has asked a deep question and to win approval for an answer which is a mistake." Humanity was allowed to question why and how, but not to question the revealed facts themselves. Augustine then asserts that he, for one, is absolutely certain that before God made the universe he did not make anything. If anyone else is surprised at God's apparent sloth, as Cicero had allowed it to be characterized, he will have misunderstood the question. Before the universe existed, Augustine explains, there was no time—time is not absolute; it is a feature of the universe. Also, for God, all time exists at once, eternally.

Within these and other investigations, Augustine advanced important philosophical ideas on the nature of time and the meaning of consciousness and the will. With Augustine, we have entered into another place in history, a place in which generation after generation will take Augustine's assumption for granted, as once the Greek gods had been taken as a fact of the natural world. But this time, thoughts of God were fraught from the beginning with the issue of belief and doubt. There was never an original time when everyone believed the same thing in Christianity, as if it were as obvious as the rocks and the trees. It always had within it the muscles of philosophy, of doubt, which it had incorporated into its doctrines back when having a philosophical justification for God was a matter of life and death for Christians. The myths of the ancient Greeks of the Archaic and Early Classical periods were not full of stories about belief and faith, and in the period when these myths were most generally believed, nobody talked about believing in them, or about how some things have to be taken on faith. There is no such period for Christianity: even in its origins, and even in the medieval period, Christians always discussed the work of belief and the problems of doubt. They did so because they had on their bookshelves remnants of a world that had already come to value rationalism, proof, and logic.

In an odd twist, Augustine praised doubt as the road to knowing anything, as long as it does not question God. It is a curious argument, not least because it preempts Descartes's *cogito ergo sum* ("I think, therefore I am"). The issue was that Stoics insist on determinism, yet Christians insist on free will, and Skeptics insist that one cannot know anything. Augustine says that he knows that he thinks, and from that he knows that he is, and that he has a will. But he puts the matter in terms of doubt: no one, he concedes, agrees on the true nature of the force behind "living and remembering and understanding and willing and thinking and knowing and judging."

> Nobody surely doubts, however, that he lives and remembers and understands and wills and thinks and knows and judges. At least, even if he doubts, he lives, if he doubts, he remembers why he is doubting; if he doubts he has a will to be certain; if he doubts, he thinks; if he doubts, he knows he does not know; if he doubts, he judges he ought not to give a hasty assent. You may have your doubts about anything else, but you should have no doubts about these; if they were not certain, you would not be able to doubt anything.[35]

And elsewhere he writes:

> I am quite certain that I am, that I know that I am, and that I love this being and this knowing. Where these truths are concerned I need not quail before the Academicians [Skeptics] when they say: "What if you should be mistaken?" Well, if I am mistaken, I exist. For a man who does not exist can surely not be mistaken either, and if I am mistaken, therefore I exist.[36]

Even within the increasingly closed system of Christianity, doubt was understood as the only way to know anything.

Augustine clearly does not reject philosophy. Instead, he furthers the argument of the early Church Fathers, claiming that philosophy is a gift of God and should be used when it is useful. His rationale for studying philosophy was put in terms of a metaphor anchored by the scene in the Hebrew Bible in which God told the Israelites to "despoil the Egyptians" as they left their bondage by taking gold and silver statuary or other works that had been pagan or profane and using the metal for their own, finer purposes. "If those . . . who are called philosophers happen to have said anything that is true and agreeable to our faith, the Platonists above all, not only should we not be afraid of them, but we should even claim back for our own use what they have said, as from its unjust possessors."[37] "Despoiling the Egyptians" has come to mean any use of another culture's art and ideas for purposes that may wholly contradict their original intention. The key to Augustine's use of philosophy is that his use of logic is beautiful, but it keeps certain propositions outside the rules. For example, if someone does not believe there can be resurrection from the dead, then they won't believe that Christ has risen. "But this consequent is false, because Christ has risen again. Therefore the antecedent is also false. . . . If there is no resurrection of the dead, then neither has Christ risen again; but Christ has risen again; therefore there is a resurrection of the dead."[38]

The other thing we need to notice is that Augustine ushers in this new period with a very mixed message, for as much as he was fighting against much of ancient philosophy, he was also fighting against the Christian heresies. At several junctures in Augustine's writings it is clear that he contributed to the exclusivity of the Catholic position: the persecution of non-Christians, nonorthodox Christians, and heretics, and the suppression, in the late fourth century, of all these people's writings. Augustine struggled to win over the Pelagians and other heretics, but ultimately he threw his weight behind violent repression. On the suppression of the philosophers, consider Augustine's letter to Dioscorus:

> If, however, in order to secure not only the demolition of open errors, but also the rooting out of those which lurk in darkness, it is necessary for you to be acquainted with the erroneous opinions which others have advanced, let both eye and ear be wakeful, I beseech you, —look well and listen well whether any of our assailants bring forward a single argument from Anaximenes and from Anaxagoras, when, though the Stoic and Epicurean philosophies were more recent and taught largely, even their ashes are not so warm as that a single spark can be struck out from them against the Christian faith.

By Augustine's time, Christianity was the religion of Rome in the West and Byzantium in the East, and it had power, money, and status to back up such calls for exclusivity. About 391, rioting Christians destroyed Alexandria's Sarapeum, a pagan temple that housed a branch of the Great Library. We don't know what was lost in that fire. It gets worse.

We are slipping now into the world that stopped speaking of doubt and lost touch with most philosophy. There was, however, a narrow corridor of continuous memory of philosophy and much of that corridor was provided by Boethius, the "last of the Romans, first of the Scholastics." Anicius Manlius Severinus Boethius lived from ca. 480 to 524 and did more to carry the old learning through the Middle Ages than anyone else. The emperor Theodoric praised him for his translations and interpretation of philosophy in no less terms than these: "From far away you entered the schools of Athens; you introduced a Roman toga into the throng of Greek cloaks. . . . Thanks to your translations, Pythagorus the musician and Ptolemy the astronomer may be read as Italians. . . . It is by your sole exertions that Rome may now cultivate in her mother tongue all those arts and skills which the fertile minds of Greece discovered."[39] It was hardly an exaggeration. Consider his work on the subject of logic alone: Boethius translated at least five of Aristotle's works on logic as well as a famous study of Aristotle by Porphyry, and he also wrote four commentaries of his own, two on Aristotle's works, one on the famous study, and one on a work by Cicero. As one historian has put it, "By his monumental achievement, Boethius guaranteed that logic, the most visible symbol of reason and rationality, remained alive at the lowest ebb of European civilization, between the fifth and tenth centuries."[40]

Boethius also wrote five independent treatises on logic and five tractates on theology and in all of them used the tools of reason and logic to elucidate Christian ideas, such as the Trinity. It is then more interesting to note that Boethius's most famous book today, *The Consolation of Philosophy*, completely ignored Christianity. Boethius had been charged with treason by the same Theodoric who had praised him, and he wrote the *Consolation of Philosophy* while in prison awaiting execution. It has often been regarded as a tremendous mystery that a man condemned to death would write a treatise that did not fall into faith, but the *Consolation* said nothing at all of Christ, Church, or doctrine. Instead, the book claims that in Boethius's despair, Philosophy showed up to comfort him. She appears to him both as a woman of normal human form and a kind of shimmering incorporeal being of such stature that her crown pierces the sky.

> I recognized my nurse, Philosophy, in whose chambers I had spent my life from earliest manhood. And I asked her, "Wherefore have you, mistress of all virtues, come down from heaven above to visit my lonely place of banishment? Is it that you, as well as I, may be harried, the victim of false charges?" "Should I," said she, "desert you, my nursling? Should I not share and bear my part of the burden which has been laid upon you from spite against my name? Surely Philosophy never allowed herself to let the innocent go upon their journey un-befriended."

Philosophy tells a heartening tale of her career, reminding him that, "though Plato did survive, did not his master, Socrates, win his victory of an unjust death, with me present at his side?" She mentions "the followers of Epicurus and in turn the Stoics" as plunderers who stole pieces of her for the masses and then wandered off. She recounts the exile of Anaxagoras, the poison drunk by Socrates, and the torture of Zeno and laments that "Naught else brought them to ruin but that, being built up in my ways, they appeared at variance with the desires of unscrupulous men." If these grow too strong, she explains, "our leader, Reason, gathers her forces into her citadel, while the enemy are busied in plundering useless baggage. As they seize the most worthless things, we laugh at them from above, untroubled by the whole band of mad marauders." Try as they may, the enemies of philosophy cannot crush reason.

Boethius had Philosophy explain that there was an intelligence to the universe, that which was once called fate, and that now we understood it to be the universal force. Philosophy thus described an ultimate governance to the universe, but Jesus' name was not mentioned. Throughout the Middle Ages, *The Consolations of Philosophy* was the most widely circulated of all early medieval writings. This rational, undogmatic, philosophical belief was a major part of the narrow path that leads continuously from the doubt of the ancient world to the doubt of the modern. When philosophy awakens in Europe, it will find Boethius near to hand.

THE JEWISH DOUBT OF ELISHA BEN ABUYAH

The early centuries of the first millennium CE produced a towering figure in the history of Jewish doubt. The Talmud tells of a heretic Jew, Elisha ben Abuyah, so violently renounced by the sages of old that his nickname is Aher, "the Other." He was one of a tiny number of rabbis excommunicated during the eight centuries of the Rabbinic period. He was more than that, though—from his time forward, in the Jewish texts, his name has been the byword of doubt. Elisha had been a beloved rabbi, renown for his insight and wisdom, and he had come to deny God and reject religion. The Talmud does not tell us much, but he was born sometime before 70 CE and died sometime after 135, and his father appears to have been a wealthy landowner of Jerusalem. We know Elisha trained in talmudic studies and became a highly respected rabbi—some of his judgments and comments appear in the classical rabbinic texts. But the Talmud tells us: "Aher's tongue was never tired of singing Greek songs."[41] He was a student of Greek and a lover of

Greek poetry. The Babylonian Talmud says that Elisha, while still a religious teacher, kept forbidden books hidden in his clothes. He apparently had some significant knowledge of horses, architecture, and wine. Elisha is also said to have walked outside the distance permitted on the Sabbath and ridden through town on a holy day. The Talmud describes the event that precipitated his excommunication. One day Elisha ben Abuyah saw a man send his young son up into the branches of a tree to perform a mitzvah: before one gathers eggs from a nest, one should shoo away the mother bird. The performance of this particular mitzvah, according to the Bible, brings long life.[42] The child did the mitzvah, then fell from the tree and died. Elisha called out: "There is no justice, and there is no Judge." From previous comments and behavior, everyone knew he meant it, and he was excommunicated.

The Talmud also tells this story: There was a group of rabbis in the years before the revolt (the last against Rome) who tried to see into the mysteries of the universe—they were likely exploring Gnosticism, and perhaps Greek philosophies as well. The rabbis were Azzai, Ben Zoma, Elisha ben Abuyah, and Akiva. They all "entered paradise" in these studies, and the result was that Azzai "looked and went mad" and "Ben Zoma died." As for Elisha: "Aher destroyed the plants"—this is generally understood to refer to his rejection of Judaism. Only Rabbi Akiva "entered in peace and left in peace." It is an important story in the history of Jewish doubt because it was ever afterward used to point to the dangers of outside learning and of mysticism (if practiced by those not prepared). The sole person to defend Elisha after his apostasy was his old student the rabbi Meir. Meir's wife, Beruriah, was the only woman to be treated as a person of learning and halakah decision in the Talmud. The daughter of a great rabbi, she was praised not only for scriptural knowledge and wisdom but for her unconventional spirit and intellectual wit. Among her keen talmudic comments: Rabbi Yosi the Galilean was once on a journey when he met Beruriah. He asked her, "By what road do I go to Lod?" She answered, "Foolish Galilean, did not the rabbis say 'Engage not in much talk with women'? You should have asked, 'Which to Lod?'"[43] Now that's comedy. We do not know if Beruriah and Elisha were partaking in a shared and perhaps common culture of questioning old beliefs, but it certainly seems possible. Beruriah stayed within the fold; Elisha's doubt and eventual disbelief haunt the texts of Judaism. The Babylonian Talmud tells a story in which a heavenly voice calls out: "Turn, O backsliding children (Jer. iii. 14), with the exception of Aher."

HYPATIA AND THE END
OF SECULAR PHILOSOPHY

We turn now to the terrible story of Hypatia, long told as the defining moment in the death of ancient philosophy at the hands of Christian orthodoxy. We learn the story from the *Ecclesiastical History* by Socrates Scholasticus, a fifth-century church historian: "There was a woman at Alexandria named Hypatia, daughter of the philosopher Theon, who made such attainments in literature and science, as to far surpass all the philosophers of her own time."[44] She became head of the Platonist school at Alexandria in about 400 CE. There she lectured on mathematics and philosophy, in particular Neoplatonism. "She explained the principles of philosophy to her auditors, many of whom came from a distance to receive her instructions. On account of the self-possession and ease of manner which she had acquired in consequence of the cultivation of her mind, she not infrequently appeared in public in presence of the magistrates." She was not frightened to go "to an assembly of men," for she was widely admired for her extraordinary dignity and virtue. But trouble came. Hypatia was friends with the Roman prefect of Alexandria, Orestes. Apparently because of her counsel, Orestes would not reconcile with the future Saint Cyril, patriarch of Alexandria (the two were locked in a struggle over church and state power). And Cyril had a gang of followers.

> Some of them, therefore, hurried away by a fierce and bigoted zeal, whose ringleader was a reader named Peter, waylaid her returning home, and dragging her from her carriage, they took her to the church called Caesareum, where they completely stripped her, and then murdered her with roof tiles. After tearing her body in pieces, they took her mangled limbs to a place called Cinaron, and there burnt them. . . . And surely nothing can be farther from the spirit of Christianity than the allowance of massacres, fights, and transactions of that sort.

She was about forty-five when it happened. We get a slightly different story from a tenth-century Byzantine encyclopedia called *The Suda:*

> Hypatia, daughter of Theon the geometer and philosopher of Alexandria, was herself a well-known philosopher. She was the wife of the philosopher Isidorus, and she flourished under the Emperor Arcadius. Author of a com-

mentary on Diophantus, she also wrote a work called *The Astronomical Canon* and a commentary on *The Conics* of Apollonius. She was torn apart by the Alexandrians and her body was mocked and scattered through the whole city. This happened because of envy and her outstanding wisdom especially regarding astronomy.

Hypatia was born, reared, and educated in Alexandria. Since she had greater genius than her father, she was not satisfied with his instruction in mathematical subjects; she also devoted herself diligently to all of philosophy.

The woman used to put on her philosopher's cloak and walk through the middle of town and publicly interpret Plato, Aristotle, or the works of any other philosopher to those who wished to hear her. In addition to her expertise in teaching she rose to the pinnacle of civic virtue.

The Suda's version leaves out the story of Cyril and Orestes. Here Cyril kills Hypatia simply because she is the learned pagan opposition:

For even if philosophy itself had perished, nevertheless, its name still seems magnificent and venerable to the men who exercise leadership in the state. Thus it happened one day that Cyril, bishop of the opposition sect [Christianity], was passing by Hypatia's house, and he saw a great crowd of people and horses in front of her door. Some were arriving, some departing, and others standing around. When he asked why there was a crowd there and what all the fuss was about, he was told by her followers that it was the house of Hypatia the philosopher and she was about to greet them. When Cyril learned this he was so struck with envy that he immediately began plotting her murder and the most heinous form of murder at that.

And *The Suda* tells a similar brutal story, adding that "The memory of these events is still vivid among the Alexandrians." It was five centuries later. We know from the correspondence of the Bishop of Ptolemais that Hypatia had admiring Christian students among her other pupils—he had been one. His many letters to her have in large part survived, and they glow with admiration and reverence. He asked her advice on the construction of an astrolabe and a hydroscope.

Nonetheless, the overall idea was not lost on anyone: the age of philosophy was over. Hypatia's father was the last director of the Museum. After she was murdered so many scholars left Alexandria that it marks the beginning of the end of Alexandria as a major center of ancient wisdom. A bribe saved the killers from prosecution, and after Hypatia's murder no non-Christian in the Roman Empire actively attempted to propagate secular philosophy. Hypatia died in 415 CE. The same Cyril who killed her was the one who attacked the Nestorian idea—also, it would seem, in a bit of a jealous snit of rivalry. Cyril went to the Bishop of Rome (the office was gaining importance, although only with Leo I [440–461]—often called the first pope—would the Bishop of Rome claim authority over the whole Church) and insisted that Nestorius's opinions had to be stomped out. They were condemned in 431. The result was that the Nestorians broke off from Rome, set up their own patriarchy in Baghdad, and spread eastward—to the Far East, even: Nestorian missionaries were in China by the seventh century. This had extraordinary importance for the history of doubt, because the Nestorians left the West when the texts and legacy of ancient philosophy were still part of an educated person's world. They would keep those texts, some in continued use, in remote Eastern monasteries while the same texts were driven out of the West and eventually forgotten there. Cyril, by the way, also drove the Jews from Alexandria.

The exile of doubt from the West reached its completion in 529 CE. That year, fearing the anger of God, the Christian emperor Justinian outlawed paganism and closed the Epicurean Garden, the Skeptic Academy, the Lyceum, and the Stoic Porch. After more than eight hundred years, they no longer existed. The graceful-life philosophies, with their beacons of doubt and rationalism, were banished from all Europe.

Pagan practices hung on for centuries longer and many were eventually stirred into Christianity. Christian bishops preached all over the empire against the Roman Feast of Kalends in January. In Carthage, on the day of that feast, Augustine himself preached for more than two and a half hours in an attempt to divert the crowd's attention, but as one historian has put it, Augustine's congregation, "though good Christians, were also loyal members of their city: they would not forgo that great moment of euphoria in which the fortune of the city, and all that was within it, was renewed."[45] By the late sixth century in Spain the clergy merely added the chant of "Alleluia" to the Feast of Kalends and jubilantly participated in the ancient rites. Christianity was steeped in Rome: saints came to be understood as members of a "celestial Senate house," invisible, but modeled on Roman "patrons" and thus *patron saints*. In the city of Rome in 495, Pope Gelasius fought against the Lupercalia, an ancient pagan ceremony that involved young men running naked through the streets. He lost and the festival went on as planned. All over the empire, people still bowed to the Sun in the morning. In Coptic Egypt the head of the great White Monastery despaired of changing the practices of the surrounding populace: "Even if I take away all your household idols, am I able to cover up the sun? . . . Shall I stand watch on the banks of the Nile, and in every lagoon, lest you make libations on its waters?"[46] Caesarius, Bishop of Arles, tried to get all Christians to refer to the days of the week numerically, starting from the Sabbath, in order to shake off the *Mars,* for example, hidden in *mar*di, *mar*tedi, and *mar*tes. Only Portugal did; their Tuesday is *terça-feira.*

By the sixth century, Christians in the West had won over the cities, but the countryside was still a place of almost endless supernatural energies, and even city dwellers saw the natural world in this spirited way. The great God of the Christians was too far away for farmers, and the Son may have been human but he was not available for watering the fields or fending off locusts. In parts of Spain the practice of leaving little piles of votive candles near springs, in trees, and on hilltops and crossroads was still so rampant that as late as the 690s dramatic Church ceremonies were staged to transfer the candles to the local churches and announce that idolatry was finally dead. What actually worked was not sermons against the enchanted natural world, but rather the reenchantment of the world in Christian terms. Gregory of Tours (538–594) was most responsible for the reinterpretation of the Christian saints as capable of helping average people in their relationship with the natural world; through them springs and crossroads once again became sanctioned places for worship. The saints brought healing, mercy, and fertility to the small places of field and hearth, and brought safety on byroads and high seas. In myriad ways, water was holy again, and trees might spring up on the graves of saints.

As the Roman Empire declined, it split: the wealthy eastern part gradually became the Byzantine Empire. Meanwhile, the West slipped away from its identity with the Roman Empire and fell into local struggles among barbarian tribes. In some cases, these tribes had converted to one or another version of Christianity in order to live comfortably with Rome; in other cases, the conversion came after Rome had receded and was more about alliances of tribes. The East and West inheritors of the Roman Empire were both Christian, but Rome had the pope—predominantly because the Bishop of Rome had been the bishop of the capital city (succession from Peter was touted later, when the position was challenged)—so, using the same argument by location, Constantinople elevated its bishop to patriarch, and the patriarch ran the Eastern Church.

In what became the Byzantine Empire, the central government remained relatively stable and consistent. Its religion was free to be religious. As historian Karen Armstrong has shown, in the Eastern Church rationalism was not taken on as a goal for religion; the Trinity, for example, was celebrated as a mystery rather than regarded as something to be interpreted until it made sense.[47] In the West it was different. The Roman government disappeared. Areas of the world that had seen no city life before Rome fell back into tribal farming or grazing. Western urban centers shrank dramatically, like puddles in the sun. What remained in these cities was generally the Bishopric and the churches, and the clergy maintained a neat hierarchy, continuously based in Rome. Although Rome's great buildings would begin to fall into ruins in this period, the steady hierarchy of the clergy did not. With the Roman Empire retreating and then gone, the Church took on secular and administrative roles. Literate education dwindled away with the suppression of the classical teachers and the shrinking of cities, and soon only the educated of the Roman Church could read and write. The language of literacy was a Latin that was frozen in its texts. Outside the monasteries, Latin transformed into the languages of modern Europe, or gave way to Celtic, Frankish, Saxon, and other spoken languages. Eventually, learning to read required learning Latin. In the West, the literature and philosophy of humanity was locked up in the Church and in its language.

As we turn from western Europe for a quick look at the Far East, we leave a tribal world that had been overlain with a sophisticated empire, then significantly overrun by new Germanic tribes, and was now slipping again into a state of kinship, farming, and warfare. This tribal world, however, was one informed by the texts of a sophisticated civilization, although those texts dwindled to just a few. The various tribes were often contentiously unified through allegiance to the borrowed tribal God of the Hebrews. Christian doctrine tended to take second place to the idea of God as bellicose, jealous, and judging. The early Middle Ages saw tremendous changes in western Christianity. The rise of monasticism for both men and women became a major factor of Christendom. Slave traders moved people around and thereby Christianized whole populations and some key individuals: Patrick, the son of a genteel family in England, was captured and sold into slavery to work in the bracing rain of the Irish coast, from which he escaped and to which he later returned—as a missionary. By the seventh century the intellectual concerns of Christianity turned to describing the world after death, detailing the delights of heaven and the torments of hell; by the early eighth century the local "mass priest" became an established feature of communities.

Charlemagne once again put the West under an empire in the late eighth and early ninth centuries. It is telling that he did not want his lands understood as a new Rome, but rather as a new Israel. His tribal people and their religious and political problems seemed more aptly reflected in the Old Testament societies and the drama of ancient Israel and her God than in the sophisticated world empire of Rome. Indeed, Charlemagne saw his empire as a theodicy, dedicated above all to stewarding the people's souls into heaven. With this in mind, in 789 he called for the monasteries and cathedrals of his empire to open schools for the exploration and proper dissemination of the faith. These schools were the intellectual centers of the West for centuries. Their subjects were always Latin, law, and theology, but laymen and clerics also came to learn medicine, or civil or ecclesiastical administration. Students also read pagan literature subsumed under grammar and rhetoric. But aside from the Bible, what were their texts?

They were limited. Beyond the translations and commentaries made by Cicero and Boethius, the Romans had never done much to translate the tremendous heritage of Greek thought into Latin, and by the fifth century, knowledge of Greek was rare in the West. The schools survived in the West on this spare diet: First, all they had of Plato was the *Timaeus*—the one book in which he gave a sustained and somewhat mythic discussion of God. Even that was only partially translated and embedded in the text of a fourth- or fifth-century commentary on it by a scholar called Chalcidius. The cathedral schools also had access to the commentary of another writer of the early fifth century, Macrobius, this one on the sixth book of Cicero's *Republic*. They had the *Natural Questions* by Seneca (ca. 4 BCE–64 CE), as well as Boethius's translations of Aristotle's elementary logical works and some of Boethius's original works on music and arithmetic. Then there were a few texts by important intellectual figures of the early Middle Ages, like Isidore of Seville and the Venerable Bede—all of which were derived from the ancient writers. In this way, and only in this way, texts by Plato, Aristotle, Cicero, and the Stoics, which were no longer available, were read about, but not read, for centuries. If we ask ourselves how the great minds of the West came to understand philosophy as a thing that stopped, necessarily, at questions of God, even though their world had already had a rationalist revolution, we have to answer by taking Charlemagne at his word: this was, to a degree, a new Israel, a tribal people, with only a strange and tangential relationship to the legacy of thought that had once flooded the land where they lived, and had since receded eastward. The West came to doubt again from within the logic of its own internal history in the eleventh century, and also from outside that logic, in the twelfth century, as the great flood of ideas came splashing back upon them from the other direction, having looped around the Mediterranean Sea.

Before we leave the early Middle Ages, we need to look farther east yet—to where the story of that other world of doubt had taken a stunning little turn.

ZEN AND THE GREAT DOUBT

In the early Middle Ages, soon after Justinian and Theodora closed the pagan schools in Byzantium (529 CE), there were some fascinating changes in Eastern doubt. From the beginning of the Common Era, the Buddhism that grew and dominated China was a theistic vision of the master's idea. For the most part, when Buddhism spread to other countries in this period, it was Mahayana Buddhism—the big raft—and it had gods. But even so, the basic idea often led, not only to theism, but also to the idea of physiological and intellectual practices that depended on theism very little and often not at all. Consider the case of Tibet, which fashioned the third great school of Buddhism through a kind of fusion and reinterpretation of Theravada and Mahayana.

The traditional story of the origin of Tibetan Buddhism is that Buddhism was introduced into Tibet in the seventh century CE by two devout Buddhist princesses, one Nepali and one Chinese, who became the wives of a Tibetan king. The new religion was actually established by one of the successors of that king who brought an Indian monk, Padmasambhava, to found a Buddhist monastery near Lhasa about 750. Tibetan Buddhism incorporated Tibet's pre-Buddhist gods as well as the gods it came with. But what most separated Tibetan Buddhism from any other was its profound incorporation of the Hindu doctrine of tantra. The tantra texts concentrate on the interconnectedness of all things and, not unlike Taoism, they advise using all the body's powers to reach nirvana. They said that by these means practitioners could speed up the process dynamically and also obtain a degree of superpower in this life. In the West we are most aware of the tantric sexual practices, but these are only a portion of the tantra. Tibetan monks sing, move, and chant. They also intensely visualize a variety of gods in order to invoke certain states. This was yet another example of the transformation that many strands of Buddhism underwent, shifting from the nontheism of the Buddha into traditions of gods that supported or personified the Buddhist journey to peace. But not all Mahayana Buddhism slipped over to the worship of gods, Buddhas, or bodhisattvas.

Zen Buddhism is Mahayana, and it is essentially early Buddhism translated through Taoism in China (and then further influenced by its later diffusion into Japan). The founder of Zen in China was called the Bodhidharma, and he came to China from India in the late fifth century CE. He taught the practice of "wall-gazing" and promulgated the Lanka-Vatara Sutra, the chief doctrine of which is "consciousness-only," which means that consciousness is real but its objects are constructed by it, and unreal. Mixing with Chinese Taoism, Zen grew in the coming centuries, and the eighth and ninth centuries were its golden age. Zen bases itself on a moment when the Buddha answered a disciple's question by holding up a lotus flower and saying nothing. The idea was that the wonder of being can only be experienced, not explained. No logical descriptions could be of any use. The only thing to do was to get oneself shocked out of the normal human assumptions about the world by asking questions that are impossible to answer within our normal human assumptions.

These questions are koans—famous ones include "What is the sound of one hand clapping?" and "What did your face look like before your ancestors were born?" The Zen master saw the adept twice a day to hear his or her latest answer, generally sending the adept back to work with a gentle prod in the right direction. Apparently, the master was hardly necessary when the right answer had finally come, because, like so many other realizations in this kind of practice, when the answer came it was as certain and as clear as any radical change in the physical universe. There was a clatter of recognition. Zen was as empirical as Siddhartha's Buddhism was: it made no claims to knowledge that could not be discovered by each individual. There was nothing to take on faith.

One way of characterizing the difference between Zen and other forms of Buddhism is to note that the Buddha offered a technique, a distinct and progressive therapy, to bring the individual follower to the awakening, whereas Zen stresses a nonprogressive vision of awakening: one suddenly snaps out of one's dream. It's the difference between, say, learning to walk again after an accident, where your daily work is cumulative and likely to take a set amount of time, and snapping out of a hysterical paralysis because of a sudden confrontation with the triggering experience. Zen works to pop you into enlightenment.

The result is that there is even less to believe in here, in a way, than in other versions of Buddhism, which at the very least give the adherent a progressive program of internal exercise. Unlike many other programs of enlightenment, Zen Buddhism specifically cultivates doubt.[48] The cultivation of doubt on such primary matters is a cultivation not of a problem, but of the great mystery. In Stephen Batchelor's words, "A problem once solved ceases to be a problem; but the penetration of a mystery does not make it any less mysterious. The more intimate one is with a mystery, the greater shines the aura of its secret. The intensification of a mystery leads not to frustration (as does the increasing of a problem) but to release."[49] That's the idea: Hum in the middle of the universe created by your mind and let yourself not understand it. Coax out the experience of not recognizing the universe as familiar.

The great teachers of Zen have urged keeping oneself in a constant state of unknowing, and they have excelled in generating an attitude of questioning that is sustained and vivid in its wonder, yet blank and unhopeful in relation to answers. Zen literature tends to be full of succinct, even terse, "case studies" of awakenings; *koan* actually means "public cases" and derives from this legal usage. Many of these hit a similar note: an adept meets a teacher, the teacher says something to him—"What is it?" is common—and nine years later (or right away, or any amount of time) the adept is awakened or enlightened by this question, and stays awake ever after. Other public cases are about extremely odd or seemingly ridiculous gestures by a great teacher. The point of strange antics—a sudden scream, a question answered with a tweak to the nose—is to unhinge the sense of normality and assumption; that is, to create a profound and transformative doubt. There is a famous Zen dictum that encapsulates the notion: "Great doubt: great awakening. Little doubt: little awakening. No doubt: no awakening." The greatness of the doubt here is said to rely not only on its magnitude but on its subject as well: for a great awakening, one must first doubt the primary issues of life, the nature of existence, and the character of meaning.[50] Note that neither scholarly knowledge, nor performing good deeds, nor prayer, nor ritual were considered of spiritual value here. There arose two main schools of Zen, the Lin-chi (Rinzai in Japan), which placed greater emphasis on the use of the koan, and the Ts'ao-tung (Soto), which emphasized sitting in meditation without expectation and with faith in one's own intrinsic state of enlightenment.

Meanwhile, the atheist doctrine of Carvaka and the materialist philosophy of Samkhya were still going strong all these centuries later. In the seventh century, the sage Purandara made an important adjustment to Carvaka's insistence that valid inference was impossible.[51] He argued that inferences based on a great many repeated experiences of things—inferences one can make about the physical world, for example—are reasonable, whereas inferences about the transcendental world, whatever that might be or not be, cannot be considered reasonable since they are not based on sense experience at all, let alone repeated sense experience. The eighth-century philosopher Sankara described the Samkhya as holding that nature made itself in the same way that "non-sentient milk flows forth from its own nature" and as "non-sentient water, from its own nature" flows along to our benefit, so the nonsentient matter of the world "although non-intelligent, may be supposed to move from its own nature."[52]

Christianity and Zen both arose out of doubt's questions. Jesus' exhortations to faith and Zen's lunging, leaping, seeking after uncertainty were both tremendously fruitful innovations in the history of doubt. They also provide some of that great history's best images, gestures, poems, howls of woe, and radical solutions. These were religions with doubt at their hearts.

SIX

Medieval Doubt Loops-the-Loop, 800–1400

Muslims to Jews to Christians

It is a common assumption that doubt died in the Middle Ages: Christianity and Islam seduced and controlled everyone in a period of darkness, disease, and discomfort. We will see, instead, that doubt had some amazing adventures in this period, in a great loop around the Mediterranean Sea. This chapter will loop-the-loop of medieval rationalism. Generally, each time doubt traveled on, it was being stamped out in the last place it left. Doubt traveled mostly by coercion at first, as the Christians chased philosophy eastward and out of their territory. If we begin in Florence, Italy, a few centuries after Jesus, we find that educated people could no longer read Greek, which had been the language of books and intellectuals in Rome. A clock metaphor may help here: Florence is at twelve o'clock, and it was midnight when the mob killed Hypatia. Soon even knowledge of the existence of much philosophy would be lost in the West. At the time, doubt was more tolerated in the eastern Roman Empire, so books and intellectuals-in-exile flowed there, and were able to thrive for a while. Then this Byzantine world closed the philosophic schools in the hopes of pleasing God. This chased doubt farther east, into Syria and Persia, where there were already a few monasteries with good Greek and Roman libraries. That would be at about two o'clock, and three o'clock. Ancient rationalism, natural philosophy, and medicine survived in Syria and Persia in these tiny flickers until the Arab Muslim state rose up along the southeast of the Mediterranean and beckoned those scholars to come and teach what they knew.

The ideas continued to spread clockwise: west, across Muslim North Africa and then north into Muslim Spain. There, Jews picked up the fever of rationalism, again transforming it, and the Christians took it up and spread it back through Europe. A great wealth of material was back in Florence by the year 1200 (it's a nice coincidence with my clock metaphor, which is about geography). Much of the ancient material had been fitted for monotheism by then. This mix of faith and philosophy had created wonderfully bizarre new kinds of doubt. So let us begin our loop-the-loop.

It is funny that Greek philosophy was first taken to Asia with Christians, but it was. Saint Ephrem founded a Christian school at Edessa, in Mesopotamia, in 363. Its faculty had studied philosophy in Athens and, along with their theology, they taught Aristotle's philosophy and the medical writings of Hippocrates and Galen. Many Nestorian scholars who went east after the condemnation against Nestorianism in 431 ended up at the school of Edessa. Indeed, the school grew too Nestorian for the emperor Zeno's taste and he ordered it closed in 489. The Nestorian professors who were chased away went to Persia and took with them Syriac translations of some of Aristotle's works; they translated these and others into Persian. Back in Syria, the school of Edessa became a new Monophysite school. Monophysites propagated the view that Jesus was not really human but only divine—of one nature, not two as orthodox Christianity held—but they continued the Nestorians' habit of studying Aristotle and also continued the work of translating Greek texts into Syriac. So by the time Justinian and Theodora shut down the Byzantine philosophical schools in 529 (at one A.M. on our clock of the Mediterranean), there were already places for the displaced philosophers to go, farther east (at two o'clock), where Nestorians had prepared the way. The scholars were welcomed into Persia in their exile. There, in these few sites, the great heritage of doubt in the Western world smoldered quietly. But not for long.

MUSLIM SKEPTICS

While the West was in its darkest days, a new faith arose in the East. Muhammad of Mecca lived from 570 to 632 and in his own lifetime saw the beginning of a great expansion of the religion he inaugurated. Within a hundred years, the Islamic world had expanded, by military conquest, from India, through North Africa, and into Spain. Islam was envisioned as a continuation of Judaism and Christianity, and Muhammad saw Moses, Jesus, and himself as the great prophets of a single God. The big difference was supposed to be location (this time the message was being brought to the Arab world) but, by and large, the Jews and Christians were not amused by the "corrections" of their faith. The Koran supposes that on Judgment Day, God will ask Jesus, "Didst thou say unto men: worship me and my mother as gods in derogation of God?" and assures its readers that Jesus will deny having done so.[1] This certainly riled up the Christians, but we see nothing new in such debate over religious detail. Muhammad's religious vision was, of course, a radical questioning and rejection of the tribal paganisms of Arabia. If anything was new in the history of doubt, here at the birth of Islam, it was in the clarity of the command to surrender to belief in God. Indeed, *Islam* and *Muslim* are variations on the same Arabic word, whose meaning is to submit. The masses may have done so, but the intellectuals of the Muslim world had access to more ancient texts than did the intellectuals of Byzantium or the west—and that made all the difference.

There were three main sects of the Muslim world. The largest were the *ahl al-hadith*, the Traditionalists. Their ideas appealed to ordinary people, in part because they believed that every person had the ability to know God, through the Koran. Priests were unnecessary and each individual had to be treated with respect. The Shia branched off in their dedication to descendants of Muhammad's daughter, Fatima, and her husband, Ali, who was Muhammad's cousin. The Shia were very concerned with succession obviously, and it was a crisis for them when the childless, twelfth-descended Imam went into hiding in 939 and was never heard from again. In response, the Shiites developed a notion that he had experienced some kind of apotheosis: the twelfth, Hidden Imam would eventually return and begin a great new Muslim golden age. In their origins, Shiites were often more interested in rational explanations of God than their mainstream coreligionists were. The third sect, the Mutazilis, were more rational still, welcoming philosophical speculation and logical proofs, and comfortable using allegory to explain away the anthropomorphism of the God of the Koran: God's "hands," for example, referred to his generosity and benevolence.

The differences that the Traditionalists, the Shiites, and the Mutazilis had with each other were about economic, political, and social issues as well, but these are outside our present concerns. What we need to know

here is that the Mutazilis came to believe that the essence of God was justice. They liked the idea for its own internal logic but also because it preserved human free will: God could do no injustice or sin, so if there is so
much injustice and sin on earth, it proves that we have the ability to make
our own decisions. This gives us responsibility and demands that we attend
to theories of ethics. The Traditionalists attacked the Mutazilis for making
God too rationalist and embraced the idea of predestination as a response:
God's justice was incomprehensible, it was all in his hands, all else was
hubris. Justice itself, they argued, was a purely human ideal, and we must
not assume that God is beholden to it.

Some Traditionalists claimed that the Mutazilis had removed all religious value from their philosophical concept of God. Traditionalists eventually declared that no rational discussion of God should be permitted. To
many, this seemed extreme. A compromise was struck when Abu al-Hasan
ibn Ismail al-Ashari (878–941), a Mutazili, dreamed that Muhammad
instructed him to turn to the Traditionalist studies. Later, after earnestly
condemning Mutazilis, al-Ashari had another dream of the Prophet and was
this time instructed not to be such a fanatic about it. A peeved Muhammad,
he tells us, appeared and said, "I did not tell you to give up rational arguments, but to support the true hadiths!" The upshot was that al-Ashari
founded the Muslim tradition of Kalam, literally *theology*, which was based
on logic but did not hold God to that logic. The idea was that Muslims
ought to use reason and logic to show that God was beyond human understanding. So there had been Muslim rationalism in the Mutazili sect, and it
had influenced a new, more rationalist mainstream. Yet, at the same time,
among the early Muslims there were a few deeply independent scholars who
doubted almost all the features of God that made him godlike, i.e., that
God was good, that he made the universe, or that he cared about humanity.
They were often referred to as atheists. There are two major figures here: the
amazing Ibn al-Rawandi and the equally surprising al-Razi. But first we
must attend to the rise of Falsafah, the dynamic philosophical movement
that appears in the Muslim world as a sudden flood of Greek texts becomes
available.

One of the reasons the Islamic world spread so far, so fast, was that the
two giant empires it demolished—Byzantium and Persia—had just exhausted
themselves fighting each other in a series of wars. Also, these old empires were
very diverse, so the Muslims took them on in small bursts, in piecemeal fashion. What's more, the considerable population of heretical Christians happily

yielded to the Muslims because, while the dominant Christians sometimes forced the heretics to convert or die, the Muslims merely demanded taxes, not conversion, from other "people of the book"—Jews and Christians. All this conquering meant that in these centuries Muslims were constantly confronted with Judaism, Christianity, Indian religion and philosophy, Manichaeism, Zoroastrianism, and some Greek and Roman ideas. It was this wildly heterogeneous world that was taken over in the Abbasid Revolution, a momentous political uprising in 750. The Abbasid dynasty reigned until the thirteenth century and brought the Muslim world its first golden age. When the Abbasids established their residence in Baghdad, Syrian scholars and doctors—many of whom were Nestorian and Jacobite (Syrian Monophysite) Christians—were invited to live, teach, and work there. When the Syrian scholars came to Baghdad, they brought their texts. Their medical treatises were the first to be translated into Arabic, but soon the focus shifted away from things of pragmatic, immediate use and toward philosophy itself.

Over the course of the ninth century the Arab world began a spectacular program of translation. There was a hunger for the ancient discussion of many areas of intellectual life. The Nestorian Christians, who spoke Greek and Arabic, first translated the books they had. Then they went in search of ancient texts preserved in the dark recesses of the libraries and monasteries of the Eastern Church, and in the monasteries that were outposts of Christianity in the Muslim world. Soon enough, there were Arabic translations of Euclid, Archimedes, Ptolemy, Hippocrates, Galen, all of Aristotle except the *Dialogues* and the *Politics,* Plato's *Timaeus, Republic,* and *Laws,* as well as the works of a few lesser-known figures. There was also something called *The Theology of Aristotle,* which was actually excerpts from Plotinus's *Enneads!* That this seemed reasonable reminds us of the degree to which Neoplatonism had been mixed into everyone's idea of Aristotle. These "new" texts were translated from the Greek directly into Arabic or, often enough, into Syriac and then into Arabic. There were some stars in this effort. A father-and-son team in the ninth century—a Nestorian court doctor named Hunayn ibn Ishaq and his son Ishaq—made more works of Galen, Euclid, and Ptolemy available in Arabic than were ever available in Greek-speaking Byzantium. The father translated the books from Greek to Syriac, and then the son translated them from Syriac to Arabic. Working with these ancient texts, the Arab world made more scientific discoveries than had been made at any previous period in history.

There also arose a philosophical humanism that celebrated the scientific and philosophical heritage of classical antiquity as a cultural and educational ideal.[2] Individualism and literary humanism was cultivated in the arts and politics. This golden age thrived in a Baghdad that had become a remarkably cosmopolitan place in terms of its arts and the urbanity of its citizenry. It was the center of both the Abbasid Empire and, after 945, of Buyid rule. It was in this context that the Faylasufs emerged, lauding Greek philosophy and adopting it as their own, with some Muslim adjustments. Their movement, Falsafah, held that the God of the Greek philosophers was identical to Allah. The Faylasufs came to believe that God was reason itself.

Yaqub ibn Ishaq al-Kindi (died ca. 870) was the first to approach the Koran through Greek philosophy, the first of the great Muslim Aristotelians. It was daring of him to do so, but he justified the act by using the old notion of philosophy as the handmaid of revelation. In a way this was made easier by a terrific interpretive error: the Arabic-speaking world in this period thought that Aristotle and Plato were the same person and were at work reconciling the contradictions in this genius's work. Plotinus's combination of the religious suggestions of Aristotle and Plato fed the error. Al-Kindi wrote that Muslims "should not be ashamed to acknowledge truth and to assimilate it from whatever source it comes to us, even if it is brought to us by former generations and foreign peoples."[3] Al-Kindi used Aristotle's proof of the necessity of a prime mover, but differed with Aristotle (and favored the Koran) in his assertion that the universe was created out of nothing. After al-Kindi, Falsafah seems to have always sided with Aristotle on this matter, despite its direct contradiction of the Koran. In general, the Faylasufs based their own knowledge in reason and disagreed with the Koran without much fanfare. They maintained that the Koran was a valid path to God for those who were incapable of finding their way to truth through reason—but for anyone who could follow the path of reason, well, that was more exquisitely true.

The rise of skepticism among Muslims grew out of the problem of prophecy. Prophecy had a central place in Islam. By the time Islam came along, the Jews had been finished with prophets and prophecy for about eight hundred years. Having broken with Judaism, the Christians were in a more imminent and malleable religious tradition; that is, they believed that new and earthshaking religious information was still being offered by God. That made them more open to prophetic types of speech than were the Jews, but they, too, did not admit contemporary prophets; everyone *knew* that the prophetic age was over centuries ago.

Before the rise of Islam, Jews and Christians did not have much of a literature on how to recognize a true prophet. There was a discussion of it in Deuteronomy, which suggested that a prophet should be attended by miracles, but that was about it. By the ninth century, however, Jews, Christians, and Muslims had each created a body of literature on the topic. As Sarah Stroumsa explains in her *Freethinkers of Medieval Islam,* it seems to have begun with Muslims defending themselves against the angry disbelief of the Jews and Christians, followed soon after by a response from those Jews and Christians.[4] The earliest of this Islamic work has been lost to time, so the prophetology we find in the ninth century was already a mature body of thought. One of the earliest important surviving works is *Proofs of Prophecy* by Jahiz, which gives much attention to the idea that the Koran was so beautiful that it was the miracle that validated Muhammad's word. He also argued that the Koran showed knowledge—knowledge of the history of the Jews, for example—that Muhammad could not have known by natural means. The Christians and the Jews argued against it, of course, but many of these believers ended up defending prophetology against deeper doubt.

Much of this was written in the form of dialog or argument. The role of the disbeliever was generally cast as either the Greek philosophers or, more marvelously, *Barahima,* the Brahmans, whom Muslims saw, essentially, as antiprophetic rationalists.[5] After a while, sometimes the doubter won.

The term *zindiq* started out meaning "secret dualist": calling someone a zindiq meant you suspected him of harboring Manichaean sympathies beneath his public Islamic piety. Manichaeism had been "Hellenized" in the four centuries since Augustine professed it; its intellectuals rationalized the belief system and read the ornate cosmology as allegory. The heresy (*zandaqa*) of Manichaeism was the way of a Muslim with a taste for rationalism. As the period of great Muslim doubters got going, *zindiq* started to mean religious doubters and *zandaqa* started to mean religious doubt. The first person to be executed for zandaqa was Djad ibn Dirham, in 742. He is said to have denied the Muslim concept of a God with attributes, saying, "God did not speak to Moses, nor take Abraham as his friend"; he was also known as a materialist. He had a following who were reputed to believe that Muhammad had lied and who denied the resurrection. Ibn al-Muqaffa was executed in 760 for Manichaeanism and for attacking Islam, its prophet, and its notion of God.

The poet Abu Nuwas was a famous doubter of that time. At his mosque one day, when the imam began to read out verse one of sura 109, "Say: O!

You unbelievers . . . ," Abu Nuwas is said to have yelled out, "Here I am!" When the police hauled him to the inquisitor, a portrait of Mani was produced and Abu Nuwas was told to spit on it. They clearly had the old idea of zandaqa in mind. Abu Nuwas demonstrated the extent of his doubt, answering their challenge by sticking his finger down his throat and vomiting on the picture of Mani. The zindiq Inb Abi-l-awja was executed in 772 for believing in the eternity of the world and denying the existence of a creator. He also had a problem with providence, asking, "Why are there catastrophes, epidemics, if God is good?"

A whole circle of zindiq poets existed in these years. The historian Ignaz Goldziher gives us this picture: "It is reported that at Basra a group of freethinkers, Muslim and non-Muslim heretics, used to congregate and that Bashshar ibn Burd did not forego characterizing the poems submitted to this assembly in these words: 'Your poem is better than this or the other verse of the Koran, this line again is better than some other verse of the Koran, etc.'" The doubters also specifically critiqued the Koran. For instance: "Al-Mubarrad tells of a heretic who ridiculed the parable in sura XXXVII.63, where the fruits of the tree Zakkum in hell are likened to the heads of devils. The critics say: 'He compares the visible with the unknown here. We have never seen the heads of devils; what kind of simile is this?'"[6]

Abu Nuwas relates a comment made by Aban, one of the freethinkers of Basra. It happened just after the call to prayer one day. "Then said Aban: 'How could you testify to [Muslim belief] without ocular demonstration? So long as I live I shall never attest anything but what I see with my eyes.'" At the mention of God having spoken with Moses, Aban supposedly said, "Then your God must have a tongue and an eye. And did He create Himself, or who created Him?"[7]

We come now to the two major authors of Muslim doubt: Ibn al-Rawandi and Abu Bakr al-Razi. We do not know much about Ibn al-Rawandi. Some scholars think he died about 860 and others that he lived until 912 or so. Some have seen him simply as an Aristotelian philosopher and others as a radical atheist. He is not generally understood within the Falsafah tradition, which was just getting under way in his era: he started out as a Mutazili but came to reject his colleagues and then to extend his critique, rejecting ever more fundamental notions of Islam, of revealed religion, and of theism itself. He wrote many mainstream scholarly works and an amazing amount of heretical work. In bold agreement with Aristotle, he supported the eternity of the world, though it meant that God did not create it. This point, which was a relatively small one in Aristotle's work, would become a major part of how everyone thought of him in the Middle Ages—and a real sticking point for all three great monotheist traditions. Al-Rawandi started it. He also took on such positions as: "against the idea that God is wise," "against the Koran," "against Muhammad," and "against all prophets"!

Ibn al-Rawandi's most important book was the very curious *Kitab al-Zumurrud,* or *The Book of the Emerald.* It survives only in detailed descriptions and some quotes in other texts. The book was a conversation between the author and his friend and mentor, the scholar Muhammad al-Warraq. In the beginning of *The Book of the Emerald,* it is the friend who is the most radical, but soon al-Rawandi's doubt goes further than al-Warraq's.

Al-Warraq was Muslim, but Muslim sources call him a Manichaean. He was certainly a doubter. Al-Warraq often referred to God as an idiot, because "He who orders his slave to do things that he knows him to be incapable of doing, then punishes him, is a fool."[8] The provocative and sarcastic role he played was purposefully disruptive, and he did not seem to mind being held in general disrepute, even courting it. Details of al-Warraq's life are vague, but we know that after a while he was persecuted, and after his death his books were banned and destroyed. We know him mostly through the excerpts found in polemics against him.

In *The Book of the Emerald,* al-Warraq explains the problems of prophecy, while his student Ibn al-Rawandi tries to defend it. Al-Rawandi insists, for instance, that Moses and Jesus both predicted the coming of Muhammad. Here's what he has the al-Warraq character say:

> Moses and Jesus did indeed predict the coming of Muhammad; any astrologer can make correct predictions. In the same way, the fact that Muhammad could predict certain events does not prove that he was a prophet: he may have been able to guess successfully, but this does not mean that he had real knowledge of the future. And certainly the fact that he was able to recount events from the past does not prove that he was a prophet [because he could have read about those events in the Bible] and, if he was illiterate, he could still have had the Bible read to him.[9]

Al-Warraq also argued, this time citing a Brahman, that if the prophets' claims support human judgment—if we are capable of figuring out, say, that it is good to be forgiving—then the prophets' claims are unnecessary. If these claims are contrary to what God's gift of intelligence reports, then we should not listen to them. This rips the rug out from under the entire notion of revealed religion. It did not praise the intellect for its capacity to know God but for the wonder of science. Al-Warraq explained that people developed the science of astronomy by gazing at the sky, and no prophet was necessary to show them how to gaze. No prophets were needed to show them how to make lutes, either, or how to play them. Humanity learned by itself that if you remove the intestines of a sheep, dry them, and stretch them over a piece of wood, it can make pleasing sounds when pounded upon. According to Ibn al-Rawandi, we figured all this out through natural intellect, study, observation, and trial and error. We can know the world on our own.

With a stubborn reliance on material explanations and human ability, Ibn al-Rawandi suggests that maybe the Koran is more beautiful than other Arab books because Muhammad was an extraordinary composer of words, or because the other Arabs were too busy fighting off Muhammad to take time to write poetry, or because the Arabs were an uneducated people. He then offers the opinion that the Koran is not that beautiful anyway. The book, he asserts, says contradictory and absurd things and is not particularly impressive to non-Muslims. *The Book of the Emerald* even points out that Muhammad's teachings themselves represent a challenge to revealed religion: he had claimed that all sorts of things that Jews and Christians believed were entirely wrong, that those things had been passed down from their prophets incorrectly. But, wonders Ibn al-Rawandi, if you cannot trust the great multitude of Jews and Christians to get the facts right, why should one trust the handful of Muhammad's followers who passed down the Muslim tradition?[10]

There were events in Muhammad's life that were explained in the Koran as having been determined by the military help of angels. Ibn al-Rawandi wrote that this helps explain why Muhammad did better than one might have expected at Badr, but it begs the question of where these angels were at Uhud, and indeed, why even at Badr they killed only seventy of Muhammad's enemies. The angels should have been able to do better. *The Book of the Emerald* actually assigns a kind of cheating to the prophets. They were not deluded or mistaken; they were actively faking, using tricks and sleight of hand to fool their audience. In a delightful praise of rationalism, the book argues that the prophets used weird and little-known natural phenomena— like magnets but less famous—to defraud their followers. *The Book of the Emerald* also criticizes prayer, concern for ritual purity, and all the ceremonies of the hajj: walking around a great stone, it claims, cannot really help any-one. It even asks why the Ka'ba "is better than any other house." Ibn bows to his friend's ideas at the end of the book.

Soon al-Rawandi's doubt was thoroughgoing. He was known to system-atically write books and then to write their refutation. He is also said to have "composed for the Jews the *Al-Basira,* refuting Islam, for the sum of 400 dirhams, which as I have heard, he received from the Jews of Samarra. Having collected the money, he contemplated writing a refutation of it, until they gave him an additional hundred dirhams, for which he abstained from writing the refutation."[11] We do not know if he was directly influ-enced by the Skeptic tradition, but his was a new voice in its canon. Among his other books, Ibn al-Rawandi wrote *Against the Koran* and a little treatise called *The Futility of Divine Wisdom.* In their content and tone, these books seem to mark a complete break with Islam. He mocks the philosopher's idea of God as "universal force" that does not know how to add two and four to get six; he concludes that if the world's events are the result of willful action, then God must be a wrathful, murderous enemy.

The writer al-Hayyat, who died about 913 CE, had this to say about Ibn al-Rawandi: "He disputed the reality of the miracles of Abraham, Moses, Jesus and Muhammad and claimed that they were fraudulent tricks and that the people who performed them were magicians and liars; that the Qur'an is the speech of an unwise being, and that it contains contradic-tions, errors and absurdities."[12] Further, he said that, according to Ibn al-Rawandi, a God who inflicts illness upon his slaves cannot be counted as one who treats them wisely, "nor can he be said to be looking after them or to be compassionate toward them. The same is true concerning he who inflicts upon them poverty and misery. Also unwise is he who demands obedience from a person who he knows will disobey him. And He who punishes the infidel and disobedient in eternal fire is a fool."[13] Al-Rawandi grew contemptuous of those who reconciled innocent human suffering with a caring, benevolent God, and that contempt got him into trouble. He was so hated that by the eleventh century it was difficult to find any of his manuscripts. Yet this vibrant and inventive voice in the history of doubt was never quite forgotten.

We cannot say that Ibn al-Rawandi was Falsafah, for they did not claim him, nor he them. He was just a little further into apostate than internal critique. Yet there was a man who was as radical in his doubt but who was tolerated by the establishment and able to help create and animate the rationalist Islamic movement: Abu Bakr al-Razi, the other star of early Islamic doubt. Before we look at him directly, let us listen to what people have said of him: al-Razi has been called "the greatest nonconformist in the whole history of Islam," "the most free-thinking of the major philosophers of Islam," "the least orthodox and the most iconoclast," "perhaps the single figure most frequently denounced and disapproved as a heretic in the subsequent history of Islamic thought," and in the words of a recent study of his work, "If, however, the sayings attributed to him are authentic, his doctrine emerges as irreconcilable with any kind of Islam, however open-minded."[14] Yet al-Razi was beloved.

The reason one was hated and the other cherished was simple: where Ibn al-Rawandi was provocative, sarcastic, and happy to play the role of the outsider, correcting the gullibility of the majority, al-Razi was a doubter who devoted himself to his community's well-being and grew famous for his generosity, intelligence, and skill. He was a doctor and has been called the most creative genius of medieval medicine. He has also been lauded as both a philosopher and a chemist and he served as the director of hospitals in his native Rayy, Iran, and at Baghdad. Al-Razi's books became classical works on Arabic medicine, and a few were well known in the West; his study of pathology and therapy was translated into Latin and long served as a teaching manual in various universities in Europe. He was often called "the Arab Galen," a name he earned not through sycophantic repetition of Galen's doctrines but by representing Galen's dedication to experimentation and observation.

Al-Razi's dates are usually given as 854–925, which means he flourished at the same time that Islamic thought was reaching maturity in various arenas: it was the zenith of Mutazilism; the first phase of the translation movement was being completed; and Islamic interest in Neoplatonism was at its peak. The period also saw the crystallization of Falsafah, as al-Razi is often considered the first true Faylasuf. The names of three of his books were *The Prophet's Fraudulent Tricks*, *The Stratagems of Those Who Claim to Be Prophets,* and *On the Refutation of Revealed Religions.* In them, al-Razi asked groundbreaking questions about prophets: "On what ground do you deem it necessary that God should single out certain individuals" by giving them prophecy, "that he should set them up above other people, that he should appoint them to be the people's guides, and make people dependent upon them?"[15] How could it be possible that a God would choose this method, since it invariably incites people against one another, spreads hostility, and increases fighting? The "most fitting" behavior of the Wise One would be to give everyone the same necessary knowledge. "He should not set some individuals over others, and there should be between them neither rivalry nor disagreement which would bring them to perdition."[16] It is a rotten situation, and no planning, compassionate God would invent it.

With such a variety of religions, al-Razi said, anyone could have foreseen that "There would be a universal disaster and they would perish in the mutual hostilities and fightings. Indeed, many people have perished in this way, as we can see." Al-Razi also wrote, "If the people of this religion are asked about the proof for the soundness of their religion, they flare up, get angry and spill the blood of whoever confronts them with this question. They forbid rational speculation, and strive to kill their adversaries. This is why truth became thoroughly silenced and concealed."[17] Here and elsewhere, his language is festooned with references to proof, soundness, questions, and rational speculation. He saw religious people as having been originally duped by authority figures and explained that they now continued to conceal the truth

> as a result of their being long accustomed to their religious denomination, as days passed and it became a habit. Because they were deluded by the beards of the goats, who sit in ranks in their councils, straining their throats in recounting lies, senseless myths and "so-and-so told us in the name of so-and-so . . ."[18]

This passage alone has been described as "the most violent polemic against religion in the course of the middle ages."[19] It may well have been, as it has a sharp blade with its smiling "so-and-so's." Al-Razi was able to get away with this ruthless religious critique because he did so much more: most people who wrote about him concentrated on his impressive body of scientific work and simply left his strange, extreme antireligious arguments to the side.

Al-Razi thought the variety of religions was a good proof that none of them had it right: "Jesus claimed that he is the son of God, while Moses claimed that He had no son, and Muhammad claimed that he [Jesus] was created like the rest of humanity." What is more, "Mani and Zoroaster contradicted Moses, Jesus and Muhammad regarding the Eternal One, the coming into being of the world, and the reasons for the [existence] of good and evil."[20] He picked apart the Christian claims about Jesus and mocked the Torah's imagining of God as an angry old man, "white-haired and white-bearded."[21] Not only that, the Hebrew God frequently asked for sacrifices of particular quality and quantity. "This sounds like the words of the needy rather than of the Laudable Self-sufficient One."

Abu Bakr al-Razi claimed that the supposed miracles of prophets were not good evidence of their validity. For one thing, "similar feats were performed by people who made no claim of prophethood," and he mentioned all sorts of feats by wonderful but purely human tricksters, from juggling and legerdemain, to dancing on spikes and walking on spears, to the rhyming speech of oracles and soothsayers. As for the Koran, it is best to hear the diatribe in al-Razi's own words:

> You claim that the evidentiary miracle is present and available, namely, the Koran. You say: "Whoever denies it, let him produce a similar one." Indeed, we shall produce a thousand similar, from the works of rhetoricians, eloquent speakers and valiant poets, which are more appropriately phrased and state the issues more succinctly. They convey the meaning better and their rhymed prose is in better meter. . . . By God what you say astonishes us! You are talking about a work which recounts ancient myths, and which at the same time is full of contradictions and does not contain any useful information or explanation. Then you say: "Produce something like it?!"[22]

Al-Razi wasn't really against ancient myths; he liked them, but he did not think the Koran qualified as a true mythology. A true mythology, he held, ought to read like a riddle, an allegory whose contemplation would lead to higher philosophical speculation, and in his view the Koran did not. He was not just trying to correct religion, though. Al-Razi made it clear that he thought all supposedly revealed religions were a disaster for humanity, because they led to bloodshed and because religious authorities tended to be cruel and despotic.

For al-Razi, human beings could negotiate existence on its most profound level without religion. How? "No soul can be purged from the turbidity of this world and escape to the next, except by contemplating philosophy. If a person contemplates philosophy and comprehends anything, be it ever so small, his soul is purged from this turbidity and is saved."²³ Advice like his could not be meant for oneself alone. Al-Razi knew a cosmopolitan doubting world and was offering comfort to its members. He was treated as a hero by them long after his death.

The Faylasufs in general were rationalists, and they belong to the history of doubt, but compared to al-Razi most were pretty tame. In fact, they came to accept prophetology: the intellect was understood as extraordinarily powerful, but for most people it was never going to be enough, and that's why the prophets were needed. The Faylasufs were accused of writing "as if" they were denying prophecy, and it was often said that their doctrine "amounted to" being against prophecy, but whereas al-Rawandi and al-Razi both denied the fundamental tenets of Islam, the Faylasufs found ways to support many of them. We should note that these were cosmopolitan, rationalist times and religious communities shared the mood. The Ismailis and their subsect "The Brethren" dedicated themselves to science and mathematics in order to discover the hidden meaning of life. The Brethren wrote that in search of the truth, we must "shun no science, scorn no book, nor cling fanatically to a single creed."²⁴ They were deeply influenced by Neoplatonic texts, even rejecting the Koran's claim of ex nihilo creation in favor of the Platonic idea of emanation. As I have mentioned, the Faylasufs also disagreed with the Koran on this issue, in favor of Aristotle's eternity. There was a lot of room for doubt.

The greatest of the Faylasufs was Abu Ali ibn Sina, known in the West as Avicenna (980–1037). Like Abu Bakr al-Razi, Avicenna was a famed medical doctor. He was born into a Shii family and was later attracted to Falsafah. After he cured the sultan of Bukhara, the sultan hired him and gave him access to a considerable library. He attributed his Neoplatonist understanding of the world to the reading this allowed him. Arguably, before Avicenna, Falsafah was an Aristotelian philosophical movement that took place within the Muslim world but was not really a Muslim movement, not a movement within the religion. Avicenna saw that Falsafah seemed elitist to people but believed it could save them from the mundane world, from fear of death, and from confusion, if only they could understand it. But even if the people could understand it, it was not human enough for them: there was no way to pray for things or talk to God or other spiritual beings, no face to imagine or arms to lean on, and there was no afterlife. Avicenna understood Aristotle in terms of the Neoplatonist idea of emanation—divine thought emanating the world into being—and declared that, in this sense, it was philosophically sound to speak in terms of a created world. It was not exactly the way it was presented in the Koran, but to some at least, it seemed closer.

This emanation idea allowed for a philosophical God who was not a personality but who was in the world and in some vague sense knew the world, since, after all, the world was an emanation of him. One could thus commune with this God in a way that was impossible if God were Aristotle's prime mover. In fact, the idea continued, sometimes this emanation is sufficiently powerful to allow for prophetic speech. But that did not quite mean this would be speech that was literally true. Thus Muhammad's abilities as a prophet allowed him to speak real truth in human terms, so that the average person could understand it. Just as important, Avicenna found a way to speak of an afterlife that also came from a Neoplatonic reading of Aristotle.

As far as he was concerned, anyone who was capable of philosophy was called to do it, and would want to—it was a moral responsibility to seek truth, but it was also the best game in town. God *was* understandable intellectually, and that was the sweetest and highest way. Intellectuals thus had access to some of the joys of revealed religion without entirely contradicting their rationalism, and mystics could support their otherwise antirationalist experiences with philosophical argument. Falsafah reigned for a couple of centuries in an efflorescence of arts and letters, learning and science, commerce and cosmopolitanism.

Islamic literary tradition often says that the three worst zindiqs of Islam were al-Rawandi, al-Tauhidi, and al-Ma'arri. We have met al-Rawandi, the philosopher; the other two were poets. Al-Tauhidi left us few overtly heretical ideas; he may have written more incendiary things that have been lost, or perhaps his interest in Greek philosophy and science was passionate enough to seem like zandaqa on its own. As for Abdallah al-Ma'arri (973–1057), he was alive at the end of the movement and was a fabulous character in the history of doubt. He was born in Syria, contracted smallpox as a child, and eventually went blind from it. We should meet him through his verse:

> By fearing whom I trust I find my way
> To truth; by trusting wholly I betray
> The trust of wisdom; better far is doubt
> Which brings the false into the light of day.[25]

Al-Ma'arri specifically lauded doubt. His criticisms also echoed Xenophane's theme that people believe what they are brought up to believe: "Our young man grows up in the belief to which his father has accustomed him. / It is not Reason that makes him religious, but he is taught religion by his next of kin."[26]

> *They recite their sacred books, although the fact informs me that these are a*
> *   fiction from first to last.*
> *O Reason, thou (alone) speakest the truth. Then perish the fools who forged*
> *   the religious traditions or interpreted them!* [27]

> *O fools, awake! The rites ye sacred hold*
> *Are but a cheat contrived by men of old*
> *Who lusted after wealth and gained their lust*
> *And died in baseness—and their law is dust.*[28]

What did he advise then? "Devout is he alone who, when he may / Feast his desires, is found / With courage to abstain."[29] Al-Ma'arri also enjoyed reminding his readers that the sacred stones in Mecca, "visited and touched with hands and lips," in actuality "are stones that once were kicked."[30] He was also interested in the cosmopolitan argument: "They all err—Moslems, Christians, Jews, and Magians"—because there are only two types of "Humanity's universal sect": "One man intelligent without religion / And one religious without intellect."[31] What's more: "The Christian, as more anciently the Jew / Told thee traditions far from proven true."

> *The Christians have lied concerning the Son of Mary*
> *The Jews also lied concerning the Son of Amran.*
> *And never the Days have brought forth new in nature,*
> *Nor ever did Time depart from his ways accustomed.*[32]

> *Religion and infidelity, and stories that are related, and a Revelation that is*
> *   cited as authority, and a Pentateuch and a Gospel.*
> *Lies are believed amongst every race; and was any race ever the sole*
> *   possessor of Truth?*[33]

He wrote that "If a man of sound judgment appeals to his intelligence, he will hold cheap the various creeds and despise them,"[34] and thought only physical punishment could have ever made human ancestors accept religion in the first place:

Had they been left alone with Reason, they would not have accepted
a spoken lie; but the whips were raised (to strike them).
Traditions were brought to them, and they were bidden say,
"We have been told the truth"; and if they refused, the sword was
drenched (in their blood).[35]

The period of Falsafah, which supported such doubt, did not last forever. It broke down at the end of the eleventh century, in part because of the actual psychological breakdown of one of its followers, Abu Hamid al-Ghazzali.

Al-Ghazzali (1058–1111) pushed Falsafah so far that he became not only a doubter in the rationalist tradition but also a doubter in the tradition of the dark night of the soul. When he arose from that dark night, it was not to bathe in the light of the intellect but to seek and cultivate a light from within. At thirty-three, al-Ghazzali was the director of a prestigious mosque in Baghdad, but he did not have a settled mind. He wanted to know God for certain, writing that he had "poked into every dark recess . . . made an assault on every problem . . . plunged into every abyss . . . scrutinized the creed of every sect" in his effort to know the truth.[36] He wrote a book, *The Opinions of the Philosophers,* that respectfully and accurately described Aristotle's thoughts on God as derived by Avicenna and others. He later wrote a book called *The Incoherence of the Philosophers.* In it he stated that he'd found twenty propositions of Aristotle that were not sufficiently demonstrated, and he spent most of his energy on these philosophical problems. But he also noted that the God of Falsafah, the God that seemed to him to be the God demonstrated by Aristotle and Plato, was not very Islamic and was not very helpful. With no bodily resurrection and with a God that does not really know individuals, there was not much hope here of a religious nature. Not only that, but he thought the emanation idea was actually another way of thinking that the world was eternal, even if the Faylasufs denied it.

This rejection of philosophy was joined by a rejection of all other conceptions of truth that he had encountered. In one historian's words, "Al-Ghazzali was as aware as any modern skeptic that certainty was a psychological condition that was not necessarily objectively true."[37] He spoke passionately against the Muslims who claimed to know God through present or hidden imams; and against the Sufis, who believed they knew God through their mystical practices; and Falsafah, who claimed to know God through their intellects—for how did any of them know that their imams, or visions, or proofs were really true? He could not stop struggling for certainty, and in or about the year 1094 he came to a crisis. His breakdown had physical dimensions: he could not swallow or move his tongue. In his *Deliverance from Error,* he recorded that he had been brought to the edge of Skepticism because of the failures of the proofs of God.

When he was an adolescent, he explained, he noticed "that Christian youths always grew up to be Christians, Jewish youths to be Jews and Muslim youths to be Muslims."[38] So what was belief? He set out to find the truth through study and concluded that the knowledge he had been given in his youth was not sufficiently grounded. Thus he found himself relying on nothing but sense perception and what seem to be "necessary truths"; but soon "Doubt began to spread here and say: 'From where does this reliance on sense-perception come?'" A shadow looks to our eyes like it never moves, yet it moves, and not all that slowly. The sun looks like it's the size of a coin, "yet geometrical computations show that it is greater than the earth in size."[39] He also asked, given the fact of dreams, "Why then are you confident that all your waking beliefs, whether from sense or intellect, are genuine? They are true in respect of your present state; but it is possible that a state will come upon you whose relation to your waking consciousness is analogous to the relation of the latter to dreaming. In comparison with this state your waking consciousness would be like dreaming!" Al-Ghazzali proclaimed that "the disease was baffling" and lasted almost two months, "during which I was a sceptic."[40]

He then came to describe philosophy as having four camps: the ones who "derive truth from the infallible imam"; the philosophers; the scholarly theologians; and mystics such as the Sufis.

> I said within myself: "The truth cannot lie outside these four classes. These are the people who treated the paths of the quest for truth. If the truth is not with them, no point remains in trying to apprehend the truth. There is certainly no point in trying to return to the level of the naïve and derivative belief once it has been left, since a condition of being at such a level is that one should not know one is there; when a man comes to know that, the glass

of his naïve beliefs is broken. This is a breakage which cannot be mended, a breakage not to be repaired by patching or by assembling of fragments."[41]

His assumption that the choices supplied by his historical moment were the only important ones is instructive. The second thought was more impressive: no one had ever recorded the psychology of doubt like this, recognizing a level of belief as necessarily unnoticed by its proponent.

Al-Ghazzali came to believe that all the various philosophers were affected by the "defect of unbelief," especially the materialists.[42] He offered this description of them as people who think the world has been eternally here "of itself and without a creator, and that everlastingly animals have come from seed and seed from animals; thus it was and thus it ever will be."[43] It was intended to dismiss the naturalists in his midst but acts as a lovely testament to their existence. Al-Ghazzali liked Aristotle best of all philosophers, "yet he too retained a residue of their unbelief . . . we must therefore reckon as unbelievers both these philosophers themselves and their followers among the Islamic philosophers, such as Avicenna." What must we read then? Well, we must read carefully. Even math can be dangerous, and again his warning gives witness to the state of doubt around him:

Every student of mathematics admires its precision and the clarity of its demonstration. This leads him to believe in the philosophers and to think that all their sciences resemble this one in clarity and demonstrative cogency. Further, he has already heard the accounts on everyone's lips of their unbelief, their denial of God's attributes, and their contempt for revealed truth; he becomes an unbeliever merely by accepting them as authorities, and says to himself, "If religion were true, it would not have escaped the notice of these men since they are so precise in this science." Thus . . . he draws the conclusion that the truth is the denial and rejection of religion. How many have I seen who err from the truth because of this high opinion of the philosophers and without any other basis![44]

It's a remarkable record of doubt. "Few are those," he concludes, "who devote themselves to this study without being stripped of religion and having the bridle of godly fear removed from their heads." His advice was that only the most wise should read this material, "Indeed, just as the snake-charmer must refrain from touching the snake in front of his little boy," the wise should not even talk about philosophy among the people.[45]

Yet, with all his ability to reject philosophy, al-Ghazzali took a long time to convince himself to leave his teaching position to search for truth with the Sufis. The memoir reads a lot like Augustine's, without the battle with sexual desire. Al-Ghazzali just wanted to stay in the world, in his prestigious position. That's when "God caused my tongue to dry up so that I was prevented from lecturing." Worldly desires tugged at him, and he worried that as soon as he gave up his possessions and positions he would wish he had them back. When he finally joined the mystics, he explained that the difference between reading about God and having an ecstatic experience of him is exactly like the difference between reading about alcohol and being drunk. He did this for two years, but because he worried about his family and other responsibilities, he came back into the world and worked in it for another ten years in an "impaired" solitude. Admitting that he had "experienced pure ecstasy only occasionally," he also said that "innumerable and unfathomable" things had been revealed to him in his periods of solitude. He indicated that there were precise stages of visions and revelations, that those stages could only be experienced; it was impossible to explain in words. Do not try to prove your worldview to anyone through miracles or reason, was the message, for "then your faith is destroyed by an ordered argument showing the difficulty and ambiguity of the miracle." Go be a mystic and prove the truth to yourself.

In *The Incoherence of the Philosophers,* al-Ghazzali also argues against the reality of cause and effect, pointing out that nothing about the fact that certain things seem to precede other things is sufficient to prove that the one causes the other—not even "quenching thirst and drinking . . . burning and contact with fire, light and sunrise, death and decapitation, . . . relaxing the bowels and taking a purgative, and so forth." He says his skeptic approach to cause and effect is meant to make room for God, who he insists is the real cause of these effects and who could easily make a decapitated person live, or not allow drinking to slake thirst. But he managed to demonstrate how all rational belief may be reduced to confusion. Al-Ghazzali was a tremendous force in medieval Islam. His biographers attribute hundreds of works to him, and his books were very popular. He had made a cogent claim for the weakness of seeking truth through philosophy, and many followed his lead. Henceforth, until modernity, Muslim theology would be based in authoritative texts and in mysticism, except for one last breath of Aristotelianism. It came so late, in this context, that it did not have much effect among Muslims—but translated into Hebrew and Latin, it was to have quite an influence on humanity over the course of many centuries.

Abu Walid ibn Ahmad ibn Rushd, known to the West as Averroës, lived from 1126 to 1198 and was to formulate a spiritualist rationalism that transformed Western theology and philosophy. Living and working in Morocco and Spain, he did much to bring philosophy back to the West. His commentary on Aristotle was so significant that when Aristotle was translated into Latin from the Arabic in which it had long existed, Averroës' response to each idea was recopied along with it. Throughout the Middle Ages in the West, for hundreds of years, "The Philosopher" and "The Commentator" meant Aristotle and Averroës. The bulk of Averroës' work was this commentary, and in physics or metaphysics Averroës differed from Aristotle freely, offering his own interpretations. But first he wanted to be sure he was starting with the real Aristotle. Averroës saw that Avicenna's Aristotle and Neoplatonism could be separated, and his commentaries on Aristotle consistently show the reader how to disentangle the real Aristotle from what was by then the culture's common understanding of him. The God that Averroës spoke of, then, was not the Neoplatonic great unity of the universe, which at least sort of jibed with Islam, but rather the Aristotelian prime mover, which did not know the world or its individuals. This last he softened a bit by suggesting that since we do not know anything about God's kind of knowledge, he may know us in some sense. There was no individual immortality. Averroës teased Avicenna for having conceded too much to Islam.

Along with his commentaries Averroës wrote a number of treatises on philosophical matters, one of which was an answer to al-Ghazzali's *The Incoherence of the Philosophers,* which he titled *The Incoherence of the Incoherence.* Another was a defense of rationalism, also essentially a counter-attack on al-Ghazzali's call for an end to philosophy. As Averroës saw it, the Koran had commanded "Reflect, you have vision" (Koran LIX, 2).[46] Based on this "we are under an obligation to carry on our study of beings by intellectual reasoning. It is further evident that this manner of study, to which the Law summons and urges, is the most perfect kind of study using the most perfect kind of reasoning; and this is the kind called 'demonstration.'" For Averroës, any Muslim who could grasp philosophy must study it diligently. "It is preferable," wrote Averroës, "for anyone who wants to understand God . . . to have first understood the kinds of demonstration" and their varying conditions of validity, along with the errors of fallacious reasoning. If someone ended up in a muddle having studied the philosophers, perhaps "owing to a deficiency in his natural capacity," it was a pity, but it "does not follow that one should forbid them to anyone who is qualified to study them." Al-Ghazzali's call to outlaw philosophy "is like a man who prevents a thirsty person from drinking cool, fresh water until he dies of thirst, because some people have choked to death on it."[47]

What if philosophy and the Koran clash? Averroës says we have to see that they do not. For instance, Aristotle said the world was eternal and the Koran said God created it—but the Koran did not say how God created it and he may have done so in a way that did not begin at any specific moment. This insistence that demonstrative truth must mesh with the Koran definitely favored demonstrative rationalism. "This being so, whenever demonstrative study leads to any manner of knowledge" the subject is "either mentioned in Scripture or not. If it is unmentioned there is no contradiction." If scripture does speak of it, "the apparent meaning of the words inevitably either accords or conflicts with the conclusions of demonstration about it. . . . If it conflicts there is a call for allegorical interpretation of it."[48]

When he goes on to explain that allegory is an expanded metaphorical interpretation of something, he asserts that the Arabic language does this of necessity, calling things and acts by things and acts that they resemble. Allegory is everywhere in thought and belief; we just have to figure out where it belongs and to what degree. After all, he says, law is derived from scripture, but everyone knows it has to be interpreted for any given case, and everyone knows it is modified to fit the times. "Now if the lawyer does this in many decisions of religious law, with how much more right is it done by the possessor of demonstrative knowledge!"[49] There could hardly be a more important observation in the history of doubt.

Averroës wrote that he was annoyed that al-Ghazzali had spoken and written about such things publicly, and said he was sorry that he had to follow suit in order to refute al-Ghazzali. Averroës thought it best if those of middling intelligence did not know anything about philosophy. But since al-Ghazzali had told the religious that philosophy was a dangerous trap, they hated it, and as a result, philosophers scorned religion in return. Al-

Ghazzali had given allegorical readings to the masses, so that they could see that scripture was true in these terms. Averroës said the masses did not need to know about allegory—the masses should simply believe the literal text. Allegories should be discussed only in "demonstrative books" so that they are "encountered by no one but men of the demonstrative class." Al-Ghazzali, he said, had shown them to everyone because "he wanted to increase the number of learned men, but in fact he increased the number of the corrupted not of the learned! As a result, one group came to slander philosophy, another to slander religion."[50] Averroës was a major early voice in the debate over philosophers' sharing their doubt with the rest of humanity.

The period of leaning on Aristotle for Muslim theology had passed, but Averroës would have considerable influence on Jews and Christians. Falsafah texts were read throughout the Muslim empire, which included a great many Jews and Christians. It was the Jews who first took inspiration from the Muslims and began a rationalist, philosophical interpretation of Judaism—in Arabic.

THE FIRST RABBI ON THE MOON

Judaism in the medieval period allowed so little doubt that the rabbis told their own Job story, leaving out the whole rebellion. The first speculative philosopher of Judaism was Saadia ben Joseph (or ibn Joseph), who lived from 882 to 942 and counted himself a Mutazili as well as a Talmudist. As a Jew born in Egypt—we are at about five o'clock on the Mediterranean Sea—ben Joseph lived among a great variety of Neoplatonists, Aristotelians, Jews, Zoroastrians, Christians, and Muslims who freely taught their beliefs, argued, and announced final refutations. Ben Joseph suggests that there was a good deal of winking going on already by his time, as people began to doubt that anyone had it right. Charmingly, he cited his own tradition's ancient doubt in the first paragraph of his *Book of Doctrines and Beliefs:* "We all seek to probe this distant and profound matter which is beyond the grasp of our senses, and regarding which it has been said by the wise king, 'That which was is far off, and exceedingly deep; who can find it out?' (Eccl.7.24)."[51] (It was still thought that King Solomon had written Ecclesiastes—which is why ben Joseph calls him "the wise king.") Before he got going with his philosophy, ben Joseph said that no one, neither the philosophers nor the champions of revealed religion, had based their system on certain knowledge. Everyone was fudging at the edges: people who argue for an infinite universe have never seen anything infinite; people who argue for atoms—for a thing that does not itself have "hot or cold, moist or dry . . . but which becomes transformed by a certain force and thus produces those qualities"—never saw anything like that. He offers his philosophy as best because, he says, (1) his arguments "are stronger than theirs," (2) he can disprove his opponents, and (3) he has the testimony of the scriptures on his side.

Although mixed with a great scattering of snippets from the Hebrew Bible, his philosophy has the feel of an Aristotelian argument, but he uses the style to support Jewish doctrine (like creation) over Aristotle. He quotes Job saying things such as "I will fetch my knowledge from afar." The mythological biblical stories have nothing to do with it, nor the biblical descriptions of God; the only true thing we can say about God is that he exists.[52]

We do not know whether ben Joseph was the first Jew to write like this, but he was the first we can read. There is evidence that ancient Jews were attracted to all kinds of Greek philosophies, from the Platonic and Aristotelian to the Epicurean, Skeptic, and Stoic, but this evidence is usually in the form of an absent enemy: traditional Jews tell us of their struggle with these people. The closest thing to secular philosophy the Jews had produced was Ecclesiastes, and that sounded more like Epicurus than Aristotle—it advised on our situation; it did not painstakingly work out proofs. Ben Joseph inaugurated the new era in which the way of the Greeks was finally allowed to occupy part of mainstream Jewish thought (this after the heirs of the Greeks had been speaking of the Jewish God for more than a thousand years). Ben Joseph saw the search for philosophical truth as a mitzvah, a religious obligation, and this idea would remain a part of Judaism—and doubt.

Along with the Aristotelians, there were Jewish Neoplatonists in the Middle Ages, too. Solomon ibn Gabriol (ca. 1022–ca. 1051) was outstanding among them. His *Fountain of Life* was pure philosophical speculation, unconcerned with theology. An odd thing about this book, which was written in Arabic, is that it was preserved only in a Latin translation made in the middle of the twelfth century. Because of the complete absence of biblical or rabbinic citations in it, medieval Christians thought the book was written by a Muslim or a Christian Arab. That this could occur is a bold indication that philosophy had a life of its own in this period. Ibn Gabriol was borrowed from and quoted approvingly by such luminaries as Albert the Great, Thomas Aquinas, and Duns Scotus. Jews, meanwhile, knew him almost only by his poetry.

The most important figure of the Jewish response to Falsafah was Rabbi Moses ibn Maimon—known as Maimonides (1135–1205). Maimonides was a talmudic scholar and a philosopher, as well as a medical doctor of great renown, who lived at the same time as Averroës, both in Cordova, the capital of Muslim Spain—about eight o'clock on the map of the Mediterranean. Like Averroës, Maimonides was thrilled by the ideas of Falsafah and wanted to study deeply to find meaning. Maimonides, however, was run out of Spain by a Muslim sect for being Jewish. The era of Muslims, Jews, and Christians living peaceably together in Spain was going to last awhile, but it was often interrupted. Maimonides' first books were about Jewish law; his *Mishneh Torah* was a code of Jewish law intended to guide Jews on how to behave in all situations just by reading the Torah and this code, without having to hunt through the Talmud for specific examples. It became a standard guide to Jewish practice and is still the basis of orthodoxy. In the Middle Ages it was said of him that "From Moses to Moses there was no one like Moses"—he was a religious man and one of the greatest Jewish legal scholars of all time, but he also offered brilliant beginnings to the philosophy of secular Judaism.

His third great work, the famous *Guide for the Perplexed,* is a key work in the history of doubt. He wrote this book specifically for people who had studied Jewish law and then studied ancient philosophy and who were upset by what philosophy suggested about the anthropomorphic God described in the Hebrew Bible. It is a notoriously peculiar book because Maimonides was living in a world in which secret knowledge was supposed to be kept secret. Thus he explains at the beginning that the book is deep and tricky, not what it seems. Only the deepest and trickiest minds would know what was being said. Not surprisingly, people have been arguing over what the *Guide* means ever since he wrote it. What Maimonides expressed was essentially a midway position between belief in prophecy and belief in rationalism. As he told it, before he came along the great sages explained our laws and beliefs as such: some have good reasons that we can see, and some have good reasons that are over our heads. Maimonides disagreed. Some of it was for health; some was for politics (such as keeping people scared of divine retribution); and some was for peace of mind, to help you cultivate your higher qualities. As for what God wants, we could not begin to wonder what God wants. Keeping the laws is a simple human thing, and there is no reason to think God cares about it. Maimonides is the first Jew we have on record as giving this kind of secular, political, and psychological explanation for the Jewish way of being. Note also that he echoed the Greek idea of the myths of religion as social control.

> Scripture further demands belief in certain truths, the belief in which is indispensable in regulating our social relations; such is the belief that God is angry with those who disobey him. . . .
>
> In some cases the law contains a truth which is itself the only object of that law. . . . In other cases, that truth is only the means of securing the removal of injustice, or the acquisition of good morals; such is the belief that God is angry with those who oppress their fellow-men . . . or the belief that God hears the crying of the oppressed . . .[53]

Maimonides was a beloved physician. He regularly worked twelve-hour days tending the crowds of patients who gathered in his courtyard, giving his last prescriptions from a daybed, too tired to stand.

What did Maimonides think about God? Well, he knew that there were "those who do not recognize the existence of God" and instead "believe that the existing state of things is the result of accidental combination and separation of the elements, and that the Universe has no Ruler or Governor."[54] These people, he says, are "Epicurus and his school, and similar philosophers." But he dismisses them with a wave, "It would be superfluous to repeat their views, since the existence of God has been demonstrated." Aristotle had this right, and there was no need to worry over it: something "thought" this world into being and put it into motion. For the history of doubt, the key point here is that Maimonides was well aware of Epicurus, of atheism, of other doubters, and of the idea of a world run entirely by chance.

Maimonides was terrific at bearing uncertainties. He was in favor of the Aristotelian idea of the eternity of the universe and also in favor of the biblical creation ex nihilo, and in the end opted for a respectful shrug on the issue. How did he find a space for biblical creation? Aristotle, he explained, had based his conclusion on an analogy between the world we know and the vast universe that is hidden from us by time, space, and our conceptual limitations. There is no reason to hope these are similar enough to bear analogy to one another. He said that Aristotle had only been offering a good guess, a fact most people missed. It is unusual for things to pop into being, yes. But maybe that is how it happened. As for creation, of course it might just be a myth, but all Maimonides held himself to accomplish was to show that Aristotle could, conceivably, be wrong on the subject. By arguing that Aristotle had made a false analogy, from the knowable world to the unknowable universe, Maimonides had accomplished this to his satisfaction. He handled Cicero's and Augustine's problem about what God did while waiting around to create the world with Augustine's answer that time came into being only with the world, as an accidental consequence of motion.

So creation was possible, yet the only good argument for it was that "prophecy" supported it. Maimonides, well informed of the Muslim doctrines reconciling Aristotle with the Koran and prophecy, believed that prophetic knowledge was real knowledge, coming from the imaginative faculty of the mind. His is an awfully human-based, philosophical prophecy, though. Speaking of the idea of God as a creator, Maimonides said, "Abraham our Father was the first that taught it, after he had established it by philosophical research."[55] Abraham the philosopher! Rationalist Jews were now not wrong to believe in creation, and these words were championed by Jews and Christians for centuries. The great love people had for the *Guide,* generation after generation, suggests that there were a lot of people who would have described themselves as "perplexed."

Maimonides developed a strange new method for speaking about God. According to good philosophy, we cannot know anything about God other than that he exists, so really, nothing can be said. Maimonides' idea was to phrase everything one says about him in the negative. "The Torah," he quotes, "speaks in the language of the son of man," and it does so to be easily understood. Actually, he explains, we know God has no corporeal body, is not involved with us, and could not be understood in terms of human traits. Of him we may only say: God is not weak, God is not strong, God is not wise, God is not great. To say that he was wise, after all, would be like meeting a king famed for his stockpiles of gold and complimenting him on having some silver. "Is this not an offense to Him?"[56]

The more thoroughly you negate attributes of God, promises Maimonides, the more you will come close to him, for you will be meditating on his unknowability. Maimonides had a lot of fun showing how one could come to know objects or phenomena, such as a ship or fire, by negative description, but here he whittled away at attributes until you knew, for instance, that you had a large, hollow, wooden thing that floats. His point about God, however, was not a clever way to figure out positive attributes. It was a purposeful communing with their absence. The ancient Temple of Jerusalem had had an empty chamber as its holiest shrine. Now the unimaginable became unimaginable in a whole new way. Maimonides said, "Do not desire to negate merely in words." You have to really mean it. For example: "It follows necessarily that He exists, but not according to the notion of the existence which is in us." As far as what we mean by existing, God doesn't.

"He who affirms that God . . . has positive attributes . . . has abolished his belief in the existence of the deity without being aware of it." If someone said that the taste of chocolate is, say, blue, you would not argue about the color, you would explain that taste is not visual. The idea of a God who could be described is completely wrong. In a great device, Maimonides asks us to imagine a man who knows that the word *elephant* signifies an animal, but nothing more. Now imagine that someone misleads him, saying that an elephant

> is an animal with one leg, three wings, lives in the depths of the sea, has a transparent body; its face is wide like that of a man, has the same form and shape, speaks like a man, flies sometimes in the air, and sometimes swims like a fish. I should not say, that he described the elephant incorrectly, or that he has an insufficient knowledge of the elephant, but I would say that the thing thus described is an invention and fiction, and that in reality there exists nothing like it; it is a non-existing being, called by the name of a really existing being, and like . . . a centaur, and similar imaginary combinations for which simple and compound names have been borrowed from real things.[57]

God is perfect simplicity. When we apply the name God to something with attributes, "we apply that name to an object which does not at all exist." That elephant that you've got there, it does not exist. Notice that, after the first wacky thing about one leg and three wings, Maimonides just references wacky things that are said about God. Recognize that nonexistent elephant now? For him the world must be the result of some singular essence with the potential for creating patterns, but that's a far cry from the anthropomorphism of the Bible. It is a pinnacle of Jewish doubt. Maimonides, by the way, quotes Ecclesiastes a lot: "For God is in heaven and thou upon the earth; therefore let thy words be few. (Eccles. 5:1)."[58] He finds other encouragement not to give one's time to prayer, such as Psalm 65:5, "Silence is Praise to thee," and Psalm 4:4, "Commune with your own heart upon your bed, and be still."[59] Rather than praising God, he advises, "it is . . . more becoming to be silent, and to be content with intellectual reflection, as has been recommended by men of the highest culture."[60]

Maimonides was a practicing Jew, but he explained the necessity of the laws, prayers, and rituals as dependent upon humanity's inability to bear an utterly abstract, lawless worship of God. People need religion for political and emotional reasons; for ideas our best options are reason, meditation, and resignation. Maimonides saw "the mass of religious people" as "the multitude who observe the commandments, but are ignorant."[61] He argues that when ancient information, either that of Aristotle or the Jewish sages, is contradicted by the growth of a scientific discipline, the ancient information must be discarded in favor of truth. When people asked him what to think about astrology, which seemed like superstition but was backed by numerous talmudic sages, he was not ambivalent. The "entire position of those who predict the future from the stars is regarded as false by all masters of science," and that, he insisted, means that it is false. He writes that if one searches the Talmud and the Midrash, one can find sages who speak of the stars having an effect on people, but, he says: "Do not regard this as a problem. It is not proper to abandon matters of established knowledge that have been verified by proofs . . . and depend instead on the teachings of individual sages who may have possibly overlooked what was essential to these matters. . . . A man should never cast reason behind him, for the eyes are set in front, not in back."[62]

Finally, note that Maimonides had an interesting take on two great Jewish doubters. Like everyone else, Maimonides scorned Elisha ben Abuyah, Aher, but he ignored the specific things that the Talmud accused Elisha of believing and doing. Instead, Elisha was characterized as one who believed and did all the things typified by heretics of Maimonides' own day, and of his Arabic world. These included believing in the eternity of the world and disbelieving the prophets. Sarah Stroumsa has made a good argument that Maimonides was referring directly to Ibn al-Rawandi: Maimonides mentions almost nothing of what the Talmud says of Elisha and instead gives a description that perfectly fits Ibn al-Rawandi.[63] It is also possible that Maimonides knew of Jews who themselves enacted these heresies. In any case, he made it clear that he knew of many Jewish unbelievers, speaking of those who "claim to be more intelligent and brighter than the Sages," and those who "repeatedly mock the sayings of the Sages." The other curious reading Maimonides offered was on Job. He devotes two chapters in his *Guide for the Perplexed* to analyzing Job, and repudiates the long-standing rabbinical habit of ignoring Job's revolt. Instead, the story was said to showcase the maturity Job gained through experience. We should feel the roar of grandeur and not worry "whether He knows our affairs or not, whether He provides for us or abandons us." Maimonides did not believe in providence or even a God who could speak and act, so he turned the Job story into an allegory for self-control and wonder.

Maimonides had tremendous stature in the late Middle Ages. As his works spread widely through Spain and southern France, they caused a burst of Jewish philosophy that lasted several centuries. The great Jewish mystical movement of Cabala took off in response to Maimonides and other Jewish rationalist philosophers. But there were forms of Jewish mysticism before Maimonides. These had a lot to do with psychosomatic practices such as fasting a specific number of days, getting into odd postures, and whispering or humming precise words and phrases. As the great scholar of Jewish mysticism Gershom G. Scholem has pointed out, mysticism depends on people feeling very separated from God: if there is naïve belief that sees God in nature, and then a period of rationalism that sees God outside nature, the period of mysticism follows only then, as an attempt to get back to this distant God.[64] We are reminded of al-Ghazzali. Unlike the philosopher, the mystic does not deny revealed knowledge, but he or she is perfectly willing to generate so much new knowledge that the old is swamped. Mysticism of this kind had a good career in Judaism, perhaps because the orthodoxy never had enough command of worldly state power to stamp it out. Cabala, the great Jewish mysticism, was a mixture of Maimonides' notion of the total unknowability of God with some ancient Gnosticism, Neoplatonism, ancient mythology from the East, and an earlier, German-Jewish mysticism called Hasidism (it's been the name of many movements because it means "the devout"). This Hasidism was a system of psychosomatic practices developed to reach states of ecstasy and get a glimpse of God, intended for a select group of seekers.

Cabala had two founders who, although very different, both came to mysticism through Maimonides and who lauded his work even after their own conversions. One was Abraham Abulafia (1240–ca. 1291). He conceived of his mysticism as adding the final step in the *Guide for the*

Perplexed, which he admired deeply. This may have been so, but it was a giant step. What he prescribed was most like Eastern religions of enlightenment. One meditated on the letters of the Hebrew alphabet using methods that were very much like music in their patterns. To cite Scholem again, "his teachings represent but a Judaized version of that ancient spiritual technique which has found its classical expression in the practices of the Indian mystics who follow the system known as Yoga."[65]

Moses de Leon, another Spaniard from the same period, transformed Cabala into the tremendous movement it became. His great book, the *Zohar,* became the central text of Cabalism and, in fact, became a canonical Jewish text. For several centuries it was ranked with the Bible and the Talmud. Here is what we know of the book's author: In 1264, when Moses de Leon was about twenty-four years old, we have a record of his going to considerable effort and expense to have a Hebrew translation of the *Guide for the Perplexed* made up especially for him. In the 1270s he became friends with a follower of Abulafia's Cabalism. We also know he read the pieces of Plotinus's *Enneads* known then as *The Theology of Aristotle.* There de Leon read of the philosopher's ecstatic rise into the world of truth.

The *Zohar* was written about 1280, at a time when the Crusades had been ripping through Europe, sometimes decimating Jewish communities along the way. Philosophy had offered a rational interpretation of Judaism, but in difficult times people long for the comforts of religion. The *Zohar* claimed that Jewish law did not need to be defended rationally at all, for all its gestures were part of the secret-knowledge rites that had to be done to fix the broken world. In this, de Leon seems to have been influenced equally by ancient Gnosticism and Neoplatonism, but such a notion never had a better world of symbols and rituals to invest with meaning. In the sixteenth century the Cabala championed by Isaac Luria would parallel this in Luria's idea of repairing the "broken vessel" of divine light through acts of kindness. Here in the thirteenth century, Cabala made the acts of Judaism come alive as thrillingly meaningful because according to the *Zohar* each individual, going through his or her lawful obligations, was mystically fixing the world. Whereas Abulafia's program was a system of ecstasy for the elite, the *Zohar's* program was explicitly for the masses, and intended to revive the doctrines of naïve popular belief that were being challenged by the philosophers.

Meanwhile, the Jewish philosophical movement begun by Maimonides also produced Levi ben Gershom (1288–1344), known as Gersonides, of southern France. We are at about nine o'clock on the Mediterranean map. By the thirteenth century the intellectual language of the Jews had switched from Arabic to Hebrew, and a Hebrew translation of Maimonides' work had profoundly shaped Gersonides' world. Gersonides himself was a scientist, an astronomer, and a mathematician. He used a camera obscura to watch eclipses and other heavenly phenomena. The image projected in one of these "dark boxes" can be traced for a precise drawing, but since Gersonides used treated paper, which he later "developed," he can be seen as having taken the first photographs. Gersonides also invented the "Jacob's staff," a pole with metal plates used to calculate angular distances with reference to the stars, which made ocean-spanning voyages possible. This combination of doing theory and technology was almost nonexistent in the Middle Ages. More than anyone else of his time, Gersonides emphasized the need for empirical observation as a basis for astronomical research, rather than just theorizing with past observations and conjecture.

He seems to have been the only person to falsify the Ptolemaic, Earth-centered model of the solar system. He may have gotten the idea from reading Maimonides, who had mentioned that there was no way to verify the odd epicycles Ptolemy described. Gersonides used his camera obscura to check if the brightness of Mars varied in ways that supported the idea of the planets moving in epicycles. It didn't. He tried to come up with an alternative model but couldn't. Not even Copernicus, Kepler, or Galileo, who would solve and resolve the riddle three centuries later, ever offered a falsification of the old model. This side of the story is almost never told, but we are interested in doubt, not merely new solutions. Gersonides put the matter into question, and knowledge of his work may have helped lead Copernicus in the right direction. In the twentieth century, these contributions were recognized by the naming of a lunar crater after Gersonides, which has earned the fourteenth-century scientist the moniker of "the first rabbi on the moon."[66]

Gersonides approached philosophy and religion in a similar way. Along with his science, he was also famous for centuries for his superscript commentaries on Averroës' commentary on Aristotle. He did not agree with the philosopher on all matters, but he came to agree that God had no knowledge of the goings-on of life. Aristotle had persuaded Maimonides of almost this much: Maimonides had lived as an observant Jew, but philosophically he did not believe that God could be thanked, praised, or petitioned. There was an essence to the universe and perhaps it had some sense of us and some benevolence; for Maimonides, this was the force at the heart of the Jewish religion, even if a lot of mythological ideas had been necessary for ancient people to believe. By contrast, Gersonides' God could be said to know of us only insofar as his pattern-full essence is what we are. "For God does not acquire His knowledge from them; rather they acquire their existence from His knowledge of them, since their existence is an effect of the intelligible order pertaining to them inherent in the divine intellect."[67] Notice that there is no divine thinking implied here.

His dedication to rationalism was a matter of personal conviction and he could cite authorities for his right to it:

> There is nothing in the words of the Prophets that implies anything incompatible with the theory we have developed by means of philosophy. Hence it is incumbent upon us to follow philosophy in this matter. For, when the Torah, interpreted literally, seems to conflict with doctrines that have been proven by reason, it is proper to interpret those passages according to philosophical understanding, so long as none of the fundamental principles of the Torah will be destroyed. Maimonides too follows this practice in many cases, as his famous book the *Guide for the Perplexed* shows.[68]

With God so distant, unknowable, and unknowing, and the Torah disprovable just up to the point that it not be destroyed, we have come rather far into Jewish doubt.

The balance that existed between Jewish mystics and Jewish rationalists was about to change. There was fatal pressure on Spanish Jews to convert, beginning with massacres in 1391, and eventually there was a population of people who were secretly still Jews, called Marranos. The Spanish Inquisition was all about ferreting out these people. When King Ferdinand and Queen Isabella finally finished kicking the Muslims out of Spain, they turned again to consider the Jews. The Jews had a rich and complex society in Spain, going back hundreds of years; Spain is as much the home of Jewry as the Middle East and Eastern Europe. In 1492 the Jews of Spain were given three months to leave or convert. How could a world leave? The departing Jews lost almost all their property. The king of Portugal offered safe haven at a price and a great many Jews took him up on it. After he got the money, he began selling the Jews into slavery or otherwise dispatching them.

After a long period of almost idyllic coexistence, the Jews of Spain were subjected to sudden and horrific persecution. Many died of starvation on the outskirts of some Christian town, in a foreign land. All lost their world, with its million ancient delights, exquisite mosaic synagogues, songs, recipes, literature, and jokes. Cabala was transformed through this awful rupture, so that it described everything in the Bible and in Jewish life in general as being about exile and return. The *Zohar* stressed acts of repentance and "savoring" the bitterness of exile. In the Cabalist image of Job in this period the fate of humanity rested on Job's ability to bear his suffering. For the exiled Jews of Spain, the idea that the pain of one person's life could have cosmic results, even if one did not believe in providence, was obviously appealing. Eventually, sometimes after generations, many Marranos made their way to better lands—they or their descendants founded the Jewish communities of Amsterdam, New York City, Hamburg, and London— where they returned to a public, common Judaism that was by then fascinatingly foreign to them. Eventually, these communities would be fertile fields for doubt. After the disaster of 1492, though, Judaism grew increasingly mystical. Doubt closed down in Jewry as it had closed down among the Muslims. But not before tipping the flame toward one more culture— the Christians.

THE SCHOLASTIC RATIONALISTS AND THE EUROPEAN RENAISSANCE

When we left Christian Europe at the end of the ninth century, Charlemagne's cathedral schools were poring over Boethius's logic. Those educated at the schools learned how the world worked through treatises on Plato, Aristotle, Cicero, and the Stoics that had been written by early medieval thinkers such as Isidore of Seville and the Venerable Bede. Finally, the cathedral schools' natural philosophy was based almost entirely on the *Timaeus* of Plato, which, as we have noted, was embedded in a commentary by Chalcidius and, moreover, was only partially translated. That is all they had; the East had the good libraries and the Arabs had the East.

Nevertheless, this world was developing its own pitched battle between rationalists and mystics, and they were somewhat influenced by the outside world. Here the rationalists were Anselm of Canterbury (1033–1109), the one with the famous proof, and Peter Abelard (1079–1142), the one with the biography that makes everyone wince. Against them was the amazing mystic and musical composer Hildegard of Bingen (1098–1179). It was an even fight. Anselm's proof of God was that we have an idea of a perfect God, but reality is always better than something that is not real, so God, being perfect, must be real. It did not take long for people to point out that, actually, we cannot make a three-course meal appear by imagining a perfect one. Still, much Christian faith was tending toward rationalism, proofs, and demonstrations. The same is true for Abelard—he was not a doubter but is of passing interest to our story as a rationalist in the service of faith. He also came to a keen understanding of Aristotle, more than had been understood in his world in centuries. Here's the wince: He is best remembered for seducing and then secretly marrying Heloise, one of his students and the daughter of an important Paris churchman. When he would not live openly with his wife, her relatives castrated him. He then went off to a monastic community and she to a convent, from which they exchanged many letters and further developed their reasoned faith.

I said above that the battle between philosophy and mysticism was a fair fight. The mystical side was well covered by Hildegard, whose career provides a demonstration of what doubt was up against. Born to a noble family in Germany in 1098, Hildegard was only eight when she was sent to a wealthy Benedictine convent. The place was run by a famed mystic abbess called Jutta, and this woman raised Hildegard herself. The girl eventually succeeded Jutta as prioress, leaving after a while to start a convent closer to the ideal of poverty, near Bingen. Hildegard's noble pedigree and her famous mentor gave her status, and she was an incredible talent. She said she had had visions from a young age, and she wrote tomes full of vibrant allegorical visions and charges of impiety.

Saint Bernard of Clairvaux, the twelfth-century monastic reformer, read her work and the two became fervent correspondents. Bernard secured the pope's approval of Hildegard's theology. He also got the pope to condemn some of Abelard's rationalist propositions. Hildegard was soon traveling all over Germany lecturing monks, clergy, and secular officials, recounting her visions and scolding them for their carnal transgressions. She also wrote well-respected medical treatises. In person, too, she had a great following, who came to her for counsel. She is often described as the most profound psychological thinker of her day. Marvelously, that's not the half of it, for she is the first composer in the West whose biography is known to us, and her music, which took the form of ethereal angelic chants, is still played today. She also invented the morality play.

So there were attractions to philosophy and there were attractions to mysticism. This balance was about to change. About the year 1000, weakness in the caliphate of Cordoba encouraged the Reconquista, the "taking back" of Spain for the Christians (the Visigoths had been there before the Muslims; they were a heretical, Christian, Germanic tribe). Over the next centuries the Reconquista brought the new ideas in Muslim and Jewish Spain to European Christians. There were major centers of learning in Toledo and Sicily, and for most of the eleventh century these were still out of reach of the Christian world, which knew the Muslims had vast knowledge far beyond its own. King Alfonso of Spain, aided by the dashing mercenary warrior El Cid, managed to take Toledo in 1085. In 1091, the Normans took Sicily. Throughout the twelfth century, scholars ran to these two destinations to find works in Arabic, and to several destinations in Italy for works in Arabic and Greek. Many of the works they got their hands on already had hairy translation histories.

It was mostly science that made the loop-the-loop around the Mediterranean, not literature. In this wave, once again, a handful of translators changed the world. Gerard of Cremona (died 1187) translated from Arabic to Latin: Aristotle, Ptolemy, Euclid, Galen, al-Razi, and Avicenna. In all he translated about seventy books. A little later, William of Moerbeke (ca. 1220–1286) translated from Greek to Latin almost all of Aristotle and Archimedes, and a great variety of commentators. He translated about fifty works. Western Europe had never been thrown that much new stuff. Historian Edward Grant says, "The impact of Euclid's *Elements* and Ptolemy's *Almagest* alone were capable of transforming the basis of science. It was as if the West had left a barren desert and moved to a richly watered oasis."[69] Grant is one of our best authorities on the Middle Ages, and his book *God and Reason in the Middle Ages* offers wonderful evidence about the prominent role of reason in these misunderstood centuries.

As Grant explains, it was the new texts of Aristotle and Aristotelianism that really changed everything. The great philosopher had written out rules and a method for all sorts of disciplines. In the culture of the Europeans of that time, these realms of thought were just a mishmash of vague and contradictory notions. Aristotle told how the Greeks had climbed out of impressionistic and mixed-up thinking and laid out the foundations for clear thinking: logic and observation. His sheer range and output were stag-gering, and by now there were numerous Greek, Latin, and Arab commen-taries that treated him as the great authority. Europe was transfixed by this work. It was the eleventh hour on the Mediterranean doubt clock. The rivalry with the mystics ended in a flash because the new, almost magically sophisticated math, medicine, and science simply blew it out of the water.

By this time, cathedral schools had been thriving in various urban cen-ters, such as they were, for a few centuries. For their own protection and to create a more coherent community, they began to get themselves *incorpo-rated*, as did almost every other distinct group in the Middle Ages. Being incorporated meant that the group had certain rights and privileges and had to abide by laws. The Latin word most commonly used for these incorpora-tions was *university*—the reason the word means "school" now is because the school-type university is the only one that made a successful transition from that time to our own. The incorporations set up the universities to do a certain amount of teaching in a certain way. So after they were incorpo-rated, they needed a lot of curriculum material, immediately. As it hap-pened, the ancient texts that were rediscovered at this time were almost all about logic, science, and math. Thus the universities, which had different specialties—law here, medicine there—all had a huge preliminary curricu-lum, which included four to six years of logic. The absence of ancient texts on theater, poetry, history, and other branches of the humanities made this a very rationalist course of study. The curriculum of the arts degree was Aristotle's natural philosophy and it remained so across half a millennium. Theology, medicine, and law were above the arts degree, being more impor-tant and later in one's education, but that meant that everyone had to first earn the arts degree.[70] Thus, logic (four to six years of it!) provided the one common ground of all these scholars, theologians, and practitioners, and for all of them it provided the foundation of their worldview.

They did just what you would think a somewhat-literate tribal world would do, if, as it was naturally developing some rational rules for knowl-edge, it suddenly found a cache of the finest thinking of an ancient world that had dedicated itself to rational wisdom. They fetishized it. They did not use it to look for new knowledge or for ways to make things, move things, or take things apart. They merely tried to make sense of the material they had found. Consider this: they never thought to understand Aristotle's works as the changing thoughts of a man across his lifetime—they simply followed the Muslims' lead and worked to posit an explanation for the "discrepancies." Since the cache of ancient wisdom they found was a very particular corner of all ancient thought, they made a world of a molehill.

The medieval scholars faced the varied genius of the new Aristotle, Euclid, Ptolemy, Hippocrates, Galen, Avicenna, Averroës, and al-Razi and blinked at the variety of claims. They met the challenge of this avalanche of opinion by writing out *yes* or *no* questions and then arranging the opinions of the great thinkers for and against. This was the gist of the Scholastic method, and those who did it were called Schoolmen; they were the Church's philosophers, the rationalist theologians. A question might be "whether there could be an infinite dimension" and then a "principal argument," which stood as a bit of a straw man, and then "Aristotle determines the opposite" or "The Commentator affirms the opposite" (Averroës, of course, being "The Commentator"), and then the author's opinion and reconciliation of any remaining problems.

I have said that the new Aristotle that came in, mostly in Arabic, was like a typhoon, overtaking mysticism. It also knocked out Boethius in one swipe. He'd had an incredible run. Boethius's labors maintained the discipline of logic for half a millennium. The reason people knew what it meant to find a whole cache of Aristotle, the reason that name cried out to Europeans when they saw the Arabic letters spelling it out, was because of Boethius.

The most important of all the new texts was Aristotle's *Sophistic Refutations*. It was a study of fallacy: how words work and what to call the ways in which they can be deceiving. It appeared in Latin about 1120 and just plain took over. For one thing, it was easier to understand than others, and for another, it offered instruction about work that was yet to be done and how to do it. The Schoolmen picked it up with glee. They started to take apart language to see how it worked so they could be sure of speaking the truth—there were treatises, for instance, on *syncategoremata,* words that cannot be subjects (such as: *every, because, and, or, if*), which made logicians suspicious of sentences with such words in them. In fact, the way that Latin words worked soon became more important than their content. What is more, although word order is unimportant in Latin, the Schoolmen assigned meanings to Latin word order and thus expanded their project. That is, the Schoolmen learned from Aristotle that this conceptual game of logic could be used to negotiate fallacies and they turned Latin into a symbolic logic system, rather than a system of communication. The Schoolmen thus used purposefully silly content, so that content would not distract them. That is why it has seemed like such hocus-pocus to the later Europeans.

Apparently a sentence was constructed so that one could argue that it was true and then argue that it was false, and the idea was that one would learn something, or hone one's truth-finding skills, by analyzing these claims.[71] Scholastic authors liked citing famous ancients in their content-nonsense texts. Consider some of their sentences:

> Varro, though he is not a man, is not a man, because Cicero is not Varro. A head no man has, but no man lacks a head. Socrates is whiter than Plato begins to be white. Socrates will as quickly have been destroyed as he will have been generated. A horse is an ass. God is not. No man lies. Some horse does not exist. What Plato is saying is false. Socrates wants to eat.[72]

A variety of exercises and games were being worked on here. Sophisms were one major project of these weird statements. Sophisms were arranged in themes. That is, they were understood to exemplify, for example, themes of signification, supposition, connotation, and insolubles. The sophism "This dog is your father" is true and intelligible because the dog is yours and the dog is a father; you own this father; this dog is your father. The best here are the insolubles, the most famous of all being: "What I am saying is false."

Scholasticism, then, was not at all religious in the sense that people think of Scholasticism as being religious. No one argued about how many angels danced on the head of a pin—people made that up later to deride the Schoolmen. True Scholasticism was not experimental, nor did it seem to be about finding any new knowledge, yet it is not right to see it as a religious game of faith-blinded theologians. These were the thinking people of the age, they were very educated, and they did not all simply take Christianity as literally true, on faith, en masse. They knew of other religions and even of a whole non-Christian past. They were curious about things other than those within the prescriptions of their religion, and working on their weird Scholastic sentences led them to some subtle and interesting philosophical questions.

When the Schoolmen's logical investigations brought them to a wall around which their faith would not allow them to peer, they usually found a way to peer anyway. To get a taste for their rationalism, consider that the Schoolmen had a whole branch of thought dedicated to "the first and last instant." Aristotle had wondered whether there was a moment in which something can truly be said to become something else, or is the continuum the only reasonable way to think about change? If so, how can we speak of things starting and stopping? The Schoolmen followed him and asked about the outside possibilities of things in the world. They asked themselves questions such as "Should a capacity such as Socrates' ability to lift things be limited by a maximum weight he can lift or by a minimum weight he cannot lift?"

They did not leap from reading Aristotle's arguments into making entirely different arguments of their own; they stayed close to his, disagreeing here and there but generally remaining pretty impressed. Indeed, they praised his empiricism without actually doing anything experimental or empirical. Yet, as we have seen, there is a way in which Aristotle was not very experimental either. Compared to Plato, who did not trust the senses and thought the world was to be perceived through ideas, Aristotle had been very empirical, it was true. Observation and experimentation were seen as the route to truth. Yet Aristotle had a definite gut feeling that the world was a noble, coherent whole: in *On the Heavens, On the Soul, Physics, On Generation and Corruption,* he was dedicated to figuring out the world honestly, but his conviction that it was all going to make beautiful sense made him look for ways for it to fit together conceptually. The Schoolmen mostly did what Aristotle had done, but with almost no actual observation of the world: they occupied themselves with discussing his observations. After all, Avicenna and Averroës and all the others had treated Aristotle (or rather the philosopher they knew as Aristotle) as the ultimate authority. That tradition went back many centuries now.

No one in Christiandom seems to have done any physical experimenting across these long centuries. What they did was to follow Aristotle's lead with a lot of thought experiments. One of the best known was about a bean. Aristotle had said that if a thing is thrown up and then comes down, there is a moment of rest in between. Here's the question the Schoolmen got hung up on: Imagine a bean thrown up just when a big stone is thrown down. If at the moment of contact the bean did in fact stop (in order to begin descending), it would have to hold up the stone, which is impossible. By thought experiments such as this, they did actually derive the "mean speed theorem" in their inquiry into the concept of variation. They also came up with impetus theory, the possibility of finite motion in a vacuum, and the idea of void space outside our cosmos. As Grant points out, "We may properly characterize medieval Aristotelianism as empiricism without observation. It was also empiricism without measurement."[73] John Murdoch has called it "natural philosophy without nature."[74] What would you expect from a world suddenly faced with advanced schools of thought from a culture not its own? Had they been recipes from another world, set in conceptual terms far beyond us, they might have been recited as poems for centuries before anyone thought to get out the pots and pans. Scholasticism was not science, but it was more rational than religious.

It was this period that saw the rise of Padua as a center of science and Averroën Aristotelianism. Albertus Magnus, Albert the Great, lived most of the 1200s and was one of the best theologians of the Middle Ages. He was educated at the university of Padua and joined the Dominicans while still a young man. When his Dominican brothers asked him to write a book explaining the "science of nature" to them, they specified that they wanted it so that they could attain at least a competent understanding of Aristotle. As Albertus explained in the introduction, "we take what must be termed 'physics' more as what accords with the opinion of Peripatetics than as anything we might wish to introduce from our own knowledge."[75] In discussing whether the heavens were made out of nothing or generated from something, Albertus made it clear that in the "principles of nature" things are always generated from other things. Thus, despite its being contrary to his faith, it was only this option that was worth considering in his *Physics*.

At this same time we find the French theologian Siger de Brabant (died ca. 1284) being called "head of Latin Averroism," which was based at the universities of Paris and Padua. In Paris, Siger taught that the individual soul had no immortality and that the world was eternal rather than created. When asked if he could still be a Christian with these notions, Siger showed that he believed he was following the Averroist idea that different intellects could handle different versions of the truth. Actually, he had perverted this notion a bit, arguing that there could be "two truths," that is, that something could be true in rational philosophy but false in religious belief.

Thomas Aquinas came on the scene in a fierce attack on Siger. Indeed, Aquinas's work was in large part an attempt to defend both Aristotle and Christianity from the stark separation that was being made here between philosophical believers and popular believers. He wrote a number of commentaries on Aristotle—on ethics, physics, politics—and these have long been considered among the finest philosophy. "Rabbi Moses," as he called Maimonides, was one of his major sources. As for his own opinion of God, Aquinas was a bit of an Aristotelian here, too, but he felt that on this one issue, revelation had made a big difference. Consider the first two questions of his most famous work, the *Summa Theologica*. Question One asked "whether, besides the philosophical sciences, any further doctrine is required," and showed how one could argue that nothing further is required. He settled in on yes, we do need revelation. Question Two was on "the existence of God" and asked whether it was self-evident. Aquinas noted that since Psalm 14 says "The fool said in his heart: there is no God," God's existence is not self-evident.

The third part of that question asked "whether God exists" Aquinas included this argument against: "If therefore, God existed there would be no evil discoverable; but there is evil in the world. Therefore God does not exist." He also included this little beauty: "It seems that everything we see in the world can be accounted for by other principles, supposing God did not exist. For all natural things can be reduced to one principle, which is nature; and all voluntary things can be reduced to one principle, which is human reason, or will. Therefore there is no need to suppose God's existence." These are an amazing pair of statements, whether or not they are offered as straw men. Aquinas's answer to them is to cite revelation—"I am Who am"—and to give philosophical proofs: we need a first cause, we need a prime mover, the world shows gradation in its creatures and things and that suggests a perfected being at the top, and finally, the governance of the natural world. The new Aristotelians had seemed hostile to the Church before Aquinas. He offered the Church an Aristotelianism that, by comparison, supported Church doctrine. It was a deal the Church would heartily defend for centuries, but it did not seem like that at first: Aquinas's work was also repressed, briefly, in the years just after his death in 1274.

After a few preliminary warnings and condemnations, in 1277 the Church became so uncomfortable with the new thinking across Europe—especially at the university of Paris—that it issued a condemnation of 219 propositions, including, for instance, that "the first cause could not make several worlds." It's funny enough to hear the Church quoting Aristotle so fully that it calls God "the first cause," but the point was that the Church had an all-powerful God and it was not going to accept Aristotle's "impossibles." Another condemned teaching was that "God could not move the world with a rectilinear motion; and the reason is that a vacuum would remain." Aristotle did not believe in vacuums, which meant God could not make one, which meant God could not move the world, because it would leave a vacuum. The Church answered that God can do anything. Aristotle had also made it clear that God could not make an accident exist without a subject—it is a logical contradiction. The Church said God could do it, and we can see why: with the Eucharist they had a substance that was transformed into God, but still seemed to have the accidents of the bread. The Church needed to keep some wiggle room around an adjective and its noun so that the accident of the eucharistic bread could exist even though the bread was gone (replaced by the body of Christ, which has no accidents).

As wild as this is, it went further. In 1277 the Church said it was now forbidden for anyone to say the following:

152. That theological discussions are based on fables.

40. That there is no higher life than philosophical life.

153. That nothing is known better because of knowing theology.

154. That the only wise men of the world are philosophers.

175. That Christian Revelation is an obstacle to learning.

37. That nothing should be believed unless it is self-evident or could be asserted from things that are self-evident.

As eminent historian Etienne Gilson put it in 1938, "The list of those opinions is a sufficient proof of the fact that pure rationalism was steadily gaining ground around the end of the thirteenth century."[76] These Averroists went much further than Averroës. "As a matter of fact," wrote Gilson, "it was like nothing else in the past, but it anticipated the criticism of the religious dogmas which is a typical feature of the French eighteenth century." Gilson further remarked, "That the so-called Revelation is mythical in its origin is everywhere suggested in Fontenelle's *History of the Oracles* (1687); Fontenelle was a very prudent man; he was merely suggesting what he had in mind; but four centuries before him, some Averroists had clearly said it." I'll discuss Fontenelle when we get to him later in the book, but it's useful to consider Gilson's conviction here. Historians tend to be very touchy about the possibility of medieval doubt. Grant writes, "It is doubtful that any natural philosophers actually incorporated such explosive and potentially dangerous articles into their written work," allowing only that "If such assertions were actually made, they were probably communicated orally around the University of Paris." We cannot know for sure, but let us think of that list again: no higher life than philosophical life; theological discussions are based on fables; Christian Revelation is an obstacle to learning. This was an active polemic, a real fight, between real opponents. There were doubters at the Sorbonne in the thirteenth century.

The same list of condemned teachings also included: "that God could not make anything new," "that there is more than one prime mover," "that eternity and time have no existence in reality but only in the mind," "that nothing can be known about God except that He is," "that God does not know things other than himself," "that after death man loses every good," "that raptures and visions are caused only by nature," "that happiness is had in this life and not in another," "that there are fables and falsehoods in the Christian law just as in others," "that a philosopher must not concede the resurrection to come, because it cannot be investigated by reason," and my favorite, "that man could be adequately generated from putrefaction." How did the Schoolmen come to all this?

We know they read Aristotle, Averroës, and Maimonides. Not only were many of the Schoolmen believers in a very philosophical cosmos, but after Aristotle had been used as a textbook for centuries it was just beginning to dawn on Europeans that Aristotle and the other ancient writers were not exactly the early texts of the Schoolmen's own, European civilization. With astonishment, it was slowly being recognized that Aristotle and Plato and the rest of them belonged to a fully other civilization that had its own answers to the big questions and that explicitly rejected a God like Jesus. Not only that, but all these ancient texts had come in from Muslim and Jewish sources, having been transformed by Muslim and Jewish sages, and the shock of confrontation with cultural diversity came from that direction as well.

The result of the Condemnation of 1277 was in part to spur rationalism in new directions. The Schoolmen's questions about how God *could*, in fact, do Aristotle's impossibles fostered among them an imaginative and bold inquiry into questions that their Christian cosmology would never have imagined alone, and that their Aristotelianism would never have imagined either. Aristotle had said that there could not be any other worlds. The Schoolmen did not believe that God had made any other worlds, but they defended the idea that he could. That made them write about the issue, intensely. After the Condemnation they generated ways that God could make a vacuum, ways that other worlds might exist, ways in which the heavens might have been animated by God at the moment of creation and never since. This last was posited against Aristotle's certainty that since the movement of the heavens was not slowing down, it must be continuously animated, and therefore the stars had souls. Dutiful Schoolmen argued that a really powerful God could do anything, including setting up a universe that runs on its own. As Grant has said, "The invocation of God's absolute power made many aware that things might be quite otherwise than were dreamt of in Aristotle's philosophy."[77] Yet, as Grant shows, few questions even mentioned God in passing, and only a tiny fraction dealt with the idea of God as a central issue.[78] Theology was becoming a mix of logic and natural science that left very little room for anything spiritual. The Condemnation of 1277 didn't do doubt much harm and it did the history of doubt a lot of good: just as with the Carvaka and the Hellenistic Jews who opposed the Maccabees, the best evidence of the existence of medieval doubt is what was preserved in the polemics of its enemies—everything else having been burned or only whispered in the first place.

In the fourteenth century, Christian theologians and scholars grew more rationalist on the one hand, and more skeptical on the other. The arts master John Buridan (ca. 1295–ca. 1360) explained that to think about the world as Aristotle did was to reject recourse to the supernatural and to try to figure things out as they appear. There were ways of imagining God doing all sorts of seemingly impossible things, he explained in one discussion, "But now, with Aristotle, we speak in a natural mode, with miracles excluded." Around 1370 the great theologian Nicole Oresme wrote:

> I propose here, although it goes beyond what was intended, to show the causes of some effects which seem to be marvels and to show that the effects occur naturally, as do the others at which we commonly do not marvel. There is no reason to take recourse to the heavens, the last refuge of the weak, or demons, or to our glorious God as if He would produce these effects directly, more so than those effects whose causes we believe are well known to us.[79]

The heavens, the last refuge of the weak! The period's other great Christian thinkers—William of Ockham, Duns Scotus, and Nicholas of Autrecourt—pushed into Skepticism, each questioning the ability of reason to describe reality. Ockham even rejected Aristotle's proof of God.

Ockham's famous question was How much can reason know faith? His answer was "not at all." His even more famous "Ockham's razor" was part of his contribution to logic, and calls for using the simplest explanation possible in all things. Along different lines, Scotus, too, forwarded the notion that reason could not penetrate beyond the world of sense experience and seemed to question whether clear thinking about metaphysics could be done at all. Nicholas of Autrecourt (born ca. 1300) went as far the ancient Skeptics, the Carvaka, and the Muslim al-Ghazzali on the question of cause and effect: we can never be certain if two rational notions are causally related, no matter how well they seem to link.

Nicholas was sentenced to burn all his writings, and did so in November 1347. Two letters to a friend survived and an excerpt from one will demonstrate the state of doubt:

> Just as you do not know whether the Chancellor or the Pope exists . . . [s]imilarly, you do not know the things of your body—whether or not you have a head, a beard, hair, and so forth. . . . I wonder very much how you can say that you are evidently certain of various conclusions which are more obscure—such as concern the existence of the Prime Mover, and the like—when you are not certain about these things which I have mentioned.[80]

A few sentences down he says, "And it seems to me, the absurdities which follow on the position of the Academics, follow on your position." Nicholas thus knew of the Academic Skeptics even if he only mentioned them to deny them, saying he himself would stick with the evidence of the senses. Elsewhere, he explains that, despite reasons to doubt one's perceptions, such as the problem of dreams, some things are more probable than others. He is often called the "medieval skeptic." We have found two Latin translations of Sextus Empiricus from this period (one late thirteenth century, one late fourteenth).[81] With a little help from the ancients, the medieval philosophers had arrived at a similar conclusion about philosophy: with it, we always get to the point where we cannot know anything. Not ourselves, not the world, and not God.

Just around the time Nicholas burned his books, big things were going on in Italy. In 1345 Francesco Petrarch stumbled upon something he had been searching for in monasteries and libraries for years: a volume of lost letters of Cicero. They astonished him. The Cicero he found there was nothing like the Cicero everyone thought they knew—the calm, unflappable philosopher. The letters themselves were a form that had not been seen in centuries; the medieval letter was very formal whereas Cicero's were personal, friendly, conversational, and meandering. We date the start of the Renaissance from this moment, this sudden and profound recognition that Cicero was not the disembodied voice of one branch of philosophy but a man with a personality, an individuality; that culture made all the difference in understanding a life, and that culture changes. The idea was powerful. The hour hand of medieval doubt, roaming around the Mediterranean, had inched up on twelve since Greek skepticism and rationalist doubt were reintroduced to Florentine culture. When Petrarch found Cicero, it was noon in Florence, and the town began to bloom. In the next chapter we will see the old Scholastic Aristotle cursed in favor of Cicero, whose casual tone and nonreligious concerns about human happiness inspired a whole new manner of thinking: Humanism. It was no attack on God, but as is clear from its name, it had a different central concern.

Doubting Muslims, Jews, and Christians had been having a very civilized conversation with each other—and with the ancient Greek and Roman philosophers—for centuries. The conversation was informed by rationalist convictions of human unity, and by the insistent relativism of a cosmopolitan experience, so it was relatively open. There were a lot of doctors, scientists, and community leaders involved.

What we learn in the rationalist loop-the-loop around the Mediterranean is that there was doubt somewhere around the Mediterranean throughout the Middle Ages. In late antiquity it passed from the crumbling world of the now Christian Romans, east to the Nestorian Christians, south to the Muslims and west across their North African world, then north to the Jews of Spain, and back to the Christians of Europe. In the next chapter, after a look at the Far East, we will see Petrarch and others damn the Schoolmen and their beloved Aristotle for being cold, arcane, and no longer relevant. That attack described Scholasticism as an intellectual game of interest only to its players. As time went on people assumed that the arcane Church philosophy was about Christian dogma. In general, the Middle Ages would increasingly be described as hopelessly entrenched in the fables of faith. We have seen that this was not the case. There were doubters, there were rationalists, and there were skeptics.

SEVEN

The Printing Press and the Age of Martyrs, 1400–1600

Renaissance and Inquisition

This chapter looks at the Far East, the new West, and the tumult of Europe in between, in a period variously spoken of as Renaissance, Reformation, Age of Exploration, and Inquisition. The time span covered is about 1350 to 1650. Our route may feel a bit hectic, but the variety of investigations will prove a point: this was a moment of contact between many worlds of doubt. If learning of other people's faith causes relativism and doubt, learning of other people's doubt redoubles it. This chapter tours the further development of Zen and Buddhism, and Eastern materialist philosophy, as well as the European Renaissance move to secularism and Protestant doubt. A section on François Rabelais tells of a great historical fight over irreligion that had its origins in Rabelais scholarship. We'll also look at a sampling of Inquisition trials for atheism or unbelief and spend a while with Michel Montaigne, one of the most entertaining doubters of all time. A section called "Uncertain Danes and the 'Debauch' of French Libertines" considers Hamlet among other undecided characters. Last, we will watch the Catholic Europeans head off to China to convert the masses and end up introducing the science of one land to the atheism of the other. It is a watershed moment in the history of doubt, yet it was typical of this period. Everywhere one looked, there was a drama of encounter between various worlds of doubt.

ZEN AND OTHER EASTERN DOUBT

We have seen that from as early as 600 BCE, the theistic Brahmanical religion of India had competition: from the materialist Carvaka and the rationalist philosophers, to the more religionlike practices of the Buddhists, a nontheist world had commanded tremendous attention for six centuries. The end of the first millennium CE saw a resurgence of the Brahmanical tradition, in its bhakti and other theistic and devotional forms, and atheism in India came under sustained attack. Buddhists lost a lot of texts in this period, and Carvaka texts may have gone then, too. What's worse, when Muslim invaders swept across India, starting in the eleventh century, they viciously attacked Buddhist monasteries. The Muslim world seems to have worked up a tremendous hatred for the monasteries, at least in part because of the atheism of the monks' project. One of the most famous monasteries, Nalanda, was said to have housed countless manuscripts and ten thousand monks—it was savagely destroyed by Turkish troops in 1197. Since Hinduism had adopted so much of Buddhism outside the great monasteries, when these monasteries were destroyed, the only specifically Buddhist practitioners in India disappeared. It began as an Indian religion, but by 1000 CE it was all but gone there. Whatever full Carvaka texts had survived up to that point were destroyed then, too.

By this time, however, the dharma was thriving throughout the rest of Asia. As we have seen, Buddhism seduced China in its Mahayana form—the big raft. The early Middle Ages were a glorious period for Buddhism in China. The Tang dynasty (618–907) was the most cosmopolitan China had known: there was a great deal of trade (especially along the Silk Road between the Middle East and China); the government took a position of religious tolerance, so there were Jews, Muslims, Zoroastrians, Nestorian Christians, and Manichaeans around; people and texts were flowing and being transformed along the trade routes. Along with economic and social stability, such cosmopolitanism led to leaps in astronomy, geography, and medicine, saw the introduction of block printing, and the flourishing of literature and the arts: calligraphy, poetry, painting, and sculpture. Sophistication and urbanity were goals: there were competing famous philosophers and authors, and a popular culture that prized learning and education. Many of the great Buddhist schools arose in this period. A central inspiration for all was Nagarjuna, who had expressed doubt that went beyond the Buddha's in a sense, because he said we could not even claim there is no self; there is not any right doctrine, there are instead meditations that can effect a ripening into enlightenment.

When the Tang dynasty declined, Buddhism was attacked as foreign and too inwardly and there was a great persecution of Buddhism in 845. The Nestorian churches were chased out of China then, too, as were the Jewish synagogues and Islamic mosques of the trade cities. A neo-Confucianism that had absorbed from Buddhism a great deal of inward-ness and concern for the individual took its place. When the dust settled, Zen emerged as the dominant Chinese Buddhism sect, partly because it was isolated in mountain monasteries and so did not get wiped out. The two main schools of Zen, the Lin-chi, which relied mostly on the koan, and the Ts'ao-tung, which concentrated on sitting in meditation, were introduced in Japan near the same time, in 1191 and 1227. In Japan they came to be known as Rinzai and Soto. Rinzai used the koan to encourage an overwhelm-ing "feeling of doubt" that leads to *satori,* awakening.[1] As in China, the Rinzai theory was "Great doubt, great awakening." Zen masters used all sorts of techniques to wear away at the normal way of being in the world, to snap out of it. The Soto thought the Rinzai got caught up in all this and missed the point—which is why the Soto chose to just sit. This was at the very beginning of a military dictatorship based on the samurai, and Zen's austerity, discipline, and aggressive approach to one's own state of mind worked well here. Zen monks were important in politics, literature, and education.

There were many forms of Zen and Zen doubt. One of the most doubt-ing of the great Japanese Zen masters was the poet Ikkyu Sojun. His vision of Zen was an intense concentration on life, not really for the sake of achieving enlightenment, but more in the insistence that enlightenment is, actually, consciousness of the pleasure of life. He counseled people to be neither angels nor demons but human beings; to be, and thus let the worry disappear.

We eat, excrete, sleep, and get up;
This is our world
All we have to do beyond that
Is to die.

Ikkyu lived from 1394 to 1481. He famously loved the pleasures of the flesh and championed the idea that Zen is not only in the eating, excreting, and sleeping, but also in Eros—even cheap Eros—and romantic love. He's remembered for putting that into his poetry, both in gleeful mentions of his

Zen trips to the brothels and in the love poems he wrote to Mori, a woman he met when he was seventy-three. Ikkyu liked his sake, too. Some people saw him as awfully licentious for a Zen master, but he thought they were the ones missing the point. His place in the history of doubt is as a sensualist, a Zen doubter, and a provocative questioner of rituals, attitudes, and customs. He was nicely explicit about God in his poem "Skeletons." In it he stated at the outset that "The original formlessness is the 'Buddha,' and all other similar terms—Buddha-nature, Buddhahood, Buddha-mind, Awakened One, Patriarch, God—are merely different expressions for the same emptiness." He then told this story of a dream: "Toward dawn I dozed off, and in my dream I found myself surrounded by a group of skeletons. . . . One skeleton came over to me and said: 'Memories flee and are no more. All are empty dreams devoid of meaning. Violate the reality of things and babble about "God" and "the Buddha" and you will never find the true Way.' I liked this skeleton. . . . He saw things clearly, just as they are. I lay there with the wind in the pines whispering in my ears and the autumn moonlight dancing across my face. What is not a dream? Who will not end up as a skeleton?" Ikkyu kept a keen sense of wonder along with his rejection of the supernatural.

Consider three short poems: "To write something and leave it behind us, / It is but a dream. / When we awake we know / There is not even anyone to read it." This feels stark, but only at first: "The vast flood / Rolls onward / But yield yourself, / And it floats you upon it." This doubt is subtle: "On the sea of death and life, / The diver's boat is frightened / With 'Is' and 'Is not'; / But if the bottom is broken through, / 'Is' and 'Is not' disappear."[2]

This encouragement to doubt was the heart of Ikkyu's philosophy, but his poetry also speaks of doubting Zen practice itself. By this point Zen had developed many habits and conventions of its own. This stanza comes from a longer poem on the theme:

Contemplating the Law, reading sutras, trying to be a real master;
yellow robes, the stick, the shouts, till my wooden seat's all crooked;
but it seems my real business was always in the muck,
with my great passion for women, and for boys as well.

Consider a few more short ones: "No one really knows / The nature of birth / Nor the true dwelling place. / We return to the source / And turn to dust." "The vagaries of life, / Though painful / Teach us / Not to cling / To this floating world." Each poem gives us another vision of tolerance for ambiguity, doubt, and emptiness. He could be gently comforting, too: "If at the end of our journey / There is no final / Resting place, / Then we need not fear / Losing our Way."

If we turn our attention back to China, we see that the ancient materialist, naturalist philosophy Samkhya was still around, but by this point there had been some Samkhya teachers who believed in some of the supernaturalism in the Vedas. In the tenth century, the philosopher Vacaspati defended the original rationalism. He wrote that while nothing should be considered impossible without investigation, the revelatory manner of reaching conclusions was unacceptable. Only reason, he protested, not authority or tradition, leads to truth. "Though there is nothing prescribed, yet what is unreasonable cannot be accepted, else we should sink to the level of children, lunatics and the like."[3] A reasonable scale of probability—what is likely—forbids believing a whole range of imaginative possibilities, even though we do not know anything for sure.

The neo-Confucianism of the Sung dynasty rejected the creative cosmologies that had become part of Buddhism. It also rejected the whole transcendental world. Neo-Confucian Shih Chieh (1005–1045) wrote: "I believe that there are three illusory things in this world, immortals, the alchemical art, and Buddheity. These three things lead all men astray, and many would willingly give up their lives to obtain them. But I believe that there exists nothing of the sort, and I have good grounds for saying so." His grounds were that no one he knew of personally had ever succeeded at any of them. Zhu Xi (1130–1200) was the greatest voice of neo-Confucianism. Taking a great deal from Buddhism in its least supernatural mode, as well as elements of Taoism and a variety of other philosophies, Xi imagined a rich metaphysics for Confucianism. According to Xi, the world was composed, on the one hand, by a pattern or principle, which is incorporeal and unchanging, and on the other, by material or force, which is physical and changeable. Xi's whole doctrine was thoroughly naturalistic. "That which integrates to produce life and disintegrates to produce death is only material force. What we called the spirit, the heavenly, and earthly aspects of the soul and consciousness, are all effects of material force."[4] When the material is gone, he explained, so are these effects.

Xi died in disgrace owing to a mundane political struggle in which he was involved, but his ideas won out for a long, long time, reigning during the Yüan, Ming, and Ch'ing dynasties. There were other contenders, across these many years, who advocated the study of the mind or the inner life, but Xi's naturalism was the orthodoxy. For some six hundred years—until the Chinese examination system was abolished in 1905—the memorization of long passages of his books was required of almost all students in China.

Doubt in medieval Asia was marvelously varied, such that even when the main focus of an established school was doubt, some came forward and doubted that method of doubting. Furthermore, while some directed their doubt expressly against the idea of God or gods, other doubters focused on questioning karma, enlightenment, authority, the sutras, any kind of life after death, or any kind of ritual—even the yellow robes. In the fifteenth century, the philosopher Aniruddha wrote, "Huge giants [or Gods] do not drop from heaven simply because a sacred verse, or competent person, says so. Only sayings which are supported by reason should be accepted."[5]

THE SHOCK OF THE OLD: RENAISSANCE AND REFORMATION

One difficulty in ascertaining popular belief is that we simply do not know how much Christianity ever penetrated the great mass of peasants and workers of medieval Europe. Christianity arose as the Roman Empire was declining in the West. Until Charlemagne's great programs began in 789, no one had had the resources to set up schools or missionaries or even bother too much about the religious ideas of the people who worked the land—the vast majority of humanity. Rural priests complained that their flocks showed up at church to gossip and play, and that when they took part they barely knew what they were saying. In the universities, people continued to labor away at Scholastic logic and natural philosophy. Outside them, there were surges in the numbers and varieties of monks and nuns, seers and mystics, sects and heretics. Average people lived in a world marked by piety, superstition, and raucous irreligion.

It is against this background that we return to Petrarch. We last saw him searching old monasteries and libraries for a copy of Cicero's letters. Surprisingly, Petrarch wrote back. Mirroring Cicero's style, his letters were full of chatty questions and comments, and Petrarch ended with an appropriate marker: "Written in the land of the living; on the right bank of the Adige, in Verona, . . . in the year of that God whom you never knew the 1345th." Godlessness found its way to the fore, even when it was not the topic of conversation.

It would be the painter Giorgio Vasari who would look back in 1550 and call the past two hundred years a "renaissance," a rebirth of ancient genius. For Petrarch it was experienced as frustration with Church theologians and their devotion to logic. A friend of his once wondered whether Petrarch was not too hard on logic, to which Petrarch replied, "Far from it; I know well in what esteem it was held by that sturdy and virile sect of philosophers, the Stoics, whom our Cicero frequently mentions," but once students learn it, Petrarch says, they should move on.[6] If anyone "begins to vomit forth syllogisms," he suggests, "I advise you to take flight."[7] In a work called "On His Own Ignorance and That of Many Others," Petrarch wrote, "I snarl at the stupid Aristotelians who day by day in every single word they speak do not cease to hammer into the heads of others Aristotle whom they know by name only."[8]

The Renaissance turned away from the syllogisms of the past several centuries. Italy had always been more interested in poetry, plays, letters, and other literature than had the more Scholastic-minded northern countries. Also from the fourteenth century on, here and elsewhere, there was a rise in books that told real-world stories, not in Latin but in the spoken vernacular of each country. Giovanni Boccaccio's *Decameron,* set in a plague year, contained ten stories supposedly told by a group of privileged young people to entertain themselves while hiding from the contagion (the book, by the way, borrowed liberally from Lucius Apuleius's *The Golden Ass*). In a personal letter to Boccaccio, Petrarch told of having had a "visit from an Averroist," who listened to Petrarch's Christian beliefs and then, as Petrarch told it: "Here he burst into a disgusted laughter and exclaimed: 'Well you are surely a good Christian. I, for my part, do not believe in such things. Your Paul and Augustine, and all the others you preach and extol, were awfully loquacious fellows. If you could only bear Averroes, you would see how much greater he is than these silly babblers of yours.'"[9] Petrarch wrote that he kicked the man out, but our interest is that he existed.

Humanism was not actually about secularism, nor blatantly about putting science above faith, but somehow everyone understood that, in a way, it was. When women of the time called for equality, they asked for rational, scientific education for girls. Christine de Pisan is known as the first woman to have supported herself by writing, and her *Book of the City of Ladies* in 1405 asserted that if daughters were "taught the natural sciences, they would learn as thoroughly and understand the subtleties of all the arts and sciences" just as well as sons. Later in the century, Cassandra Fedele would echo the notion, claiming that "man is rightly distinguished from the beasts above all by [the capacity for] reason" and woman deserves her share as well.

In 1417 a manuscript of Lucretius's *On the Nature of Things* was discovered by Poggio Bracciolini. A Latin translation of Diogenes showed up in 1430. Lorenzo Valla, court historian to King Alfonso of Naples in 1435, became devoted to Epicurus. Valla also revolutionized the historical study of languages, and for linguistic reasons refused to believe that the Apostles' Creed had been composed by the twelve apostles. His book *On Pleasure* championed Epicurus, advocating a life spent in prudent delights: quiet wisdom, decent virtue, and good times. For his tastes, Valla explained, the Stoics were too interested in controlling the passions. The Inquisition found him heretical on eight counts, including his defense of Epicurus, but in a close call, King Alfonso stepped in and saved him from the stake.

In 1453 the Muslim Turks took Constantinople, killed the emperor, and thus ended the eastern Roman Empire, long called the Byzantine Empire. That collapse shook a lot of texts out of the old Byzantine shelves; good texts—in their original Greek or at least translated only once. In western Europe, also in 1453, Gutenberg's press turned out the first printed book, a Bible, and presses immediately began to pop up all over Europe, reproducing all the ancient material that was pouring in from the East. By 1473 the new presses had published *On the Nature of Things*. Only twenty years had elapsed between the first mechanically printed Bible and the first mechanically printed Lucretius!

In Italy, humanists noticed that they were living among the ruins of a real civilization. They launched campaigns to study the ruins and started to build in their style, or with lessons learned from them. Architecture was the first art to be transformed by the times: it was the great physical thing that the ancients had left behind. Then, of course, there were all the statues. Renaissance artists started copying those, too. It took a while: it is only in the High Renaissance that we see the first freestanding nude since ancient times, Michelangelo's *David* (1501–1504). But let us take a moment to note the boy's origins—they are the heart of the Renaissance mixture: the Hebrew story, in a Greek art form, informed by dissection, which was part of the new experimental mood that was just beginning and that would change the world. Despite the secularism of this outpouring of great art, the Italian popes tended to see it as a way to bring glory back to a weary Rome.

The ancients' paintings had not survived, but eventually classical ideas were transferred from the other arts. Painting in the Middle Ages had been decorative or instructional, and the human figures props for the story. There had been many paintings of the Last Supper throughout the medieval period, for example, but as beautiful as many of them were, they were pictograms of a sort. In Leonardo's *Last Supper* (1498), by contrast, each man in the painting responds to Jesus' announcement of betrayal according to the nature of his heart. This is still a biblical story, but it is now about the psychology of its human characters.

The Church was confident, but not utterly at its ease. By as early as the mid thirteenth century, a movement of Latin Averroism was centered in Paris and Padua. In 1513 Pope Leo X issued a condemnation of any teaching that concluded that the human soul was mortal. It was aimed at what was going on in Padua, and the condemnation was not effective. The great Paduan professor Pietro Pomponazzi (1462–1524) published all his books in the years just following it, and they all concluded that the soul is mortal. Pomponazzi had himself been educated at Padua, and he recounted that his fellow students had pressed their teachers for a rationalist evaluation of Averroës and Saint Thomas. Scholars in Florence were already generating Neoplatonic, mystical answers for such questions, but they were not to Pomponazzi's taste. Pomponazzi said that he, too, wrote of such matters because one of his students demanded a straight answer on the question of the soul, "leaving aside revelation and miracles, and remaining entirely within natural limits."[10] The straight answer was that he agreed with Aristotle and Averroës that the independent soul of a human being needs its body, and it exists only in its body.

Pomponazzi rejected the idea that people need threats of heaven and hell in order to be moral, reminding his readers of the heroic virtues of animals and even noting that self-interest could create such virtues as patriotism. Ghosts he rejected as a mix of vapors from the charnel house and human imagination. Demons and angels were not real. Possession and demonic prophecy were delusional states brought on by sickness or madness. What's more: "It is likely that the whole world is deceived in this common idea of immortality, for if we assume that there are three major religions—Christ's, Moses', and Muhammad's—either all of them are false and the whole world is cheated or two are wrong and the greater part of mankind is deceived."[11] He mentioned as atheists Epicurus, Lucretius, Diagoras, and some lesser-known figures, and counted these men as the few who had managed not to be taken in. Although he was sometimes issued warnings, Pomponazzi lived a full life, became a professor of philosophy at Bologna, and was considered the greatest Aristotelian of Italy.

Pomponazzi was doubt's philosopher, but the man people thought of as the great unbeliever was Niccolo Machiavelli. His book, *The Prince,* appeared in 1513, having been written as a job application—to display Machiavelli's abilities as a political counselor. It shocked Europe because the general political idea had long been that a leader should be a moral guide for people to follow into heaven—political philosophy still tasted of Charlemagne's tribalist theocracy. Great power had been seen as crucially linked to great goodness. A good ruler, Machiavelli said, had to lie sometimes, precisely because everyone else did it, even popes. Machiavelli was not the conniving politico his name implies nowadays, but he did critically question the relationship between high morality and worldly success. People of his time generally believed that Machiavelli had admired pagan religions above Christianity.[12] Among the last words of *The Prince* one finds a discussion of Christian principles: "These principles seem to me to have made men feeble, and caused them to become an easy prey to evil-minded men, who can control them more securely, seeing that the great body of men, for the sake of gaining Paradise, are more disposed to endure injuries than to avenge them." Religion should be in service of the state, he explained, and Christianity was not the best religion for the task.

Pomponazzi and Machiavelli were part of the ebullient, philosophical doubting that was going on in Italy in the prosperous culture of the Renaissance, in the bath of fresh texts from the Byzantine Empire. Italy was not entirely alone in this mood: in France the philosopher Girolamo Cardano celebrated Pomponazzi, whom he called "the great new Averroist of Padua."[13]

Yet for the most part, outside Italy, doubt in Europe in this period was about doubt in the Church. The Church had become a huge international bureaucracy and was often described as decadent. There were bishops who collected several salaries and lived in none of the dioceses that paid them, clergy with wives and children, surprisingly ignorant priests. Indulgences were on the rise, too. These were pieces of paper that one bought in order to help the soul of a dead loved one, or oneself, to attain passage to heaven. The slogan "As soon as the coin in the strongbox rings, a soul from purgatory springs!" surely did not come from the pope, but this was often the mood. There were different prices for the rich and the poor, and the Church was not shy about using the money for self-beautification rather than, say, charity. This all seemed like materialist foolishness to some; to other people, indulgences seemed a reasonable way to contribute to the

Church and to the glory of God and thereby do a good deed and help oneself at the same time.

Desiderius Erasmus laughed and lectured about all this in the same
spirit that he rejected Scholasticism, the Church's primary intellectual
world. His 1509 *Praise of Folly* made fun of the Scholastics to the point of
suggesting ridiculous new questions for them: "Could God have taken on
the form of a woman, a devil, a donkey, a gourd, or a flintstone? If so, how
could a gourd have preached sermons, performed miracles, and been nailed
to the cross?"[14] Erasmus complained that the Scholastics were "so busy night
and day with these enjoyable tomfooleries," that they haven't had time to
read the Gospel "even once through."[15] And elsewhere, "Who could have
imagined, if the savants hadn't told him, that anyone who said that the two
phrases 'chamber pot you stink' and 'the chamber pot stinks'. . . are equally
correct can't possibly be a Christian."[16] He used this foul metaphor to
remind his readers that theology had become extremely worldly in its concerns. When it came to the truth of it all, Erasmus shrugged. Listen to him
on what we can know of truth: "Human affairs are so complex and obscure
that nothing can be known of them for certain, as has been rightly stated by
my Academicians, the least assuming of the philosophers."[17] "Academicians"
meant Skeptics, the Middle Academy, and it is useful to note that Erasmus
based his opinion on the soundness of their conclusion that we could know
nothing clearly, but that he also preferred their mood. He also published an
edition of Lucian's *Dialogues*.

We come now to one of the great paintings in the history of doubt. In
1508 Pope Julius II's architect, Donato Bramante, recommended Raphael
Sanzio to paint for the pope's library. Libraries of Imperial Rome had had
portraits of great poets, and the practice was being revived in late-fifteenth-
century Italy. What Raphael did, though, was new. The fresco he painted
was called the *School of Athens* and in it were arranged, in meaningful poses,
all of doubt's old friends: Plato at center, holding the *Timaeus* and pointing
up; Aristotle holding the *Ethics* and gesturing toward our physical world;
Socrates at left, explaining something to Alexander the Great (in fancy soldier dress); and Alcibiades, another student Plato described in such scenes.
Xenophon listens in from under his black-and-white hat. The Stoic Zeno is
the old man with green cowl and book at far left, near where Epicurus
leans, looking well fed. He wears a laurel wreath and seems to be getting a
bit of a rubdown as he writes. Diogenes the Cynic, who lived in the street
and rejected the whole human effort against the universe, is the figure read-

ing on the steps, stretched out in the sunshine. Heraclitus is at the front left, seated and leaning, and is in fact a portrait of Michelangelo. Averroës is nearby in a turban. Ptolemy and Zoroaster chat on the right, holding a sphere of stars. Euclid and Pythagoras are also there. The architecture around them is yet another portrait: Raphael had seen Bramante's drawings for Saint Peter's. As for Raphael, he's in the front, far right, a twenty-seven-year-old looking directly at us.

So the Catholic Church had room for Erasmus's deep skepticism, as well as images of materialist philosophers and casual rejecters of all faiths. Erasmus's call for reform was just that, and certainly was not about "believers" versus doubters. That's where Martin Luther came in. By the time he found himself rejecting the selling of salvation through the indulgences (a new, jubilee indulgence had just been announced to help finish Saint Peter's), on a church door in Wittenberg in 1517, Luther had already come through great agonies of doubt. He could not convince himself that he had done enough to deserve salvation. Luther was near despair when he turned to the works of Saint Paul and found his answers.

It was the other great shock of the old. Erasmus and Luther returned to the ancient Christian texts just as Petrarch and Valla had returned to the ancient philosophers. The shock of Paul's claim of "justification by faith alone," and of his suggestions of predestination, convinced Luther: Christianity should be about faith alone. Paul had thrown out the laws of Moses. Fifteen hundred years later, the tense situation of belief being more important than acts was purposefully revived. The Bible text was described as literally true and the community who joined Luther challenged itself simply to believe it, as Paul had asked believers to accept the resurrection and as Jesus had asked the crowd to believe his miracles and that God's kingdom had come. Life within such challenges of belief is an inner battle to be certain; it can be ecstatic and immediate when it's good, and full of nail-biting worry when it is bad. The new Protestant doubt drew freely upon the image of Augustine in his garden, wracked with struggle.

Luther wrote *Against Scholastic Theology,* a list of ninety-seven points asserting, for instance, that "It is an error to say that no man can become a theologian without Aristotle. This in opposition to common opinion."[18] In fact, "no one can become a theologian unless he becomes one without Aristotle." It's not just the ancient philosopher Luther is rejecting: "In vain does one fashion a logic of faith." We should not even want rational proofs to believe, let alone speculative doubting. "If a syllogistic form of reasoning holds in divine matters, then the doctrine of the Trinity is demonstrable and not the object of faith." In case we missed his point, he says, "Briefly, the whole Aristotle is to theology as darkness is to light. This is in opposition to the Scholastics." What Luther rejected in the Reformation was not just Church corruption. It was the primacy of Aristotelian logic that had dominated education and intellectual culture.

Luther also scolded Erasmus's tactic of Skeptical obedience. Doubt was not Luther's cup of tea. Answering Erasmus, he wrote, "[Y]ou foster in your heart a Lucian, or some other pig from Epicurus's sty who, having no belief in God himself, secretly ridicules all who have a belief and confess it."[19] Luther was dazzling with a harsh word. He continues, "Permit us to be assertors, to be devoted to assertions and delight in them, while you stick to your Skeptics and Academics till Christ calls you too." Then he delivers one of the great lines in this history: "The Holy Spirit is no Skeptic, and it is not doubts or mere opinions that he has written on our hearts, but assertions more sure and certain than life itself and all experience."[20] The idea that the Holy Spirit might be a Skeptic is Luther's own flight of fancy; Luther's gift to the history of doubt.

Of course, the Catholics saw the reformers as the real skeptics and atheists, for they had ditched the common gestures of worship and disavowed the authority of the Church. This generated a lot of uncertainty. Luther said the Church had no authority and that individuals could feel what the truth was: easy, perhaps, when the subject was priestly sexual improprieties or the hawking of indulgences, but on more subtle questions, like the usefulness of confession, the sense of moral obviousness was gone. The whole thing begged the question: If we do not trust the Church to know the truth, why should we trust ourselves? How can we know? It might have precipitated a skeptical crisis by itself, but since Cicero and Sextus were becoming increasingly available, that was not necessary.

In the next generation, Calvin pushed Luther's justification by faith a step further: Luther said we were all too sinful to be saved by acts; only faith and God's grace could save us. That left room for some gestures on one's own behalf; one could struggle to believe. Calvin said omniscient God knows before we are born who is capable of such belief, so we are each damned or blessed before we cry our first tear. The struggle here becomes to know that one has been saved and is among the Elect, by finding oneself to be sufficiently full of belief. In this exceedingly anxious situation, all of the churches, the Roman church and the various Reformed churches, regularly accused one another of atheism. A word that had not been much used since ancient times was suddenly everywhere. There were countless books devoted entirely to discussing the varieties of atheism. In Geneva, when Jacques Gruet, a prominent member of the bourgeoisie, protested Calvin's increasing domination of the city's public life, Calvin had Gruet burned as a "speculative atheist," that is, one who intellectually rejected the faith. Calvin insisted that papers found at Gruet's house proved the matter, but they were burned with him, so we do not know more. Calvin had a man named Monet beheaded for being a "practical atheist," that is, one who behaved as if there were no God. Monet had a book of pornographic pictures that he called "my New Testament."[21]

Increasing numbers of the ancient authors were accused of atheism, and an expanding cast of relatively contemporary writers as well. It is striking to see the word suddenly all over so many manuscripts and letters. Frequently, the people being accused were more reasonably seen as zealots of one sort or another than as unbelievers, yet this new use of the word is also part of the history of doubt. Lucian was generally understood as an atheist by Renaissance and Reformation Europeans. When they accused each other of atheism, they delighted in spicing up the indictment by calling their target a "slave of Lucian" or a "student of Lucian." This public worry over atheism happened because authority and sufficient belief were such tense issues, suddenly, again. Many who were themselves more certain thought their opponents had rejected God's decrees, and that seemed like atheism; many people accused others of atheism because doubt troubled them; and some people were reporting on real, serious doubt.

Meanwhile, in 1543 the Pole Nicolas Copernicus published a book called *On the Revolution of Heavenly Spheres.* He had been sent to Italy to study medicine so he could serve as his powerful uncle's private doctor. There he picked up some Neoplatonism, which was carrying on Plato's reverent equation of the sun with the ultimate good. It seems to have influenced his sense that the sun was worth going around. He did not publish his big idea until he was on his deathbed and he claimed he did not believe in the system he described, but merely found it useful for calculations. By the old system, to make the planets' observed orbits match the worldview of an Earth-centered solar system, you had to have the planets going around their orbits in wacky epicycles. Copernicus found that if you pretend the sun is at the center, you find the planets almost exactly where they would seem to be from Earth. That is, you could predict the path of the planet through our sky. Most people were first aware of or convinced of heliocentrism by the work of Galileo in the next century. Here, note that henceforth some people were confronting the possibility that humanity is not the center of anything—heaven no longer above us and demons no longer below.

Amid all this, Ignatius of Loyola, a Spanish nobleman, experienced a doubt similar to that of Luther and decided to answer it with blind dedication to the pope. He set up the Jesuits as "shock troops of the papacy" dedicated to converting people all over the world to Catholicism and thus making up for the pope's losses in Europe. Meanwhile, the Inquisition was killing Protestants, Protestant leaders were killing various dissenters, spontaneous massacres arose in which neighbors killed neighbors, and there were full-on Wars of Religion. It was a bloodbath.

THE HISTORICAL PROBLEM CALLED RABELAIS

A center of fresh doubt in these years was at the court of Margaret of Navarre. She lived from 1492 to 1549, right through the tumult of the new explorations, new religious movements, and the height of the Renaissance. A near-contemporary included the following story in his *Lives of Illustrious Women:* When one of her maids was fatally ill, Margaret sat by her bedside with such strange concentration that other ladies asked her "why she looked with so much attention on that poor dying creature." Indeed, the queen "never stirred from [the girl's] bed-side, as long as she was agonizing, looking her earnestly in the face, without interruption, till she was dead." The reason, replied the queen to her ladies, was that she had heard many learned men assert that the soul left the body the moment it died. Queen Margaret wanted "to see if there came from it any wind or noise, or sound on the removal and going out of the soul," but in the end, she reported, "she could perceive nothing like it." It was a serious matter: "she added, that if she were not well settled in her faith she should not know what to think." As it was, though, "she would believe what her God and her Church commanded her to believe."[22]

As we started to see, at this time people thought of Italy as the breeding ground of atheists, and northern France was getting a bit of a reputation as well. Navarre, on the border of France and Spain was about to join them, and this had a lot to do with our Margaret of Navarre. A writer of stories, plays, and poems, she was best known for her *Heptaméron* (1558), an original collection of stories in the genre established by Boccaccio's *Decameron.* Her brilliant court was frequented by literary men, among them the famous writers Etienne Dolet and François Rabelais, both of whom have been thought of as atheists for centuries. As queen consort of Navarre and sister of King Francis I of France, Margaret could do as she pleased to some degree. She was a strong supporter of religious liberty and mild church reform. Did she doubt? Did Dolet and Rabelais?

There is a historical fight over disbelief that centers on these two men. When Rabelais was alive, a lot of people called him an atheist, an "ape of Lucian," and a drunk. He wrote books about two giants he invented—a father and son, Gargantua and Pantagruel—and he gave them lives full of messy carousing, complete with unceremonious sexual adventures, excretions, and general muck. These books were funny then and they are funny now. In them, many people make fun of priests, scripture, Church hierarchy, and ritual. What historians have gone back and forth on is what Rabelais himself actually believed.

At the end of the nineteenth century, a historian of great prestige, Abel Lefranc, known as "the commanding general of an army of Rabelais experts," claimed that Rabelais was an aggressive atheist, barely hiding behind his literary veil.[23] Lefranc had terrific evidence, not least that Rabelais's colleagues called him an atheist, and vociferously! Lefranc's book became the last word on the subject for a generation. Rabelais was an atheist. Then along came Lucien Febvre, who read the classic book and, he tells us, found the thesis absurd. In 1947, in his celebrated *The Problem of Unbelief in the Sixteenth Century: The Religion of Rabelais,* Febvre claimed that Rabelais not only had not been an atheist but *could not have been* an atheist. The great critique of Febvre's day was against history that impressed modern values on the past. He was saying that historians had been doing this to the question of irreligion, reading our modern, secularist ideals into history. Febvre proceeded to show that Lefranc had made a lot of errors reading his sources. Back then, Febvre explained, simply everyone used the word *atheist* to insult one another. Furthermore, upon careful consideration of complicated texts, Febvre was able to show that most of the jibes Lefranc had thought were aimed at Rabelais were actually being hurled at Etienne Dolet, Rabelais's friend at the court of Margaret of Navarre.

To support his claim that the "mentality" of the times could not allow atheism, Febvre got a lot of his information from tinkers, tailors, and tanners, as well as the usual theologians and kings. The "mentalities" idea and the attention to popular opinion helped shape the way history was done in the twentieth century. But what Febvre claimed here was pretty extreme. He held that the mentality of the time was such that there were no intellectual replacements for the answers theism offered, so no one could have rejected them. In fact, he claimed that no one could have been an atheist until Descartes made it conceptually possible.

What then of Rabelais's friend Dolet? Febvre himself offered good evidence that contemporaries saw Dolet as an atheist. He was clearly a doubter, and he was killed for it—hanged and burned at age thirty-seven. He would be called the "Martyr of the Renaissance" in the sixteenth century and a "Martyr of Free Thought" in the nineteenth, but Dolet had actually tried to stay out of trouble, writing that he hated persecution but that in response to it: "I play the part of a spectator. I grieve over the situation, and pity the misfortunes of some of the accused, while I laugh at the folly of others in putting their lives in danger by their ridiculous self-will and obstinacy."[24] He was no fanatic.

Yet he had some things to say. Dolet put at the end of one of his books, "Death? Let us not fear its blows. It will either grant us to be without feeling or it will allow us to enter a better world and a happy state—unless our hope of Elysium is entirely groundless." Contemporaries understood what it meant to consider that death may "grant us to be without feeling." A contemporary wrote of Dolet, "Sneer away, you ape of Lucian. . . . To deny the existence of a God in Heaven who wished his son to die for the salvation of men . . . to deny the Last Judgment and the punishments of Hell—this is madness."[25] In the end, as I said, he was burned for it.

Dolet alone should have thrown the "no real doubt" thesis into turmoil. But there's more: Febvre himself, only a few years after publishing his book on Rabelais, found a text by Bonaventure des Périers called *Cymbalum mundi*. It had been published in 1537, and it struck Febvre as the real thing: unbelief. He dealt with it by claiming that des Périers had been unusual, an exception, because he happened to have access to ancient texts that sufficiently augmented his mentality. The thing was that Febvre's *The Problem of Unbelief* was so impressive in its use of sources and reconstruction of cultural oddities, and his invective against Lefranc was so mocking and bilious, that even after "mentalities" history was no longer new and dominant, historians have not wanted to come out in favor of unbelief in this period.

This almost unconscious fear of the subject has kept scholarship away from it for the last sixty years. Some historians studied the deep rebelliousness within popular forms of religion and some studied the various heresies, and that nuanced the picture, but did not changed it. The dominant understanding remained that real, serious doubt in God was impossible in this period.[26] In the last few decades, however, a few historians, especially in Europe, have stood up and said that the whole thing was ridiculously overdrawn. Scholars like Susan Reynolds, Michael Hunter, and David Wootton have cried foul that half a century has gone by and this essential subject, in one of the most productive periods of historical scholarship ever, has been bizarrely ignored. Maybe all that evidence of people calling each other atheists actually meant some people were atheists. This revision, however, has not yet reached far, so most people still tow Febvre's line. Even the fine religious historian Karen Armstrong in her *History of God*, in 1994, writes that "As Lucien Febvre has shown in his classic book *The Problem of Unbelief in the Sixteenth Century* the conceptual difficulties in the way of a complete denial of God's existence at his time were so great as to be insurmountable . . ." No one could be an atheist "[u]ntil there had formed a body of coherent reasons, each of which was based on another cluster of scientific verifications. . . . Without this support, such a denial could only be a personal whim or a passing impulse that was unworthy of serious consideration."[27]

But things are changing. The medieval historian Susan Reynolds argues that the idea of "the Age of Faith" persists "more or less unnoticed rather like a shabby old chair in our mental sitting-rooms. . . . It is rather rickety, and goodness knows where it came from, but it looks all right now that we have smartened it up with the new loose covers of heresy and 'popular religion.' All the same, the old chair is still there underneath. It is constructed out of the assumed credulity, the incapacity for atheism, of the medieval mentality."[28] Reynolds argues that there is no reason to be certain that even in completely "untouched and traditional societies" there are not some people who doubt. There is certainly no reason to believe in a complete lack of doubt in a culture so variously reminded of other religions and philosophies, ancient and foreign. Reynolds asks why else people like Anselm were worrying about proving God's existence: "They clearly knew about unbelief and regarded it as dangerous, even before the thirteenth century brought so much more of Aristotle as well as other classical pagans and some Jewish and Islamic philosophers to their notice."[29] Most important, Reynolds writes that we cannot keep explaining away evidence of atheism based on the idea that there could not have been any, which is itself dependent on the idea that we cannot find any. Reynolds's lecture on the subject in 1990 is the most colorful attack on the "mentalities" idea of medieval faith, but there are others like it, many collected in Michael Hunter and David Wootton's *Atheism from the Reformation to the Enlightenment*.[30]

The problem had always been thought of as such: "no one" in this period called *himself or herself* an atheist; indeed, many who have been called doubters or even closet atheists for these past centuries specifically denied it. Febvre had even mocked historians who called such figures atheists because in so doing, the historians were calling them liars and cowards. After all, he said, other heretics were willing to die for their beliefs. Well, yes, but first of all, one does not die for atheism as one dies for a god. Second, atheists have a long history of pride in silently smiling and going about their business with respect for the believers and compassion toward their needs. What is much more, however, is that indeed there *were* people who announced that they did not believe in God. We have seen some explicit suggestions of it in condemnations as early as the thirteenth century. The real mother lode, however, is in the sixteenth-century documents from the Inquisition. In the past few decades, historians have been at work finding these documents and interpreting them. They reveal actual men and women who explain how they came to doubt.

Before looking at the Inquisition papers, we should note first that the passages in Rabelais that Abel Lefranc read as atheist, and Febvre read as "not atheist," were part of the canon of doubt, either way. How much of a doubter was Rabelais? Febvre tells us not to take the scatological humor and the bawdy jokes as evidence of doubt, and certainly not to read things into it—one historian had suggested that when one of the giants farted out little men and women it was a critique of virgin birth, and we may agree with Febvre that this is a bit of a stretch. Then again, we may not. In either case, the world we are shown by Rabelais is not one of rational unbelief, but it also is not one of piety or traditionalism. This is how the book begins:

> MOST noble boozers, and you my very esteemed and poxy friends—for to you and you alone are my writings dedicated—when Alcibiades, in that dialogue of Plato's entitled *The Symposium*, praises his master Socrates, beyond all doubt the prime of philosophers, he compares him, amongst other things, to a Silenus.[31]

A *Silenus,* he tells us, is a fancy painted box that holds serious drugs. Socrates liked to drink and carouse, we are further told, and Plato warned that many people missed the real interior meaning of his teacher because of this, so he calls him a Silenus. Likewise, warned Rabelais, one should endeavor to see the deeper meanings of his seemingly comic books.

Some of Rabelais's material is more outrageous than anything in any teen movie today. In one famous story, some men from one island visit another during a holiday; they get silly and one of them responds to a paraded picture of the pope by making an obscene gesture—called "showing a fig." A few days later, the insulted hosts avenge themselves, arriving at the offenders' island, killing many, and offering the remainder a choice: death or, if they are willing, a task: a fig is placed in the "private parts" of a donkey and anyone willing to both remove that fig with his teeth and then replace it again would be granted his life. Rabelais tells us that some died and some went for the fig. Those that performed the fig maneuver were called Popefigs ever after.[32]

Amid the gross stories, written in tavern slang, there is a letter from Gargantua to his son, Pantagruel, written in high style and with exquisite prose. This letter has always been taken as a singularly immediate commentary on Rabelais's historical moment—it's one of the great descriptions of the Renaissance as seen from within. It has also been read as the closest thing to a straight confession that Rabelais, or his literary character, did not accept the Christian afterlife. The setup is that Pantagruel is away at school, and the letter from his father is intended to hearten the boy toward his studies. First Gargantua explains that parents care so much about a child's education because of the realities of their own mortality.

> [W]e can, in this mortal state, acquire a kind of immortality and, in the course of this transitory life, perpetuate our name and seed; which we do by lineage sprung from us in lawful marriage. By this means there is in some sort restored to us what was taken from us by the sin of our first parents, who were told that, because they had not been obedient to the commandment of God the Creator, they would die, and that by death would be brought to nothing that magnificent form in which man has been created.
>
> But by this method of seminal propagation, there remains in the children what has perished in the parents. . . .
>
> When, at the will of Him who rules and governs all things, my soul shall leave this mortal habitation, I shall not now account myself to be

absolutely dying, but to be passing from one place to another, since in you, and by you, I shall remain in visible form here in this world, visiting and conversing with men of honour and my friends as I used to do.[33]

Gargantua also says to Pantagruel: "to see my grey-haired age blossom afresh in your youth . . . I shall not now account myself to be absolutely dying."

The letter mentions God, as we see above, but no use is made of him. One deals with the fact of death by mortal, human means. Gargantua's letter gives his son a lot of philosophy, and God is on the sidelines there, too. Instead, the thing to do is study. When Gargantua was himself a young man, his education was poor because, he reports, "the times were still dark." Nowadays, the humanist's life is available to all, and his son must embrace it.

> Now every method of teaching has been restored, and the study of languages has been revived: of Greek, without which it is disgraceful for a man to call himself a scholar, and of Hebrew, Chaldean, and Latin. The elegant and accurate art of printing, which is now in use, was invented in my time, by divine inspiration; as, by contrast, artillery was inspired by diabolical suggestion. The whole world is full of learned men, of very erudite tutors, and of most extensive libraries, and it is my opinion that neither in the time of Plato, of Cicero . . . were there such facilities for study as one finds today. . . . I find robbers, hangmen, freebooters, and grooms nowadays more learned than the doctors and preachers were in my time.
>
> Why, the very women and girls aspire to the glory and reach out for the celestial manna of sound learning.[34]

It is not an antireligious letter, but it is worldly. Since Rabelais started his book with a whisper to read it for unspoken claims and hidden messages, the book has been experienced by many as pointedly irreligious. Rabelais's place in the history of doubt is as a sage of secular life, and as a shocking jester, as the bulk of his work is outrageous enough to break traditional barriers of decency in any age. The quiet moments of contemplation he created were in the mode of the graceful-life philosophies: confronting death as an absolute, and making some meaning based on loving one's family, having a carousing good time now and again, drinking deeply of ideas, letting that metaphor end in a hangover now and again, and doing the work of wisdom—reading, writing, and meditation on the truth. The French poet Joachim Du Bellay, a contemporary of Rabelais, wrote him this poem:

> *Who can with so much learning write*
> *That now we have at last in France*
> *Democritus reborn, whose lance*
> *Cuts through the Schoolmen's murky light.*[35]

Democritus reborn, no less. After Rabelais died in 1553, Du Bellay wrote another poem, this time from the great jester's point of view. Thus we have the posthumous Rabelais with these words in his mouth: "Sleep, gluttony, wine, women, jest and jibe: these were my gods, my only gods, when I was alive." Maybe so, maybe not. In either case, along with Socrates and Ikkyu Sojun, Rabelais was one of doubt's most prominent jesting, drinking sages.

THE INQUISITION

It was just after the publication of Rabelais's books that the Inquisition really got going. We should know a few things. The Inquisition had been started to keep an eye on hertics, Jews, and Muslims and then turned its tongs and talons on the Protestants. That led to a lot of documentation about what people actually believed. These records are sketchy, but fascinating. The voice of doubt among common people in Europe had been beyond our hearing, and with these trials it is suddenly audible. These common doubters will draw on many of the sages and poets of doubt that we have seen arise out of the aging world: Greek, Hebrew, Eastern, Roman, and Arab. Yet, we would do well to listen to the independence of thought experienced by the people themselves. A woman from Montaillou was asked where she got her doubts about hell and the resurrection and she said that she got them from no one: she thought of them for herself.[36]

Information is sometimes very thin, but we get an idea after only a few examples: In 1497 a Gabriele di Salo was tried at Bologna for claiming that Christ's miracles were natural phenomena.[37] In 1533 the Venetian Council of Ten complained that the friars of San Fermo were responsible for many "filthinesses" with the nuns of the Magdalene and "do not want to live under the rule of their founder, but as sons of iniquity . . . as Epicureans and Lutherans."[38] The weird combination of accusations against the friars is probably to be explained because, this early, the term Lutheran was often used for anyone who questioned religious doctrine.

In 1550 an Anabaptist (radical Protestants best known for rejecting infant baptism) religious council in Venice declared that the impious die with their bodies and the elect sleep until the day of judgment. They believed in no immaterial beings and explicitly denied the divinity of Christ.[39] In 1558 Catherine de Medici's cousin, Pietro Stozzi, a mathematician who translated Caesar into Greek, was dying, having been wounded in the siege of Thionville. On his deathbed he plainly renounced God and denied immortality. The evening before, it was reported, he had asserted that the scriptures were a fiction.[40] Others fought the trend, but still reported on it. Consider the revival of ancient Skepticism as described by Guillaume Budé (1467–1540), the French humanist responsible for starting both the Collège de France and the Bibliothèque Nationale:

> Oh God, Oh Savior, misery, shameful and pitiless fault: we believe Scripture and Revelation only with difficulty. . . . Such is the result of frequenting cities and crowds, mistresses of all errors, which teaches us to think according to the method of the Academy and to take nothing for certain, not even what Revelation teaches us concerning the inhabitants of heaven and hell.[41]

It's true that the city can give one a case of cosmopolitan relativism, and it is nice to hear him tell us how far it went. A Professor Galland at the Collège de France wrote this:

> All of the other sects, including even that of Epicurus, busy themselves with safeguarding some religion, while the Academy strives to destroy all belief, religious or otherwise, in men's minds. It has undertaken the war of the Titans against the gods. How would he believe in God, he who holds nothing as certain, who spends his time refuting the ideas of others, refuses all credence to his senses, ruins the authority of reason! If he does not believe what he experiences and almost touches, how can he have faith in the existence of the Divine Nature which is so difficult to conceive?

We've seen that Diogenes was already in print, and in 1562 Sextus Empiricus was published for the first time in the Renaissance. By 1569 all his work was in print and it was continuously reprinted and translated after that date. This ancient Skepticism would have tremendous impact.

By late in the sixteenth century, tensions that had begun earlier that century among the various versions of Christianity had erupted into unprecedented violence. To take one prominent example: In these years, hundreds of noble families, many of them related to one another, lived at the royal palace in Paris, each installed in their own apartments. Some were Catholic and some Reformed, and these were mixed within families. Before dawn on August 24 in 1572, some of the Catholic noblemen, accompanied by the king's guard, went around and killed about a hundred of these Reformed Church friends and family members. The king had feared a Protestant coup and meant to stop it; that is, it was a political issue specific to these particular Protestants. But, hearing of it, people felt this was an encouragement to get rid of French Protestantism, and in the next three days Catholic Parisians murdered about three thousand of their fellow townsmen. As the news spread throughout France, several thousand more Protestants were killed. Forever after in the history of doubt, the Saint Bartholomew's Day Massacre (named for the feast day on which it began) would be cited as evidence that religion, especially any religion that thinks it has a monopoly on truth, does more evil than good.

Returning to the Inquisition, in a 1573 "denunciation" text we learn that Matteo de Vincenti, a woodworker in Venice, left a sermon about the doctrine of the "Real Presence" on Palm Sunday, saying, "It's nonsense, having to believe these things—they're stories. I would rather believe I had money in my pocket."[42] Geoffroy Vallée was executed in 1574 for denying God. He cited the unbelief evident in Ecclesiastes and Psalm 1.[43] More people than ever could now read the Bible, and have access to a copy of it, and could thus avail themselves of Ecclesiastes' graceful-life philosophy. As for the first psalm, we recall that it specifically mentions "the counsel of the ungodly," the behavior of "the ungodly," estrangement of "the ungodly," and the ultimate punishment of "the ungodly."[44] It was enough to suggest to Vallée that some people did not believe in God. On trial, he said that he picked up these opinions from conversations with learned men in his foreign travels.[45] Vallée complained that believers recited "like parrots" the irrational views that they had memorized before they left the cradle. Instead, he asserted, one should believe only what one could learn by the senses, and those ideas for which one could show rational proof. The inquisitors asked him if he disbelieved everything for which he had no direct evidence, and it was recorded that he "did not know how to respond."[46] That is understandable. As historian Nicholas Davidson notes, it is one thing to have a gut sense of what rational or reasonable proof might be, but it is something else again to be able to parse probabilities for various hypotheses, recite evidential criteria, and then compare the strengths of various claims. This was tricky stuff and Vallée burned.

Also in 1574 the Venetian Inquisition received a denunciation against Commodo Canuove of Vincenza, accusing him of saying "we have never seen any dead man who has returned from the other world to tell us that paradise exists—or purgatory or hell; all these things are the fantasies of friars and priests, who wish to live without working and to pamper themselves with the goods of the Church."[47] In 1575 a physician named Pietro Sigos was denounced for claiming that images cannot produce miracles—"it's simply not possible; it's all an invention of the priests to get more money."[48]

In 1579, in Venice, a craftsman named Pietro was said to deny God.[49] In 1580 Alvise Capuano was sentenced as an atheist, having been denounced to the Inquisition in 1577. The trial documents summarize his confessions, saying he believed

> the world was created by chance . . . that when the body dies the soul dies also . . . that Christ was the adopted son of the Madonna, born as other men are, and that angels and demons do not exist . . . that there are no true witches, and that belief in witchcraft arises from melancholic humors . . . the world has neither beginning nor end . . . Christ's miracles were not true miracles but natural acts . . . and that the only law that must be obeyed is the law of nature.[50]

We do not know where Capuano got his ideas, but we know now that they were in circulation in a variety of cultural conversations.

In the 1581 confession of Evangelista de Vintura to the Venetian Inquisition, there is something we have not heard before. He confessed in writing that since the plague of 1555 killed his mother, his brothers, and his sisters and separated him from all his property, he had "lived far away from the true and correct Christian way of life."[51] He explained, "I doubted that God could have any providential control over events, since he had treated me so badly." This was a Job who said no to God, and that is how it stayed for some thirty years. Perhaps such sorrow only ignited doubt in a century in which talk of doubt was available to any reader, or perhaps it was always common to doubt providence when you knew that you had demonstrated more moral character than the world had shown you. The Inquisition notes let us hear it.

In the later sixteenth century, Noel Journet wrote two manuscripts showing inconsistencies in the Bible story and concluded that the whole of the Good Book was a fable.[52] He asked his readers how Moses could have written Deuteronomy when the book tells of Moses' own death. Journet also thought Jesus had been fully human and thus an impostor. Since Journet thought people needed religion, but that Christianity and its idea of God were wicked, he set out to do a better job by starting a new religion. In 1582 the Inquisition burned him on a pyre with his books. We have no copy of them, and know what we know of his thoughts only from the trial documents and from a contemporaneous refutation of one of his destroyed books.

At last we come to a piece of peasant doubt that can be fleshed out a bit: it is the tale of a miller in Italy. The year 1584 saw the first trial of Domenico Scandella (1532–1599), known as Menocchio. Menocchio's story has been carefully recomposed by the historian Carlo Ginzburg, and was offered in part as a response to Lucien Febvre's thesis of the impossibility of unbelief in the sixteenth century.[53] Menocchio was neither well off nor extremely poor—he rented two fields, looked after a large family, and scraped by milling and selling grain. By the time of his trial he had served as mayor of his village and had been administrator of a parish church. Witnesses at his trial said that they liked him a lot, and that so did most people, but that "He is always arguing with somebody about the faith just for the sake of arguing—even with the priest."[54] He was fifty-two years old at the trial, and friends said he'd been intense and vocal about religion for thirty years.

Here are a few of the things he was reported to have said: "The air is God . . . the earth is our mother" and "Who do you imagine God to be? God is nothing but a little breath, and whatever else man imagines him to be."[55] He also said: "What did you think, that Jesus Christ was born of the Virgin Mary? It's impossible that she gave birth to him and remained a virgin. It might very well have been this, that he was a good man, or the son of a good man."

At his trial Menocchio said, "I believe that the law and commandments of the Church are all a matter of business, and they make their living from this."[56] But this was not all about economics. A version of Menocchio's cosmology that ended up in the Holy Office read:

> I have said that, in my opinion, all was chaos, that is, earth, air, water, and fire were mixed together; and out of that bulk a mass formed—just as cheese is made out of milk—and worms appeared in it, and these were the angels. The most holy majesty decreed that these should be God and the angels, and among that number of angels, there was also God, he too having been created out of that mass at the same time and he was made lord, with four captains, Lucifer, Michael, Gabriel, and Raphael.[57]

Where did he get this stuff? There were Anabaptists around, and some secret Lutheran groups, but when asked about predestination and justification, Menocchio did not know the meanings of the words. Anyway, those ideas have little in common with the cheese-and-worms theory.

The cheese and the worms seem to have belonged to a peasant culture that was many centuries old and that had never really been much more than adjusted to make room for Christian doctrine. Menocchio himself vociferously declared, "Sir, I have never met anyone who holds these opinions; my opinions came out of my own head."[58] Yet books mentioned by him at the trial show there were influences from learned culture: he had borrowed books from a wide group of people, including a woman. One, surprisingly, seems to have been the Koran. One was the *Travels* of Sir John Mandeville, which he said upset him a good deal. After describing a land where cannibalism at death was an act of supreme respect and love, Menocchio wrote, "From there I got my opinion that when the body dies, the soul dies too, since out of many different kinds of nations, some believe in one way and some in another."[59] Elsewhere he says that "what I have said came from that book of Mandeville that I read," which he later calls "that book of Mandeville about many kinds of races and different laws that sorely troubled me."[60] Mandeville ends his book with a plea for tolerance of all these peoples. Some of them, he says, possess our best virtues without knowing anything of God or of the Bible. Explaining Menocchio, Ginzburg writes that Mandeville's *Travels* was "an echo of medieval religious tolerance, [that] reached even the age of the wars of religion, of excommunications, and of the burning of heretics."[61]

In his trial Menocchio told the story of the three rings: A king told his three sons that whoever was given his ring would be his successor. Before the king died, he made copies of his ring and gave one, in confidence, to each of them. So, Menocchio explained, did God do with his children: the Christians, the Jews, and the Turks. Said he, "I do believe that every person considers his faith to be right, and we do not know which is the right one: but because my grandfather, my father, and my people have been Christians, I want to remain a Christian, and believe that this is the right one."[62] He'd gotten the story from Boccaccio. Menocchio had borrowed the *Decameron* from a painter friend of his. It was a prohibited book.

Ginzburg did some historical sleuthing and found a path from the University of Padua and its Averroist circles to this village miller. A childhood friend of Menocchio was a priest and was himself prosecuted for heresy (in part for being a "whoremonger and ruffian") a few years before Menocchio's trial. At this trial we learn of a Paduan professor with whom the accused had had some chance, meaningful contact. We hear of a few other possible conduits as well, but Menocchio had a vernacular Bible and that could have been enough. "As for the things in the Gospels, I believe that parts of them are true and parts were made up by the Evangelists out of their heads, as we see in the passages that one tells in one way and one in another way."[63] His fellow villagers said he used to repeat, in words close to those in Ecclesiastes, "when man dies he is like an animal, like a fly . . . and when a man dies his soul and everything else about him dies."[64] At other times he appears independent of all books, musing that "If he really was God Almighty he would let himself be seen."[65]

The Reformation and the printing press made Menocchio's philosophy possible by supplying him with the books that he used to support his popular knowledge and his own ideas. At the very least, they gave him courage and loosened his tongue. Of Reformation specifics, Menocchio seems to have thought Luther was a man who had brazenly questioned doctrine and had overthrown the priests, that Lutherans were other people who bravely doubted, and that Geneva was a town that was a haven for such people, where one lived in religious freedom. He once joked with a friend that at his own death, "Some Lutherans will learn of it and will come to collect the ashes."

In his first trial, although he tried to hold his peace, the examiners goaded Menocchio and he had a wonderful time giving in; for pages and pages he sounded more like he was lecturing than confessing. The result was nasty solitary confinement for years at the expense of his family, which fell apart as a result. His wife and the son to whom he was closest both died.

Menocchio finally got out, swearing that he had repented. He fantasized a bit about going to Geneva, but he was not a man to go far from his village—a good thing, too, as Geneva was not at all the place he imagined. He was more careful now, but a few injudicious remarks brought him back before the Inquisition. The twenty years between his trials had been full of turmoil and the Protestant Reformation had become connected with many economic and political upheavals. It was against a sense of social, intellectual, and political free fall that tolerance for doubt began to disappear. Established interests of the Reformation Church and the state governments all wanted to crush that aspect of the movement that encouraged the questioning of authority. Menocchio said less this time, but by now each word meant more, and he was put to death.

The trials keep on coming. In 1586 the Inquisition tribunal in Venice, interrogating Girolamo Garzoni, concluded "that you have not believed anything in our faith and that you are consequently an atheist—that is, that you do not believe there is any God in this world and that the world was created by chance."[66] Pomponio Rustico was executed in Rome in 1587 for many doubts, including the view that "the stories described in the Bible . . . are worthy only of derision." There was also some great doubt among more public personalities in this period, and no story was more powerful than that of Giordano Bruno.

It is best to hear the highlights of Bruno's life straight through, so that the excommunications chime like a musical scale. Born in 1548, Bruno was only fifteen when he entered the Order of Saint Dominic. Nine years later, in 1572, he was ordained a priest. By 1576 a formal accusation of heresy had been brought against him. He went to Rome, and soon enough the accusations were renewed against him at the Convent of the Minerva. Within a few months of his arrival, he left the city and the order. In 1579 he seems to have been in Geneva and to have converted to Calvinism. He later denied having done it (at a Catholic tribunal in Venice!), but we know he was excommunicated by the Calvinist Council and asked to leave Geneva. He stayed out of trouble in France for a few years and in 1583 he went to England and for a while was a favorite of Queen Elizabeth. But when Oxford's theologians mocked his ideas and would not hire him, he attacked the professors in print and was back in hot water. In 1585 he returned to France. He went to Germany in 1587 and was there excommunicated by the Lutherans. In February 1593 Bruno was sent to Rome, where he was thrown in a dungeon for six years. He was burned at the stake in 1600.

How could anyone get into so much trouble? Well, let's check his tone first. Here is a snippet of what he wrote about his interview with the Oxford theologians:

> They spoke Latin well, [were] proper men, . . . of good reputation . . . fairly competent in learning but mediocre in education, courtesy and breeding . . . , for 'tis *yes my master; yes my Father, or my mistress; yes sir forsooth;* . . . elect indeed, with their long robes, clad in velvet. One wore two shining gold chains about his neck while the other, by God, whose precious hand bore twelve rings on two fingers, had rather the appearance of a rich jeweler. . . . Did they know aught of Greek? Aye and also of beer. . . . One was the herald of the idol of Obscurity and the other the bailiff of the goddess of Presumption.[67]

They had treated him badly when they found he could not speak to them in their own Scholastic terms. Bruno disagreed with the Church on countless points. The one that was most shocking was his complete conviction that the universe was infinite and filled with many other suns like ours. Bruno was one of the earliest supporters of the Copernican model. He believed that planets like ours go around these other suns, and that these planets were populated and busy with life.

Bruno read Lucretius and he read Copernicus. He also knew some pre-Socratic philosophy and a good deal of Neoplatonism. He put it all together with one terrific insight: the Copernican heliocentric system—and its destruction of the stable and contained Earth-centered universe—gave weight to Epicurus' claim that our world might be one among many. Adding the Neoplatonism and the pre-Socratic philosophers, he believed that the universe was all one great divine singularity, and the sun at the center of our system seemed a reasonable direction for one's admiration. It added up to materialistic pantheism: God and the world were one. God did not make the universe, which was eternal; rather, God was the universe. Christianity, then, was meaningless. Bruno believed that Christ was not God but merely an unusually skillful magician, and he rejected the virgin birth and the resurrection.

His scientific theses were so outrageous that he began his *On the Infinite Universe and Worlds* of 1591 with a preface that insisted he was not joking. The preface also praised Democritus and Epicurus for arguing "that everything throughout infinity suffereth renewal and restoration," and for "alleging a constant and unchanging number of particles of identical material that perpetually undergo transformation one into another."[68] Many future doubters would find Democritus and Epicurus through the attention thrown on them by Bruno. The main text is presented as a conversation. In one wonderful scene for the history of doubt, we see a character called Burchio listen to the idea of an infinite universe. Says Burchio, "Even if this be true I do not wish to believe it, for this *Infinite* can neither be understood by my head nor brooked by my stomach." His friend Philotheo replies that, while infinity does seem bizarre to us, we must remember that "we have experience that sense-perception deceiveth us concerning the surface of this globe on which we live, much more should we hold suspect the impression it giveth us of a limit to the starry sphere." Anyway, for Philotheo "it appeareth to me ridiculous to affirm that nothing is beyond the heaven" as Aristotle would have it. There is drama in his next words:

> Thus let this surface be what it will, I must always put the question, what is beyond? If the reply is *Nothing,* . . . let us now see whether there can be such a space in which is naught. In this infinite space is placed our universe (whether by chance, by necessity or by providence I do not now consider). I ask now whether this space which indeed containeth the world is better fitted to do so than is another space beyond?

They go on, talking out the issue, but we pause at that remarkable parenthetical phrase. He has been remembered best for such frank little statements of doubt.

In another text Bruno offers relativism in a close paraphrase of Epicurus. As Bruno puts it: "There is no absolute up or down, as Aristotle taught; no absolute position in space; but the position of a body is relative to that of other bodies. Everywhere there is incessant relative change in position throughout the universe, and the observer is always at the center of things."[69] That is a powerful idea and Bruno's revival of it would have tremendous influence, despite his usually being seen as decidedly quirky. Bruno's science was so full of fantastic imaginings, and his verse so full of passion for the natural world, that in every age he has always been a bit of an oddball. Without making any real discovery or passing down a coherent, lasting philosophy, Bruno made a big impact on the history of doubt.

Bruno said he was not an atheist, because even though God, for him, was the same thing as the universe, the world we know is somehow the result of this God-universe he posited, so they were not exactly the same thing. Bruno's God was not the creator or even the first mover, but the soul of the world. When Bruno was sentenced to death by fire, he famously said, "Perhaps you, my judges, pronounce this sentence against me with greater fear than I receive it." He was invited to repent but would not. As the flames got going, someone offered him a crucifix, but Bruno pushed it away with an expression of sharp disdain.

Another great doubter of the late sixteenth century was Lucilio Vanini. Although he is not as famous as Bruno today, in the centuries that followed his own, Vanini was often cited by doubters as an influence. He was born in 1585, was educated by the Jesuits, joined the Carmelites, earned a doctorate in law, and was in Padua after 1608. Rumors of his unorthodoxy led to his leaving Italy for England and travel in northern Europe, during which time he wrote two books, published in 1615 and 1616 in France. The first book, in which he was already showing the influence of Pietro Pomponazzi and Girolamo Cardano, protested that God could not have personality and struggled with the problem of evil, but it still argued from within the Christian tradition. The second book was called *Of the Marvelous Secrets of the Queen and Goddess of the Mortal Ones, Nature* and it went further. *Of the Marvelous Secrets . . . of Nature* was set up as a debate between a student character called Alexander and a teacher character called Julius Caesar. Since Lucilio had changed his name to Julius Caesar Vanini, the teacher character was plainly identifiable with the author.[70] From the start he casts doubt on every aspect of Christianity, wondering whether an immaterial God could make a material world and whether an immaterial spirit of any sort could speak to human beings, and concluding that the only true worship is the worship of nature.

What of creation then? Vanini believes in the eternity of matter, makes jokes about the idea of creation, and even proposes that the animals and human beings had come into being through some kind of putrefaction—the same way life always arises from rotting things. He insists that there are no real nonmaterial beings, so neither ghosts nor spirits nor the independent human soul exists. He even writes that all religions, including Christianity, are human inventions, fictions cooked up by kings and clergy for the sake of power. All miracles associated with prayer are just coincidences or have natural explanations; all inner knowledge is just reason; and immorality is simply the result of illness or bad diet. It was far-reaching doubt. He was burned to death at the age of thirty-four, in 1619. The charge was blasphemy and atheism.

Paolo Sarpi started out as the Venetian government's theological consultant, but only a year later, in 1607, he was excommunicated. He published descriptions of the universe as he saw it that left no place at all for God. He also wrote that "the end of man is to live, just like any other living being."[71] Sarpi also rejected all supernaturalism. Most notably, he argued that the human sense of God comes from our ignorance and frailty—that is, it is about human needs—and that these needs can be transcended. Intelligent people could see past the myths and live moral lives without fear of God or death. When the Church investigated Sarpi, Rome came to believe he was the "head of an atheist company" located in Venice, and Sarpi was reported to have said that there were many atheists in Venice.

MONTAIGNE'S SHOCK OF THE NEW WORLD

Montaigne lived his life in a changeable age. In his lifetime whole new worlds had been found by explorers, and in those worlds there were new plants, new animals, and most surprising of all, new cultures. The ancient world had become vividly present as an alternative reality. Politics in his time was also pure upheaval: the government and official religion of England had changed several times during his own life. Also, the intellectual world was coming to know of the Copernican model of the solar system and was beginning to take it seriously, and that model called for an unparalleled imaginative shift. We should also note that Montaigne was reared Catholic but was the son of a Catholic man and a "New Christian" woman; that is, his mother's family had been Jewish.

Montaigne invented the genre of the *essay;* the word means "try" in French. We now have his work in one volume called *Essays.* The overall project of the book was new: Montaigne had set out to describe his personality; to get to know himself psychologically. In his time—one normally labored to keep oneself and one's imperfections hidden, especially in a text. He confessed he had had a best friend for four important years of his young life, and then the guy died. Montaigne seems to have opted out of affective ties at that point. He did not seem to think of his wife as a companion, and he did not find true friends again until the last years of his life, when he had stopped looking. Montaigne decided instead to make a friend out of his essays. One of them, "The Defense of Raymond Seybond" (1576), is one of the greatest works in the history of doubt. It was early vintage and is less interested in psychological self-exploration and the voice of the best friend than most others (it's also much longer than most). Still, although this was a published book, it had some of the tone of a private conversation.

"The Defense of Raymond Seybond" was written at the request of a princess friend of Montaigne's, Margaret. Margaret was married to Henri of Navarre (a great-grandson of that other Margaret we saw leaning over her dying maid and raising a glass with Dolet and Rabelais). Montaigne had translated a book by Seybond because his father had asked him to, and the princess had read the resulting work. Now the princess wanted help understanding Seybond so she could use him to defend Catholicism against those who were rejecting it at her court. Montaigne's answer dismissed Seybond early on though it said he was "bold and courageous" for having tried to "prove against the atheists all the articles of the Christian religion" by means of human reason.[72] Montaigne even attested that he knew "a man of authority, nurtured in letters, who confessed to me that he had been brought back from the errors of unbelief through the medium of Seybond's arguments," and Montaigne conceded that these arguments will probably do—against the average attacker. But he could not say he believed it. Instead, Montaigne answered the princess's questions more broadly: what did he really think was true, and what is the best Catholic defense? The answer was a marvel in the history of doubt.

Montaigne set out his great bombshell immediately: Custom and law defined religion, not some inner knowledge of truth or any rational argument for truth. Sense experience and reason tell us nothing of God, and so we should simply believe. Of course, if we cannot know anything about God through reason; we cannot know anything about religion either. We have a religion because "we happen to have been born in a country where it was in practice, . . . we regard its antiquity or the authority of the men who have maintained it, [and] . . . we fear the threats it fastens on unbelievers."[73] If we lived someplace else, we would believe other things.

Ancient Skepticism was shocking stuff to the Renaissance doubter. These sophisticated philosophical texts announced that the questioning of religion in favor of rationalism led to belief in nothing and the acceptance of death as real. Montaigne agreed that philosophy does lead to believing nothing for certain and accepting death. That was, however, a little hard to take. According to Montaigne, God seemed to have communicated himself to a few ancient individuals so that we would all know of him. That is all we had to go on. Why should anyone be willing to burn for details, when they were so unknowable? We should all just do as the ancients suggested and follow the religion of tradition, in this case Catholicism. If both sides could accept that all was ignorance, the Catholic versus Reformed Church debate would fall to his side, the Catholic, by the advice of the ancients.

Montaigne's essays, including this one, are all festooned with quotations from ancient writers set off with indentations or, in the later ones, italicized within the text. The style has a kind of family resemblance to the Scholastic practice of questioning and answering the ancients, but the voice of the sixteenth-century individual has grown chatty and uses the ancients a bit like pearls decorating a bodice. It is worth noting that especially in this essay, but in many ways throughout the *Essays,* Montaigne cites Lucretius's Epicurean *On the Nature of Things* more than anything else. Cicero—mostly from his *On the Nature of the Gods*—comes in a close second. Indeed, with these two books, things go further: there are times in Montaigne's argument when he casually starts speaking to these two works, mentioning Balbus and his questions, for instance, and arguing or agreeing point by point. Montaigne also quotes Lucian saying things such as "They fear their own imaginings."[74] The essay also mentions many mechanistic understandings of the world: Archelaus, Socrates' teacher, said that "both men and animals were made from a milky slime squeezed out by the heat of the earth." Whatever he believed, Montaigne laid out centuries of doubt's ideas, heroes, and poetry for young modernity to read.

Montaigne did not want to join the argument over philosophy and religion; he wanted to escape it through Skepticism. "What good can we suppose it did Varro and Aristotle to know so many things? Did it exempt them from human discomforts? Were they freed from the accidents that oppress a porter? Did they derive from logic some consolation for the gout?"[75] They were famous for knowledge in the time when knowledge was greatest, "but we have not for all that heard that they had any particular excellence in their lives." An inset quote from Horace reads: "Does the illiterate's tool stand less erect?"[76] Montaigne knew how to use a joke from Horace to lighten things up, but the point was to ask what smartness gets you. Even if philosophy "would actually do what they say, blunt and lessen the keenness of the misfortunes that pursue us, what does it do but what ignorance does much more purely?"[77] Pyrrho had told of a time on a storm-tossed ship when all were panicking but himself and the swine on board. Pyrrho used it as an example of natural calm. Montaigne notes the story and says if the learned and the ignorant are both calm and happy, why not take it easy with the pigs?

Montaigne forgets about Seybond early in the essay, pushing on to rehearse every ancient argument against God or the gods, against life after death, and against religion. When Montaigne tells about Anaxagoras's claim that the sun was a glowing stone, he teases that "he did not stop to think that fire does not blacken those whom it looks on."[78] It was a nice little catch about needing ultraviolet rays to get a tan. Xenophanes was a favorite of ancient doubt for Montaigne, and he brings the old philosopher's metaphor to life: "Xenophanes," he writes, "used to say wittily that if the animals make gods for themselves, as it is likely they do, they certainly make them like themselves, and glorify themselves as we do. For why shall a gosling not say thus: . . . I am the darling of nature?"[79]

Montaigne complains that the tricks for living dreamed up by Stoics and Epicureans, tricks such as not having desires or not attending to one's pain, are just not that great: one wants to want things and feel things, and usually one cannot help it anyway. He accuses the Stoics of faking their calm as they suffered inside, and he laughs at "the other advice that philosophy gives, to keep in our memory only past happiness," chiding, "as if the science of forgetfulness were in our power."[80] Of the Cynics, he tells a story of Aristippus walking by Diogenes one day wearing a robe given to him by Dionysius the Tyrant. Diogenes, washing cabbages, shouts out to Aristippus, "If you knew how to live on cabbage you would not pay court to a tyrant." Aristippus answers, "If you knew how to live among men, you would not be washing cabbages."[81] Not a bad point on both sides, which, of course, was what Montaigne meant by telling the story: "See how reason provides plausibility to different actions."

Montaigne then sets out a wonderful and wandering explanation of Skepticism. The first problem of knowledge is the relativism of judgment. Of beauty, "the Indies paint it as black and dusky, with large swollen lips and a wide flat nose . . . in Peru the biggest ears are the fairest," and in Basque country the women are more beautiful when they shave their heads and "plenty of other places."[82] As for pure forms, Plato loved the circle, but Epicureans liked the pyramid or the square. There is also the problem of the interpretation of someone else's judgment: "some have considered Plato a dogmatist, others a doubter."[83] Then there is the matter of madness, which Montaigne described as a heretofore ignored topic in Skepticism and a fertile one.[84] How can we ever be certain of our sanity and its conclusions? Then there is the issue of perception, that wine tastes different in the mouth of a sick man. "Do you think that the verses of . . . Sappho smile to an avaricious and crabbed old man as they do to a vigorous and ardent young man?"[85] Sexual desire, to which Montaigne admitted never having been very susceptible, could make some people befuddled. For his own part, when he was dragged into sex by some seduction, he was surprised by how single-minded he became until the urge was relieved. While he was aroused, he lost his values and approved only of arguments that encouraged him to become satisfied: "I would see my soul regain another kind of sight, another state, and another judgment."[86] Yet, that, too, would disappear fast once desire was sated.

Montaigne's general phlegmatic disinterest in sex is its own little lesson in doubt, for the most extreme torture Augustine ever felt was of having to put aside sex, and here is our pious Skeptic, free to indulge but never having had much taste for it. There are different types of people, he reminds us, and what is more, any given person changes constantly. "He who last night was invited to come to dine this morning, today comes uninvited, seeing that he and his hosts are no longer themselves: they have become others."[87] We change our minds, fast sometimes. No ideas, opinions, desires, or perceptions stay put. Even time "is a mobile thing, which appears as in a shadow together with matter, which is ever running and flowing, without ever remaining stable or permanent."[88] We disagree not only with each other, but with ourselves across our lifetimes. How can we bear to trust our own opinions over and over again? "Has it not happened to me, not once, but a hundred times, and every day, to have embraced with these same instruments, in this same condition, something else that I have since judged false?"[89] After all, "if my touchstone is found to be ordinarily false . . . Is it not stupidity to let myself be fooled so many times by one guide?"[90] Inset, Montaigne has the Epicurean poet whisper:

> *The latest find*
> *Kills prior things and spoils them in our mind.*
> LUCRETIUS

Worse yet is the problem of mood. Montaigne declares himself to be moody—now sad, now happy, now angry—and says that the world looks wholly different to him in these various states. Hunger can be enough to change one's attitude toward everything from art to ethics. "Preachers know that the emotion that comes to them as they talk incites them toward belief." Some people will believe things just because they reject the "pressure and violence of authority." In fact, pride or reputation has "sent some men all the way to the stake to maintain an opinion for which, among their friends and at liberty they would not have been willing to burn the tip of their finger." A terrific observation. An investigation of social and personal pressures sometimes makes nonsense of convictions—even fatal ones.

Montaigne may have rejected the ancient philosophies, quoting chapter and verse from Sextus, but he was also ardently respectful of them.

> The writings of the ancients, I mean the good writings, full and solid, tempt me and move me almost wherever they please; the one I am listening to always seems to me the strongest; I find each one right in his turn, although they contradict each other. The facility that good minds have of making whatever they like seem true, and the fact that there is nothing so strange but that they undertake to color it enough to deceive a simplicity like mine, shows evidently the weakness of their proof.[91]

Unable to prove any of the various brilliant voices wrong, he knocks their heads together: with so many, none can be true. The shock of Skepticism must have been intense. The Greeks, after all, developed philosophy slowly, coming slowly to the sense that the schools of philosophy were problematically numerous, varied, and sure of themselves. Eventually they developed a great deal of circumspection about philosophy altogether. Skepticism counted Socrates as one of its founders, in his doubting that we can know, but it was only with the later Pyrrho that it became a doctrine that took seriously the disagreements among thinkers. When Montaigne found the Pyrrhonists, he was smacked with a wallop of relativism. He tells us that he had been moving toward this relativism before he read Sextus. When he found him, Montaigne carved quotations from Sextus into the rafters of his study.

Explaining the Skeptics' position, Montaigne said that if you say snow is white, they say it is black, if you say it is neither black nor white, they argue that it is both. If you say you are sure you don't know, they will say you do. "Yes, and if by an affirmative axiom you assure them that you are in doubt about it, they will go and argue that you are not." It is "by this extremity of doubt that shakes its own foundations" that they reject even opinions that support their own claims. All they ask you to agree with is their right to it. Says Montaigne, "Why, they say, since among the dogmatists one is allowed to say green, the other yellow, are they not also allowed to doubt?"[92] People usually come to their beliefs by the "custom of their country," or by parental upbringing, "or by chance—as by a tempest, without judgment or choice, indeed most often before the age of discretion," and then "to such or such an opinion, enslaved, and fastened as to a prey they have bitten into and cannot shake loose"; bizarrely, to "*whatever doctrine they have been driven, as by a storm, to it they cling as to a rock.*" The italicized phrase is Cicero's. Isn't it better, Montaigne asks, to free oneself from certainty and thereby glide above the fray?

How do we live in such a situation? Well, we cannot trust our senses, nor can we believe the world to be as it appears to us. But we can accept things as they come to us and simply enjoy them. For this, he paraphrases the Skepticism of Ecclesiastes: "'Receive things thankfully,' says the Preacher, 'in the aspect and taste that they are offered to thee, from day to day; the rest is beyond thy knowledge.'"[93] Montaigne had these words inscribed on the ceiling of his library. He was a man who liked to live around his words. He worked up a motto and symbol for himself and had that carved into the wall, too, among the Sextus and Ecclesiastes: the image was of a scale, its two plates at empty balance. The motto was *Que sçais-je?* (What do I know?) Not a bad symbol for the history of doubt.

Montaigne's answer to the princess was through Sextus: We cannot know anything—the only evidence for even God, let alone any dogma, is ancient hearsay—so we might as well stick with the Catholic Church, just as the ancients advised.[94] This sort of approach to religion is called fideism—explicitly blind belief—but it is hard to know how much faith Montaigne really had. On another page he casually mentions that his argument tends against belief in God, and that religion is a human invention for the sake of social peace:

> How could that ancient god [Apollo] more clearly accuse human knowledge of ignorance of the divine being, and teach men that religion was only a creature of their own invention, suitable to bind their society together, than by declaring as he did, to those who sought instruction . . . that the true cult for each man was that which he found observed according to the practice of the place he was in?[95]

Montaigne says his own custom-based idea of religion clearly accuses humanity of no real knowledge of God.

The ancient world brought colossal doubt. As for the modern, we might think that the Copernican revolution would have felt like a great new certainty, but it did not:

The sky and the stars have been moving for three thousand years; everybody had so believed, until it occurred to Cleanthes of Samos, or (according to Theophrastus) to Nicetas of Syracuse, to maintain that it was the
earth that moved, through the oblique circle of the Zodiac, turning about
its axis and in our day Copernicus has grounded this doctrine so well that
he uses it very systematically for all astronomical deductions. What are we
to get out of that, unless that we should not bother which of the two is so?
And who knows whether a third opinion, a thousand years from now, will
not overthrow the preceding two?[96]

Hard to say, but his take on this is stunning doubt. With so much written
history, cosmopolitanism now had a deep temporal dimension.

Before the principles which Aristotle introduced were in credit, other principles satisfied human reason, as his satisfy us at this moment. What letters-
patent have these, what special privilege, that the course of our invention
stops at them, and that to them belongs possession of our belief for all time
to come? They are no more exempt from being thrown out than were their
predecessors.

We have to keep in mind the magnitude of the sense of weirdness that
hung over people in this period, and the sense that they had learned their
lesson about reality: we must not take anything for granted. The very earth
had unfurled another inhabited world alongside their own:

Behold in our century an infinite extent of terra firma, not an island or one
particular country, but a portion nearly equal in size to the one we know,
has just been discovered. The geographers of the present time do not fail to
assure us that now all is discovered and all is seen,

For what we have at hand always seems best of all.
LUCRETIUS

The question is, if Ptolemy was once mistaken on the grounds of his
reason, whether it would not be stupid for me now to trust what these
people say about it; and whether it is not more likely that this great
body that we call the world is something quite different from what we
judge.[97]

When he wrote of the amazing peoples in this other world, he could be hostile to religion. He explained that among their customs we find "coincidences between a great number of fabulous popular opinions and savage customs and beliefs, which do not seem from any angle to be connected with our natural reason. The human mind is a great worker of miracles." The Brazilians, he said, died only of old age because they were so serene that they were ultra-healthy, because they spent their lives "in admirable simplicity and ignorance, without letters, without law, without king, without religion of any kind."[98]

Montaigne says we may not have enough senses to know the world. Perhaps one needs eight or nine senses to get a decent view of the universe, and we are hopelessly missing crucial information. But look where he takes it: "The properties that we call occult in many things, as that of the magnet to attract iron—is it not likely that there are sensory faculties in nature suitable to judge them and perceive them, and that the lack of such faculties causes our ignorance of the true essence of such things?"[99] It is a nice insight, and he got there not by discovering refraction patterns or X rays or infrared vision but by engaging his faculties of doubt.

Montaigne tells the princess that "for an ending to this long and boring discourse" he will quote Seneca's line that man is a vile thing unless he ascends above humanity, but he quickly despairs that we cannot "make the handful bigger than the hand."[100] Montaigne's advice is that life is good, but in order to live well one must study one's own psychology with patience and intensity. His aggressive yet humble doubt was immediately popular. The Catholic Church in France adopted it as a credo; others took it as a guide for further challenge to religion. Intellectually, it supported widespread skepticism, and someone was going to have to find a way to know things again.

Before we leave Montaigne, we can note that this cranky man had a few years of real happiness toward the end of his life, when two brilliant fans showed up. In fact, in his final essay he advises fellow writers never to put off publication until they're older out of fear of trouble, because there is too much possible pleasure to be lost: publication brought intellectual companionship. The two fans were Marie de Gournay and Pierre Charron, and Montaigne came to call them his adopted daughter and adopted son. For the next several centuries, Charron's name would be well known to the history of doubt, and I'll consider his impact when it happens, a little later in this chapter. As for Marie de Gournay, she was born in 1566 and like most reasonably well-born women of her time, she had been given just a little education. Yet even as a child she was thrilled by the new ideas of her time. To take part, she taught herself Latin by comparing French and Latin texts and studied Greek as well, eventually becoming a teacher. In 1580, when she was only about fifteen, she read Montaigne's essays and was swept away by their ideas and their style. She wrote of wanting to meet him from that time on, and finally did in 1588, in Paris. At twenty-three she was much his junior, but their friendship—which by all attestations was never romantic—seems to have been a great delight for Montaigne. The first work Marie de Gournay published was about their discussions and included a novel of hers, followed by poems and some passages she had translated from the *Aeneid*.

She and Montaigne met only four years before he died in 1592, but he left her full control of his work. She became the principal editor of the *Essays*, writing the preface to the 1595 edition and editing the 1598 and the 1641 editions. Her own work was fiercely Skeptical, and her approach to religion followed suit. Skepticism was infinitely preferable to the arrogance of the reformers. "Who can also suffer these new Titans of our time . . . not wanting anything as true if it does not seem probable to them."[101] Nothing seemed probable to her. The only thing reasonable was to act respectfully, but to doubt everything. Marie de Gournay was treated abysmally by so many male intellectuals that she ended up spending much of her creative energy on sexism; her great works on the subject were *The Equality of Men and Women* (1622) and *Grief des dames* (1626). She also wrote about religion and philosophy and translated Virgil and Ovid and, although ridiculed by the university men, her work sold well. After Christine de Pisan, she was one of the first successful, professional women writers. She was a crucial figure in the history of doubt: a keen skeptic who lived a philosophical life, wrote about real-world problems, as well as poetry, and also edited and introduced the *Essays,* the magnum opus of Renaissance skepticism.

UNCERTAIN DANES AND THE "DEBAUCH" OF FRENCH LIBERTINES

The year Bruno burns, Hamlet first broods. The young Dane asks whether or not to be. It was a question about whether, and how, to stay in a world where people could be impostors, where truth could be contradictory and unavailable to reason, where mood and madness might be responsible for our experience, where authority had gone so far awry that one could not tell if a given act was a gesture of submission or rebellion. Think of Hamlet reducing the universe to a quintessence of dust as he reels in uncertainty, cursing the inconstant world. There is something dryly secular and loosely skeptical about Shakespeare's whole project. Think of *Macbeth:*

> *To-morrow, and to-morrow, and to-morrow,*
> *Creeps in this petty pace from day to day*
> *To the last syllable of recorded time,*
> *And all our yesterdays have lighted fools*
> *The way to dusty death. Out, out, brief candle!*
> *Life's but a walking shadow, a poor player*
> *That struts and frets his hour upon the stage*
> *And then is heard no more: it is a tale*
> *Told by an idiot, full of sound and fury,*
> *Signifying nothing.*

Is God not that idiot, or is there no God? Scholars have noted that prayers for help or safety often directly precede death and disaster throughout these plays, and many characters are given irreligious speech. *Twelfth Night* tells us: "What is love? 'tis not hereafter. Present mirth hath present laughter." Since we are dealing with fiction, such specifics do not make as strong an argument as does the character of the entirety. Think of *The Tempest:*

> *Our revels now are ended. These our actors,*
> *As I foretold you, were all spirits and*
> *Are melted into air, into thin air:*
> *And, like the baseless fabric of this vision,*
> *The cloud-capp'd towers, the gorgeous palaces,*
> *The solemn temples, the great globe itself,*
> *Ye all which it inherit, shall dissolve*
> *And, like this insubstantial pageant faded,*
> *Leave not a rack behind. We are such stuff*
> *As dreams are made on, and our little life*
> *Is rounded with a sleep.*

The grace of Shakespeare is that there is always another side to things; there is always doubt.

Meanwhile, in 1601, Montaigne's "adopted son" Pierre Charron published *Of Wisdom,* which was a description of Montaigne's philosophy. Montaigne's *Essays* wove his intellectual ideas in with his private thoughts on all manner of things. Even within a given essay, one found little structure and many digressions, and the collection of them is a hodgepodge of its own. Charron, by contrast, was a great organizer and laid out Montaigne's modern skepticism in a nice theoretical style. He added a bit of Machiavelli and hammered on the points about cultural relativism: things are done differently everywhere, and if you were born there, you'd do it that way, too. Custom is more powerful than nature. *Of Wisdom* is also important in the history of doubt because it speaks straightforwardly about how to write a book that seems pious enough, but that wise readers will recognize as sending a different message indeed.[102]

For centuries the book would be described as a seminary of atheism.[103] Right away, it was a triumph: it was put on the Index of Forbidden Books fast, in 1605, but it sold like hotcakes right through the century. Charron even came out with a mini version of it the next year. Here he dealt with his surprise that people could find doubt uncomfortable. Charron said of doubt:

> It alone can provide true repose and security of our spirits. Have all the greatest and most noble philosophers and wise men who have professed doubt been in a state of anxiety and suffering? But they say: to doubt, to consider both points of view, to put off a decision, is this not painful? I reply, it is indeed for fools, but not for wise men. It is painful for people who cannot stand freedom, for those who are presumptuous, partisan, passionate and who, obstinately attached to their opinions, arrogantly condemn all others. . . . Such people, in truth, know nothing. They do not even know what it is to know something.

It was a claim that doubt can make you happy, can ease your pain, and can be a home. It may have been the first time anyone in modernity spelled it out like that.

People loved it. The new Pyrrhonism fueled two movements: the French Catholic Counter-Reformation, as I have mentioned, and the *libertin erudits.* Of the first, I offer only the foremost example, Cardinal du Perron: On the occasion of a dinner with King Henry III, du Perron had regaled the table with a series of proofs of the existence of God. When the king praised him, du Perron said, "Sire, today I have proved by strong and evident reasons that there is a God. Tomorrow, if it pleases Your Majesty to grant me another audience, I will show you and prove by as strong and evident reasons that there is no God at all."[104] Du Perron, by the way, was a close friend of Marie de Gournay and had read Charron. As the story goes, the king threw the cardinal out on his ear. The joke, though, was on the king, who had shown his ignorance of the surprising new French orthodoxy. The French clerics of the Catholic Counter-Reformation saw Skepticism as a broom to sweep away rationalism, such that faith, or at least calm obedience, would have to take its place. For his help in the cause, Sextus was never put on the Forbidden Books index, and neither was Montaigne until 1676.

The other major inheritors of Montaigne's Skepticism were the "Libertines," sometimes called the "Erudite Libertines," a French phenomenon of the early seventeenth century. They read Pomponazzi and other doubting Italian philosophers. It was an interesting crowd. There was François de La Mothe le Vayer (1585–1672), who was such a skeptical fideist that he thought all scientific research was both silly and impious in its arrogance and became an actual defender of irrationalism. He wrote of "the divine Sextus," which is fun for us, given that we have already seen Epicurus "divinized" by Lucretius. La Mothe le Vayer himself was often called "the Christian Skeptic" or an "Epicurean unbeliever." Then there was Gabriel Naudé (1600–1653), whom Cardinal Bagni took to Rome to be his librarian. He became a medical doctor at the University of Padua and was King Louis XIII's physician. Richelieu hired Naudé to be his librarian, but the cardinal died and Naudé stayed on as Mazarin's librarian (both were important French cardinals and statesmen). Naudé held that a library should not be without Sextus Empiricus. For several centuries now, Naudé has usually been remembered as an atheist. When Cardinal Bagni asked him what were the best books in the world, he said, the Bible first, then Charron's *Of Wisdom,* and the story goes that the Italian cardinal blinked unknowingly at the brazen choice and promised to read it.[105] These anecdotes all suggest that the clergymen choosing faith through classic Skepticism knew that the idea seemed outrageous.

Guy Patin, who was a scholar, medical doctor, and rector of the Sorbonne medical school was another member of these Libertines. Pierre Gassendi (1592–1655), a leading mathematician and influential figure in science and philosophy, is among the best known of them today. Isaac la

Peyrère pioneered an extraordinary new kind of Bible criticism that questioned every aspect of the text, including that the Five Books were written by Moses. His favorite critique could be summed up in what he called his "pre-Adamism"—his belief that there were people in the time before the Bible story begins.[106]

The Libertines were seen as having a kind of bacchanal of irreligion, and we have been left with the image of a period overrun with "Skeptical banquets" and "Pyrhonnian debauches." Here's a letter written by Patin in which he explains that "M. Naudé, librarian of Cardinal Mazarin, intimate friend of M. Gassendy, as he is of mine," had arranged for all three of them "to go and sup and sleep in his home at Gentilly" the following Sunday,

> provided that it will only be the three of us, and that there we will have a débauche; but God knows what a débauche! M. Naudé regularly drinks only water and has never tasted wine. M. Gassendy is so delicate that he would not dare drink it. . . . As for me, I can only throw powder on the writings of these great men. I drink very little, and nevertheless it will be a débauche, but a philosophical one, and perhaps something more. For all three of us, being cured of superstition and freed from the evil of scruples, which is the tyrant of consciences, we will perhaps go almost to the holy place. A year ago, I made this voyage to Gentilly with M. Naudé, I alone with him. There were no other witnesses, and there should not have been any. We spoke most freely about everything, without scandalizing a soul.[107]

They were an odd group, but still, the idea of doubters forming a delicious, hidden resistance is plainly in full force here. Whatever they said, they were experiencing a pleasant frisson from the scandalous talk—the atmosphere is charged.

In the 1620s Marin Mersenne claimed that there were fifty thousand atheists in Paris—this at a time when the whole city held only four hundred thousand. Did he really mean "atheist" when he said atheist? Did he really mean to say that one in eight Parisians entirely ignored religion? It's hard to know, but it certainly suggests a good deal of doubt. Mersenne was the hub of science in his time: new ideas were brought to him and he disseminated them. He had long friendships with Galileo, Hobbes, Descartes, and Gassendi, to mention only the best known. In the book in which fifty thousand atheists are claimed, Mersenne's catalog of atheists included our Bonaventure des Périers (Febvre's *Oops I found one*), Charron (Montaigne's adopted son), Machiavelli (who was said to have demolished moral order), Vanini (whom we saw burn), and Bruno (also lost to the stake). Whether or not we agree that all these characters were atheists, the list was a fine little history of doubt. In a later book wonderfully called *The Impiety of Deists, Atheists, and Libertines of These Times,* Mersenne narrowed his sights upon three chief opponents: Giordano Bruno, Pierre Charron, and Girolamo Cardano (the French champion of Pomponazzi). For Mersenne, the Skeptic problem was not a problem—and he seems to have been the first to articulate this—because we are free simply to investigate the phenomena that our senses present to us, whether or not we trust our senses in some ultimate fashion. Skepticism becomes irrelevant as soon as you stop asking how we can know the real truth about the anthill or the atom, and just ask what we can determine about phenomena as they appear to us. That's a major answer to the Skeptical question and it lets us out through a wormhole in an otherwise claustrophobic little corner of thought.

Pierre Gassendi was also very much responsible for this first conceptual leap past Skepticism and into empirical science. His scientific work was mainly in the fields of astronomy and cartography, but in 1647 he published his *On the Life and Character of Epicurus* and by 1649 had followed it up with two more works on Epicurus. Through these works, Gassendi showed that atomism explained how the world could have made itself, by itself—but, he assured his readers, God made the atoms. Owing to this innovation, for the first time atomism was not necessarily atheism. Gassendi's interest, however, clearly did not stop at the atoms; he lauded Epicurean ethics as well, and generally took Epicurus as his hero. Gassendi also knew his Sextus. On the Skeptic question, like Mersenne, Gassendi explained that if we are worried because our senses now tell us honey is sweet, and later tell us it is bitter, we must not give up the project of knowledge, but rather must come to understand honey and taste buds on such a minute level that we can fathom their variation. Whatever else they did, Mersenne, Gassendi, and the rest of the Erudite Libertines brought many ancient and later doubters to the consciousness of a new generation.

JESUITS IN CHINA

In the sixteenth century nothing remained of the Christian communities founded in China by the Nestorian missionaries in the seventh century, or by the few Catholic monks in the thirteenth and fourteenth centuries. If they made an impact, it was not exactly Christian, especially the ones who learned to fit in, met a nice Chinese woman, and disappeared into China's history and into its hills. Now that the Protestant Reformation had taken so many Europeans from the Roman Church, though, the Church set out to make up for it with converts in Asia, Africa, and Brazil. Off sailed the Jesuits, who had come into being in 1540 for just such purposes.

Matteo Ricci was the founder of the Catholic missions of China. At the Jesuit school in Rome he had studied mathematics, cosmology, and astronomy, as well as philosophy and theology. In 1577 he asked to be sent on the missions in farthest Asia, and he embarked from Lisbon the next year. Ricci knew the Chinese would not tolerate an overt conversion mission (they had already rebuffed a few attempts), so his plan was to awe them with science and technology first. Once they trusted Europe to be possessed of great truths—then he would start talking about Jesus.

Ricci took with him mathematical and astronomical instruments, glass prisms and other optical toys, large and small clocks, musical instruments, oil paintings and prints, and diagrammed works of architecture, cosmology, and geography. He also had a big, beautiful map of the world that impressed people a great deal. It was actually based on European *and* Chinese sources and Ricci first used it just to show his route, but everyone wanted a copy. Some of the math and science was much more profoundly surprising, but even so the Chinese were merely impressed and interested, not swept away into a state of shocked obedience. Furthermore, when it became clear that Ricci's real goal was religious, people were annoyed at the subterfuge, and there is a line of Chinese hostility to the West that is always traced from this deception.

Ricci's superior in this adventure died early on, yet Ricci continued his mission. He was soon able to dismiss his interpreters and began to dress in the style of the educated Chinese. He'd been working his way to the capital, starting small flocks in large cities, when he was summoned by Emperor Wan-li in 1601. Ricci brought his library, museum, music, and scientific devices; and many educated Chinese flocked to learn everything they could. Ricci himself wrote books of Christian morals, such as *On Friendship,* in Chinese, and these seem to have deeply impressed them.

His *T'ien-chu-she-i,* or *True Doctrine of God,* was a brief rundown of the philosophical proofs of God's existence, the creation of the world, the immortality of the soul, and the coming reward or punishment in a future life. All were demonstrated as philosophical propositions; the Europeans offered no burning bushes. The *True Doctrine* also included proofs against the worship of idols and the belief in the transmigration of souls. Confucians liked this part, as compared to the fanciful metaphysics and practices of many sects of Chinese Buddhism. In general, however, the Christians seemed more like the Buddhists with their priests and monks, concern for the afterlife, and talk of God or gods.

Ricci may have had the most impact on the history of doubt in the matter of ancestor worship and ritualized respect toward the image of Confucius. Ricci could tell there was no way any of the Chinese were going to stop tending their family shrines anytime soon. It would have been illegal, for one thing, but it also seemed like a bizarre request. Ricci decided that these were purely civic rites and gestures, and so there was no need to stop them.

> The honor they pay to their parents consists in serving them dead as they did living. They do not for this reason think that the dead come to eat their offerings or need them. They declare that they act in this manner because they know no other way of showing their love and gratitude to their ancestors. . . . Likewise what they do, they do to thank Confucius for the excellent doctrine which he left them in his books, and through which they obtained their degrees and mandarinships. Thus in all this there is nothing suggestive of idolatry, and perhaps it may even be said that there is no superstition.

Nicely done, but it had its problems. It meant that he was able to claim a nice batch of Christians (two thousand baptized in 1608), but that he had defined all Confucians as atheists. The Jesuits liked this idea in general because it made the Chinese look more sophisticated and therefore a more valuable conversion. It is funny to see the religious of Europe impressed with a sober civil rite while mildly disgusted by the genuflections of a pagan religion.

Meanwhile, in defense of the argument of universal agreement (there must be a God since all people know of him), Dominicans and Franciscans argued that the Chinese customs were actually religious rites. It was the Jesuits who won the day, and China was increasingly thought of as the land of atheism.[108] Contact with China brought doubt to Europe along a number of lines; for instance, the French libertine Isaac la Peyrère asserted that the Bible must be wrong because, after all, Chinese history goes back ten thousand years whereas biblical history has the world in existence for only about six thousand years. The Jesuits' insistence on the country's atheism, however, was the biggest factor by far. Ricci told new converts that they could speak to Jesus at the family shrine, so Catholics were a lot like anybody else in China, but with a few crosses. Another problem was what to call God. Having studied ancient Chinese texts, Ricci concluded that they spoke of *T'ien*, heaven, and *Shang-ti*, sovereign lord, in the same way Europeans spoke about God—as the source of all power and order. Meanwhile, modern Chinese scholars of Ricci's time used T'ien and Shang-ti to apply to the actual sky in the material world. Ricci knew this, but he argued that this material interpretation did not do justice to the ancient Chinese texts. So Ricci had them call God a name that meant *sky*, which lent many prayers and doctrines a distinct flavor of natural science. Ricci made all these decisions by the book, getting the authorities to sign off on policy, but in the centuries to come he was blamed for having introduced Jesus to China as a force small enough to fit into a world of polytheism and atheism without much of a ripple.

Ricci's three famous, elite converts (the "Three Pillars" of the early Christian church in China) were all fascinated by science and math. One, Xu Guangqui (1562–1633), worked with Ricci on translations of European books on astronomy, hydraulics, geography, and math, including a modern study of Euclid's *Elements of Geometry*. Xu rose to the position of grand secretary, one of the highest political offices in the empire. Li Zhizao (died 1630) held out for a while over the issue of his concubine—the one Christian issue Ricci was strict about was monogamy. Li wrote on astronomy as well as God, and his cleverly titled *First Collection of Writings on Learning about Heaven* told of the shock of introduction both to the European cosmology and to a God who was explained using Plato and Aristotle. It must have been a wondrously strange experience. The last of the three pillars was converted by Li just after Ricci's death. He was Yang Tingyun (1557–1627). He, too, struggled before managing to send away his concubine, the mother of his two sons; and he, too, did as much to disseminate European science as he did to disseminate Christianity. He also represents well many of his contemporaries in coming to see the world in relativist terms, in which the Chinese way of life was only one of many valid human cultures.

Things went awry when the Ming dynasty fell, because some Jesuits fled with the court while others remained in Beijing and served the new court.

The educated Han Chinese never forgave them for this easy shift in alliance. Meanwhile, the upshot of all the new knowledge was not only a golden age of studying Western science but also a simultaneous revival of the rationalist sciences of China's own history. It was the Jesuits, the shock troops of the papacy, of all people, who brought European rationalism to China and brought news of a world of atheists to Europe.

Just as Buddhism was born in India but died there just before it took off in China, Zen was born in China but had its greatest life in Japan. As various ideas of doubt swirled around Asia, they tossed up marvelous materialist philosophers and poets, and new nontheistic psychophysical enlightenment programs. Europe was at this time beginning to laugh openly at the posturing of its major religion, such that Rabelais and his giants sloshed beer and, when things got serious, philosophized. Meanwhile, in the same moment that Vasco da Gama and Christopher Columbus were providing a relativist shock to educated Europeans such as Montaigne with their voyages of discovery, the great medieval Mandeville's *Travels* was catching up with villagers like Menocchio, and providing an equal jolt of the strange. The Inquisition allowed us to hear the voice of common doubt; its place in the larger story is that, henceforth, the Christian churches will be constantly criticized for their cruelty and intolerance, a theme that we had not seen much of until now. The coming Enlightenment will be fueled by it—and by new solutions to the problem of Skepticism.

EIGHT

Sunspots and White House Doubters, 1600–1800

Revolutions in the Authority of Reason

This chapter begins in a faraway world and ends at the doorstep of modernity. We will see bold developments in Skepticism, and a rising struggle between science and religion. Yet, the loudest voice of doubt will come from people furious over the religious persecution that was still being carried out by the churches. For now, we are back in a world that is breaking its head over Skepticism: it seemed that no one had the right to claim anything, and there was no solid ground from which any knowledge could build. It was into this unmoored world that René Descartes loosed a few new tricks.

THE SCIENTIFIC REVOLUTION

In 1628 the young mathematician and scientist René Descartes was invited to the home of Cardinal Bagni (for whom the libertine Naudé would work) for an intellectual party. It would be one of the red-letter evenings in the history of doubt. A chemist named Chandoux was to entertain a group of leading thinkers of the day with a lecture against the old Scholastic philosophy. The listeners were skeptical fideists: subscribers to the view that we know nothing at all and are therefore free to blindly embrace the idea of God and even the specifics of Catholicism. Mersenne, he of the "fifty thousand atheists in Paris" line, was there. There was warm applause afterward; Descartes alone was peeved—and visibly. When asked what the problem was, he explained that, although glad for the negative attitude toward

Scholasticism, he could not abide resting all reality on a good guess suspended above blank uncertainty. He agreed that we could not trust information brought in from the senses and interpreted by the mind, but claimed he had found another way to truth, one that did not depend on the senses or tricks of the mind. He could offer a proof of God. When he explained himself, one of the cardinals there insisted that the young mathematician henceforth devote himself to working out this new philosophy.

What Descartes came up with was this: to determine truth, we must find out if there is one thing that we can know for certain, and build from there just as in ancient geometry. But how? The answer for Descartes was doubt, just as it had been for Augustine; but Descartes's doubt does different work. His boldly titled *Meditations on the First Philosophy in Which the Existence of God and the Distinction Between Mind and Body Are Demonstrated* (1641) had six parts; the first was "Of the things which may be brought within the sphere of the doubtful," and in it he walks his reader through a Skeptical crisis. In this first meditation, he explains that it was "now some years since I detected how many were the false beliefs that I had from my earliest youth admitted as true and how doubtful was everything I had since constructed on this basis."[1] He had become convinced that his senses deceived him. To explain, he says that as he writes this he is seated by the fire, in a dressing gown, holding paper in his hands, "and how could I deny that these hands and this body are mine?"[2] Yet, there are people in the world who "think they are kings when they are really quite poor . . . or [think they] are nothing but pumpkins or are made of glass." Leaving aside madness, had he himself not dreamed of sitting by the fire in a dressing gown? Perhaps he was dreaming now. What if God was actually playing with him? he wonders. What if some evil spirit had "employed his whole energies in deceiving" him?[3] His every experience might be a lie, down to the idea of extension, and body, and time.

Descartes decides he will fight the evil spirit: he will force himself to believe that everything—the sky, the earth, colors, numbers, sound—is an illusion. Furthermore, "I shall consider myself as having no hands, no eyes, no flesh, no blood, nor any senses." This was an awful struggle, since it felt like he did have them. "I shall remain obstinately attached to this idea," he announces, even though just as a slave does not want to awaken from a dream of freedom, his mind did not want to awaken from the illusion that all its information was reliable. Yet he pledges to do it. The next chapter's meditation begins with the confession that "The meditation of yesterday filled my mind with so many doubts that it is no longer in my power to forget them"; he feels as if he has fallen into deep water and cannot find the bottom or the top. He needs one Archimedean point from which he can right his world. What he finds is this: he is a thing that thinks. He knows that does not answer everything, even at its own level—for example, "I am, I exist, that is certain. But how often?"[4] But he runs over this delightful notion to get to the key point: "Am I not the being who now doubts nearly everything . . . ?" Even if he were dreaming, he decides, it would be he who was dreaming. Even if none of his perceptions correspond to a real thing in the world, they are still his perceptions. Meditation III was "Of God: that He exists," and in it Descartes begins from his point of certainty, "I am a thing that thinks, that is to say, that doubts, affirms, denies . . . , that wills, that desires, that also imagines and perceives." So there we have it: Descartes's famous axiom, *Cogito Ergo Sum*—I think, therefore I am—is perhaps more accurately expressed as *Dubito Ergo Sum*—I doubt, therefore I am.

He then claims that the certainty with which he knows this first point, that he exists, has taught him to recognize certainty elsewhere, and (and here's the leap) he claims he is certain that his perception of God came from God. He also perceives by inner knowledge that this God is good and therefore could not be generally deceiving him. We may trust the world and our senses, having confirmed their basic veracity against an inner certainty that was discovered without recourse to those senses. We have seen Stoics, Jews, and Christians claim that the magnificence of the world is evidence that God exists. Skeptics questioned every aspect of our ability to know that outside world. Descartes flipped the argument: it wasn't that the magnificence of the world proved God exists, it was that inner knowledge of God could prove that the world exists. Consciousness is suddenly esteemed higher than the universe.

Descartes performed this flip in order to protect belief, not erode it, and that is how it was taken by the Church: after an early hostile response, the Church quickly came around. Seeing that Scholasticism was besieged, and that skeptical fideism was too uncomfortably based on knowing nothing, the Church took Descartes's solution. Yet his work ended up forwarding the history of doubt in a big way. What he had done was to take God completely out of the world, and this allowed the new science a free rein. It's how religious scientists ever after conceived their project—and they avoided science of the mind. Meanwhile, this God known by inner sensation helped support the validity of scientific inquiry, since our senses could now be trusted.

The *Meditations* was not Descartes's first work. The first was a scientific treatise called *The World,* which he was about to publish in 1634 when he learned that Galileo had been condemned by the Church for teaching Copernicanism. Since *The World* was Copernican, and since Descartes valued his liberty above this particular point, he prevented the book's publication. The story helps us remember the timing. Born in 1564, Galileo Galilei had been trained in the old Aristotelian astronomy but at the University of Padua learned the Copernican system and was converted. Improving on the lens strength of the telescope, he was able to show a universe that was not at all perfect orbs and crystalline spheres: the sun had spots, Jupiter had moons, and the surface of our own moon was rough and raw. The Church asked him to stop teaching that the Copernican system was anything other than a mathematical device, but he would not, because he was convinced. He first published his support of the theory in 1613, about two decades after he came to Padua. The Church banned him from teaching the theory in 1616, and in 1633 he was brought to trial, shown some instruments of torture, and asked to take back his claim that the earth was careening around the sun. He did, but the story goes that on his way out he whispered under his breath, "E pur, si muove" (But still, it moves!). He was an old man by the time the authorities placed him under house arrest at his villa in Florence, and he died five years later, in 1642, at age seventy-eight.

Galileo was one of doubt's mechanics, one of its great students of the physical world, and he did unravel cherished notions of the faith. Yet, for his part, Galileo seems to have been just trying to get religion right. As early as 1615 he had written his famous "Letter to the Grand Duchess Christina" in which he explains his attitude toward scripture. The letter begins, "Some years ago, as Your Serene Highness well knows, I discovered in the heavens many things that had not been seen before our own age." This "stirred up against" him many professors, "as if I had placed these things in the sky with my own hands in order to upset nature and overturn the sciences." It is hard not to think of Anaxagoras pointing at the meteorite that had fallen in 467 BCE and wondering what the angry authorities wanted him to say. There it was. Bringing on the first-ever indictment for atheism, Anaxagoras's claim that the stars might be hot rocks rather than gods has a kind of direct descendant in Galileo's claims, despite vast differences between the two offended religions involved. What Galileo said to the duchess in the rest of his letter sounded like a man who trusted both his senses and the divine origin of the Bible, even though they contradicted each other. To Galileo it was obvious that this schism meant the scriptures had been understood too literally. God wouldn't have cared whether the science in the background of the Bible's lessons was correct. God, Galileo said, had let the Bible speak in terms that matched the simplest human perception of things.

The passage in the Bible that gave Galileo's contemporaries the most trouble was the one in which Joshua bids the sun stop its movement in the sky to give his warriors more time.[5] Galileo said that the Bible was just describing how things looked. He then cited the ancient philosophers who had considered heliocentrism; mentioned that Copernicus had been a cardinal and beloved by the Church; and offered quotations from Augustine advising the Church to avoid making decrees about the physical world, lest they be overturned by new knowledge. Citing an unnamed "eminent" ecclesiastic, Galileo quipped that "The intention of the Holy Ghost is to teach us how one goes to heaven, not how heaven goes." After all, if all interpreters of the Bible spoke by divine inspiration there should not be differences among them, but there were. It was this brash attitude, more than his scientific claims, that angered people.[6]

Galileo said he did not study Epicurus directly, but he wrote a work called *Saggiatore* in 1623 that made so much use of atomism that it was referred to as his Epicurean book.[7] By this time, Gassendi had loosened the connection between Epicurean atomism and atheism, but the Church attacked *Saggiatore* because it left no room for the doctrine of transubstantiation. The two-millennium connection between atomism and atheism was obviously also significant. From 1632 forward, Jesuit professors were repeatedly warned not to teach atomist physics.[8]

Galileo was not the only one in this period getting in trouble for reimagining the universe. When the Spanish and Portuguese Jews were exiled in 1492, a lot of them settled in Amsterdam, one of the most open and progressive cities in Europe at the time. Baruch Spinoza was born there in 1632. His family had been Jewish, then Marrano (converted to Catholicism but secretly still Jewish), and then Jewish again. The mother tongue of Amsterdam's Jewish community was still Portuguese; Spinoza later learned Spanish, Latin, and Dutch. As a young man he ran with a crowd who were admirers of Descartes, and he personally greatly admired Bruno. At some point Spinoza began telling people that the Torah was not literally the word of God, that Jews were not God's chosen people, and that there is no immortal human soul. Spinoza would not back down and was excommunicated from Judaism in 1656. He was in his early twenties and he never took part in the religion again, or any other. What is more, his reply to the excommunication grew into his great work, the *Tractatus Theologico-Politicus*. The opening of the book's preface announced that people would never be superstitious if the world was orderly, or if they were always lucky, "but being frequently driven into straits where rules are useless, and being often kept fluctuating pitiably between hope and fear," people "are consequently, for the most part, very prone to credulity." Here's a keen observation on doubt:

> This as a general fact I suppose everyone knows, though few, I believe, know their own nature; no one can have lived in the world without observing that most people, when in prosperity, are so over-brimming with wisdom (however inexperienced they may be), that they take every offer of advice as a personal insult, whereas in adversity they know not where to turn, but beg and pray for counsel from every passer-by.
>
> No plan is then too futile, too absurd, . . . if anything happens during their fright which reminds them of some past good or ill, they think it portends a happy or unhappy issue, and therefore (though it may have proved abortive a hundred times before) style it a lucky or unlucky omen. . . .
>
> Signs and wonders of this sort they conjure up perpetually, till one might think Nature as mad as themselves, they interpret her so fantastically.

Although Spinoza never said he was an atheist, he was known as one in his own century, and ever afterward. Descartes had said that the material world was one kind of substance and mind was something else. Yet on a grander scale, Descartes said that the only substance and the only thing one could know for certain was God. Spinoza let this idea inflate until it curled around itself: God and everything were the same. God's thought did not make the world, God *is* his thought, and the God-thought *is* the world. Spinoza's argument followed the model from geometry, taking one axiomatic idea at a time and building on it. As he had it, the universe was a self-caused substance with infinite attributes; human beings are "nodes" of that substance; and he called that substance "God or Nature." This is pantheism, although the word did not yet exist. It is no wonder that the Jewish fathers had a problem with it; it left little room for Jewish history, cosmology, or theology. Since the world was the lawful unfolding of the cosmic event, the idea went, everything was predetermined. Each thing led to the next, mechanically, so unavoidably. No one, not human beings, not God, could have free will. That is one of the reasons that even though Spinoza always talked about God, people saw him as an atheist: a god that has no free will and is not separate from the unfolding universe is not much of a god. After Spinoza was excommunicated, he changed his given name to its Latin equivalent, Benedict, and lived out his life as a philosopher, a lens grinder, a maker of fine telescopes and microscopes, and a tutor.

Spinoza resolved to look hard at revealed religion and see whatever was there to be seen. Taking the lead from Isaac la Peyrère, he wrote about a host of logical impossibilities, concluding that the unphilosophical masses needed scripture but that the real truth was more subtle. God did not have purposes. Nature was self-causing and unfolded according to necessary law. There were no miracles. Galileo had viewed Joshua's miraculous stopping of the sun from the viewpoint of a heliocentric world and declared that whatever happened it accorded with science. Spinoza got further in his doubt because Jewish tradition had precedent for this kind of thing: Maimonides had said that the miracle was extended local daylight, without there actually having been any change in the movement of heavenly spheres. Gersonides, "the first rabbi on the moon," had said the miracle was that the victory was so fast it took place in that brief time during which the sun seems stopped at the top of the sky, i.e., the whole thing was a literary device. We should not be too surprised, then, to hear Spinoza say in his *Tractatus*, "Do we have to believe that the soldier Joshua was a skilled astronomer . . . or that the sun's light could not remain above the horizon for longer than usual without Joshua's understanding the cause? Both alternatives seem to me ridiculous."[9] Spinoza was willing to allow that something weird had been perceived, perhaps owing to "excessive coldness of the atmosphere," and that Joshua had given it a supernatural explanation out of simple ignorance. But he also argued that the book might just be untrue. Spinoza was convinced that the Bible had multiple authors and he rejected divine authorship altogether. His understanding of the Bible was similar to Carneades's view of the Greek pantheon; its supernaturalism did not have to be rationalized—it could simply be dismissed.

Spinoza's biblical criticism was lively and memorable, and his indictment of religion could be blunt:

[T]he man who endeavors to find out the true causes of miracles, and who desires as a wise man to understand Nature, and not to gape at it like a fool, is generally considered and proclaimed to be a heretic and impious by those whom the vulgar worship as the interpreters both of Nature and the gods. For these know that if ignorance be removed, amazed stupidity—the sole ground on which they rely in arguing or in defending their authority—is taken away also.[10]

Privately, he recommended a life the ancients would have recognized: study, wine, good food, the beauty of green things, theater, and sports. His encouragement of such sensualism was expressly set against religious self-denial: "nothing forbids our pleasure except a savage and sad superstition."[11] Above all perhaps, he was known for his advocacy of virtue for its own sake and for the earthly rewards it brings. His ethics were seen in relation to his doubt. A contemporary described Spinoza as offering his work "for public utility," in order to encourage citizens to "live honestly and obey their magistrates" and to keep themselves virtuous, "not in the hope of a compensation after death, but simply for the excellence of virtue itself and for the advantages for virtuous people in this life."[12] The same contemporary mentioned that in order to do this, Spinoza's work had as its main objective "the destruction of all religions, particularly the Jewish and Christian religion, and to introduce atheism, libertinism, and freedom in all religions."

The two great figures of atheism in the seventeenth century were Spinoza and Hobbes—although neither ever described himself as an atheist. Hobbes is best known today for the political science of his masterwork, *Leviathan,* which claims that without authoritarian government people's lives would be "solitary, poor, nasty, brutish, and short." It was a support for the monarchy of his time, but the book was at least as important for its role in the history of doubt. Hobbes said we do not know anything about God other than that he exists. His biblical criticism treated the Bible like any mixed-up historical text; he teased apart its different authors on the basis of literary and historical analysis, much as Spinoza did. The truth about religion, as Hobbes explained it, is that it had been formed and sustained by people in power, to control their subjects. He allowed that religion was good for people but said there was no reason for the priesthood ever to have power above the monarchy, since the clergy have no special information on God. They just operate the cult. Hobbes understood the world as a machinelike thing that runs itself. He also claimed that our souls are mortal (he cites Job saying so), but that the saved will be revived at Judgment Day while the others simply will not.[13] Hell, he said, was just a fantasy to control people. Foolish people, "they that make little or no enquiry into the natural causes of things," are driven by anxiety about their future and make up fanciful relationships between events and "powers invisible," and end up "in awe of their own imaginations, and in time of distress . . . invoke them, and as also in the time of an expected good success, to give them thanks, making the creatures of their own fancy, their Gods."[14]

Hobbes said people believe religion as an explanation for why good and bad things happen. When someone "cannot assure himself of the true causes of things (for the causes of good and evil fortune for the most part are invisible), he supposes causes of them, either such as his own fancy suggesteth, or trusteth to the Authority of other men."[15] Hobbes knew that "some of the old Poets said that the Gods were at first created by human Fear," but "acknowledging of one God eternal, infinite, and omnipotent may more easily be derived from the desire men have to know the cause of natural bodies" than from the fear of what was to befall them in time to come.

Causality leads one to the conclusion of a first mover, "a first and eternal cause of all things which is that which men mean by the name of God."

> It was the same with that of the soul of man; and that the soul of man was of the same substance with that which appeareth in a dream, to one that sleepeth, or in a looking-glass to one that is awake, which men not knowing that such apparitions are nothing else but creatures of the fancy think to be real and external substances and therefore call them Ghosts.[16]

Hobbes summed up religion as derived from four mistakes: belief in "ghosts, ignorance of second causes, devotion towards what men fear, and taking of things casual for prognostics," and from these errors, "different fancies, judgments, and passions of several men, hath grown up into ceremonies so different, that those which are used by one man, are for the most part ridiculous to another."[17] Many people have believed Hobbes was an atheist and that when he made statements that left room for God he was just saving his neck. Then again, why offer one's own idea of resurrection, dangerously heretical in itself, as part of a believing smokescreen? We cannot know. What we can know is that he argued against religion, and against any conception of God beyond the simplest statement that God exists, and many were unconvinced that he meant that. The effect on his world was monumental.

There is a sense that Hobbes's reliance on absolute government, despite its abuses, was necessary precisely because the idea of governance by God did not seem likely to have much of a career left to it. People had to be made to be good somehow. Naudé, too, had believed in religious rebellion and political submission, but Hobbes was easily doubt's greatest political conservative since the Stoics. Still, people were scandalized by his doubt: a bill against atheism was introduced to the English Commons in 1666 that mentioned *Leviathan* by name. Another great wave of the plague had hit England and it was suggested that this was punishment for harboring an unbeliever; a cry went up to burn Hobbes in hope of turning away the wrath of God. Hobbes, however, was a well-protected man by then, and he died at the ripe old age of ninety-two. For centuries, despite the important political meanings of Hobbes's work, to be called Hobbesian was to be called an atheist. We have the witness of a contemporary observer, one Richard Bently, who wrote in a letter to his professor, in 1692, "There may be some Spinosists . . . beyond the seas; but not one English Infidel in a hundred is any other than a Hobbist; which I know to be rank Atheism in the private study and select conversation of these men; whatever it may appear to be abroad."[18]

To hear another witness: in 1669 David Scargill caused much ado with his "Recantation," wherein he confessed that he had "lately vented and publicly asserted" at the University of Cambridge "divers, wicked, blasphemous and Atheistic positions," including "that there is a desirable glory in being, and being reputed, an Atheist; which I implied when I expressly affirmed that I gloried to be an Hobbist and an Atheist."[19] Scargill thus testified to atheism and to reading Hobbesianism as atheism; that he professed these ideas at Cambridge; and that he had been gloriously proud.

It's good to keep Scargill's Hobbesian pride in mind as we sidle up to Blaise Pascal's famous wager. Pascal knew Mersenne, too, but late; he was a youth in these conversations when the rest of the skeptical fideists were aging. In his 1670 *Pensées,* Pascal wrote that we are incapable of knowing whether God exists or not, but we have to guess. He proposed that our choice should be influenced by the various results:

> "Either God is or he is not." But to which view shall we be inclined? Reason cannot decide this question. Infinite chaos separates us. At the far end of this infinite distance a coin is being spun which will come down heads or tails. . . .

> Which will you choose then? Let us see: since a choice must be made, let us see which offers you the least interest.[20]

In *Pensée,* Pascal was speaking these questions to an unbeliever, someone who says "I am so made that I cannot believe."[21] Pascal's answer is that life is full of affliction and death is always threatening the "appalling" possibility of annihilation, and hence "the only good thing in this life is the hope of another life."[22]

Pascal claimed that if humanity had to live without God it would be pure misery, and if God did exist, it would be bliss—the various values of the outcomes (rather than the probability of the thing in question, which he insisted was fifty-fifty) changed the guess. This idea that preference for a particular outcome can, in some cases, reasonably influence our choice, was a watershed in decision theory. It was a bit odd that he thought the chance of God's existing was one in two just because there were two choices; we may be struck by lightning or not, but that doesn't make it a fifty-fifty proposition. Anyway, the key factor here, as historian Richard Popkin has put it, is that "What Pascal decried as the misery of man without the Biblical God, was for Spinoza the liberation of the human spirit from the bonds of fear and superstition."[23] Hobbes inspired similar pride: the bad outcome for Pascal was not the bad outcome for all others. For some, there was liberation in it.

While the seventeenth century offered doubters who are famous to this day, such as Spinoza and Hobbes, and doubting believers such as Pascal, there were some hugely popular doubters who have been forgotten. The anonymous book *Theophrastus Redivivus,* published in about 1650, was famous for more than a century. It was a compendium of old arguments against religions and belief in God, and it precipitated a cultural explosion in discussions of unbelief. As historian Tullio Gregory has shown, the author was familiar with the Paduan Averroist, Pomponazzi; the Renaissance secularist Machiavelli; as well as Julius Caesar Vanini. The author had also read the Skeptics Montaigne and Charron, and the Libertine Naudé.

Some of these authors—Pomponazzi and Vanini, for example—seemed to say just what was on their minds. Others—Machiavelli, Montaigne, Charron, and Naudé—capped their doubting treatises at either end with protestations of faith. To some people, that meant that they believed; to others, it meant that these writers were smuggling their arguments past the censors. Whatever the intention, these writers had smuggled a great deal of doubt into the hands of the reading public. *Theophrastus Redivivus* was testimony that readers were savvy enough to recognize such smuggling. The book was a compilation of the most antireligious, atheist, and variously doubting positions of all these doubters. The book was also famous for having proclaimed that all philosophers, always, had been atheists. The tone and certainty suggested that the author knew other atheists and felt part of a special community, across time.

At the end of the seventeenth century, there was a surge in the persecution and aggressive suppression of unbelief in Italy. The Church now began to silence not only hot blasphemy but also cool materialist philosophy. In Pisa in 1670, Cardinal Leopoldo de' Medici discreetly warned the local scholars who supported atomism that the notion would no longer be tolerated.[24] The trials in Pisa were stepped up soon after, in 1676. The authorities caught wind of a whole flock of offenders in 1688: eleven men and women who were accused of denying the existence of God and the doctrines of creation, the afterlife, and Christ's divinity. They believed instead that the world was eternal and best explained by atomism. Across Italy, Inquisition trials increased for atheism and related crimes. From the late 1680s to the early 1690s, in Naples, a de Cristofaro lingered in prison six years for teaching Lucretian atomism.[25] There was to be no lull in the campaign against unbelief until the 1720s. Not surprisingly, science went north.

In 1642, the year that Galileo died while under house arrest in Florence, Isaac Newton was born in Lincolnshire, England. Newton profoundly advanced the new concept of the world as understandable and regular. He invented the calculus (primarily, the mathematics of change), explained that white light was made up of colored light, and discovered the law of gravity—all, by the way, by the age of twenty-three, although he waited to publish. His *Principia* (1687) drew on Descartes's law of inertia, Galileo's ideas on acceleration, and Kepler's laws, and brought it all to a mathematically expressed synthesis that made the world strangely intelligible. One did not need Aristotle's spheres to hold objects in the heavens, and the problem of the planets' continuous motion disappeared. *Principia* was also influenced by Gassandi, and atomism was a cornerstone of Newton's thinking. In the early drafts for the second edition of *Principia,* Newton included ninety lines from Lucretius's *On the Nature of Things* in association with his concept of inertia.

Newton accomplished a revolution in worldview, yet he spent as much of his life on alchemy and mysticism as he did on physics. He believed in God. For several centuries now, Newton has been called the first physicist and the last magician—and all along historians have wondered how he balanced belief in, say, miracles with having concocted the physics that mechanized our minds and took us to the moon. For the answer we need to note that in the sixteenth and seventeenth centuries, miracles became extremely important in the Protestant world. For Catholics the problem of discovering true miracles had almost always been concerned with evidence of sainthood. Protestants were after bigger quarry. For them miracles are humankind's crucial evidence that Christianity is true. The Protestant physicist and chemist Robert Boyle (1627–1691), for example, wrote that the evidence of miracles "is little less than absolutely necessary to evince . . . that the Christian [religion] does really proceed from God."[26] Boyle thought that since miracles were real, scientists would be the best people to identify them.

Newton believed in some miracles. He said that the way our solar system was set up seemed a little unstable, and he thought either it would need adjustment, or that it would fall apart eventually and God would need to set it up again. Recall that Spinoza had already said there could be no miracles, that miracles did not even make sense; God was the universe and its laws. It was known that there were English "deists" by now who agreed that scripture could not always be reconciled to truth. The Newtonians disagreed. There were a few real miracles, where God stepped in and changed things, at creation, in the period of Jesus, and again at the end of time—which Newton associated with the time when our solar system would be destroyed.

Otherwise, the Newtonians reconciled some miracles with their lawful universe but let them maintain a divine component. Take the flood story, a saga of punishment sparing Noah's family, his ark, and his animals. The Newtonians believed God knew in advance that this period would be sinful and so he built into the pattern of the universe a series of events that led to a natural flood to wipe out the sinners. Even prayer worked that way, so that a god who never intervenes can be said to have heard you in advance and set up a natural chain of events in your behalf. The *Cosmologica Sacra* (1701), by Nehemiah Grew, which was supposed to be a defense of religion against Spinoza's way of interpreting the Bible (i.e., that some of this stuff just didn't happen), suggested that the plagues of Egypt had resulted from "sundry Natural Causes." When Moses turned the Nile to blood, Grew proposed, it was really because "all the Fish, small and great, with the Hippopotamus, Crocadile, and other Amphibious Creatures were seiz'd with a Dysenterick Murrain."[27] That is, a huge bout of nasty marine dysentery had transformed the river into a stream of bloody effluvia.

All this science and philosophy reached the average person through the popularizing work of Pierre Bayle (1647–1706). He was born in France, but the wars of religion had ended with the dominance of a singular authority: Louis XIV, who wanted France Catholic and revoked the tolerant Edict of Nantes. Bayle spent the bulk of his life in Holland, whose cities were joining Athens, Alexandria, Rome, Padua, and Paris as some of doubt's great centers. Bayle's was Rotterdam, and he was the leading member of a vibrant intellectual community there. His *Miscellaneous Thoughts on the Comet of 1680* argued that the comet was a natural phenomenon and did not presage disaster; but this was not the point of the book. What the work really represented was the first-ever all-out defense of the morals of an atheist. Chapter headings include avowals that "Atheism does not necessarily lead to corruption of morals"; that belief in God does not make people honest; that "In a society of atheists there would be laws of propriety and honor"; that "Belief in the mortality of the soul does not prevent people from desiring to immortalize their name"; and that, historically, atheists have "not been especially conspicuous by the impurity of their morals." In this last he praised the upright life and unimpeachable moral fiber of Epicurus and his followers, the Jewish sect the Sadducees (who, Bayle reminds his readers, "frankly denied the immortality of the soul"), as well as the atheist Julius Caesar Vanini and a number of lesser-known and anecdotally well behaved unbelievers. Epicurus and Vanini themselves had made it explicit that doubters could be even more moral than believers, for the former are motivated by reason. For that matter, so did Socrates, Plato, and Aristotle. Still, the sense that the common people needed God for morality was strong. Only after Bayle's aggressive argument do we begin to see the disappearance of an old reason for self-censorship: doubters had chosen to be silent or at least to whisper (even when they did not fear for their lives) because of their fear that without a belief in God and the afterlife, the masses would run wild in the streets. Fear of other people's behavior, should they lose religion, had always been a reason to stay quiet. Now it was less so.

Bayle's 1695–1697 *Historical and Critical Dictionary* is one of the weirdest books ever written and was, in historian Thomas M. Lennon's words, "the philosophical blockbuster of all time," reaching more of its potential audience than even Plato.[28] Through the eighteenth century, this book was on more shelves than any other. Even after the hugely popular Locke and Voltaire were on the scene, they trailed behind Bayle by leagues. Indeed, they and other philosophers used his tremendous cache of arguments so often that Bayle became known as the "arsenal of the Enlightenment."

The reason his book is so strange is that for every page of text there are about twenty pages of lengthy footnotes, and even the footnotes have footnotes. In Bayle's editions the notes to the notes ran in the margins; in modern editions, the footnotes take over pages after the original entry, and the secondary notes take up the foot of these pages. The entries, nearly three thousand of them by the time he was done, were almost all biographies. Bayle would say who the person was in a brief and often ancient sketch and then use a note to leap off on a tangent. The information was already being offered in the least thematic form possible—alphabetically—but when it is combined with the footnotes, the busyness is giddy. What hides in this puzzle box of a book? A good deal of sex and a great deal of philosophical Skepticism. Both could show up anywhere. The entry on Jupiter is festooned with notes on the god's incestuous proclivities that roamed over several generations. Bayle also lets the notes from Jupiter's entry carry him off to discuss Skepticism through Carneades, Anaximenes, Epicurus, and Cicero. The entry's last note finishes up as follows:

> Those who claimed to be best acquainted with the doctrines of theology showed, when they expressed themselves clearly, that they did not acknowledge any other gods than air, the stars, and the like. This was at bottom true atheism. It amounted to converting the necessity of nature into god. I have noticed a passage in Euripides in which Jupiter is invoked without actually knowing what he is. It is admitted that he governs all things justly by occult methods, but he is found extremely difficult to know, and one cannot tell if he is the necessity of nature or human intelligence. What a faith! A Spinozist would just about agree to this.[29]

Then he cites the passage, from the *Trojan Women,* that he mentioned above: "O vehicle of earth, residing on it, / Whoever thou be, inscrutable to us, / Necessity of nature, or men's minds, / O Jupiter, I invoke thee."

Consider Bayle's entry for Simonides. The ancient poet's life and work are summed up in a paragraph, within which we read, without introduction of any kind, "The answer he gave to a prince who asked him for the definition of God is very famous."[30] This notion gets two footnotes, which go on for fifteen pages. They explain that Simonides continually asked for more time, finally answering that the more he thought about it, the less he knew. Bayle then brings in Cicero's skeptical comment on this piece of Skepticism (the Roman guessed the Greek, too, had "despaired of all truth"), and then merrily discusses Skepticism for pages, bringing in Thales, Descartes, La Mothe le Vayer, Aristotle, and Charron. The odd style of texts and subtexts puts us in an ever more private conversation. Bayle speaks of the question of

God's gift of free will, allowing man to sin, in terms of a mother who would allow her daughters to go to a ball although she knew for a fact they would give in to the enticements of some gallant gentlemen there and "part company with their virginity."

His entry on Spinoza—a relatively long one at a few pages, and the footnotes go on for thirty more—is an attack on the idea of God as indistinguishable from the universe, which Bayle called clear atheism. The second footnote includes an incredible story. It is offered as witness to the contention that Spinoza's atheism was not new and could be evidenced in many places, including China. Bayle tells of a Chinese teacher, called Foe, who had gone into the desert at the age of nineteen and studied there until he was thirty. He then emerged "to instruct men," and he "represented himself as a god" and attracted eighty thousand disciples.[31] At seventy-nine and near death, he confessed it was time to tell them the truth. "'It is,' he said, 'that there is nothing to seek, nor anything to put one's hopes on, except the nothingness and the vacuum that is the principle of all things.'" At death Foe "began to announce his atheism." Afterward, his disciples "divided their doctrine into two parts, one exterior, which is the one that is publicly preached and taught to the people, the other interior, which is carefully hidden from the common people."[32] The exterior doctrine is "only like the frame on which the arch is built and which is later taken away when one has finished the building." Listen to his description of the false exterior doctrine:

> It consists in teaching: 1) that there is a real difference between good and evil, justice and injustice; 2) that there is another life in which one will be punished or rewarded for what one will do in this one; 3) that happiness can be attained by means of thirty-two figures and eighty qualities; 4) that Foe is a deity and the savior of mankind, that he was born out of love of them . . . that he expiated their sins, and that by this expiation they will obtain salvation after death, and will be reborn happier in another world.[33]

By contrast, the interior doctrine "that is never revealed to the common people because they have to be kept in their place by the fear of hell," is that there is nothing; "that our first parents came forth from this vacuum and that they returned there after death."[34]

In Bayle's entries on such figures as Pyrrho, Xenophanes, and Pomponazzi, he walks his reader through all sorts of religious doubt, sometimes contesting a point, sometimes reporting calmly. There were doubters in

Bayle's dictionary who became Skeptical heroes after Bayle published: Julius Caesar Vanini was once again known—Bayle calls him western Europe's great martyr for atheism.[35] He dubbed Bruno "the knight-errant of philosophy," helping to romanticize that freethinker. Bayle was so positive toward Judaism that some have posited that he was secretly of that faith—in any case, he preached tolerance for all religions. He had an entry on Muslim Skeptics that, although tiny, included a long footnote on whether all philosophers end up as atheists. He notes there that most people thought the association was true:

> There are those who maintain that the Arabian philosophers were only Mohammedans in appearance, and that they actually made fun of the Koran because they found things in it that were contrary to reason. You cannot get a great many people to stop believing that Descartes and Gassendi were as little convinced about the Real Presence as about the fables of Greece.[36]

Bayle credits Gassendi for making Sextus Empiricus known to his contemporaries. Others who make appearances in Bayle's notes include Democritus, Confucius (with whom Bayle argues regarding the nature of nothingness), Montaigne, and Naudé. Bayle also discusses English Socinianism, a movement that rejected the divinity of Jesus but believed in God: Faustus Socinus (1539–1604) left the Roman Catholic Church when, influenced by the writings of his uncle, Laelius Socinus, he came to deny the Trinity. Faustus left Italy to spend his life in the comparatively tolerant Poland. His doctrine spread from there to England through John Biddle (1615–1662). Bayle also spoke of Jansenism, a kind of reformed Catholicism that grew up in these years. Bayle sympathetically argues with Maimonides throughout—indeed, Richard Popkin has written of Bayle's *Dictionary* as an attempt to provide a new *Guide for the Perplexed.*[37]

The *Dictionary* became the bible for doubters. Bayle upheld Montaigne's insistence that religious claims are not confirmed by any inner knowledge, but instead were fed to us in our childhoods. Since all is custom, there is no point in religious intolerance. Bayle also combines the ideas of the ancients with the attitude of the new science to produce a new Skeptical empiricism, but he does not describe himself as a Skeptic. He argues for sensible judgments (on the basis of experience and probability) and the great new world of modern science. "It is therefore only religion that has anything to fear from Pyrrhonism. Religion ought to be based on certainty. Its aim, its effects, its usages collapse as soon as the firm conviction of its truths is erased from the mind."[38] He made excellent Skeptical arguments and then said such ideas destroyed religion, but for himself said that the urge to do good rather than evil might be a satisfactory sign of God. More than that we can only guess, and many have guessed that Bayle did not believe. His advice to believe seemed a cold choice compared to his heated critique of religion. As Popkin puts the question, "Was Bayle, in his forceful, skeptical way, trying to lead people to faith, or was he secretly trying to destroy it, as Voltaire and many others have since suspected, by making it so irrational, so lacking in morality, and so ridiculous?"[39]

As we leave Bayle, consider his entry on Geoffroy Vallée. Vallée had said he found unbelief in Ecclesiastes and the first psalm and was executed in 1574 for denying God.[40] Bayle made much of Vallée as evidence that unchecked reason always leads to unbelief. Bayle wrote as if he were warning against that unbelief, but that is not how it read to his tremendous readership, and it gave Vallée, who had been drifting out of memory, a proper place in the history of doubt. Yet, along with looking to the past, Bayle's book was fresh: impatient with the variety of philosophies, confident in science, and resigned to ignorance in matters of religion.

In the early eighteenth century a new popular compendium of doubt appeared on the scene, infamously titled *The Three Imposters*. The impostors in question were Moses, Jesus, and Muhammad. Under this title, the text was the most widely available of all the clandestine manuscripts that circulated during the eighteenth century.[41] It was first published in 1719 under the title *The Life and the Spirit of Spinoza,* and in fact there were two texts here, one called "The Life" and the other "The Spirit." "The Life" was a biography of Spinoza, written in about 1678 by a devoted disciple, which drew him as an exceedingly virtuous genius, following truth. "The Spirit" was itself a meld of two things: the first part was the three-impostors legend, which was traceable to Pomponazzi and Averroës and which held that each of the great religious leaders was tricking his followers and was not an agent of the divine; the idea had been recently revived by the Libertines and most recently by the *Theophrastus Redivivus.* The second part was much more modern: texts of Spinoza that changed a few words here and there, rendering it fully materialist. It was composed sometime between 1702 and 1712 and published anonymously. Historian Silvia Berti has offered evidence for its authorship: Jan Vroesen of Rotterdam, a friend of Bayle's and founder of The Lantern, the most important philosophical society in town and famous as a company of heretics and freethinkers. The Lantern met in Vroesen's house.

"The Spirit" also highlighted Spinoza's discussion of the making of the Bible, including the talmudic story of a council of rabbis who almost removed Ecclesiastes from the Bible but decided for it, finally, because it praised the law of Moses.[42] More generally, what "The Spirit" took from Spinoza was an array of spicy quotations critiquing religious ceremony and literal belief in the Bible. "The Spirit" also offers psychological reasons for the development of religion: "Men, feeling that they are capable of wishing and hoping, falsely conclude that this is all that is required to make them free."[43] We fall into this error because wishing would be an efficient way to get what we want, so we hardly even think about what might actually work.

Then, believe it or not, the editors who brought out *The Life and the Spirit of Spinoza* added whole chapters to Vroesen's text. One new chapter was an entire section of Spinoza's *Ethics,* which actually represented the first French translation of any of it. That was particularly significant since— though popularizations spread the *Ethics'* ideas widely—a full French translation was not published until 1842! In the opening passage of the section of *Ethics* included here, Spinoza gave his account of God acting through the necessity of his nature, but that had been cut, removing even Spinoza's odd idea of God. The stand-in paragraph was close to one from Hobbes's *Leviathan,* where "the chimerical fear of invisible powers" was described as the origin of religion.[44] Another added chapter was the intensely irreligious chapter of *Leviathan* titled "Of Religion." Still another new chapter was lifted from Charron's *The Three Truths,* and one from his *Of Wisdom.* Four chapters relied on Julius Caesar Vanini (executed exactly a century earlier) and the Libertines Naudé and La Mothe le Vayer, both dead for more than a half-century and in need of a new audience. In Berti's words, the editors' purpose "was to construct and disseminate the first portable philosophical compendium of free thought."[45] They succeeded, and it was incredibly popular.

ENGLISH DEISTS

English deism starts with Edward Herbert of Cherbury (1583–1648), a soldier, poet, and philosopher. He was also ambassador to France from 1618 to 1624 and got to know Mersenne and Gassendi, hung around with the Erudite Libertines, and circulated with those known to discuss Sextus

Empiricus.[46] In his masterwork, *The Truth,* published in 1624, he presented to the English world of letters the problem of Skepticism and his refutation of it. The refutation was not much (it had to do with our innate knowledge of truth), but Herbert had gotten the conversation going. He gave deism its name and its first tenets: there was a supreme deity who should be worshiped and who metes out justice in this life and after it; everything the churches added to this was bunk.

English deism was much enlivened with John Locke (1632–1704). Locke studied at Oxford, where he became friends with Newton and Boyle—friendships that persisted all their lives. Locke was also much influenced by Gassendi's work. He went to France, where he met the bright new minds of French philosophy and science. When he returned to England he was soon suspected of radicalism by the government, and took off for Holland, where he wrote his *Essay Concerning Human Understanding* (1690). Many philosophers had thought our basic ideas about the world (ideas of God, of virtue, of truth) were innate in the human mind or soul. Locke said when we are born the mind is a blank slate, a *tabula rasa,* and that all knowledge, ideas, images came from sense and experience. To improve human life, do not ask God for help, but change experience and thereby improve people. Locke appreciated the Skeptical critique but thought that sensory knowledge was more manageable than that: we can break it down, following the kind of demonstrations of evidence made by Newton. One can start by saying things have primary qualities like solidness or size, and also secondary qualities like taste and sound. The first are all about the object; the second are about the way objects impact our sense organs—the world outside our heads cannot be said to have taste or sound. This was real progress in dealing with Skepticism.

Locke had digested the doubt of the ancient philosophers (there were no anthropomorphic gods) and had digested the doubt of the ancient Skeptics (nothing could be known at all). He also understood how Christian theology had transformed the idea of knowledge (by seeing religion as the new source of it, through revelation), and he saw how the Renaissance Skeptics such as Montaigne had brought the whole thing down again. Locke did not agree with Descartes, because Locke noticed that "I think, therefore I am" is a bit of a leap (as the Buddha might have happily pointed out); that "I think, therefore thinking happens" is pretty much all you can get. Locke did believe in God, as a deist, because there was thinking in the universe. He considered it possible that thinking could be a property of organic matter but decided that, probably, thinking could not have appeared in the universe if it had not come from some great thinking something.

Locke's ideas spawned one of the history of doubt's most illustrious of believers. The philosopher George Berkeley (1685–1753) said that Locke's primary and secondary qualities were *all* products of our minds. Berkeley's doubt of the reality of the material world went further than that of any other philosopher in the West. It took as true what was, in the old Skepticism, only the most extreme of the various possibilities. For Berkeley, the world was only in our minds and the mind of God.

By 1725 J. F. Reimann's *Universal History of Atheism* included as atheists: Thomas Hobbes, John Toland, Count Charles Blount, and Anthony Collins. I have spoken of Hobbes. The other three came a half-generation later and, though less known today, were cutting-edge in their own time. Toland had met and become friends with Locke and joined forces with him in the public fray. Yet Toland's own thought was much influenced by Lucretius and Bruno. Toland was often said to have started the "deist controversy" when he published, in 1696, *Christianity Not Mysterious,* which read the Bible in the new critical mode and incorporated the most doubting aspects of Locke's philosophy.[47] The book was burned in Ireland in 1697. In England, an Act of 1696 was aimed at any who said there were more gods than one or denied the truth of the Christian religion. Thus, Toland's choice to attach his name to the second edition of *Christianity Not Mysterious* meant that he spent most of his life in financial difficulty. His biography of John Milton caused a scandal for doubting the authenticity of the New Testament. He fought for full rights for Jews in England and Ireland in a lifelong campaign for religious toleration. Jonathan Swift called him the "great Oracle of anti-Christians," and Locke coined the term *free-thinker* in reference to him.[48] Toland also may have been one of the editors involved in the production of *The Three Imposters.*

In Toland's *Letter to Serena* (1704), written to Sophia, Queen of Prussia, he coined the term *pantheist.* He used it again in the title (only) of *Socinianism Truly Stated, by a Pantheist* (1705); and in 1710, in a letter to the philosopher Gottfried Leibniz, he offered a definition: "the pantheistic opinion of those who believe in no other eternal being but the universe." Toland worked it all out in *Pantheisticon* (1720), which proposed a civic religion with meetings, community rituals, and a secularist liturgy. Although modeled on Masonic lodges (cultish nonreligious citizens' clubs), Toland's imagined civic religion was the first of its kind. He clearly wanted more community for doubters. At least he had friends from the past: in 1698 he bought Queen Elizabeth's bound copy of four dialogues by Bruno, which must have cost him a great deal, and we know he had a copy of Lucretius's *On the Nature of Things* near him in the room where he died.

Toland was an atomist and believed that Aristotle's problem with the motion of the world had been proved foolish by the new physics: if matter had motion intrinsically, that is, if it turns out that matter is in motion by its very nature, the hypothesis of a prime mover becomes unnecessary. Newton saw that force is not a thing outside matter that pushes on it, but rather that pieces of matter exert forces on each other. Today we think less of forces at all, and instead think of the pieces of matter having a field of interaction. Still, Toland had noticed that Aristotle had been tricked by friction and gravity into thinking that things needed outside force in order to move. The Prime Mover was unnecessary after all.

Toland did the historian of doubt a remarkable service. He wrote on the hidden or "internal" doctrines of ancient writers—how they had exoteric teachings that were public to all and esoteric teaching that they kept only for special students—and he taught how one could parse a single text and find both the public and the secret, embedded meanings. He was writing about the ancient philosophers, but Toland admits, "I have more than once hinted that the external and internal doctrines are as much in use as ever."[49] He further notes, "considering how dangerous it is made to tell the truth, 'tis difficult to know when any man declares his real sentiments of things."[50] In this way we are told outright that, when criticizing religion can get an author beheaded, a lot of authors hide the true meaning of their books. It's an obvious point, but historians try hard not to read anything into texts unless there is very good evidence to support it. Toland's contemporary recipe for subterfuge lets us be a little bolder in our interpretation of the texts of his times. The first step in disguising a thesis, Toland explains, is to start off your work with "a bouncing compliment" to whatever it is you are going to attack. Next, mention that you are not dissembling—it reminds people of the possibility of dissembling. Insisting that you are not a wolf in sheep's clothing (indignantly!) can subtly influence the innocent reader, while serving as nothing less than a secret handshake to those in the know. Of course, he continues, the best technique for smuggling doubt to the reading public is simply to offer short, lousy arguments for the orthodox view (to which you have offered your "bouncing compliment") and excellent, extensive arguments for what you really believe.[51]

Anthony Collins (1676–1729) was friends with Toland and Locke and was much influenced by Spinoza. He said that there was a God, but that God was the world; thus everything is the only way it could have been, and therefore God is not free. Berkeley knew Collins, and Berkeley swore that someone he called "our Diagoras" had announced that he had found a proof against the existence of God. Historian David Berman has done a nice job of showing that this Diagoras was Collins.[52] Their contemporary, Samuel Johnson, offered a bit more information when he wrote that Collins had announced finding a demonstration against God and then "soon after published [it] in a pretended demonstration that all is fate, and necessity."[53] That is, Collins told his select crowd about his Spinoza-influenced atheism, but in print he said that everything is mechanistically predetermined, without mentioning that this meant God was not free and was thus not anything more than the universe itself. A contemporary wrote, "I am told by those, who are very capable of informing me, that the modern Atheist [in England] has given up the system of Epicurus as absurd and indefensible and adheres to that of the Fatalists."[54] For many of these people, Fatalism meant not only Spinoza but also ancient Stoicism. Listen to Collins on immortality: "The immortality of the soul was nowhere plain in the Old Testament, was denied by the Sadducees, the most philosophical part of the Jewish Nation . . . was thought doubtful by most sects of the Greek philosophers and denied by the Stoics, the most religious sect of them all; had never, according to Cicero, been asserted in writing by any Greek extant in his time . . . and was first taught by the Egyptians."[55] The soul he emphasized was a new, transplanted idea, from the land of mystery cults and animal worship.

The first to translate sections of Spinoza's *Tractatus* into English was Count Charles Blount (1654–1693), in his *Miracles No Violations of the Laws of Nature* (1683). He combined it, paragraph by paragraph in places, with selections from Hobbes's *Leviathan*. Blount was a close friend of Hobbes's and was devoted to ancient Stoicism. One of his favorite aphorisms was Seneca's line: "After death nothing is, and nothing Death." Blount put out a scholarly edition of a classical text that camouflaged a compendium of ancient, Renaissance, and contemporary sources of doubt: Pliny, Cicero, and Lucretius; Pomponazzi, Vanini, Montaigne, and Machiavelli; and then

Hobbes, Spinoza, and Isaac la Peyrère. Blount also wrote *Oracles of Reason—"Lucretius Redivivus."* He seems to have been an important figure of a thriving world of freethinkers. Blount grew ill and, in the end, killed himself—long cited as a proof of his commitment to Stoic over Christian ideas. He has also been remembered as an atheist.

Also part of this group was the third Earl of Shaftesbury, Locke's patron, like him a deist and champion of Plato, Epicurus, and Marcus Aurelius and of general free thought. The jurist and deist Matthew Tindal was also important, arguing that true religion was universal and all particular dogma was nonsense, based on unverifiable hearsay. There were fights among these thinkers: Bernard Mandeville's *Free Thought on Religion* (1720) was understood by many as the great text of English deism. It favored the Epicureanism of Hobbes and Gassendi and Montaigne's approach to moral problems, and rejected other deists as maintaining frankly Christian ideas in their supposedly rational religion. Amid the squabbles, though, Shaftesbury in particular brought a concentration on politeness and manners to philosophy and this tended to open the conversation to anyone who could learn the civil niceties.

So doubt was thriving among philosophers. What of the rest of the population? Consider the case of young Thomas Aikenhead. The events took place in Scotland and "made a great noise" in London at the end of the 1790s.[56] Aikenhead was a medical student at the University of Edinburgh, charged with blasphemy in 1796. He was accused of saying that theology was "a rhapsody of feigned and ill-invented nonsense" and that the scriptures were "so stuffed with madness, nonsense, and contradictions, that you admired the stupidity of the world in being so long deluded by them."[57] His accusers said he described the Old Testament as "Ezra's fables, by a profane allusion to Esop's fables [*sic*], and that being a cunning man he drew a number of Babylonian slaves to follow him, for whom he had made up a feigned genealogy." Ezra, he guessed, showed the genealogy to the Persian king Cyrus and thereby convinced the king to set them up in Israel. Aikenhead was also accused of calling Jesus "the impostor Christ" and saying Jesus must have learned some magic in Egypt and then "picked up a few ignorant blockish fisher fellows, whom he knew by his skill in physiognomy, had strong imaginations."[58] Jesus then played "pranks" on them with his supposed miracles. Aikenhead was also accused of saying that Moses, "if ever . . . there was such a man," had, like Jesus, "learned magic in Egypt, but that he was both the better artist and better politician than Jesus."

Lectures at the University of Edinburgh were filled with discussions of Hobbes and Spinoza so that the students could learn to refute them. Aikenhead might have gotten the idea of Ezra's inventing pre-Babylonian Jewish history from Spinoza. Further, the library there had Blount and Toland, among other doubting texts. Faced with these charges, he confessed all and begged forgiveness, reminding the court that he was still a minor, only twenty years old. These were harsh times, though. The world was changing fast, and the old world tried to hang on. In the same year that the Privy council initiated the last major witch-hunt in Scotland, it also decided, by one vote, to execute Aikenhead. This was done on January 8, 1697, about three months before his twenty-first birthday. He died holding the Bible. Our best collection of materials on Aikenhead, by the way, was meticulously preserved by John Locke.

MORE JESUITS IN CHINA

With the fall of the Ming dynasty (which lasted from 1368 to 1644; it was the dynasty that built the Wall) and the rise of the Qing (1644–1911), the Jesuits' fortunes rose in China. We saw that their mission had been in decline. Now the new Qing emperor, who was avid for Western science and math, would change everything. The Jesuits there responded to his interest by sending for experts and technicians of many types. They even built a harpsichord and started giving lessons. The court seems to have become a campground for science, math, and music. Intellectuals and courtiers scheduled lessons for themselves all day long. They built European fountains, studied geometry, played music, and made windmills.

For all the variety, astronomy was still the Jesuits' most impressive offering. Soon after the transfer of power in 1644, Jesuits began serving as directors of the imperial calendar, an important state office. Remarkably, however, the Europeans did not want to tell the Chinese about Copernicus and Galileo and heliocentrism. The Jesuits actually taught the Chinese a system proposed by the astronomer Tycho Brahe, in which the earth is stationary and the other planets go around the sun, which in turn orbits Earth. The deception is striking: Brahe's system had never been accepted in Europe and was now known to be false. But, with the right mathematical corrections, it allowed prediction of the heavenly motions and a version of the old cosmology. The Chinese found the betrayal a little stunning when they learned of it. That happened in 1760 when the missionary Michel Benoist at last brought heliocentric theory to the Qianlong emperor's attention. By then the Chinese scholars had grown less impressed with the European system anyway, because it was complicated and the calculations were off.

Yet despite the tragic comedy of lying to millions about the nature of the universe for over a century, and the suspicion that it generated, the science that the Jesuits brought into China vitalized Chinese astronomy and math. Wang Xishan (1628–1682), Mei Wending (1633–1721), and Xue Fengzuo (died 1680) were the first scholars in China to respond to the exact sciences brought in by the Jesuits after 1644, and they were responsible for a scientific revolution. Through their work, for example, geometry and trigonometry largely replaced numerical algebra. They convinced Chinese astronomers that mathematical models were not only useful tools for predicting phenomena but were actually explanatory, thus changing the whole tenor of the experience: the work became investigative. This burst of intellectual activity renewed interest in Chinese philosophers, and in historian Joanna Whaley-Cohen's words, "part of this movement involved a repackaging of the sages of antiquity as initiators of Chinese technology."[59] As she also points out, however, scientists like Mei Wending claimed that one had to disagree with the ancient sages if new scientific truth contradicted them.

The movement that arose around that idea was called *kaozheng,* or "evidentiary research." It championed the rationalist sages of the Chinese past as a way of connecting even specifically Western scientific knowledge with the authority of Chinese antiquity. In fact, as Whaley-Cohen puts it, "To encourage serious attention to the new knowledge, eminent scholars created a myth that Western mathematics had evolved out of ancient Chinese ideas."[60] A mid-eighteenth-century kaozheng scholar insisted that "In ancient times . . . no one could be a Confucian who did not know mathematics. . . . Chinese methods [now] lag behind Europe's because Confucians do not know mathematics." So the great kaozheng movement was keenly inspired by European math and science, but it folded this energy back into China and made itself a profoundly Chinese ideal.

Meanwhile, the effect from the other side was considerable. We saw that Ricci had created a strange brand of Chinese Christianity in the seventeenth century, but missionaries there did not see anything too odd about it, immersed as they were in the local Chinese culture. In the early eighteenth century, new missionaries arrived who knew little about China or the history of Ricci's decisions, and they were fairly shocked. It looked as if

Ricci and his followers had gone mad: the cross was but one of the symbols on family shrines, the word for God was essentially "sky," and Christians merrily performed rituals at Confucian shrines. The new missionaries ran to Pope Clement XI and said that the entire Catholic flock in China was praying to Confucius, and to their own dead, and praising the sky for sustaining them.

In response, the Jesuits redoubled their claim that all this ritual was civil and secular. The pope decided against them this time, but the idea of widespread atheism was further popularized; it was soon posited elsewhere, too. The concept of universal belief was shattered. When Clement XI condemned Chinese ritual in 1715, Catholic converts were no longer allowed even to watch Confucian rites, "because to be a bystander in this ritual is as pagan as to participate in it actively."[61] This stung and amazed the emperor Kangxi; and missions were banned from China. The Emperor's Decree of 1721 responded to the pope: "To judge from this proclamation, their religion is no different from other small, bigoted sects of Buddhism or Taoism. I have never seen a document which contains so much nonsense. From now on, Westerners should not be allowed to preach in China, to avoid further trouble." In England, the idea of an atheist China appealed to many deists. Among Collins's references to atheists we find "the literati in China" and the "followers of Spinoza."[62] Tindal argued that Jews and Christians cannot be the only people allowed to perceive truth since there were three hundred million atheist Chinese who, he said, were in some ways better off with their Confucianism.

ENLIGHTENMENT

Reason to doubt was not the only thing the Europeans brought back from their travels; they also returned with something to drink while you talked about the problem. Bars or "public houses" had always been the place to find radical talk, because people gathered there and because alcohol loosens lips. In the eighteenth century, suddenly, tea and coffee flooded into Europe. It was an immediate and lasting craze, and it gave rise to teahouses and cafés that could serve the same function as bars, but on a level both more sober and more refined. When Lord Shaftesbury wrote about politeness and philosophy, he was advising on the rules of conversation of the London coffeehouse. In Paris, salons were the great theaters of philosophy and conversation. When the Enlightenment philosophers called themselves "French Lord Shaftesburys," it was in reference not only to his doubt but also to his commitment to the new politeness that kept serious conversation from ending up in a brawl, or from being dominated by the elites.[63]

The French salons got started when a few upper-class women of early-eighteenth-century Paris sought to get an education in the new sciences and philosophies of the day. They had been barred from real learning, and the culture of their time was noted for its levity—they were hungry for substance. These parties started for the purpose of self-improvement, bringing together the most interesting scholars, scientists, and authors for a splendid supper and a few hours of talk. The *salonnières* who started all this, in the first decades of the century, were Claudine-Alexandrine Guérin de Tencin and the Marquise de Lambert, Anne-Thérèse de Marguenat de Courcelles. In the next generation, the three salonnières who created famous, world-changing salons of the mid-century Enlightenment were Marie-Thérèse Geoffrin, Julie de Lespinasse, and Suzanne Necker.[64] These were respectable events, but they were also a bit on the fringe. Tencin had escaped from a convent and forced vows and then had a baby out of wedlock—and it was no secret. But she was not the biggest scandal around: her guests were such men as Montesquieu, the radical political philosopher, and Bernard Fontenelle, whose great contribution was as a popularizer of modern science, especially astronomy. Fontenelle defended Copernicanism, critiqued miracles, and satirized religious wars. His work was thought of by many as the first discussion of the Bible as myth.[65] It was Geoffrin who can be said to have invented the Paris salon by making the dinner early, at one P.M., so that the company could talk afterward (and stay alert), and by making each salon fall regularly every week. She held a Monday salon for artists (she was the most prominent collector of contemporary French painting of her age) and a Wednesday salon for writers.[66] The other salonnières arranged their salons on various days, so that each could attend the others. For several decades, if invited, a person could attend a salon every day of the week. Some did.

The ideas that were discussed in these salons were important in the history of doubt, but so was the social and political role that these events came to play. At the end of the wars of religion, authority had won: the monarchy grew in strength over the nobles. Nobles responded to the growing power of the state and its religion by inventing the notion of a Republic of Letters, a kind of state within the state where equality reigned and the monarchy was not the judge of all things. The new judge was, first of all, the salonnière. She shaped the conversations in her rooms as she thought interesting and convincing. The men who competed to get letters of invitation so that they could attend, and came back week after week for decades, extolled the salonnières as directors of conversation and as the source of connection to employment and publication. Some Italians returning from Paris wrote to one of the great salonnières and said they had tried to meet on their own, at home, but "There is no way to make Naples resemble Paris unless we find a woman to guide us, organize us, *Geoffrinise* us."[67]

If the salonnières themselves provided the first answer to who should judge intellectual ideas in the Republic of Letters, they also helped to create the second part of the answer, which was public opinion. The salons generated a plethora of publications intended to bring salon conversation and salon news to as wide an audience as possible. Bayle started the first journal of the *Republic of Letters,* calling it just that. As such journals proliferated, their open nature not only involved average people in the world of ideas but went further and asked all readers to contribute. All this was conceived in terms described by the salonnières: the Enlightenment was to be a social act of mutual education.

It was in salons and journals that Voltaire (1694–1778) showed off his famous wit. Voltaire was not atheistic, but he probably inspired more people to reject their childhood religion than anyone else at that time. As a taste, consider his entry on "the Divinity of Jesus" in *The Philosophical Dictionary.* Voltaire writes: "The Socinians, who are regarded as blasphemers, do not recognize the divinity of Jesus Christ. They dare to pretend, with the philosophers of antiquity, with the Jews, the Mahometans, and most other nations, that the idea of a god-man is monstrous," and he goes on to explain how the idea of Jesus as God had grown over the course of Christian history.[68]

Violent persecution, though tapering off, was still around and Voltaire was a great leader in stopping it. In the 1760s, when he was already a phenomenally well known author, Voltaire embarked on three great campaigns against the current cases of suppression. The first was the Calas affair. The Protestant Jean Calas and his wife, Anne-Rose, were rearing four sons and two daughters in largely Catholic Toulouse. One son had recently converted to Catholicism; another had expressed the desire to do so but did not follow through. One day in 1762 this son was found hanged, and Catholic neighbors suggested that the boy had been killed to keep him from becoming a

Catholic. As a result, the whole family was arrested, except for the daughters, who were away—when they returned they were forced to enter a convent. The jury in Toulouse found Jean Calas guilty. He was sentenced to torture "ordinary and extraordinary," was then broken on the wheel and finally burned at the stake. The brothers ran off to Geneva. Voltaire met the remaining family and took up the cause of exonerating Jean: By 1765, a Parisian tribunal unanimously pronounced Calas innocent. What was left of the family's property was returned to them, and gifts from average people and the king himself allowed what was left of the Calas family a decent material life. This seems to be the first time a man of letters conjured up a wave of public opinion in support of a cause. His essay on tolerance emerged from this experience.

Voltaire's second great cause of this period was the Sirven affair, which stretched out across nine years. Here again, the Sirven family was accused of killing one of its own—a daughter this time—to prevent her from converting to Catholicism. The third cause centered on a nineteen-year-old youth named La Barre who refused to take his hat off to a religious procession and also mutilated a wooden crucifix. He was tortured and executed for blasphemy in 1766. Voltaire was particularly stung that La Barre's death might have been precipitated because the boy was known to possess a copy of Voltaire's *Philosophical Dictionary.* Voltaire rose up against "this sentence so execrable, and at the same time so absurd, which is an eternal disgrace to France." This is why Voltaire's works are peppered with the injunction to "Ecrasez l'infame!" (Crush the infamous thing!)—which meant crush the Church, make it stop. Consider his entry on "Martyrs" in his *Philosophical Dictionary.* Here he has just explained how Christianity overstates the role of martyrdom in its own history:

> Do you want good well-authenticated barbarities—good and well attested massacres, rivers of blood which have actually flowed—fathers, mothers, husbands, wives, infants at the breast, who have in reality had their throats cut, and been heaped on one another? Persecuting monsters! Seek these only in your own annals: you will find them in the crusades against the Albigenses, . . . in the frightful day of St. Bartholomew . . .

He laughs at the idea of Marcus Aurelius being called a "monster of cruelty" by "you who have deluged Europe with blood and covered it with corpses, in order to prove that the same body can be in a thousand places at once, and that the pope can sell indulgences!"[69] It's amazing that the young La Barre had this book in his hands, and that his death was in part caused by his possession of it: the cruelty was committed in punishment for the accusation of it.

Voltaire had a lot more to say on the subject, but we will settle for a last word from the "Optimism" entry in his *Dictionary*. Shaftesbury had followed the idea that this world is a self-running mechanism and since God began it without any intention of interfering with it, it must be that it is a good world. What seems like evil is but one side of a balance. Leibniz did the most with this. Voltaire didn't like it. As so many have noticed—and Voltaire refers to Epicurus in his discussion—there is an awful lot of pain in the world, and it does not seem well distributed. That's the joke of *Candide:* the naïf hero's teacher preaches that this is the best of all possible worlds, as life whittles him down to a nub. Here the main attack is broader, but it narrows down to the same point:

> What! to be chased from a delicious place, where we might have lived for ever only for the eating of an apple? What! to produce in misery wretched children, who will suffer everything, and in return produce others to suffer after them? What! to experience all maladies, feel all vexations, die in the midst of grief, and by way of recompense be burned to all eternity—is this lot the best possible?

Voltaire thought the existence of the world was proof of a creator, but of no more. "The question of good and evil remains in remediless chaos for those who seek to fathom it in reality." His was a very moral critique of religion. Voltaire repeated the ancient Greek smirk over the foolishness of myth, and the ancient Hebrew holler about providence, and added his own howl at religious intolerance. That last one opened people's minds to the others in this period and brought on the Enlightenment's efflorescence of secularism. Motivated by this rebellion against the stake, even calmer, more old-style doubters, such as the scientists and historians, were growing political teeth.

Denis Diderot (1713–1784) and his friend the mathematician Jean d'Alembert (1717–1783) created the Enlightenment's most famous project, the *Encyclopedia,* a compendium of knowledge and know-how showing the old secrets of the guilds, the latest technology, and the most scandalous new ideas. It was considered extremely antireligious. It is worth noting that d'Alembert was the illegitimate child of Claudine-Alexandrine Guérin de Tencin, mentioned in the discussion of the salons.[70] Diderot is often written of as an atheist, although others note that in some cases he pulled back from the idea that the universe had nothing like a spirit to it. In any case, he was a world-class doubter.

His 1777 "Conversation of a philosopher with the Maréchal de—" was actually the story of a conversation with Madame la Maréchale. A man is waiting for the Marshall to return and the great man's wife asks, Is he "M. Crudeli . . . then it's you who believes in nothing?"[71] When he says yes, she expresses surprise that his morals are the same as a believer's. Diderot explains that one does not disbelieve in order to secure license for oneself. An honest person is honest without threats or supervision, and many a believer is dishonest. He also points out that "if it suddenly took the fancy of twenty thousand Parisians to conform strictly to the Sermon on the Mount ["Blessed are the poor . . . ," etc.], there would be "so many lunatics that the police wouldn't know what to do with them."[72] You cannot subject a nation "to a rule which suits only a few melancholic men." For a better morality, Crudeli suggests, "Make it so that the good of individuals is so closely tied to the general good that a citizen can hardly harm society without harming himself."[73] Madame la Maréchale has many questions. She asks him, sincerely, "Does the idea of being nothing after death not distress you?" He agrees he "would prefer to exist."

> MADAME LA MARÉCHALE: If, . . . the hope of a life to come seems sweet and consoling even to you, why deprive us of it?
> CRUDELI: I haven't got this hope because the desire hasn't in the least concealed from me the emptiness of it, but I deprive no one of it.[74]

But you can keep it, he says, only if you can believe you will see without eyes, hear without ears, think without a head, love without a heart, feel without senses, exist without being anywhere, and be something without place or size. She responds: "But this world of ours, who made it?"

> CRUDELI: I ask you that.
> MADAME LA MARÉCHALE: God made it.
> CRUDELI: And what is God?
> MADAME LA MARÉCHALE: A spirit.
> CRUDELI: And if a spirit makes matter, why could not matter make a spirit?

MADAME LA MARÉCHALE: But why it should it make it?
CRUDELI: Because I see it do so every day.

People invent gods (and mindless things combine to create creatures with mind). Doubt was starting to have fun, and in public. Parts of the above dialog were so risqué that Crudeli felt he had to get up and whisper in the lady's ear—for instance, that religious morality makes it more admissible to damage a virtuous woman's reputation than to piss in a sacred vessel, although the first, he said, was a great issue of civil life and it regularly goes unpunished, while the second was harmless nonsense but it would get you killed.[75] Like Bayle, Diderot had a touch of Rabelais in him, and, although he played at whispering his impiety and usually denied proselytizing, in truth it all helped broadcast his news. Diderot also said humanity would not be free until the last king is strangled in the entrails of the last priest.[76]

Claude-Adrien Helvétius (1715–1771) was another of the Encyclopedists. His *Essays on the Mind* (1758) was publicly burned at Paris. He himself was tolerated in part because his wife, who was friends with Voltaire's mistress, the influential Newtonian author Emilé du Chatelet, was also friends with Queen Marie Antoinette. Anyway, by now condemnation did not always do a book harm: *Essays on the Mind* was translated into most of the European languages and is said to have been a major best-seller. It was materialist in its conception of the universe and convinced that a nonreligious morality is what really guided most people's virtue. For at least a century, doubters would credit Helvétius as a major influence.

When people think of the origins of modern doubt, now as then they tend to think of the literary and the comic—it was the light works of Voltaire and Diderot that became emblems of enlightenment. They reached a wide audience and changed the way ordinary people thought. Yet the most innovative philosopher of doubt of the period was the Scotsman David Hume (1711–1776). It is difficult to say whether Hume believed in God. In the penultimate chapter of his *Enquiry into Human Understanding* (1748), he has a character take up the role of defender of the tenets of Epicurus, which deny "a divine existence and consequently a providence and a future state." In this way Hume presented many arguments against contemporary believers but exclaimed "O Athenians!" every once in a while to remind readers that he was really just talking to the ancient Greeks. It reads remarkably like a man finished with God, not least because he talks about how useless the concept became once we had agreed that we could not know anything about him. Also, if the order of things as it is, is God, what does it add to say that God exists? The tone of his voice on questions of providence, life after death, and miracles was pretty direct. In the same way, Hume explains that everyday morality is based on the simple fact that doing good brings you peace of mind and praise from others and doing evil brings rejection and sorrow. We don't need religion for morality, and what is more: religion itself got its morality from everyday morality in the first place.[77] That's a smart little notion. As for adding a God to the system, Hume asks why bother, especially concerning ethics? After all, roars Hume, *"Are there any marks of a distributive justice in the world?"* [78]

Hume's *Dialog Concerning Natural Religion* of 1779 created a great furor. The conversation was between Philo, Cleanthes, and Demea. Where did he get the names? The answer is that the whole book was a love letter to Cicero's dialog *The Nature of the Gods,* and that is where the first two names came from. In Cicero's dialog, Cotta was the Skeptic and his teacher was named Philo; Balbus was the Stoic and his teacher was Cleanthes (Velleius offered the Epicurean view). Also, Cicero himself introduced and concluded his dialog, yet he was clearly also the Cotta character; and although he favored Cotta's Skepticism all along, in the end he declared Balbus and his bland Stoic theism the winner. In Hume's dialog, he takes Cicero's introduction and reinterprets it a bit but retains the plea that this question of God is one we must address with forthright honesty. He also maintains the conceit that he himself had been present for these arguments and, again, too young to participate. Yet Hume is also the Philo character. This Cleanthes (whose name stood for the Stoic) is a deist philosopher. Demea, named for a famously strict father character of the ancient world, is described as a priest of "rigid inflexible orthodoxy."

Cleanthes offers the argument from design: houses and machines are made by an intelligent creator, therefore the universe, which is also complex and ordered, must also have one. Demea argues the need for an ultimate cause to explain the great series of cause and effect that we see in the world. Hume presents these as the best arguments for God, and then has Philo take them apart. Philo, named for the ancient Skeptic, here also represents the Epicureans' insistence that the world is perfectly capable of making itself. Philo was a brilliant and very modern creation. Against the argument from design offered by Cleanthes, Philo explains that the world has its own internal logic and that, in any case, a lot of the order we perceive is actually in our heads—it's just the way we see things. Against the ancient argument of a first cause, Hume brings in the equally ancient idea that there is no reason to believe in cause and effect in the first place. This doubt had been posed by doubters from the Carvaka to al-Ghazzali to Nicholas of Autrecourt; Hume brought it to modern Europe. Yet even if we do accept it, Hume asks, why should the great stream of cause and effect have to start from some outside force and get its meaning therefrom? Perhaps it started itself and means exactly what it is.

When the conversation first begins, deist Cleanthes teases Philo about his Skepticism: "In reality, would not a man be ridiculous, who pretended to reject Newton's explication of the wonderful phenomenon of the rainbow, because that explication gives a minute anatomy of the rays of light; a subject, forsooth, too refined for human comprehension?" Hume says yes, there is a "brutish and ignorant skepticism" that supports the vulgar in their prejudice against what they do not easily understand: "They firmly believe in witches, though they will not believe nor attend to the most simple proposition of Euclid." But the refined and philosophical Skeptics, meanwhile, doubt only our ability to know metaphysics, and believe that we can know the world of daily life. Hume's position was midway: rational investigation of the world and ourselves is only what it is, but it can be done and is deeply worth doing.

Then turning on the priest Demea, Cleanthes laughs that as soon as reason began making headway against faith, all the arguments of the "ancient academics" were adopted by the Catholic Church, marked now by "all the cavils of the boldest and most determined Pyrrhonism." A different kind of doubt was building now, one that combined reason and skepticism: Locke, he said, was the first to say that religion "was only a branch of philosophy," and that just as on any other subject, we are always revising toward truth on the question of God. As Cleanthes tells it, the skepticism of "Bayle and other libertines" had "still further propagated the judicious sentiment of Mr. Locke," such that it is now "avowed by all pretenders to reasoning and philosophy, that Atheist and Skeptic are almost synonymous."

Cleanthes adds a polite, "And as it is certain that no man is in earnest when he professes the latter principle, I would fain hope that there are as few who seriously maintain the former," and that's when Philo, Hume's character, breaks in here. Philo sets it up, saying that, "having mentioned David's fool, who said in his heart there is no God," Bacon observed "that the

Atheists nowadays have a double share of folly; for they are not contented to say in their hearts there is no God, but they also utter that impiety with their lips, and are thereby guilty of multiplied indiscretion and imprudence. Such people, though they were ever so much in earnest, cannot, methinks, be very formidable."

That is, Hume suggests, the people who are formidable atheists are the ones who are careful what they say. The wise hold their peace—foolish doubters speak. But, says Hume, even though this may cause you to "rank me in this class of fools," he cannot quite help himself from telling Cleanthes his impression of "the history of the religious and irreligious skepticism with which you have entertained us." In "ignorant ages . . . such as those which followed the dissolution of the ancient schools," people were more trusting, so the priests thought that the best protection against "Atheism, Deism, or heresy of any kind" was to carefully defend a set dogma.

> But at present, when . . . men . . . have learned to compare the popular principles of different nations and ages, our sagacious divines have changed their whole system of philosophy, and talk the language of Stoics, Platonists, and Peripatetics, not that of Pyrrhonians and Academics. If we distrust human reason, we have now no other principle to lead us into religion. Thus, skeptics in one age, dogmatists in another; whichever system best suits the purpose of these reverend gentlemen, in giving them an ascendant over mankind, they are sure to make it their favorite principle and established tenet.

Cleanthes' response is that it is natural for the Church to use whatever works against "Atheists, Libertines, and Freethinkers of all denominations." So the conversation ends with Hume's Philo promoting a quiet atheism, while the believer admits that belief uses any argument in fashion.

Hume offers a few other doubting observations that are too pretty to miss.

> The Deity, I can readily allow, possesses many powers and attributes of which we can have no comprehension: But if our ideas, so far as they go, be not just, and adequate, and correspondent to his real nature, I know not what there is in this subject worth insisting on. Is the name, without any meaning, of such mighty importance? Or how do you mystics, who main-

tain the absolute incomprehensibility of the Deity, differ from Sceptics or Atheists, who assert, that the first cause of all is unknown and unintelligible?

Thus "those who maintain the perfect simplicity of the Supreme Being" are "in a word, Atheists, without knowing it." A perfect simplicity could not have thought, he explains. "A mind, whose acts and sentiments and ideas are not distinct and successive; one, that is wholly simple, and totally immutable, is a mind which has no thought, no reason, no will, no sentiment, no love, no hatred; or, in a word, is no mind at all. It is an abuse of terms to give it that appellation."

In another lovely argument, Hume has Philo say that nothing on earth could tell us much about the universe. The wonder of human thought should not make us expect thought elsewhere: "What peculiar privilege has this little agitation of the brain which we call thought, that we must thus make it the model of the whole universe? Our partiality in our own favor does indeed present it on all occasions; but sound philosophy ought carefully to guard against so natural an illusion." In a formulation that would be much quoted, Hume says that "Epicurus' old questions are yet unanswered. Is he willing to prevent evil, but not able? then is he impotent. Is he able, but not willing? then is he malevolent. Is he both able and willing? whence then is evil?" For him the universe is neither good nor evil. Philo also defends "the old Epicurean hypothesis," that the world came into being by chance, that "matter can acquire motion without any voluntary agent or first mover." Philo enjoys this in the same way Toland did, with yet more sophistication, saying that gravity, elasticity, and electricity all move things without volition, adding, "Besides, why may not motion have been propagated by impulse through all eternity, and the same stock of it, or nearly the same be still upheld in the universe." The conservation of momentum helps explain where all the new energy is coming from—it's not new energy.

So how did the animals, human beings, and everything else come into being? Hume says it was trial and error. An animal without the right internal order dies and the stuff of it will reform into another thing—even with collections of objects the ordered ones persist longer, so when you look around you see a lot of order. Even if things were totally random, we could expect accidental order to persist longer than any given example of chaos. Yet he also suggests that patterns of order may be an innate feature of all material things, and these patterns may generate the order that we perceive.

In the end of his *Dialog,* Hume bluntly declares Philo the Skeptic wiser than the orthodox Demea, but in an almost exact quote from the last line of Cicero, Hume's last line claims that the arguments of deist Cleanthes brought us nearer to the truth. Just as with Cicero, it has been assumed that the gesture was for the sake of safety.

So imagine Hume at age fifty-two visiting Paris. It is 1763 and the city is alive with vibrant salons and a bold Republic of Letters. He has already written many of his great works and is the toast of the town. It was on this trip that Hume visited the house of the Baron d'Holbach (1723–1789) and gave the history of doubt another of its legendary dinner parties. Along with Diderot, d'Holbach was among the most famous of the band of younger, more daring Enlightenment philosophers—by now Diderot and d'Holbach were fifty and forty, respectively, while Voltaire was sixty-nine, for example, and Montesquieu seventy-four. The gathering was a philosophical dinner, modeled on the salons, and Diderot, who was there, relates the story:

> The first time that M. Hume found himself at the table of the Baron he was seated beside him. I do not know for what purpose the English philosopher took it into his head to remark to the Baron that he did not believe in atheists, that he had never seen any. The Baron said to him: "Count how many we are here. We are eighteen." The Baron added: "It is not too bad a showing to be able to point out to you fifteen at once: the three others have not made up their minds."

David Berman has argued that the reason Hume took it into his head to say he had seen no atheists was "that Hume's opening gambit was rather like a Masonic handshake: an attempt to elicit a response from, and communicate with, someone whose secret identity he guesses."[79] It is a persuasive argument. The anecdote is clearly written from the perspective of the French group, designed to show how outrageous they were. Hume comes off as a bit of a chump, and the anecdote has been used to prove that Hume himself was not an atheist. However, given what Hume had written by then, and that he was already famous throughout Paris for it, the scene makes sense only if it is read the way Berman suggests—without the smirk Diderot added. In any case, there is something brazen in this conversation coming down to us in the way it has: it is doubt's great coming-out party.

This was the first group of actual avowed atheists; no dissembling, no caveats, just no gods, no God, nothing like it. For the first time, doubters were silenced neither by fear of being killed or exiled nor by fear of how the masses would behave if they became convinced there was no God and no hell. This crowd believed morality was available to anyone through reason. The central text here was Baron d'Holbach's *System of Nature*. D'Holbach was steeped in several traditions of the history of doubt (his children's tutor, for instance, published a new translation of Lucretius).[80] It is not insignificant that the book's preface, by d'Holbach's friend Naigeon, was equally brash in its evangelical atheism. It meant that the author did not stand alone even in his most extreme position. Naigeon opened with the assertion that the idea of God was nonsense and the barrier to all human progress. D'Holbach took over and just kept swinging. He'd gotten the idea that matter did not need outside force, let alone a prime mover, from Toland; and he borrowed from Collins as well. Diderot edited the book and added footnotes referencing doubters from Cicero to Hobbes. The text never hedged its claim that there is no God at all.

> Is there anything more frightful than the immediate consequences to be drawn from these revolting ideas given to us of their God, by those who tell us to love him . . . and to obey his orders? Would it not be a thousand times better to depend upon blind matter, upon a nature destitute of intelligence, upon chance, or upon nothing, upon a God of stone or of wood, than upon a God who is laying snares for men, inviting them to sin, and permitting them to commit those crimes which he could prevent, to the end that he may have the barbarous pleasure of punishing them without measure, without utility to himself, without correction to them, and without their example serving to reclaim others?[81]

This moral critique about burning in hell was matched by the moral critique about burning on earth, at the stake. No God at all makes human behavior along these lines seem even more horrible. "Abandon your chimeras," he advised, "occupy yourselves with truth; learn the art of living happy; perfect your morals, your governments, and your laws; look to education, to agriculture, and to the sciences that are truly useful; labor with ardor." Enjoy pleasures and multiply them, and "If you must have chimeras, permit your fellow-creatures to have theirs also; and do not cut the throats of your brethren, when they cannot rave in your own manner."[82] Finally, he sighed, "if the infirmities of your nature require an invisible crutch, adopt such as may suit with your humor, select those which you may think most calculated to support your tottering frame," but do not let these "imaginary beings" upset you or let you forget your real duty, to sustain the real people around you. This was the origin of the "crutch" indictment. No one had been quite so brashly dismissive of belief since the ancient world.

Edward Gibbon's (1737–1794) *Decline and Fall of the Roman Empire* inaugurated the development of a new, careful, secular history. Before it, the accepted understanding was that Christianity took over the Roman Empire because it was ordained by God, that the horrible Roman emperors martyred the early Christians in droves, and that Christians prevailed through the power of God. In this model, Rome was the best humanity could do until we had further revelation, then God stepped in, fixed the world, and Rome sank away. Not only did Gibbon tell the story without any intervention from God, he also told the story of Rome's decline as due to the disease of Christianity's spreading through the Roman Empire and rotting it. This was powerful stuff. To get a quick dose of his mood, listen to Gibbon on Christian universality and exclusivity—the insistence that only Christianity be practiced: "These rigid sentiments, which had been unknown to the ancient world, appear to have infused a spirit of bitterness into a system of love and harmony."[83] Gibbon quoted Tertullian to let us see Christians as the cruel ones: "You are fond of spectacles," the Church Father wrote,

> except the greatest of all spectacles, the last and eternal judgment of the universe. How shall I admire, how laugh, how rejoice, how exult when I behold so many proud monarchs and fancied gods, groaning in the lowest abyss of darkness . . . liquefying in fiercer fires than they ever kindled against the Christians; so many sage philosophers blushing in red hot flames, with their deluded scholars; so many celebrated poets trembling before the tribunal . . . so many tragedians . . . so many dancers.[84]

Then Gibbon said, "But the humanity of the reader will permit me to draw a veil over the rest of this infernal description which the zealous African pursues in a long variety of affected and unfeeling witticisms."[85]

Before he published, Gibbon sent some of these pages to Hume in 1776, and when Hume wrote back praising Gibbon's style and scholarship, he added, "It was impossible to treat the subject so as not to give ground of suspicion against you, and you may expect that a clamor will arise."[86] It did.

Gibbon also explains that persecution of Christians in the Roman Empire was piecemeal and never consistent and widespread. He tells the story not as a battle between good and evil, but rather as the disturbance made by a fanatical cult that would not pay respect to the symbols of the state. There were not that many martyrs anyway, wrote Gibbon, announcing "a melancholy truth which obtrudes itself on the reluctant mind," that "even admitting" all the Christian martyrdom history has recorded, "or devotion has feigned . . . it must still be acknowledged that the Christians, in the course of their intestine dissensions, have inflicted far greater severities on each other than they had experienced from the zeal of infidels."[87] The number of Protestants "executed in a single province and a single reign far exceeded that of the primitive martyrs in the space of three centuries and of the Roman empire." Gibbon's name would henceforth represent religious doubt.

There were many lesser-known doubters who popularized these works. I note only one of the best of them, the Marquis d'Argens, Jean-Baptiste de Boyer (1704–1771). His book, the *Chinese Letters,* pretended to be written by the well-traveled Sioeu-Tcheou, which allowed d'Argens to issue a sustained criticism of European customs as if from an innocent outsider. Montesquieu had already proved the technique, but d'Argens took the device to the question of doubt, including the contemporary debate between atheism and deism. The fictional Chinese narrator saw such debates as universal, and all peoples as animated by the same concerns. In the public imagination, doubt existed, had various levels and personnel, and was the concern of peoples all over the world.

FRANKLIN, PAINE, JEFFERSON, AND ADAMS

Benjamin Franklin was born in Boston in 1706. When he left town as a teenager it was in part because, in his words, "my indiscrete disputations about religion began to make me pointed at with horror by good people as an infidel or atheist." He had been brought up piously Presbyterian but soon became a doubter:

But I was scarce fifteen, when, after doubting by turns of several points, as I found them disputed in the different books that I had read, I began to doubt of Revelation itself. Some books against Deism fell into my hands. . . . It happened that they wrought an effect in me quite contrary to what was intended by them; for the arguments of the Deists, which were quoted to be refuted, appeared to me much stronger than the refutations. I soon became a thorough Deist.

Franklin joined the earliest Masonic lodge in America and in 1734 became its president. When on state business in Edinburgh, Franklin stayed with Hume; in his later years he knew and encouraged the young Thomas Paine; in his still-later years he lived in Paris for a while, where he proposed to, and was refused by, Anne-Catherine de Ligniville, widow of the great doubter Helvétius. Franklin's doubt was Enlightenment vintage. The next generation was more revolutionary in its claims.

Thomas Paine (1737–1809) is an extraordinary personality in our story. Having profoundly influenced the move toward American independence, he also was elected to the French National Convention during France's Revolution. While in France he wrote his major treatise on religion, *The Age of Reason,* and smuggled it out of Paris to be published in 1794. He was almost not so lucky himself, coming very close to losing his life in the revolution. When he returned to the United States in 1802, he was befriended by Jefferson.

Paine was bold: "Every national Church or religion has established itself by pretending some special mission from God. . . . Each of those Churches accuses the other of unbelief; and for my own part, I disbelieve them all."[88] On the evidence of revelation, Paine said that if someone did hear the voice of God, "it is revelation to that person only." It is a good point. Paine clarified that "When he tells it to a second person" and that person tells it to others, "it is revelation to the first person only and hearsay to every other; and consequently they are not obliged to believe it." We would expect some literary pizzazz from the man who had convinced some of England's far-flung farmers to declare themselves independent of their king. Paine does not disappoint: "When also I am told that a woman, called the Virgin Mary, said, or gave out, that she was with child without any cohabitation with a man, and that her betrothed husband, Joseph, said that an angel told him so, I have a right to believe them or not." The claim reminds us that, in the past, ordinary people and figures of popular culture had attacked the mythic ideas of Jesus' life, but the philosophers had left him alone and instead wrestled with the God of the philosophers. Paine did both. He weakened the idea of a prime mover but also attacked the mythic aspect of Christianity. Paine found a nice hero in the history of doubt, writing: "It appears that Thomas did not believe the resurrection and as they say would not believe without having ocular and manual demonstration himself. So neither will I: and the reason is equally as good for me and every other person as for Thomas."[89] Thus Doubting Thomas was finally picked up as a positive icon of doubt.

Paine also used history to tremendous effect: "The best surviving evidence we now have respecting this affair is the Jews. They are regularly descended from the people who lived in the times this resurrection and ascension is said to have happened, and they say *it is not true.*"[90] He reminded his readers that Jesus was born Jewish and stayed Jewish; and that the statue of Mary developed from statues of ancient goddesses. "The deification of heroes changed into the canonization of saints," he explained, and over time, "the Church became as crowded with the one as the Pantheon had been with the other and Rome was the place of both."[91] Christian theory is "the idolatry of the ancient mythologist, accommodated to the purposes of power and revenue," and now it is "to reason and philosophy to abolish the amphibious fraud."

Paine cautioned that his argument was not against the real Jesus: "He was a virtuous and an amiable man. The morality that he preached and practiced was of the most benevolent kind, and, though similar systems of morality had been preached by Confucius, and by some of the Greek philosophers . . . and by many good men in all ages, it has not been exceeded by any."[92] Paine wrote that Jewish priests felt threatened by him, and he probably "had in contemplation the delivery of the Jewish nation from the bondage of the Romans." Thus between the worry of the Jewish priests and the ire of the Romans, "this virtuous reformer and revolutionist lost his life." He spoke of Jesus as if he were, well, as if he were Tom Paine. Paine thus thought he was doing the best work that could be done—he was a virtuous reformer and revolutionist—and drew a direct line between his life's devotions and those of a secular Jesus. In modernity, traditionally religious impulses will be translated into politics for many people.

For a mellower founding doubter, consider Thomas Jefferson (1743–1826), the third president of the United States of America. In a private letter to a friend, Jefferson wrote:

As you say of yourself, I too am an Epicurean. I consider the genuine (not the imputed) doctrines of Epicurus as containing everything rational in moral philosophy which Greece and Rome have left us. Epictetus indeed, has given us what was good of the stoics; all beyond, of their dogmas, being hypocrisy and grimace.[93]

Epictetus was the Stoic who most emphasized the brotherhood of humanity. With "hypocrisy and grimace," Jefferson was making fun of the way some Stoics emphasized hiding one's pain. But he cautions against judging them too harshly. The letter continues (with no break): "Their great crime was in their calumnies of Epicurus and misrepresentations of his doctrines; in which we lament to see the candid character of Cicero engaging as an accomplice." That's Jefferson getting annoyed at Cicero for his rough treatment of the Epicurean in *The Nature of the Gods*—and we may recall how silly Velleius was made to appear. Jefferson continues, still on Cicero:

Diffuse, vapid, rhetorical, but enchanting. His prototype Plato, eloquent as himself, dealing out mysticisms incomprehensible to the human mind, has been deified by certain sects usurping the name of Christians; because, in his foggy conceptions, they found a basis of impenetrable darkness whereon to rear fabrications as delirious, of their own invention. These they fathered blasphemously on him who they claimed as their founder, but who would disclaim them with the indignation which their caricatures of his religion so justly excite.

Now there's a presidential indictment. Jefferson here scolds Plato for inventing an unintelligible mystical idea that had since been made into a god by the so-called Christians, who found that their crazy made-up stuff was well supported by Plato's. Then Jefferson says all this was wrongly hooked on Jesus, who would reject all Christianity were he to know of it. Elsewhere, again speaking against Plato's inventive mysticism, Jefferson delivered one of the greatest parenthetical statements in history:

(Speaking of Plato, I will add, that no writer, ancient or modern, has bewildered the world with more *ignes fatui* [misleading influence], than this renowned philosopher, in Ethics, in Politics and Physics. In the latter, to specify a single example, compare his views of the animal economy, in his

Timaeus, with those of Mrs. Bryan in her *Conversations on Chemistry,* and weigh the science of the canonized philosopher against the good sense of the unassuming lady. But Plato's visions have furnished a basis for endless systems of mystical theology, and he is therefore all but adopted as a Christian saint. It is surely time for men to think for themselves, and to throw off the authority of names so artificially magnified. But to return from this parenthesis.) [94]

Margaret Bryan was a natural philosopher and a schoolmistress. Among her works, in 1797 she published a *Compendious System of Astronomy,* with a portrait of herself and two daughters as a frontispiece. In 1806 Mrs. Bryan published *Lectures on Natural Philosophy* (thirteen lectures on hydrostatics, optics, pneumatics, acoustics), with a notice in it that "Mrs. Bryan educates young ladies at Bryan House, Blackheath." *Conversations on Chemistry* came out the same year. Jefferson's praise of Mrs. Bryan over Plato is a puzzle whose key is the context of doubt.

In this same letter Jefferson's friend, William Short, must have asked about moralists, because Jefferson seemed to be making a list, saying Socrates was a great one, but we have nothing he wrote, and that we cannot trust Plato; and Seneca was good.

But the greatest of all the reformers of the depraved religion of his own country was Jesus of Nazareth. Abstracting what is really his from the rubbish in which it is buried, easily distinguished by its luster from the dross of his biographers, and as separable from that as the diamond from the dunghill.

Jefferson says that if we had a clean text, one that included only the words of the real Jesus, it would help conquer bigotry and fanaticism.

I have sometimes thought of translating Epictetus (for he has never been tolerably translated into English) by adding the genuine doctrines of Epicurus from the *Syntagma* of Gassendi, and an abstract from the Evangelists of whatever has the stamp of the eloquence and fine imagination of Jesus. The last I attempted too hastily some twelve or fifteen years ago. It was the work of two or three nights only, at Washington, after getting through the evening task of reading the letters and papers of the day. But with one foot in the grave, these are now idle projects for me. My business is to beguile the wearisomeness of declining life, as I endeavor to do, by the delights of classical reading and of mathematical truths, and by the consolations of a sound philosophy, equally indifferent to hope and fear.

Jefferson's devotion to Epicurus helps explain "the pursuit of happiness" line in the Declaration of Independence. It was Jefferson who put together the bill for establishing religious freedom in the United States, grounded in the belief that a person's opinions cannot be coerced. This was his great contribution and he fought for it his whole life.

One can read hundreds of wonderfully imaginative lines in which Jefferson publicly says that religion is entirely the business of each person, and his or her God, and not at all the business of anyone else. But to friends, here is his advice:

> Fix Reason firmly in her seat, and call to her tribunal every fact, every opinion. Question with boldness even the existence of a God; because, if there be one, he must more approve the homage of reason than of blindfolded fear. . . . Do not be frightened from this inquiry by any fear of its consequences. If it end in a belief that there is no God, you will find incitements to virtue in the comfort and pleasantness you feel in its exercise and in the love of others which it will procure for you.[95]

An amazing piece of American history. Also, Jefferson explains, "the superlative wisdom of Socrates is testified by all antiquity, and placed virtue in the comfort and pleasantness you feel in its exercise and in the love of others which it will procure for you."[96] When Plato tells us more mystical things, we can trust that this was just "whimsies of Plato's own foggy brain."

In a letter to Short, Jefferson defends his conception of Jesus, saying that if Jesus claimed the things people said he claimed, he was an impostor, so secularizing Jesus saves him from this ignominy. It is easy, he assures Short, to separate the religious foolishness from the philosophy. After all, when historians "tell us of calves speaking, of statues sweating blood, and other things against the course of nature, we reject these as fables not belonging to history."[97] Jefferson also said Jesus had wanted to make rational the Jewish religion, and therefore was dedicated to discarding "idle ceremonies, mummeries and observances" that were "of no effect towards producing the social utilities which constitute the essence of virtue," instead preaching "philanthropy and universal charity and benevolence." Jefferson's ancient

Jews sound like Christians in the Age of Religious Wars, and Jesus sounds like, well, like Jefferson: "The office of reformer of the superstitions of a nation, is ever dangerous. . . . They were constantly laying snares, too, to entangle him in the web of the law. He was justifiable, therefore, in avoiding these by evasions, by sophisms, by misconstructions and misapplications of scraps of the prophets . . ." Here Jefferson makes Jesus sound like Toland, honorably dissembling in order to look more religious to the censors! His finale is singular:

> That Jesus did not mean to impose himself on mankind as the son of God, physically speaking, I have been convinced by the writings of men more learned than myself in that lore. But that he might conscientiously believe himself inspired from above, is very possible. The whole religion of the Jews, inculcated on him from his infancy, was founded in the belief of divine inspiration. . . . he might readily mistake the coruscations of his own fine genius for inspirations of an higher order. This belief carried, therefore, no more personal imputation, than the belief of Socrates, that himself was under the care and admonitions of a guardian Daemon. And how many of our wisest men still believe in the reality of these inspirations, while perfectly sane on all other subjects.

There you have it: Jefferson was enough of a doubter to insinuate that belief was insane. It is not strange that the man is best known for a declaration of independence. Most charmingly, like so many other doubters, Jefferson noted that what he said about Jesus "is no more than is granted in all other historical works." Doubters often mention that much of the best evidence for doubt is already lying around any well-stocked library. When we look at a nickel or gaze on Mount Rushmore, we should remember the great doubter before us.

John Adams was less a doubter than Jefferson, who seems to have gone all the way, but they had a long correspondence mutually critiquing the religious. Adams seems to have doubted everything until he found the Unitarians, an American sect that grew out of English Socinianism that also embraced the ideal of God but no dogma. As the second president of the United States, Adams signed into law the Treaty of Tripoli (1797), which declares that the "the government of the United States is not, in any sense, founded on the Christian religion" and is in no way an enemy to Muslims. On this basis "it is declared . . . that no pretext arising from religious opinion shall ever produce an interruption of the harmony existing between the two countries." Further, "The United States is not a Christian nation any more than it is a Jewish or a Mohammedan nation." It was carried unanimously by the Senate.[98]

In one of his last letters to Jefferson, January 23, 1825, Adams wrote, "We think ourselves possessed, or, at least, we boast that we are so, of liberty of conscience on all subjects . . . yet how far are we from these exalted privileges in fact! There exists, I believe, throughout the whole Christian world, a law which makes it blasphemy to deny or doubt the divine inspiration of all the books of the Old and New Testaments, from Genesis to Revelation." In America as in Europe, he wrote, there are still such laws on the books. If they were still enforced, "Who would run the risk of translating Dupuis? . . . I think such laws a great embarrassment, great obstructions to the improvement of the human mind." Dupuis was Charles François Dupuis (1742–1809), a historian who explained Jesus' story as classically mythic: born at winter solstice, beset with difficulties, reborn from beneath the earth in spring. Both Adams and Jefferson died on the Fourth of July of the following year; Jefferson at eighty-one, Adams at ninety.[99]

TWO GERMANS ANSWER A QUESTION

The philosopher Moses Mendelssohn (1729–1786), grandfather of the great Romantic composer Felix Mendelssohn, was given a strictly talmudic education in his youth. Then at age thirteen he found and read Maimonides' *Guide for the Perplexed.* (The *Guide* was written in 1190, and in 1305 a Jewish ban had forbidden the book to anyone under age twenty-five.) The book changed his life. At that young age he got permission to go to Berlin to study, although he'd have to go alone; he was still a boy and it was not a friendly place. Jews were not allowed to live in Berlin outside their ghetto and had to use the livestock gates to come and go. Here Mendelssohn began the traditional Jewish intellectual labors, which he augmented by first learning German (Jews there spoke Yiddish), French, and English and reading widely. As the years passed, Mendelssohn became known for his learning and he became tutor to the children of a "protected Jew," meaning one who could live in Berlin itself. He himself got a letter of protection years later, with the help of the Marquis d'Argens (the Enlightenment popularizer), with whom he had become friends. The marquis's letter to King Frederick, requesting that Mendelssohn be allowed to live in Berlin, read: "A philosopher who is a bad Catholic, hereby begs a philosopher who is a bad Protestant, to grant a favor to a philosopher who is a bad Jew. There is too much philosophy in all this for reason not to side with my request."[100] It worked.

Mendelssohn wrote a range of philosophical works, but it was his 1767 *Phaidon* that caused a craze in Europe. The book started out with a beautiful translation of part of Plato's *Phaedo* and then Mendelssohn's discussion of it. Here are some phrases he gave Socrates: "He who has taken care of his soul on earth by pursuing wisdom and cultivating both virtue and a sense of true beauty has surely every hope of proceeding on the same path after death."[101] His audience had forgotten that philosophical language could be so reverent about the idea of another world. The book was translated into every major European language, and Mendelssohn was soon called "the German Socrates." He married, and he and his wife, Fromet Gugenheim, began what would become the greatest salon in Berlin. They had three daughters and three sons, of whom we will hear again. Mendelssohn's writing was at first not specifically Jewish, but because he was a famous philosopher and also a practicing member of a vilified group, there was pressure on him to defend his choices. He resisted it, but one day a German deacon, Johann Lavater, publicly challenged Mendelssohn to either defend Judaism or give it up. Mendelssohn's response was quiet and composed, and challenged the idea of converting people. The various truths of various people were not at odds: "Why should I convert a Confucius or a Solon?" he asked. "As he does not belong to the congregation of Jacob, my religious laws do not apply to him, and on doctrines we should soon come to an understanding. Do I think there is a chance of him being saved? I certainly believe that he who leads mankind on to virtue in this world cannot be damned in the next."[102]

Note that, out of all humanity, he chose Confucius and Solon, two men widely known for having proclaimed rules of conduct without bringing in God or revelation. Lavater replied with newfound respect but it wasn't about him anymore: the encounter started a hurricane of pamphlets and published letters about Jews that soon forced Mendelssohn to give an Enlightenment account of his people. The result was *Jerusalem, or On Religious Power and Judaism* (1783), and it is this small book for which he is remembered. The second title is more informative: the text was really two little books, one on religious and political power and one on Judaism. In the first, Mendelssohn chides Hobbes for championing authoritarian government and freethinking:

"In order to retain for himself the liberty of thought, of which he made more use than anyone else, he resorted to a subtle twist."[103] The "twist" was that Hobbes thought all ideas were fair game so long as one lived within the law. Mendelssohn preferred less inner rebellion and more public freedom. No church should have any legal power at all. How could anyone try to legislate the relationship between God and a human being? Mendelssohn laments that people have always known these obvious principles but never act as if they were true. "Happy will they be if in the year 2240 they cease to act against them."[104] It is a comfort that he gave us more time.

Mendelssohn also says that the "smallest privilege which you publicly grant to those who share your religion" is actually a bribe. You can call it "privilege" if you like, but he cautions: "To the linguist such a notation may be useful, but to the poor wretch who must do without his rights as a man because he cannot say: I believe, when he does not believe, who will not be a Moslem with his lips and a Christian at heart, this distinction brings only a sorry consolation."[105] Mendelssohn rejects "atheism and Epicureanism" as undermining social civility. In this he was not in sync with many of his time, and he knew it. "Let Plutarch and Bayle inquire ever so much whether a state might not be better off with atheism than with superstition." Still, the state should take note of this "only from a distance."[106]

Mendelssohn had a great deal of sympathy for Spinoza, with whom he did not agree, but whom he had always admired. Mendelssohn mentions Judaism only once in part one of *Jerusalem,* and it is on this issue: "Reader! To whatever visible church, synagogue, or mosque you may belong! See if you do not find more true religion among the host of the excommunicated than among the far greater host of those who excommunicated them."[107] Great doubters are often more invested in religious questions than is the average believer. Mendelssohn rails that to expel a dissident from the church is like barring a sick person from the pharmacy. "By the magic power of sympathy one wishes to transfer truth from the mind to the heart; to vivify, by participation with others, the concepts of reason, which at times are lifeless, into soaring sensations."[108] It is a nice description of the good that religious society can do—with or without belief.

Jerusalem's second part is about Judaism. It explains that Judaism can be held by a deist. "It is true that I recognize no eternal truths other than those that are not merely comprehensible to human reason but can also be demonstrated and verified by human powers." But, Mendelssohn writes, when his detractors think this means he has renounced the faith of his fathers, they are "misled by an incorrect conception of Judaism."[109] "To say it briefly: I believe that Judaism knows of no revealed religion in the sense in which Christians understand this term. The Israelites possess a divine legislation."[110] These laws and rules of life "were revealed to them by Moses in a miraculous and supernatural manner, but no doctrinal opinions, no saving truths, no universal propositions of reason. These the Eternal reveals to us and to all other men, at all times, through *nature* and *thing*, but never through *word* or *script*."[111] After all, "Why must the two Indies wait until it pleases the Europeans to send them a few comforters to bring them a message without which they can, according to this opinion, live neither virtuously nor happily?"[112] Whenever Mendelssohn mentions atheism, Epicurus is there. We saw that Jewish thinkers had taken his philosophy seriously from its first appearance; now they also mention Lucretius, Helvétius, and Hume.[113]

Mendelssohn had a new reason that truth could not have been revealed or proved by miracles, even those in the Old Testament. Eternal truth was not revealed on Sinai:

> In reality, it could not have been revealed there, for who was to be convinced of these eternal doctrines of salvation by the voice of thunder and the sound of trumpets? Surely not the unthinking, brutelike man whose own reflections had not yet led him to the existence of an invisible being that governs the visible. . . . Still less [would it have convinced] the sophist whose ears are buzzing with so many doubts. . . . He demands rational proofs not miracles. And even if the teacher of religion were to raise from the dust all the dead who ever trod the earth, in order to confirm thereby an eternal truth, the skeptic would say: The teacher has awakened many dead, yet I still know no more about eternal truth than I did before. I know now that someone can do, and pronounce, extraordinary things; but there may be several suchlike beings, who do not think it proper to reveal themselves just at this moment.[114]

The real showstopper is this: "Among all the prescriptions and ordinances of the Mosaic law, there is not a single one which says: You shall believe or not believe. They all say: You shall do or not do." It's fascinating because, while it is true, stating it brings the possibility of unbelief within the flock into the realm of the probable. In Mendelssohn's conception, "faith," the Christians' "easy way" in contrast to the rigorous Jewish laws, had finally backfired. Finally, it was easier to do the laws than to believe. The Jew could step up now and say, All God asked us to do is not eat pork, and so on, and we are free to do science and face truth. "Nowhere does it say: Believe O Israel, and you will be blessed; do not doubt, O Israel, or this or that punishment will befall you." The Jew had only actions, to do or to avoid. "Belief and doubt, assent and opposition, on the other hand, are not determined by our faculty of desire . . . but by our knowledge of truth and untruth."[115] Mendelssohn's *Jerusalem* gave a lively tour of the history of doubt. It praised anyone who had Socrates' ability to say *I do not know;* reprinted a selection from a book on Hinduism; and explained the Cabalist obsession with numbers as a skeptical refusal of all that could be spoken—which it was.[116]

Yet Mendelssohn was sure Jews should keep up the laws. His counsel was, "Adapt yourselves to the morals and the constitution of the land to which you have been removed; but hold fast to the religion of your fathers too. Bear both burdens as well as you can!"[117] To the Christians he says, Do not ask us to convert anymore, we couldn't stop living this way any more than Jesus could have. He wouldn't have converted or have married you either. He wouldn't even have eaten with you. Mendelssohn's solution echoes the line of the Mishnah: "Better that they [the Jews] abandon Me, but follow My laws."[118] Mendelssohn also combats the growing idea of a universal religion, one with innocuous, agreed-upon rituals derived from all the existing religions. As a practicing Jew, Mendelssohn wasn't going to like the idea. His response to it was fun; for one thing, "Woe to the unfortunate who comes a day later and finds something to criticize. . . . To the stake with him!" But the real argument against rationalizing religion so that we could all be one congregation was deeper: it wouldn't work. "In reality, everyone would then attach to the same words a different meaning of his own, and you would pride yourself on having united men's faiths, on having brought the flock under a single shepherd? . . . Brothers, if you care for true piety, let us not feign agreement where diversity is evidently the plan and purpose of Providence."[119] It's a nice plea for pluralism. The book's last words are "Love truth! Love peace!"[120]

In 1783 a prominent Berlin newspaper called for essays in response to the question "What is enlightenment?" Mendelssohn answered, calling it a process, an education of humanity through the use of reason, and explaining that it had only just begun. Immanuel Kant answered, too. For Kant, enlightenment was "man's release from his self-incurred immaturity." He famously cheered: "*Sapere aude,* have the courage to know: this is the motto of the Enlightenment."

Was Kant, then, a doubter? Well, he had doubt, yes, but faith, too. The truth is, Kant transformed the whole conversation about doubt, but was himself a believer. Kant liked to say that reading Hume knocked him out of his "dogmatic slumber." Bowled over by Hume's sophisticated Skepticism, which studied how much our minds create time, space, and shape, Kant set out to solve the problem of knowledge. In the opinion of many, he succeeded.

Kant agreed that the senses deceive, and madness and dreams make us question our certainties. He asked that we follow this deeper, as Hume had begun to do, and accept that it would be very coincidental if we small, fleshy organisms were equipped with sensory-gathering abilities that would provide a good understanding of reality. Our minds, Kant explained, project all the basic categories of human understanding onto the world, so that time, space, and extension are all coming from us. He believed there are real objects in the world—this is not George Berkeley; we are not dreaming all this—it's just that we have no access to the real stuff because all access is through perception, which changes everything. The world we cannot perceive is the real world, the noumenal. The world we know, the one we live in and snack on, is the phenomenal world. The Skeptical problem of perception was taken so far that it disappeared as a problem. We can't know "things-in-themselves" at all. But we are free to know this phenomenal world through science, the science of how things seem to us. Philosophy was for awakening to reality.

Kant so fully demolished the last remaining philosophical proofs of God that Mendelssohn called him "the all destroyer," and the name stuck. Kant would be a major figure of doubt, but the man himself was a believer. He thought that moral feelings were a hint from the unknowable world, and because of our total lack of knowledge of that real, noumenal world, one might as well choose to believe that there is a God out there. He revolutionized philosophy without having to shift much in his pew.

THE FESTIVAL OF REASON

An early priority of the French Revolution, which began in 1789, was to nationalize and democratize the French Catholic Church and its vast holdings of French land and wealth. By 1790 considerable amounts of Church property had been seized for redistribution. Moreover, the "Civil Constitution of the Clergy" proclaimed that all clergy now worked for the French state, not Rome, and they had to swear an oath of allegiance. As the Revolution escalated to its most radical phase, the attack on religion also escalated, so that it was no longer an attack on clerical abuse or wealth—that is, no longer a belated Protestant-like break from Rome—and became instead an attack on religion and Christianity, and for a while, the existence of God. It was in January 1793 that the revolutionaries guillotined the king. France was at war abroad and in turmoil at home, yet that October, the elected body governing France—the Convention—made a priority of adopting a new calendar to shake loose the Catholic week with its holy Sunday and the Catholic year with its constant saints' days and feasts.

Just as the medieval Catholic Church tried to shape the lives of Europeans through the calendar, the Revolution sought to infuse daily life with its values. The years would no longer be counted off from the birth of a Jewish preacher in Palestine. The new calendar called 1792 the year one. Each season had three months with similar names: Germinal was in Spring, Brumaire (foggy) in autumn, and Thermidor overlapped July and August (both named by Caesars). Months were all thirty days long, divided into three ten-day weeks, with each day named for its number. It was just as the Catholic Church had requested centuries earlier in order to get rid of the pagan gods of the weekdays! But because every day in Catholic France had had a saint, which was often how people referred to the days, the revolutionary calendar also gave an individual name to every day of the year, and these names were for animals, vegetables, and herbs. I am writing this on Potato, Firstday, the fifth of Wineharvestmonth, Year 211 (that *Primidi* of *Vendémiaire* was Friday, September 27, 2002). The leftover five or six days at the end of each year were holidays called *Sansculottides,* "workers' days," and included the Day of Virtue, the Day of Work, and the Day of Reason. While it lasted (Napoleon brought back the Gregorian calendar in 1806— the new one was tough on anyone doing business outside of France), it was an earnest attempt to change the signposts of everyday life from religious to secular and natural.

In the fall of 1793, an atheist campaign got under way. Its three leaders were the self-proclaimed atheists Pierre Chaumette, Josephe Fouché, and Jacques Hebert. In the department of the Nièvre, Fouché imposed social leveling through taxes and an assault on wealth and privilege that stood out in its severity even amid the Revolution. Fouché stripped the churches of their gold and statues and sent the loot to the national treasury. He also established the cult of the goddess of Reason and ordered an inscription made for the gates of the town's cemeteries that read: "Death is an eternal sleep." In September, Fouché and Chaumette met at Nièvre and, soon after, introduced the cult of Reason in Paris. It caught on. Hebert, leader of the sansculottes (after Marat's murder), was also well known as an atheist. Soon one Parisian section after another was turning its churches into Temples of Reason, marrying off its priests, renaming its streets and its children.

On November 10, 1793, Chaumette and Hebert led the climax of this atheist revolutionary cult by throwing a grand Festival of Reason, gutting Paris's great Notre Dame to do it. A woman dressed up in white portrayed "reason," and they paraded her through the streets of Paris with much new-fashioned ritual—based mostly on rites of the ancient world. All Parisian churches were officially closed beginning November 22. Meanwhile, Robespierre was growing in power and decided he could not tolerate the threat Hebert posed. In March of 1794, he had Hebert and his major followers killed by the cartload. Then in April he got rid of the last figure with whom he shared power, the more moderate George Jacques Danton. Fully in charge now, Robespierre repudiated the atheism of Chaumette, Fouché, and Hebert, claiming that people need to believe there is a god and an after-life. Thus, on June 8, Robespierre threw the Festival of the Supreme Being. He ran this one. It was a celebration of a deist god, of the state, and of state-ordained violence, all without a hint of Christianity. It didn't last long. The Revolution turned and Robespierre was guillotined in July.

Napoleon Bonaparte would choose to make peace with the papacy when he took over, though he famously did not let the pope crown him emperor. Bonaparte had a scientific worldview, and was not shy about it, but warned against anyone trying to separate the masses from their supersti-tions. Echoing Critias's idea, the emperor believed government must make use of religion to keep order and morality—in particular, for Bonaparte, to keep the poor from murdering the rich.

ZEN IN THE EIGHTEENTH CENTURY

Our last look at the eighteenth century returns us to the great doubting reli-gion of Japan, Zen Buddhism. By now Zen had a long tradition of taking doubt as its central vehicle. Takasui, an eighteenth-century[121] Japanese Zen master, laid out the idea like this: "The method to be practiced is as follows: you are to doubt regarding the subject in you that hears all sounds. All sounds are heard at a given moment because there is certainly a subject in you that hears. Although you may hear the sounds with your ears, the holes in your ears are not the subject that hears. If they were, dead men would also hear sounds." He says, "You must doubt deeply, again and again, asking yourself what the subject of hearing could be." Ignore the thoughts that come to you. "Only doubt more and more deeply," concentrate, "without aiming at anything or expecting anything" and "without intending to be enlightened and without even intending not to intend to be enlightened; become like a child in your own breast." It is striking to note how, across these centuries, people in the East and the West have been waking themselves up to the great fact of doubt by focusing on it, fighting the common trance of life. Takasui continues his instructions:

> But however you go on doubting, you will find it impossible to locate the subject that hears. You must explore still further just there, where there is nothing to be found. Doubt deeply in a state of single-mindedness, . . . becoming completely like a dead man, unaware even of the presence of your own person. When this method is practiced more and more deeply, you will arrive at a state of being completely self-oblivious and empty. But even then you must bring up the Great Doubt, "What is the subject that hears?" and doubt still further, all the time being like a dead man. And after that, when you are no longer aware of your being completely like a dead man, and are no more conscious of the procedure of the Great Doubt but become yourself, through and through, a great mass of doubt, there will come a moment, all of a sudden, at which you emerge into a transcendence called the Great Enlightenment, as if you had awoken from a great dream, or as if, having been completely dead, you had suddenly revived.[122]

It is wonderful that such different adventures in doubt and enlightenment have swirled around the planet at the same time and across time. What happens next is equally dramatic. This chapter covered the time in which modern doubt came out and pronounced its own name without a negation or a wink. The next century's doubters will be downright evangelical about it.

NINE

Doubt's Bid for a Better World, 1800–1900

Freethinking in the Age of Science and Reform

The nineteenth century was easily the best-documented moment of widespread doubt in human history: there were more doubters writing and speaking where they could be heard than ever before, and many more had come to hear them. The big new element was the reformists. They begin the century demanding an end to religious persecution and end it in defiance of religious support for political injustice. Indeed, many of the famous calls for reform—for an end to slavery, for women's rights, for free speech—were made by doubters. Many of these figures experienced religious doubt first and considered their other battles a continuation of this prior and fundamental revolt. Doubters thus established the terms of democracy. Quite a few of these reformist doubters were female, and that's one of the nice things about this part: we get to hear from more of the women. In philosophy and poetry, doubters also embraced art as a source of natural transcendence, and created the modern idea of the artist. There are more brazenly irreligious philosophical doubters than ever before, and they are often marked now by a celebration of Eastern doubt. Throughout this century, people will speak about the old idea of replacing religion with science or philosophy, but now they call for it out in the open and sign their names. Now some will also speak of replacing religion with politics or with art.

In this century, doubters of all stripes, from Elizabeth Cady Stanton, to John Keats, to Karl Marx, were committed to their doubt and interested in figuring out what came next. In part because of this, interest in the traditional history of doubt is at a low point: there is less talk of Epicurus, Cicero, Averroës, Pomponazzi, Hobbes, Spinoza—or any of the great line. The story drops out a bit because doubt is now eloquently espoused by common people whose educations lacked much philosophy; they'd never heard of any of these characters. Also, even the best-educated doubters felt that the time for doubting religion was over: it was time to start building something in which one could truly believe, a happy new world. They guessed that it would be a better world because the money and energy once given to religion would be devoted to generating food, clothing, medicine, and ideas. They also thought they might see farther than ever before, now that their vision was mended.

Along with reform and art, there was science. We have long followed the doubters' suggestion that humanity somehow developed naturally, and in this century Darwin helps the doubters win the point. We have also long followed the doubters' insistence on atomism; that moves into the mainstream now, too. And just wait until you find out what happened to ancient Skepticism! The nineteenth century also has a pretty dramatic story of the Jewish Enlightenment and reform movement. It's told here under the heading "Mendelssohn's Daughters," although they are more its cautionary tale than its emblem.

We are in a very cosmopolitan world here, and that led to secularization and doubt, as it always does. The century was overrun with change. Capitalism, on the rise since before Adam Smith wrote of it in 1776, broke a million traditions with the past, and because growth and change were an implicit part of it, even the new traditions were now constantly broken to sell something new. Industrialization pulled people out of family-based work and toward the cities. Colonial attacks furthered an explosive intermixing of the peoples of the planet; the shock of culture clash was not from a story in a book but from people around you. Vastly expanded elementary education and new, cheaper methods of printing brought the curious creation of the Enlightenment, "public opinion," to an ever-wider public. Democratic government made a monarch of public opinion; the real beliefs of individuals now meant a lot. The century also saw terrific new distractions from the duties and pleasures of religion. New leisure activities appeared. More people lived in the cities than ever before, and the city often had a secularizing effect, but in most cases, the measurable rites of religion (church attendance, number of people keeping kosher) declined in the countryside as well as in the cities.

There were many types of doubt, but there were many doubters who partook in more than one of these and saw them as united by doubt.

One work that had colossal impact on both politics and art was Ludwig Feuerbach's *The Essence of Christianity* (1841). It was a touchstone of conversion to doubt for a century. Feuerbach noticed that if we have sensed the divine all along, and it turns out there is no God, then what we called divine was coming from us in the first place. For Feuerbach, if God is our projection of ourselves onto the heavens, we are divine:

> Eating and drinking is the mystery of the Lord's Supper;—eating and drinking is, in fact, in itself a religious act; at least, ought to be so. Think, therefore, with every morsel of bread which relieves thee from the pain of hunger, with every draught of wine which cheers thy heart, of the God who confers these beneficent gifts upon thee,—think of man! But in thy gratitude towards man forget not gratitude towards holy Nature! Forget not that wine is the blood of plants, and the flour the flesh of plants, which are sacrificed for thy well-being![1]

In his hands, everything about God was real; our only mistake was in thinking it came from outside us. Feuerbach argued that we could learn about ourselves by studying religious urges outside the context of belief. This doubter thought to take religious myth seriously.

Yet, amid this modern doubt, people still remembered the religious wars and Inquisitions; it was all fresh enough to inspire fury against the Church. At the beginning of the century Francis Horner, one of the founders of the *Edinburgh Review,* told his heirs to preserve documents on the burning of young Aikenhead as well as "similar documents from century to century, by way of proving, some thousand years hence, that priests are ever the same."[2]

MENDELSSOHN'S DAUGHTERS

Moses Mendelssohn had three daughters, Brendel, Recha, and Henriette. Moses had most carefully educated his sons, and it was his sons' successes in banking that brought the Mendelssohn family its wealth. Having taught the girls philosophy, literature, math, and religion, Moses insisted on arranged marriages for Brendel and Recha. He had decided on the ratio of secular and sacred and presented it to his children as a finished product. This does not always work. Brendel, first of all, changed her name to Dorothea. As Felix Mendelssohn's biographer Peter Mercer-Taylor has written, "her forthright manner and staggering erudition" made her extraordinarily popular among Berlin's intellectual elite.³ Her salons were at the center of the German Enlightenment. She and her sister Recha both left their husbands, Dorothea amicably: her first husband let her have the children and helped support them even though she ran off with the scholar Friedrich Schlegel. Dorothea then converted to Protestantism so she and Schlegel could marry, and the two went off to Paris where they ran another important salon. Later, the two moved to Vienna for a job and converted to Catholicism.

It was Schlegel who firmed up the idea of Romanticism, called out its members, and energetically promoted it. Romanticism was an important voice of doubt as well as spiritualism. It was a rejection of rationalism in favor of feelings, individual experience, and passion. It supported a spiritualism that posed a challenge to traditional religion, but also energized it. On the other hand, the valuing of individual experience encouraged people to wander away from traditional communities, roles, and duties. Schlegel saw Romanticism as fundamentally about freedom and breaking down stultifying cultural mores. He wrote a surprisingly overt novelized treatment of the relationship between himself and Dorothea, she wrote a novel in response, and it was hers that became famous—she henceforth supported them with her writing. His great work was that he introduced the study of Indo-Aryan languages in Germany, and he became a publisher in order to print the Bhagavad-Gita and the Ramayana. Schlegel and Dorothea were fully engaged in religious questions but stepped away from Judaism, from tradition, from Christianity, from the constraints of religious morality.

Henriette, the youngest Mendelssohn sister, was bolder still. She never married. Instead, she went to Paris, opened a school for girls, ran a famous evening salon, and at the age of thirty-six was given a plum job overseeing the education of the daughter of an extremely important French count— Catholic, of course. Although she had been upset over Dorothea's conversion, soon after taking up residence with her new charge, Henriette became a Catholic. She had an exciting life among the elite of France, and her thirteen-year tutorship came with a comfortable life pension. Moses had died in 1786, before either woman converted; Henriette waited to do it until just after her mother's death in 1812.

One is tempted to regard the conversions as the legacy of Moses Mendelssohn's rationalist religious attitude, but thousands of German Jews converted in the first years of the nineteenth century. In the French Revolution, Jews were recognized as citizens of France for the first time.

Later, as Napoleon's victories exported French laws, Jews across Europe were recognized as citizens in the countries in which they lived. Ghettos were abolished, as were mandatory badges; Jews could dress as they chose, live anywhere, and work at any job. In no time they had settled outside Jewish areas, learned the language of the country they were in, settled their children into public schools, such as there were, and generally become comfortable. After Napoleon's defeat in 1815, the old monarchies were reinstated and the laws protecting Jewish rights were scrapped; in many countries, Jews lost citizenship with all its rights to schools, universities, a multitude of jobs, and much else in public life. Faced with either losing their livelihood or undertaking a brief conversion ceremony, many who saw themselves as Enlightenment Jews converted to Christianity in these years.

Later, Abraham Mendelssohn (the middle son) and his wife, Lea, also converted, to Protestantism. Several members of Lea's family had done it, too, one taking the Christian name Bartholdy, which is why their son Felix, the composer, is sometimes called Mendelssohn-Bartholdy. That first Bartholdy had written coaxingly to Abraham and Lea saying: "You may remain faithful to an oppressed, persecuted religion, you may leave it to your children as a prospect of life-long martyrdom, as long as you believe it to be absolute truth. But when you have ceased to believe that, it is barbarism."[4] The modern sense of pride in cultural, secular Jewishness was strikingly missing from the equation. Consider how Abraham counseled his own daughter to understand the family's conversion:

The outward form of religion . . . is historical, and changeable like all human ordinances. Some thousands of years ago the Jewish form was the reigning one, then the heathen form, and now it is the Christian. We, your mother and I, were born and brought up by our parents as Jews, and without being obliged to change the form of our religion have been able to follow the divine instinct in us and in our conscience. We have educated you and your brothers and sisters in the Christian faith, because it is the creed of most civilized people, and contains nothing that can lead you away from what is good, and much that guides you to love, obedience, tolerance, and resignation, even if it offered nothing but the example of its founder, understood by so few, and followed by still fewer.

By pronouncing your confession of the faith, you have fulfilled the claims of society on you, and obtained the name of Christian. Now be what your duty as a human being demands of you, true, faithful, and good . . .[5]

It is an amazing statement. Note first of all that he quickly dismissed the turf-war aspect of Judaism and Christianity (by sticking the pagans in between, he makes the battle seem distant and cold) and instead seems harder pressed to explain why any religion at all is necessary. Then there comes the marvelous note on Christianity's "founder," a plain suggestion that the true thing Christianity had to offer—the example of an ancient Jewish philosopher—was actually missed by most Christians.

The Jewish poet Heinrich Heine also converted, in 1825, to secure his rights as a German citizen. Wrote Heine, "In dark ages people are best guided by religion, as in a pitch-black night a blind man is the best guide; he knows the roads and paths better than a man who can see. When daylight comes, however, it is foolish to use blind, old men as guides."[6] He had a big interest in the history of doubt, and great facility with an image: "All our modern philosophers, though often perhaps unconsciously, see through the glasses which Baruch Spinoza ground."[7]

It was in this atmosphere of rampant conversion that a coherent, rationalized, even secularized, "official" form of Judaism was born. It came to be called Reform Judaism. The reformers wanted to stop the hemorrhaging—they also wanted to make Judaism useful and attractive for themselves, their families, and the future. In the 1820s, many ordinary Jews and not a few rabbis started to drop some of the traditional rules, at first at home, but then in synagogue, too. They saw themselves as no less Jewish, but as merely cleaning out the arcane annoyances of the Jewish way of life. Abraham Geiger (1810–1874), the great German reform rabbi and scholar, came of age in this world. He studied the Torah critically, in secular terms, and concluded that the ever-shifting document, with its many authors and obvious flights of imagination, could not possible be binding on the lives of adult men and women in a modern world. Leopold Zunz (1794–1886) was one of the other crucial figures: probably the greatest Jewish scholar of the nineteenth century and the champion of what he called Scientific Judaism. The movement was most popular with German Jews, but it had a wide following.

In France, Samuel Cahen (1796–1862) reported in his Jewish reformist journal, "We are asked what reforms we support. Our response is: reforms of our ritual wherever it stands in contrast to our actual habits. We support reforms which our sages would have instituted were they living in 1840."[8] Another great leader of this movement was Rabbi Samuel Holdheim (1806–1860). Holdheim saw true Judaism as a commitment to monotheism and morality. Almost every aspect of Jewish law, ritual, and custom was seen as ancient history, no longer relevant in the modern era.

When the great rabbis arranged a meeting in Berlin to talk about reforming Judaism in 1845, many of those who showed up had already been running reformed services in their various communities for a few decades. These rabbis were also very influenced by Friedrich Hegel. Hegel dominated philosophy in the period after Kant and introduced the idea of history as the unfolding self-revelation of the world-spirit. This philosophy helped the rabbis see change and development as positive and progressive rather than as a mark of decline. But what, the reformists wondered, should change? With much debate, the movement called for worship in the language people speak in their country; the reintroduction of organ music in worship; equal parts for men and women in everything; dropping general observance of the minor holidays; and rejection of the dietary laws—usually completely. The injunction to keep one's head covered was generally let go as well. They spoke of circumcision as barbaric and useless, and many average Jews did not circumcise their sons in this period—although, eventually, the rabbis decided to retain circumcision. Intermarriage was acceptable. Lots of reformers also suggested taking Sunday as the Sabbath, since there was school and work on Saturday in most countries, but this one never took.

Another change that was generally repudiated by the second half of the twentieth century had to do with the ancient land of Israel. The reformers rejected the long-standing notion of Jews being in exile and called instead for them to devote themselves to the countries in which they lived. These reformers tended to be explicitly anti-Zionist. The new idea of Judaism was to celebrate the Diaspora, for through it the Jews were able to bring their moral monotheism to the world: in this way, these Jews counted Christianity and Islam as the success stories of Judaism, her "daughter religions." They got rid of the "second holiday," a habit of celebrating every holiday again the next night so that Diaspora Jews were sure to be celebrating at the same time Jerusalem Jews did. Jews of Jerusalem had never celebrated the second holiday, and now "reformed" Diaspora Jews would not either, as exile was no longer their situation. Synagogues, the places of worship far from the Temple, which later became the only places of worship, were now to be called temples. Even the idea of bringing music into the service was about choosing finally to stop mourning the Temple's destruction—for that was when instrumental music had gone out of the Jewish service.

In response, the orthodox defenders actually said things such as "The Torah teaches . . . 'Hear, O Israel' not 'Think, O Israel.'" That wasn't much to work with. In France the lay members of the community asked the rabbis "to be the first to raise the standard of reform"; without reform, they said, they could offer the next generation only arcane ceremonies choked with unimportant detail and had no "defense against the invasion of irreligion."[9] Something had to change. "Everywhere there is doubt," they wrote, and they asked for a Judaism they could agree on because it was minimal, rational, and egalitarian.[10] To enliven historical memory on the changeability of religious law, they spoke of acting "as in the days of Ezra."[11]

Many of the reformers, Geiger among them, were dedicated to the equal participation of women.[12] A rabbinical report of 1846 called for total equality for women in Judaism, explicitly rejecting the humiliations women had been subject to at the hands of the religion; the rabbis joked that, cruel as the male Jews had been to women, at least it was not as bad as the Christians in the Middle Ages "debating whether women had a soul at all!"[13] The benediction wherein each man thanks God that he was not born a woman was to be abolished, both sexes were to have a communion at age thirteen, and women would count in the minyan.[14] The rabbis admitted that while they were trying to be fair to women, they were also hoping to make them want to run Jewish homes.

The first modern rabbinical seminary in Europe was in Padua, the great old town of doubt. Samuel David Luzzatto (1800–1865) taught there. He was a friend of Zunz and an admirer of his Scientific Judaism. Luzzatto followed Mendelssohn's idea that belief was secondary in Judaism, and even took it to another level. For him, rationalism could not stop at denying the biblical story of creation and history, letting habit protect the "revealed" idea of the laws. "I call the belief in revealed Judaism supernaturalism, not in the sense of the dogma restraining freedom of thought, but certainly in the sense of the dogma which admits the supernatural meaning of events that happen contrary to the usual order of nature, like miracles and revelation."[15] If one is a rationalist, one may choose which laws are appropriate for modern Jews—like keeping the Sabbath and eating matzo instead of bread on Passover—and enjoy them as a mark of culture and community.

In one of the few acts of the early Reforms that created rather than negated ritual, Rabbi Michael Silberstein in 1871 helped lead the progressive Jews to take up the festival of Hanukkah, which became the great holiday of secular Judaism (hence the peculiar experience of Jews explaining to others that though it is the best-known Jewish holiday, it is not really a major Jewish holiday). His reason for championing the holiday was to stop Jews from celebrating Christmas: "It is a known fact that unfortunately a misuse has arisen in Jewish families, namely, the observance of the Christmas holy day as a day of Jewish sanctity." He told his fellow rabbis to make Hanukkah popular not only in synagogue but "also in the schools," and to "point out to the parents that the festival of Hanukkah should be turned into a family celebration."[16]

The holiday was celebrated by modern, progressive Jews because it fell at the same time as Christmas, but it commemorated the victory of the Maccabees, which was actually an attack on Jewish cosmopolitanism and progressivism. It seems ironic, but in another way, it is a perfect fit: the story of Hanukkah is in the Apocrypha, usually published with Catholic Bibles, never in the Hebrew Bible. This slightly-late, almost-disappeared little text affords us our clearest window into ancient Jewish doubt. Whatever else it is, Hanukkah is a time to remember Miriam with her sandal, striking the Temple and calling it a consuming wolf—not to mention all those Jewish boys who worked out in the Greek gyms, learned some philosophy, practiced a public profession, went to the theater, and read Greek poetry with their wives. Again, that's not why the Reform Jews picked it. For them, it was a matter of finding a way to make Judaism fit with the modern world, so that Judaism would persist in the world. When we remember the Mendelssohns, we can see their point. Sometimes having your own party—with presents and feasts and candy—can really help.

There were big changes in European Jewry: when the new German Empire was declared in 1871, its constitution gave equal civil rights to Jews throughout the empire. Meanwhile, Reform Judaism, which had begun in Germany, came up against limitations there. Across Europe the official leadership of religious communities was controlled by the secular governments, and these tended to think of the Orthodox as the legitimate leaders and appointed Orthodox rabbis to lead synagogues. Things could move much faster in the United States. In 1885 the Pittsburgh Conference (which met as a continuation of the German conferences of the 1840s) set out a "Declaration of Principles" to define Reform Judaism.[17] The rabbis' declaration began with cosmopolitanism: "We recognize in every religion an attempt to grasp the Infinite." It went on to declare that "the modern discoveries of scientific researches in the domain of nature and history are not antagonistic to the doctrines of Judaism," and that the Bible's miracles reflected "the primitive ideas of its own age." The laws of Moses were declared over, except for the moral laws; they would now observe only the rituals that "elevate and sanctify our lives," rejecting those "not adapted to the views and habits of modern civilization." They also announced that they were "no longer a nation," but a religious community, and rejected Zionism. As for the supernatural, they wrote that Judaism is a progressive religion, "ever striving to be in accord with the postulates of reason," rejected the ideas of heaven and hell, but reasserted the doctrine of the immortal soul. Finally, they announced that in the tradition of Mosaic equality for the rich and poor, "we deem it our duty to participate in the great task of modern times," to bring justice to "the contrasts and evils of the present organization of society." Reform Judaism is a reformation of Judaism, but it is also Judaism in the service of social reform.

RINGLETS AND BEARDS

Anne Newport Royall was born in 1769 and at age eighteen became a maid at the home of Major William Royall, a widower and veteran of the Revolutionary War and a freethinker with a good Enlightenment library. She found that when "reason is cultivated and our minds enlightened by education," we can reject "knavery, bigotry and superstition."[18] Ten years later they were married and they had a happy fifteen years, but when William died his children fought the will and Anne was left with very little. Royall began to travel and found she could support herself publishing books about her trips: both about where to get a good meal or a quiet room and about the mores of the various young states. The discussions were political, too: against slavery, for public relief for widows, in defense of the Native Americans, and most of all, against the missionaries.

In her *Black Book* (1828), she scorned the missionaries swarming "like locusts" across America, stumping for cash, and getting it, often from the poorest and most sadly superstitious people.[19] She also warned that if the champions of a national religion managed to "get two-thirds of the states to alter the Constitution . . . then let the people get their throats ready. May the arm of the first member of Congress who proposes a national religion drop powerless from his shoulder; his tongue cleave to the roof of his mouth and all the people say amen."[20] Royall was angry about new crimes perpetrated in the name of religion, but she remembered old ones. Do they think, she asks, we have forgotten how the "orthodox" used power when they had it?

Do they think we have forgotten how they drenched England in blood, created a civil war, (what they are in a fair way to do here) and, when they could no longer retain the power of killing there, came over to this country, and began it afresh—dipping their hands in the blood of a harmless, unresisting people? . . . Do they think we have forgotten how they put innocent men, women, and children to death, in cool blood, under the pretense of witchcraft? . . . Children of ten years of age were put to death; young girls were stripped naked (by God's people, the ministers) and the marks of witchcraft searched for, on their bodies, with the most indecent curiosity.[21]

She knew how to keep an audience. She also wrote letters to a friend, Matthew Dunbar, which became *Letters from Alabama* (1830). In it an 1821 letter asked: "What think you, Matt, of the Christian religion? Between you, and I, and the bed post, I begin to think it is all a plot of the priests. I have ever marked those professors, whenever humanity demands their attention, the veriest savages under the sun."[22]

In this period, if a widow wanted to receive her husband's veteran's pension, she had to petition Congress for it. When Royall got to Washington, instead of stopping at her own case she began lobbying Congress to change the law. Royall was now both radical in her doubt and educated in the ways of politics. In 1827 when a Reverend Ezra Stiles Ely began campaigning for Americans to elect only Protestants to government, Royall became the first person to lobby Congress regarding the separation of church and state. She went on to investigate all aspects of government for religious increepings, and her writing on religious rituals at West Point resulted in a congressional investigation of the school. Also, although she failed, she fought hard against the Sabbatarian campaign to stop Sunday mail delivery: "Supposing for argument sake, that it is a sin to carry the mail on Sunday, what is it to them?"[23] When invited to speak, she was often either mobbed or refused admittance to the town; over the years she was arrested, fined—ten dollars in 1829—and once pushed down a flight of stairs. On the other hand, she was so admired that President Andrew Jackson showed up to pay the ten dollars for her, only to find that he had been beaten to it by one of the witnesses for the defense, Secretary of War John Eaton. Later in life she founded and ran two newspapers, declaring her motto "Good works instead of long prayers" on the masthead and supporting what she'd written in the *Black Book:* "All priests are dangerous when clothed with power."[24] Royall died in 1854 and was remembered as an important American figure well into the next century. As her biographer George Stuyvesant Jackson wrote in 1937: "She was nationally known, liked, feared, ignored, detested; but she would be heard whatever the reactions." He characterized her as a cross between Voltaire, Carry Nation (a fiery temperance leader), Joan of Arc, and H. L. Mencken.[25]

In the early nineteenth century, there were a lot of people around who believed that religion had misdirected human energies and thought that because we had finally realized this, it was time to find a better basis for morality and fix this misshapen world. Jeremy Bentham's Utilitarianism took ideas on secular ethics from Helvétius, Diderot, Voltaire, Locke, and Hume and suggested we forget about parsing "good and evil" and work logically to minimize pain and increase pleasure; the greatest happiness for the greatest number. Bentham was friends with the like-minded philosopher James Mill and helped educated Mill's son, John Stuart—who soon came into his own as a philosopher as well. The Mills, Bentham, and many of their followers were all doubters. John Stuart Mill mentioned in his autobiography that he was "one of the very few examples in this country, of one who has not thrown off religious belief, but never had it"; in fact, he said, he looked upon the modern religion "exactly as I did upon the ancient religion, as something which in no way concerned me."[26] John Stuart described how his father rejected revealed religion as contrary to reason, and after much thought found "no halting place in Deism" and "remained in a state of perplexity," until concluding that nothing at all could be known about the origins of things. And, Mill argued, "the grounds were moral, still more than intellectual." His father's idea of religion "was of the same kind with that of Lucretius: he regarded it with the feelings due not to a mere mental delusion, but to a great moral evil."[27] He also called his father's standard of morals "Epicurean, inasmuch as it was utilitarian, taking as exclusive text of right and wrong, the tendency of actions to produce pleasure or pain."[28] In character he was a Stoic, Mill continued, and in his rejection of most worldly pleasure, a Cynic.

Much of John Stuart Mill's great work was the result of collaboration with Harriet Taylor. We need to know about women and doubt, so we will trace Taylor's role for a moment. Mill met Harriet Taylor in 1833 and the two worked closely together thereafter; when her husband died in 1849, they married. Starting with their 1851 *The Enfranchisement of Women,* they wrote a series of works that became foundational texts of modern liberal democracy. By her request, even those works written mostly by Taylor bore only Mill's name. He variously credited her as coauthor, insisting that some of his books were her ideas delivered by his pen. The book that drew the most controversy in their own lifetimes was *On Liberty* (1859). This great call for freedom of individual consciences arose from letters they had been writing to each other about conformism. Respectability had taken over society at the expense of creativity, freedom, adventure, and a "pagan individualism" (as opposed to Christian self-denial).[29] *On Liberty* held that government should interfere with citizens only if they are hurting others: people should be able to smoke opium if they wanted (the British opium wars of 1839–1842 had just popularized the drug) or, if they wanted to risk it, walk across dangerous bridges. Yet the heart of *On Liberty* was a call for religious freedom and an attack on the calcification of custom: "The mere example of non-conformity, the mere refusal to bend the knee to custom, is itself a service. Precisely because the tyranny of opinion is such as to make eccentricity a reproach, it is desirable, in order to break through that tyranny, that people should be eccentric."[30] It is a stunning contrast to the ancient doubters' bow to the state's religion. To speak your own truth was now a virtue.

Consider another of doubt's Harriets, the Englishwoman Harriet Martineau (1802–1876). Martineau's family was Unitarian, which she later called a "wonderful slovenliness of thought," but she counted herself a believer as a child.[31] She wrote a few religious books in her early twenties, and at thirty-two published a study of American women that was as highly celebrated as Tocqueville's work on America. Her best-selling books brought her financial independence, and in 1846 she took a trip, touring the Mideast in order to study the great religions. Thereafter she wrote as an impassioned doubter. Consider the tone: "There is no theory of a God, of an author of Nature, of an origin of the Universe, which is not utterly repugnant to my faculties; which is not (to my feelings) so irrelevant as to make me blush."[32] Martineau was meanwhile becoming a famous abolitionist and crusader for women's rights. People in these movements often wished she would quiet down about atheism, which, they rightly observed, could do harm to her other causes. Martineau's answer was that her primary dedication was to freedom of belief, "the very soul of the controversy, the very principle of the movement."[33] She also called for children's books for "the Secularist order of parents."

In her *Autobiography* (1877) Martineau wrote that she "certainly never believed" in the idea of "God as the predestinator of men to sin and perdition. . . . I never suffered more or less from fear of hell. The Unitarianism of my parents saved me from that."[34] Yet she was astounded by "how late on in my life" she had still believed in an afterlife—even after she no longer believed religion: "But at length I recognized the monstrous superstition in its true character . . . and found myself with the last link of my chain snapped, a free rover on the broad, bright breezy common of the universe."[35] She called herself the "happiest woman in England." Experiencing a "still new joy of feeling myself to be a portion of the universe, resting on the security of its everlasting laws . . . how could it matter to me that the adherents of a decaying mythology . . . were fiercely clinging to their Man-God, their scheme of salvation . . . their essential pay-system, as ordered by their mythology?"[36] She did not miss it. Martineau mused that Christianity "fails to make happy, fails to make good, fails to make wise," so there was not much loss. What is more, "To the emancipated, it is a small matter that those who remain imprisoned are shocked at the daring which goes forth into the sunshine and under the stars to study and enjoy, without leave asked, or fear of penalty."[37]

She did good work under those stars: William Lloyd Garrison wrote that "the service she rendered to the antislavery cause was inestimable," and Florence Nightingale wrote that Martineau "was born to be a destroyer of slavery in whatever form, in whatever place."[38] Diagnosed with a fatal heart disease in 1855, Martineau wrote her autobiography, concluding:

> I have now had three months' experience of the fact of constant expectation of death; . . . And now that I am awaiting it at any hour, the whole thing seems so easy, simple and natural. . . . The case must be much otherwise with Christians. . . . They can never be quite secure from the danger that their air-built castle shall dissolve at the last moment. . . . I used to think and feel all this before I became emancipated from the superstition. . . . But now the release is an inexpressible comfort; and the simplifying of the whole matter has a most tranquilizing effect. I see that the dying . . . desire and sink into death as into sleep. . . . Under the eternal laws of the universe, I came into being, and, under them, I have lived a life so full that its fullness is equivalent to length . . .[39]

It is somewhat new to find a doubter who believed everyone doubted. The deathbed scene was usually imagined as the sweetest surrender of the believer, so it is a strong claim. Martineau, for her part, surprised everyone and lived another twenty-one years, dying in 1876 at age seventy-four.

Frances Wright, usually called Fanny, was one of the best-known American reform women who championed religious doubt. Her work had a terrific appeal. Thomas Jefferson's personal journal contains seven pages filled with passages from one of Wright's books. He invited her to visit him at Monticello, which she did, along with Lafayette. She was also invited and stayed with Andrew Jackson at his home, the Hermitage; and she met Monroe as well. When she visited Europe she became friends with Jeremy Bentham. Walt Whitman wrote of Fanny Wright, "We all loved her; fell down before her."[40] She and her sister Camilla were orphaned young and later inherited a fortune from an uncle, which allowed them a great deal of freedom. At eighteen Fanny had a literary and philosophical club at which members delivered essays to one another: Fanny gave one about Epicurus, including a discussion of Leontium, "Epicurus's first female disciple." It was later published as *A Few Days in Athens,* augmented with even stronger doubt: "Surely the absurdity of all other doctrines of religion, and the iniquity of many, are sufficiently evident. To fear a being on account of his power is degrading; to fear him if he is good, ridiculous. . . . I see no sufficient evidence of his existence; and to reason of its possibility I hold to be an idle speculation."[41] She traveled in the United States (she also wrote a play that was produced on Broadway to good reviews) and wrote that American Unitarianism was so rapidly taking over that it could outstrip Calvinism. The Calvinists were furious, "but fortunately Calvin could no longer burn Servetus, however much he might scold at him."[42]

She was influenced by Utilitarianism, but also by what would later be thought of as utopian socialism. The French produced most of its leaders: Henri de Saint-Simon, Charles Fourier, and Pierre-Joseph Proudhon, for example, were all famous for their schemes for the perfect community, including a great deal of gender equality, free love, and of course, free thought. Proudhon, in particular, attacked the Catholic Church. The great utopian socialist of Britain, Robert Owen, made a fortune in textiles and then sought to find a way in which industrialization could avoid exploiting the worker. He bought twenty thousand acres of land in the United States and invited workers to come join him in a socialist venture; eight hundred showed up, and they called the place New Harmony. What did Owen think of religion? "All the religions of the world are based on total ignorance of all the fundamental laws of humanity. . . . Fully conscious as I am of the misery which these religions have created in the human race . . . , I would now, if I possessed ten thousand lives and could suffer a painful death for each, willingly thus sacrifice them to destroy this Moloch."[43]

When Owen invited the like-minded Fanny Wright to give the July 4 address at New Harmony, she became the first woman in America to give a lecture to an audience of both men and women. Her speeches in general spoke of the failure of religion and called for each community in America to form a Hall of Science with an auditorium, a school, a museum, and a library. In 1829 she bought the Ebenezer Baptist Church on Broome Street in New York City and renamed it the Hall of Science. It seated about twelve hundred people, and there were lectures and debates throughout the week, with a special event on Sundays. The bookstore sold works by Paine, Shelley, Owen, and Wright. It also sold birth control tracts, as birth control information was a point of attack in the fight for freedom of speech. At one lecture she told the crowd:

> The halls of science are open to all. . . . She says not to one, "eat no meat on Fridays"; and to another "plunge into the river"; to a third "groan in the spirit"; to a fourth "wait for the spirit" . . . and to nine hundred and ninety-nine thousandths of the human race "ye were born for eternal fire." Science says nothing of all this. She says, only, "observe, compare, reason, reflect, understand"; and . . . we can do all this without quarreling.[44]

She mentioned Galileo, pretty much alone among heroes of doubt, but her thought on doubt was sophisticated: "A necessary consequent of religious belief is the attaching ideas of merit to that belief, and of demerit to its absence. Now here is a departure from the first principle of true ethics." The only true ethics was "beneficial action."[45] Wright was deeply appreciative of the new home doubt had in the United States. She mentioned it when countering the religious idea of the "innate corruption" of man: "Think of his discoveries in science—spite of chains, and dungeons, and gibbets, and anathemas! Think of his devotion! . . . Think of the energy . . . with which he fought, and endured, and persevered throughout ages until he won his haven of liberty in America! Yes! he has won it. The noble creature has proved his birthright. May he learn to use and to enjoy it."[46] It is a stirring speech for the history of doubt, but for her, free thought had to lead to a more responsible world, and she could be stirring here, too: "I will pray ye to observe how much of our positive misery originates in our idle speculations in matters of faith, and in our blind, our fearful forgetfulness of facts—our cold, heartless, and I will say, insane indifference to visible causes of tangible evil?"

There are three more world-changers who need be mentioned here: Ernestine Rose, Karl Marx, and Elizabeth Cady Stanton. With Ernestine Rose the tone of doubt became just a bit more ironic than it had been before. Rose was born in Poland in 1810, the only child of a rabbi. He taught her to study the Torah in Hebrew, and the same freedom that allowed him to break custom and instruct her seems to have infused her whole personality. She later wrote that she had rejected the Bible by age fourteen. Early an activist, by twenty-four she was invited by Owen to give a speech to a workers' meeting. The speech was doubt's new favorite form of communication and she was good at it. Rose married, moved to America, and began campaigning against slavery and for women's rights and religious freedom. She was part of the movement to create nonreligious holidays, throwing a Thomas Paine celebration on the 113th anniversary of his birth, January 29, 1850. Because Ernestine Rose wore her hair in easy-to-care-for ringlets, many freethinking women adopted the style. At mid-century, describing a woman as "in ringlets" meant she was a freethinker and a reformer.

At the 1856 Seventh National Women's Rights Convention, an audience member scolded that the Bible submitted women to men. Rose said, "Do you tell me that the Bible is against our rights? Then I say that our claims do not rest upon a book written no one knows when, or by whom. . . . Books and opinions, no matter from whom they came, if they are in opposition to human rights, are nothing but dead letters."[47] The connection between women and religion was also financial and practical. "Sisters," she enjoined, "when your minister asks you for money for missionary purposes . . . for colleges to educate ministers, tell him you must educate woman, that she may do away with the necessity of ministers, so that they may be able to go to some useful employment."[48] Rose became a famous abolitionist, women's rights advocate, and atheist lecturer. In an 1861 lecture in Boston, "A Defense of Atheism," she argued philosophically against God but also joked that instead of saving Noah and the rest, God should have "let them slip also, and with his improved experience made a new world."[49]

What did she make of the world without a creator? One believer had told her that an eyeless fish living in a cave in Kentucky proved that there was a creator, since this showed design. Rose explained, "He forgot the demonstrable fact that the element of light is indispensable in the formation of the organ of sight, without which it could not be formed." This reminded her of a preacher who had proved the existence of God by noting that someone had placed the rivers near large cities. Rose believed the world could make itself, by its own logical patterns. This was 1861. Darwin had published in 1859 and she did not mention him—here it is enough to note that Rose had a notion of how some doubters have always understood the world: "The Universe," she wrote, "is one vast chemical laboratory, in constant operation, by her internal forces. The laws or principles of attraction, cohesion, and repulsion, produce in never-ending succession the phenomena of composition, decomposition, and recomposition."[50] Nature suffices as explanation, and nature is the only thing that can claim universal consent. Wrote Rose, "We are told that Religion is natural; the belief in a God universal. Were it natural, then it would indeed be universal; but it is not."

Rose had a strong belief in the coming of a wonderful new world based on the energies of freethinking people: "The Atheist says to the honest conscientious believer, Though I cannot believe in your God whom you have failed to demonstrate, I believe in man; if I have no faith in your religion I have faith, unbounded, unshaken faith in the principles of right, of justice, and humanity. Whatever good you are willing to do for the sake of your God, I am full as willing to do for the sake of man." She added that "the monstrous crimes the believer perpetrated," on account of difference of belief, would never be committed by the atheist, "knowing that belief is not voluntary, but depends on evidence."[51] The twentieth century was not able to bear out this optimism, and I cite this in part because it clangs with such power against that coming lesson, that secular states can make vicious decisions, too. For her, whatever believers would do in hope of heavenly reward, "the Atheist would do simply because it is good."[52]

If Ernestine Rose associated doubting women with ringlets, Karl Marx connected doubt and beards. His grandfather was a rabbi in Prussia; his father, Heinrich Marx, was a modern man, a deist, and did not practice Judaism. Just before Karl was born, in order to keep his post in the Prussian civil service, Heinrich had himself baptized; the children were baptized when Karl was six; Karl's mother waited until her own father, a rabbi, had died, and then was baptized, too. In his youth Karl was a bad poet; in 1837 he was converted out of poetry by the philosophy of Hegel. Hegel saw the world as being a result of minds thinking about it. It was a kind of pantheism, seeing the universe as God, with the mind of God coming into being as the minds of his creatures. Hegel then posited a "spirit" in history, which needs to go through a sequence of epochs. From the Enlightenment on, people had considered the idea of secular history having its own purposeful progress, but Hegel took it to grand levels. Human progress is the developing self-consciousness of the cosmos-God itself. By the time Marx was reading these ideas, Hegel was dead and a group calling itself the Young Hegelians was arguing over whether the philosopher had been for or against Christianity. It was hard to tell: Hegel thought Christianity was the best of the religions, yet he offered a secular morality in which community replaced God as the arbiter of good and evil. Marx went to the café the Young Hegelians frequented and there met Bruno Bauer, one of the fieriest atheists of the period.

Between 1839 and 1841 Marx wrote his doctoral thesis. It was on Epicurus and Democritus. His introduction growled, "Up to this time there has been nothing but repetition of Cicero's and Plutarch's rigmarole." Meanwhile, "Gassendi, who freed Epicurus from the interdict laid on him by the Fathers of the Church and the whole of the Middle Ages—that age of materialized irrationalism"—doesn't offer much. "It is more a case of Gassendi learning philosophy from Epicurus than being able to teach us about Epicurus's philosophy." Marx says Gassendi's attempt to make Epicurus cohabitate with the church "is like throwing the habit of a Christian nun over the exuberant body of the Greek Lais."[53] He also cited David Hume saying that philosophy should not have to answer to religion. Then came the coup de grâce: "Philosophy, as long as a drop of blood shall beat in its heart, being absolutely free and master of the universe, will never grow tired of throwing to its adversaries the cry of Epicurus," that the blasphemous person is not the one who scorns the God of the masses, but the one who blindly embraces him. "Philosophy makes no secret of it."[54] Note that there is nothing here about communism; Marx was an old-school atheist before anything else. In the summer of 1841, he and Bauer began to edit a journal called the *Atheist Archives*. It didn't pan out. Then they wrote an atheist pamphlet that got Bauer fired and made it impossible for Marx to find work in academia.

Marx then found Feuerbach and was struck by his idea of religion as a human creation, and his claim that by studying it we could learn about ourselves. Thereafter, Marx traded Hegelian idealism for the philosophical materialism he would ever after proclaim. In 1843 Bauer published an argument that the problem of the Jews—still at a civil disadvantage in

England and Germany—should be solved by Jews and Christians alike giving up religion. Having read Feuerbach, Marx found he could no longer agree with Bauer. Religion was not some crazy nonsense that could be swept away and beneath it would be a better world. Now Marx saw religion not as an independent problem, but as a symptom of a cruel economic world: people had religion because their lives were rotten; make their lives better and religion will melt away. In an 1844 paper (on Hegel), Marx wrote: "Religion is the sigh of the oppressed creature, the heart of a heartless world, just as it is the spirit of a spiritless situation. It is the opium of the people. The abolition of religion as the illusory happiness of the people is required for their real happiness."[55] The task of history, therefore, "once the world beyond the truth has disappeared," is to "establish the truth of this world."[56] For Marx it is social revolution, not science, that will finally dissolve religion. After millennia in which doubters had noted that religion seemed designed to control the masses, Marx said yes, let's do something about that. Also, tradition had it that well-fed, educated, cosmopolitan people often wander away from religion whereas their hard-scrabble neighbors thank God for their crumbs. Socialist doubt helped change that image.

Religion was the palliative that had to be removed in order to wake people up to the pain of life as it was, but there did not have to be a concerted effort against the palliative, because it would fall out of use as soon as things got better. And that's it. As historian Owen Chadwick has remarked, Marx "wrote so little about religion that some readers have doubted whether it was important to him."[57] Even in the *Communist Manifesto* (1848) there was not much. Friedrich Engels, with whom he wrote the manifesto, came to his own atheism through Feuerbach, Bauer, and the Young Hegelians, and had spent two years in England studying Robert Owen's work. Still, all the *Manifesto* said about religion was contained in a few lines. Of the proletariat: "Law, morality, religion, are to him so many bourgeois prejudices, behind which lurk in ambush just as many bourgeois interests." Perhaps the key statement of doubt in the *Manifesto* is the first line: "A specter is haunting Europe—the specter of communism." Marx and Engels nowhere mention that this specter was replacing another specter that once haunted Europe, but it was, and the allusion was not that vague. Many would note that with its savior and martyrs, symbolism, festivals, and dreams of paradise, Marxism took on many of the characteristics of a religion.

Of all the great doubters among American reforming women, the greatest were Elizabeth Cady Stanton and her comrade in arms, Susan B. Anthony. Anthony is more famous now because it was she who could travel and lecture—she had no children and Stanton had seven. Stanton was also a much more vocal doubter, although their ideas seem to have been similar: there did not appear to be a just God and there was no evidence of any other kind either. Anthony had seen, from close up, how Ernestine Rose's public atheism had hurt her strength in the movement for women's property and voting rights. She had defended Rose with vigor but chose a more discreet way for herself. Stanton took the vocal doubting role of Rose; for Stanton, doubt was fundamental. Here's how she told Anthony about the arrival of her second daughter: "Well, another female child is born into the world! Last Sunday afternoon, Harriet Eaton Stanton—oh! the little heretic thus to desecrate that holy holiday—opened her soft blue eyes on this mundane sphere."[58] In an 1860 address, "Antislavery," along with the central issue, she called for those enslaved by religion to be "born into the kingdom of reason and free-thought."[59]

Stanton spoke out on myriad church-and-state issues (a campaign she led managed to keep the World's Fair open on Sundays) and initiated feminist biblical criticism, pointing out how man "can stand in the most holy places in the temples, where woman may never enter," and that, throughout the Bible, "there is a suspicion of unworthiness and uncleanness" regarding women. She commented, in her wry tone, that you can't even sacrifice a female goat to God. But it wasn't really funny. As she proclaimed in 1882: "According to Church teaching, woman was an after-thought in the creation, the author of sin, being at once in collusion with Satan. Her sex was made a crime, marriage a condition of slavery, owing obedience, maternity a curse, and the true position of all womankind one of inferiority and subjection to all men; and the same ideas are echoed in our pulpits to-day."[60] Stanton did not have many heroes in the history of doubt, but she did tell the Galileo story, with its dramatic "Still, it moves."[61] Closer to home, she mentioned that "Harriet Martineau said that the happiest day of her life was the day that she gave up the charge of her soul," and she agreed that the happiest period of her life had been since emerging from the "shadows and superstitions of the old theologies."[62] Stanton and Anthony both praised Ernestine Rose for having helped women "to do their own thinking and believing."[63] She praised Paine as a major forerunner, too, but said that the most influential for her had been the abolitionist Lucretia Mott:

I found in this new friend a woman emancipated from all faith in man-made creeds. . . . Nothing was too sacred for her to question. . . . It seemed to me like meeting a being from some larger planet, to find a woman who dared to question the opinions of Popes, Kings, Synods, Parliaments, . . . recognizing no higher authority than the judgment of a pure-minded, edu-cated woman. When I first heard from the lips of Lucretia Mott that I had the same right to think for myself that Luther, Calvin, and John Knox had, and the same right to be guided by my own convictions, and would no doubt live a higher, happier life than if guided by theirs, I felt at once a new-born sense of dignity and freedom; it was like suddenly coming into the rays of the noon-day sun, after wandering with a rushlight in the caves of the earth.[64]

It is a nice story of awakening. For more detail on where she ended up, consider a letter of 1873. A formidable feminist, Isabella Beecher Hooker, was disappointed that Stanton did not discuss the afterlife as a part of women's salvation. Wrote Stanton to a friend, "To suppose this short life to be all of this world's experiences never did seem wholly satisfactory, but at the same time I see no proof of all these vague ideas floating in Mrs. Hooker's head."[65]

Stanton's famous address "The Solitude of Self," delivered before the U.S. Senate Committee on Woman Suffrage on February 20, 1892, is a plea for civil rights on the basis of metaphysical need. In it, Stanton gently states that the essential reason for women having equal rights is that women, like men, live and die alone, under a perhaps godless sky.[66] Economics and poli-tics were important, but this was about losing superstition and getting some philosophy. Elsewhere, Stanton worked to help fix religion. Of the Bible she joked, "Disraeli said that the early English editions contain 6,000 errors in the translation from the Hebrew. . . . It is fair to suppose that at least one-half of these errors are with reference to woman's position."[67] Again, she was a wit, but serious: "We do not burn the bodies of women today, but we humiliate them in a thousand ways, and chiefly by our theologies."[68] In 1895 at a celebration for her eightieth birthday at New York City's Metropolitan Opera House, before an audience noted to be a few thousand feminist and freethinking fans, Stanton noted that clergymen were "still preaching ser-mons on the 'rib origin'" and excluding women from church government. "We must demand that the canon law, the Mosaic code, the Scriptures, prayer books and liturgies be purged of all invidious distinctions of sex."[69]

The first volume of her most scandalous work, *The Woman's Bible,* came out two weeks later. When people got mad, Anthony defended Stanton just as she had defended Rose. *The Woman's Bible* was a best-seller, going into seven printings in the first six months.

PHILOSOPHERS OF DOUBT

Arthur Schopenhauer (1788–1860) is the person who, coming upon Kant's philosophy, noticed that the God therein was not attached to anything and shook the text till the God fell out. Suddenly nothing rattled, and it was a hell of a hush. Schopenhauer called himself a pessimist and seemed pessimistic to a lot of people, yet on the page there's an odd cheerfulness to his language. Here's a charming line on believers: "For if we could guarantee them their dogma of immortality in some other way, the lively ardor for their gods would at once cool; and . . . if continued existence after death could be proved to be incompatible with the existence of gods . . . they would soon sacrifice these gods to their own immortality, and be hot for atheism."[70] It may be the funniest statement in the history of doubt.

Schopenhauer agreed with Kant that our minds project time and space *and* inference and causality onto the world around us, so we cannot know anything about the real world, about the reality of things outside our perception of them. Kant thought we could have some intimations from the real, noumenal world, but Schopenhauer saw this as an error: since time, too, is an idea of the mind, we cannot even imagine what an intimation from the noumenal world would be since our thoughts are arranged in sequence, in time. If we can imagine it, it is from the phenomenal world. The other world was real then, more real than this one, just as Plato had said all those years ago.

There was no need for God in this understanding of the universe, and, indeed, Schopenhauer was an avowed atheist. As he saw it, there is no God, nothing made the world, we are accidental animals, and our way of knowing creates the world as we know it. Philosophy had "proofs" of God until Kant, but, as Schopenhauer put it: "Kant first suddenly wakened it from this dream; therefore the last sleepers (Mendelssohn) called him the all-pulverizer."[71] Both Kant and Mendelssohn had accepted that philosophical arguments for God had been pulverized, but they believed anyway. Schopenhauer didn't, but he was not happy about it. Some people like doubting and don't mind dying; some doubters don't like it and choose to believe: think of the nice woman Diderot portrayed, Madame la Maréchale, or think of Pascal. But some people find doubting painful and do it anyway. The Preacher of Ecclesiastes saw no justice and advised a melancholy acceptance of it. Schopenhauer took this to a new level:

> Many millions, united into nations, strive for the common good. . . . Now senseless delusion, now intriguing politics, incite them to wars with one another; then the sweat and blood of the great multitudes must flow. . . . In peace . . . inventions work miracles, seas are navigated, delicacies are collected from all the ends of the earth, the waves engulf thousands. All push and drive, some plotting and planning, others acting; the tumult is indescribable. But what is the ultimate aim of it all? To sustain ephemeral and harassed individuals through a short span of time, in the most fortunate case with . . . comparative painlessness (though boredom is on the lookout for this), and then the propagation of this race. . . . With this evident want of proportion between the effort and the reward, the will-to-live . . . appears . . . as a folly, or . . . as a delusion. Seized by this, every living thing works with the utmost exertion of its strength for something that has no value.[72]

Schopenhauer was aware of the worst: the struggle among animals was most obvious. Given how complex the natural world is, with its arrangements of perfect-prey for perfect-hunter, you would think it all added up to something sublime. "Instead of this we see only momentary gratification, fleeting pleasure conditioned by wants, much and long suffering, constant struggle, *bellum omnium,* everything a hunter and everything hunted, pressure, want, need and anxiety, shrieking and howling; and this goes on . . . until once again the crust of the planet breaks."[73] One may demur that Schopenhauer hardly mentions the equal cacophony of birth, joy, and satisfaction and note that his only critique against pleasure is that it is fleeting (a judgment best left to each individual creature). Still, his lament is compelling. Optimism for him "seems to me to be not merely an absurd, but also a really wicked, way of thinking, a bitter mockery of the unspeakable sufferings of mankind."[74]

Schopenhauer said Kant had missed the final step of perfecting his own philosophy. We know our sensory equipment is not up to the task of knowing the world, but does that mean that our experience is the only real thing, as Berkeley said, or are there real things that we simply can't access? Kant thought there are "things-in-themselves" and we just can't know them. Schopenhauer said there are not *things,* there is only one great "thing-in-itself," the universe. The world is one field, which burbles along in a way we will never be able to access. Schopenhauer got to this by a kind of evolutionary assumption: human senses and human mind are devices suited to keep us alive, fed, and reproducing. It is a tough world, and the human body is suited to survive in this world, not to seek truth. We are deeply hypnotized by our needs and desires; that is, the world is actually fabricated by want, by hunger, by will. In this he presaged Darwin. Schopenhauer discussed desires that remain hidden for decades, working in some covert way, active all along. In this he presaged Freud.

In 1813, just after he published his first great philosophical work "completing" the work of Kant, Schopenhauer discovered Buddhism and Hinduism. The relevant texts were making their way into Germany, and he was shocked to find in them all sorts of claims that strikingly paralleled what he had published. His most beloved find was a Latin translation of a Persian translation of the Upanishads, and ever after discovering it, we are told, he read a few pages of the book each night before bed. He once wrote that "With the exception of the original text, it is the most profitable and sublime reading that is possible in the world; it has been the consolation of my life and will be that of my death." An Eastern doubt had arrived to comfort a Western one. No wonder, then, that in his masterpiece, *The World as Will and Representation* (1818), Schopenhauer excitedly points out the similarities between his philosophy and the atheist religions of India. He says that Kant had found his own way to the doctrine of Plato, especially in the metaphor of the cave, and to the doctrines of the Hindu Vedas and Puranas. "Plato and the Indians," marveled Schopenhauer, had somehow perceived the unreal nature of the world but presented it "mythically and poetically." Kant then "made of it a proved and incontestable truth" by realizing the extent of the limitations of our cognitive apparatus.[75]

Schopenhauer did not think there was any danger that the English missionaries flooding India would be successful; it was "as if we fired a bullet at a cliff . . . the ancient wisdom of the human race will not be supplanted by the events in Galilee. On the contrary, Indian wisdom flows back into Europe, and will produce a fundamental change in our knowledge and thought."[76] He believed Christianity had brought to Europe the true values of Asia: "contempt for the world, self-denial, chastity, giving up of one's own will, that is, turning away from life and its delusive pleasures. Indeed, it taught one to recognize the sanctifying force of suffering; an instrument of torture is the symbol of Christianity." Schopenhauer found a few heroes in the Old Testament, too: he liked Swift's custom of celebrating his birthday by reading Job.[77] Schopenhauer also memorialized the martyrdoms of Socrates and Giordano Bruno, along with "many a hero of truth [who met] his death at the stake at the hands of the priests."[78] And when he cursed all philosophers, from Augustine all the way to Kant, for upholding "the prevailing national religion over philosophy," he quickly noted that "Bruno and Spinoza are to be entirely excepted," because they saw the world as one, and because they suffered horribly for truth. Wrote Schopenhauer, "The banks of the sacred Ganges were their true spiritual home."[79] The comment forces us to imagine that ebullient Italian hothead standing next to the excommunicated Dutch lens grinder on a warm afternoon by the Ganges. Connections between Eastern and Western doubt were growing strong enough to provide imaginary homes.

Schopenhauer's main point in this work was not religion, but when he mentioned it (mostly in footnotes), his tone was sharp. He wrote that believers convince themselves their religion's myths are somehow connected to its ethical code and thus "regard every attack on the myth as an attack on right and virtue." Almost comically, "this reaches such lengths that, in monotheistic nations, atheism or godlessness has become the synonym for absence of all morality." Due to this confusion, explained Schopenhauer, priests get away with murder. In Madrid alone, he reports, "The inquisition in three hundred years put three hundred thousand human beings to a painful death at the stake, on account of matters of faith. All fanatics and zealots should be at once reminded of this whenever they want to make themselves heard."[80] This was tucked in a note. In another stunning observation, also in a note, he wrote that the endless battles over the contradiction "between the goodness of God and the misery of the world," and between free will and "the foreknowledge of God," all miss one thing:

The only dogma fixed for the disputants is the existence of God together with his attributes, and they all incessantly turn in a circle, since they try to bring these things into harmony, in other words, to solve an arithmetical sum which never comes right, but the remainder of which appears now in one place, now in another, after it has been concealed elsewhere. But it does not occur to anyone that the source of the dilemma is to be looked for in the fundamental assumption, although it palpably obtrudes itself.[81]

Schopenhauer added, "Bayle alone shows that he notices this." He loved the history of doubt. He quoted Lucretius, too; and he crackled with bright psychological insights, such as: "The prayer 'lead me not into temptation' means 'Let me not see who I am.'"[82]

Schopenhauer did not believe in God, but he did not believe in science either. To him, trying to learn about reality by figuring out the laws of nature (as they appear to us) is doomed. Yet he believed there was a worthy pursuit of truth, through art. He said people think individual examples are just data, and that the real truth is some overall concept. People prefer concepts because they can be communicated, but we all know concepts are of use only if you can cash them back in as helpful, in a given "real case." What we really need is to know real cases. Schopenhauer wrote, "If perceptions were communicable there would then be a communication worth the trouble; but in the end everyone must remain within his own skin and his own skull, and no man can help another. To enrich the concept from perception is the constant endeavor of poetry and philosophy."[83] His influence on Romanticism was tremendous. As the Schopenhauer scholar Bryan Magee has put it, he helped elevate the arts "into something approaching a religion and this so suffused the general mental climate that in the remainder of the century most cultivated Europeans, and not only the romantics, attributed an unprecedented importance to art in the total scheme of things."[84] Some turned from religion to science; some turned from religion to art.

Schopenhauer's *Dialogue on Religion* addressed doubt head on. Where Hume's 1779 *Dialog Concerning Natural Religion* had borrowed its structure from Cicero's dialog *The Nature of the Gods*, Schopenhauer's *Dialogue on Religion* borrows from them both—although the relationship is looser. The characters are now Philalethes and Demopheles and they embody, respectively, the voice of philosophy and the voice of the people. Demopheles defends religious belief as "the metaphysics of the masses."[85] Religion, he argues, rouses average people from their "stupor" and points "to the lofty meaning of existence." And, he says, even if it isn't true in the same way philosophy is true, there's not much to be done about it. "For, as your friend Plato has said, the multitude can't be philosophers, and you shouldn't forget that. Religion is the metaphysics of the masses; by all means let them keep it."[86] Demopheles says this popular metaphysics is also a guide in life and a comfort in suffering and death, and even goes so far as to say that "it accomplished perhaps just as much as the truth itself could achieve if we possessed it." He then scolds his friend, saying, "Don't take offence at its unkempt, grotesque and apparently absurd form; for with your education and learning, you have no idea of the roundabout ways by which people in their crude state have to receive their knowledge of deep truths." It is "shallow and unjust," he says, to attack them.[87]

Philalethes' answer has sat at the heart of doubt ever since: "But isn't it every bit as shallow and unjust to demand that there shall be no other system of metaphysics but this one, cut out as it is to suit the requirements and comprehension of the masses?" Should these doctrines "be the limit of human speculation"? This entails "that the highest powers of human intelligence shall remain unused and undeveloped, even be nipped in the bud, in order that their activity may not thwart the popular metaphysics."[88] All the while, the members of this dominating folk-metaphysics constantly lecture a morality they do not practice: "Isn't it a little too much to have tolerance and delicate forbearance preached by what is intolerance and cruelty itself? Think of the heretical tribunals, inquisitions, religious wars, crusades, Socrates' cup of poison, Bruno's and Vanini's death in the flames! Is all this to-day quite a thing of the past?" It is good to see our old friend Julius Caesar Vanini remembered in the nineteenth century! (In another work Schopenhauer cited Vanini, with the clause: "Vanini, whom his contemporaries burned, finding that an easier task than to confute him.")[89] Getting back to his point, Philalethes says that "genuine philosophical effort, sincere search after truth" has to struggle against a system of metaphysics that has a state monopoly, "the principles of which are impressed into every head in earliest youth so earnestly, so deeply, and so firmly, that, unless the mind is miraculously elastic, they remain indelible."[90] That, he sighed, has a serious effect on the capacity for original thought and unbiased judgment, which is already weak enough without this extra handicap. It was the first real argument that doubt should be encouraged in the masses.

Demopheles would not be convinced, but neither would Philalethes: "We won't give up the hope that mankind will eventually reach a point of maturity and education at which it can on the one side produce, on the other receive, the true philosophy." No philosopher had quite afforded all humankind this respect before, but Demopheles' reply was the more usual one: "You've no notion how stupid most people are." To that Philalethes merely counters, "I am only expressing a hope which I can't give up. If it were fulfilled . . . the time would have come when religion would have carried out her object and completed her course; the race she had brought to years of discretion she could dismiss, and herself depart in peace: that would be the euthanasia of religion."[91] Philalethes also says that humanity would find real truth faster if we were all working on it. It is presented as the stronger argument, but Schopenhauer did not give it the palm of victory—nor did he follow the example of Cicero and Hume and crown a false king. Instead, he lets the two characters agree to disagree: "Let us . . . admit that religion, like Janus, or better still, like the Brahman god of death, Yama, has two faces, and like him, one friendly, the other sullen. Each of us has kept his eyes fixed on one alone."[92]

Søren Kierkegaard's is a doubt that yearns to believe. The dominant philosophers of the period were Hegelians, and Kierkegaard was enraged with what he saw as the complacent conformism of this crowd. He was more passionate than they in both his doubt and his belief. In *Fear and Trembling* (1843) Kierkegaard explains his doubt through the story of Abraham and Isaac. It is a lovely aspect of the history of doubt that interest in Abraham's doubt, faith, and actions threads through the centuries, and that here in this late period the interest is in this later period of Abraham's life. Kierkegaard said that if anyone found a man today who was taking his son someplace to murder him because a voice told him to do it, we would attempt to stop him and we would despise the fellow. If Abraham was to be lauded as the father of faith (as the Hegelians did), Kierkegaard said they must see that what he did was in fact publicly indefensible. "Humanly speaking he is insane and cannot make himself understood to anyone."[93] So if this was moral, morality cannot be merely what is communally approved. Kierkegaard said that what Abraham believed when he was sharpening his knife was that God would restore Isaac to him; otherwise, we would think him the father of resignation, not faith. How could Abraham believe such a thing, when no suggestion of it has even been made by God? Kierkegaard's answer is: "on the strength of the absurd."

Now Kierkegaard did not say that he, himself, had faith on the strength of the absurd. Instead, he said again and again that he was not capable of it. But at least, he argued, he did believe in faith. He was interested in it, he longed for it, he was sure that, although he was "happy and satisfied," those who have faith are happier. He, the doubter, defended faith against those who said they had it but were really just talking about civic politeness. Listen to Kierkegaard's wistful pride: "I have seen horror face to face, I do not flee it in fear but know very well that, however bravely I face it, my courage is not that of faith and not at all to be compared with it. I cannot close my eyes and hurl myself trustingly into the absurd, for me it is impossible, but I do not praise myself on that account." It is a powerful new formulation of the problem. He continued on to an explanation of his experience of doubt: "I am convinced that God is love; this thought has for me a pristine lyrical validity. When it is present to me I am unspeakably happy, when it is absent I yearn for it more intensely than the lover for the beloved; but I do not have faith; this courage I lack."[94] Further on: "When learning how to make swimming movements, one can hang in a belt from the ceiling . . . likewise I can describe the movements of faith but when I am thrown in the water . . . I make other movements."[95]

Toward the end of *Fear and Trembling* Kierkegaard casually drops clauses like "Nowadays, when indeed all have experienced doubt . . ."[96] And he advises caution "when one sometimes judges a doubter severely for speaking." In his opinion, even "if things go wrong, then a doubter, even if by speaking he should bring all manner of misfortune upon the world, would still be far preferable to these miserable sweet-tooths who try a taste of everything and would cure doubt without being acquainted with it, and are therefore as a rule the immediate cause of outbreaks of ungoverned and unmanageable doubt."[97] His notion of the absurd opened up a new way to imagine faith and hooked up the notion of the absurd to the problem of doubt.

Doubt was at the center of things now, for many people, and they thought it was only a matter of time before doubt changed the world. Nietzsche's famous line "God is dead" is in the "Madman" story in *The Gay Science*. It begins: "Have you not heard of that madman who lit a lantern in the bright morning hours, ran to the market place, and cried incessantly, 'I seek God! I seek God!'" People laugh at him because "many of those who did not believe in God were standing around just then." Someone asks, "Why, did he get lost?" Another says, "Did he lose his way like a child? . . . Or is he hiding? Is he afraid of us? Has he gone on a voyage? Or emigrated?" People jeered. "The madman jumped into their midst and pierced them with his glances. 'Whither is God' he cried. 'I shall tell you. We have killed him—you and I . . .'" As a result, value was meaningless:

> Is there any up or down left? Are we not straying as through an infinite nothing? Do we not feel the breath of empty space? Has it not become colder? Is not night and more night coming on all the while? . . . God is dead. God remains dead. And we have killed him. How shall we, the murderers of all murderers, comfort ourselves?[98]

When the madman's listeners merely stare at him in astonishment, he throws down his lantern and declares that he has come too soon—they do not yet realize the importance of their moment.

Nietzsche said there is no God and that in his absence we ought to look at the whole religious tradition as a farce. Most doubters throughout the West considered Judeo-Christian morality to be deeply valuable even in a secular world. Nietzsche thought the Judeo-Christian morality was inferior to that of the ancient world. It advised meekness, humility, and subservience—it was a slave religion. Machiavelli had said the same thing. Nietzsche pointed out that Christianity at first spread among the poor. It was suited to them. Nietzsche proposed a new morality of the "superman," the person who steps outside the civil bonds of the moment and transcends, through knowledge and training, to the ranks of the great of all time. One more word from Nietzsche, on doubt itself:

> Christianity has done its utmost to close the circle and declared even doubt to be sin. One is supposed to be cast into belief without reason, by a miracle, and from then on to swim in it as in the brightest and least ambiguous of elements: even a glance towards land, even the thought that one perhaps exists for something else as well as swimming, even the slightest impulse of our amphibious nature—is sin! And notice that all this means that the foundation of belief and all reflection on its origin is likewise excluded as sinful. What is wanted are blindness and intoxication and an eternal song over the waves in which reason has drowned.[99]

It's nice to hear him speak of doubt as our "amphibious nature."

ATOMISM AND ANTHROPOLOGY

Science forwarded the nineteenth-century belief that doubt had turned a corner. Modern atomic theory began with work John Dalton published in 1808: elements are composed of atoms that are specific to them, identical in size and weight, and different from atoms in all other elements; they then unite in simple numerical ratios to form compounds. By 1808, atomism was up and running without its metaphysics—that is, without its history of doubt. At the other end of the century, Marie Curie's demonstrations of radioactivity furthered atomic theory. Her first Nobel Prize came in 1903. Her father was a Polish freethinker and, although she was reared by her

Catholic mother, she left the church in her late teens. When she and Pierre married, it was a civil ceremony, which she explained as such: "Pierre belonged to no religion and I did not practice any." Across the century, atomism was no longer considered Epicurean, but it still offered an explanation of the world as self-creating that seemed wonderfully complex, but sensible. The theory lost its connection to Epicurus and Lucretius in part because that was a requirement of its being widely accepted, and in part because once atomism found a functioning mechanism, its proponents no longer felt they needed to quote the ancient authorities. It is a long jump from the idea of atoms to a thesis specific enough that it offers experimental predictability. Still, when moderns credit the ancients for atomism at all, they mention Democritus, who did, after all, make it up. But what gets missed is that for more than two thousand years atomism was thought of as the crazy/brilliant idea of Epicurus and Lucretius and their followers, embraced precisely because it explained the world as self-creating. It was a doubter's doctrine.

Beyond atoms, throughout history doubters had guessed that the world made itself, through the same kinds of repetitions, accidents, and patterns that we see around us every day. Charles Lyell's *Principles of Geology* first appeared in 1830 and argued that presently observable geological processes were enough to explain geological history. With enough time it is possible that rain, sea, volcanoes, and earthquakes could explain everything. It was not only the world that was suspected of evolving. During the French Revolution, Jean-Baptiste Lamarck came up with an idea for how the species might have done it: "the inheritance of acquired characteristics." The idea was celebrated by the revolutionaries because it said self-improvement and social change were natural, and also because it explained life on earth making itself, without God. In France and England, before Darwin, there were lively groups of political radicals, deists or atheists, who believed in some kind of evolution: "natural transformism" in France, "animal transmutation" in England.

Pre-Darwinian evolutionism was very much about politics and religion. Those who believed in transformism tended to be on the side of the doubters, and on the left politically. The new hot spots for such doubt and republicanism were the medical schools of Europe. In France, after the Revolution and Napoleon, with the restoration of the monarchy, the great anatomist Georges Cuvier took over establishment natural science, rejected Lamarckianism as revolutionary nonsense, and promoted the idea that each species was fixed in its God-given place. The deist Etienne Geoffroy Saint-

Hilaire arose as the opposition, describing a materialist evolutionary determinism.

In *The Politics of Evolution,* historian Adrian Desmond has detailed the pre-Darwinian evolutionary beliefs of various groups of English materialists, atheists, deists, and social reformers of every stripe. He tells us that Geoffroy Saint-Hilaire was ignored by clergy and gentlemen naturalists, while the medical schools, notorious for freethinkers, invented courses in order to use his books.[100] Benthamites and utopian socialists often promoted Lamarckianism. Leftists like George Jacob Holyoake praised Lamarck as supporting the "evolution" toward republicanism. Authority, too, now saw doubt as very much to do with biology: the royalist philosopher Louis de Bonald cursed both the "insane" system of d'Holbach and the species transformism of Lamarck.[101] In 1844 there was a bit of a surprise: Robert Chambers's *Vestiges of the Natural History of Creation* mixed things up by championing an idea of evolution directed by God. There were religious people and scientific people who fulminated against it as contradicting the Bible and being laughably bad science, but everyone read it. So just before Darwin, there were attempts to take on the idea of evolution even among the religious. This did not stop doubters from seeing Chambers's work as more evidence that the world made itself.

One of the most outspoken pre-Darwinian materialist transformists in England was Robert E. Grant, and Desmond shows that Grant took Darwin under his wing at medical school in Edinburgh. Grant's transformism was very much concerned with spontaneous generation, that is, with proving that life could get started with no God, and very much concerned with radical politics. Darwin rejected his mentor's transformism in those early days. When he changed his mind, it was under the influence of the economic thesis of Malthus, which stated that as long as people have more than one child per parent, there will be more people than can be supported—some will always die. Darwin saw that some creatures, too, will always die or otherwise fail to reproduce themselves, and that this was a mechanism by which nature could choose a trait and favor it, just as human beings had long done in their selective breeding of pigeons, horses, and dogs.

Famously, after Darwin saw this, he waited twenty years to publish his theory of evolution. In fact, it was only when Alfred Wallace showed up with the same theory and asked the much more established Darwin what he thought, that the shocked Darwin moved to avoid being trumped for all history. We suppose he waited because he feared the reaction. The wider world had already heard of transformism and knew of it as part of a politically radical worldview. Darwin's mother and wife were Unitarians—doubt in the myth of Genesis would not be a problem for them; it was not a big step for them to incorporate evolution into their religious world. But they would not have wanted to be associated with transformism's usual associates. When he had to publish, he went out of his way to disassociate himself from transformism's past. Darwin was able to forward the transformist revolution not only because he had figured out a mechanism by which change happened, but also because he went out of his way to be markedly conservative. To separate himself from the earlier, atheist believers in species transformism, he kept the argument away from spontaneous generation, and now and again mentioned "the Creator." Of course, he nowhere claimed evolution was a basis for socialism. Indeed, it supported the capitalist competition of the Industrial Revolution and the notion that whoever is surviving best is the fittest. Wallace, who was a well-known freethinker, socialist, and feminist, has been comparatively forgotten.

Darwin was careful what he said about God in public, but in his notebooks we find such items as "Love of the deity effect of organization, oh you, materialist!—Why is thought being a secretion of brain, more wonderful than gravity a property of matter? It is our arrogance, our admiration of ourselves."[102] There were also cautionary notes to himself: "To avoid stating how far, I believe, in Materialism, say only that emotions, instincts, degrees of talent, which are hereditary are so because brain of child resembles parent stock."[103] That is, stay off the issue of mind. The equation of brain and mind was a standard of freethinkers by the 1830s, not only in debates about transformism, but in the phrenology craze as well. Phrenology, the study of bumps on people's heads, also meant atheism to a lot of people (practitioners as well as opponents) because it was based on the idea that the mind and the brain were the same thing. That brain matter defined personality was one of the century's favorite arguments against God. In any case, atomism and the anthropology of the origins of humanity had historically been mentioned most often in letters, books, and speeches that were about doubting religion. The evidence in both cases got overwhelming, and those who brought these ideas to their new status kept comments about philosophical materialism in private letters and notebooks.

The reception of Darwinism varied. In Germany *On the Origin of Species* was translated quickly and well by Heinrich Georg Bronn, a man with an excellent reputation who did not believe the theory. He omitted the one sentence of the final paragraph that stood as the only mention of humanity in Darwin's big book on pigeons, horses, and dogs ("light will be shown on the origins of man").[104] In France *On the Origin of Species* was translated rather late and, oddly, by Clémence Royer, a woman who was a Lamarckian from long before she had ever heard Darwin's name. She explained in her extensive preface that transformism was a settled fact and that it proved there was no God. Yet even where Darwinism was not introduced by an evangelical atheist, soon enough, some evangelical atheists appeared to preach the new evolutionary gospel.

The century had produced a new breed of doubter, not as educated as the philosopher considering the universe, but not as ignorant as the villager balking at myth and razzing the portrait of the pope. The heroes of these new doubters were Galileo, Voltaire, and d'Holbach, and their mentors were Bentham, Mill, and the utopian socialists. Many were angry from past and present religious cruelties. When Darwin's book hit this crowd, it was like Christmas for the ex-Christians.

In Germany there were Vogt, Moleschott, and Büchner. Karl Vogt was a geology professor and materialist who lost his post in Germany, got a new one in Geneva, and there in 1858 translated Chambers's *Vestiges* into German. When Darwin's theory came out in 1859, Vogt immediately saw that he now had a mechanism for the material creation of the biological world, and he toured Europe's lecture circuit with the news. He gained fame with lines such as "Thoughts come out of the brain as gall from the liver, or urine from the kidneys."[105] Jakob Moleschott was a professor of physiology and the son of a Dutch freethinker. He read Feuerbach, started writing about the material basis of humanity, and got famous for the comment "no thought without phosphorus"; then got famous again when he advocated cremation so that bodies could return to nature. His name became a catchword for science and doubt. Vogt was the public preacher, and Moleschott a powerful symbol, yet it was Ludwig Büchner's *Force and Matter* that, for many people, was the century's key book on science and belief. It was published four years before *On the Origin of Species,* but later editions incorporated Darwinism into the argument. The point of it, in all its editions, was that there are only force and matter: the universe is eternal, infinite, and self-propelling; and thought is entirely dependent on matter. He heated things up by asserting that the universe has no purpose, and that it will eventually be destroyed. This was all good, soothed Büchner, for it was better to be "a proud and free son of Nature" than a "humble and submissive slave of a supernatural master."[106]

In France, a group of anthropologists played this role. The group had actually met as a secret freethinkers' society, and when they read Clémence Royer's introduction to *On the Origin of Species* they became anthropologists to use this new weapon against the Church. Along with Royer, this atheist group joined Paul Broca's Society of Anthropology and replaced its general positivist credo with an insistence on materialism and atheism. Royer and the other freethinking anthropologists—Gabrielle de Mortillet, Charles Letourneau, André Lefèvre, Eugene Véron, and Abel Hovelacque were the key figures—learned anthropology as they went along, and made some contributions to the field, but the promotion of atheism and materialism was always their chief aim. As I've explored in my *End of the Soul,* the greatest example of this was the Society of Mutual Autopsy.[107] They wanted to show the church that there was no soul by proving a direct relationship between a person's material brain—its shape, form, and weight—and his or her personality and ability. To that end, they donated their brains to one another and carried out the autopsies over a period of some thirty years. Broca, who was a freethinker himself, had found the first lasting mind-brain connection: damage to a particular spot on the brain correlates with particular speech problems—such impairment is still called Broca's aphasia. While hunting for more connections, the Society of Mutual Autopsy created a secular, even atheistic, version of Catholic death rituals, including a materialist deathbed scene, the keeping of relics, and a chance for unbelievers to confer their very bodies, after death, to science instead of religion.

Meanwhile, in England Thomas Huxley was so loud and tenacious in his support of the theory of evolution that he was nicknamed Darwin's Bulldog. He was also the person who coined the word *agnosticism.* The term *naturalism* was not good enough, he said, because a naturalist could espouse materialism, idealism, determinism, or libertinism. Huxley showed his knowledge of the history of doubt in his rejection of the term, saying that someone known to profess naturalism "may be a pure empiric, or a believer in innate ideas; a Platonist or an Epicurean. Doctrines as widely different as the pantheism of Spinoza and the so-called atheism of the Buddhist are forms of 'Naturalism.'" Agnosticism, Huxley explained, was less certain than any of these doctrines. What he then described was Skepticism. Where did he get the idea? As he put it: "Before now, I have had occasion to speak of the pedigree of Agnosticism; and I have vainly endeavored to placate its enemies by showing that it is really no child of mine, but that it has a highly respectable lineage which can be traced back for centuries."

He explained that he first heard about it in a passage by Sir William Hamilton (published in 1829, but not read by Huxley until 1840), "which, so far as I am concerned," wrote Huxley, "is the original spring of Agnosticism." The quoted passage was this: "Philosophy . . . is impossible. Departing from the particular, we admit that . . . our knowledge, whether of mind or matter, can be nothing more than a knowledge of the relative manifestations of an existence, which in itself [we] recognize as beyond the reach of philosophy." Huxley explained:

> When, long years after these words had made an indelible impression on my mind, I came across the *Limits of Religious Thought* (which I really did read, though the fact that I once unfortunately spelt Mansel with two l's has been held by a candid critic to be proof to the contrary), I said to myself "Connu!"; and the thrill of pleasure with which I discovered that, in the matter of Agnosticism (not yet so christened), I was as orthodox as a dignitary of the Church, who might any day be made a bishop, may be left to the imagination.[108]

Henry Longueville Mansel was a student of Hamilton, the chief philosophical Skeptic of the period, and Mansel applied that Skepticism to religion in his famous *Limits of Religious Thought*. So it really was straight out of Skepticism that agnosticism sprang! In that first generation it had Skepticism's strict denial of judgment. Huxley said that if you've never met any creatures from Saturn, and have no indication that they exist, you may not necessarily believe they don't exist either. Agnostics "totally refuse to commit" to the denial of the "supernatural." But he insisted on his right to doubt: "the future of our civilization . . . certainly depends on the result of the contest between Science and Ecclesiasticism which is now afoot."[109] Huxley celebrated Descartes as the first to train himself to doubt: "The enunciation of this great first commandment of science consecrated Doubt. It removed Doubt from the seat of penance . . . to which it had long been condemned, and enthroned it in that high place among the primary duties."[110] Like Taylor and Mill, Huxley finds doubt a duty. He celebrated other historical doubters, too, writing a book on Anthony Collins, who he called a "Goliath of Freethought." Speaking of priests and miracles, Huxley wrote, "That true man of letters, Lucian, had something to say about these people and their dupes which is well worthy of modern attention."[111]

Huxley was moderate in comparison to some. The nearly forgotten scientist and doctor Henry Bastian's overt atheism in the debates over Darwinism and spontaneous generation marginalized him in his profession and in our historical memory of him. A recent book of documents compiled and explained by historian James Strick follows his dramatic story.[112] Bastian insisted that the understanding of Darwinian evolution that was taught and accepted should include the idea of spontaneous generation to cleanly announce that science dispelled the need for God. Darwin stayed out of it. It was Huxley, working hard to keep evolution respectable, who was chiefly responsible for the eventual rejection of Bastian's materialism.

Many learned of agnosticism from the Social Darwinist and early sociologist Herbert Spencer. Spencer liked Huxley's term: he was an agnostic and his famous approach was "Contempt before investigation." On the matter of God, he believed there is nothing to investigate so the question is best ignored. Spencer had a Benthamite, anticlerical uncle who influenced him a good deal, but we may also note that he was the eldest of nine children and the only one of them to survive infancy—that parade to tiny graves could make anyone a little circumspect of anything unproved.

THE SECULAR STATE

Especially toward the end of the nineteenth century, there were activists whose primary or even sole interest was the freedom of doubt, and secular movements arose all over the world. We'll take a quick tour starting with France, for it was France in the second half of the century that raised up the greatest din of anticlerical rebellion the world had ever known. The Frenchman Auguste Comte (1798–1857) did something that had not been done in a long while: he pitched a populist, deeply secular, antireligious "religion." It was called positivism and it came to dominate European (especially French) attitudes for much of the century. The idea was that human history comprised three successive stages. They were the Theological stage, "in which free play is given to spontaneous fictions admitting of no proof"; the Metaphysical stage, "characterized by the prevalence of personified abstractions or entities"; and last, the Positive stage, "based upon an exact view of the real facts of the case."[113] The hubris is kind of funny, but to Comte's mind, that's the end of history: "The third is the only permanent or normal state."[114] Comte was magisterial in his conviction that these three stages are the meaning of all history. He insisted we needed a science of society and coined the word *sociology* and came up with a few of the field's early tenets. Comte became a hero to generations of people because he furnished a secular credo: positivism would replace religion.

Comte rejected atheism. In his *General View of Positivism* he wrote that "The fact of entire freedom from theological belief being necessary before the Positive state can be perfectly attained, has induced superficial observers to confound Positivism with a state of pure negation."[115] He then said that atheism had at one time been "favorable to progress" but was not anymore. "Atheism," he said, "even from the intellectual point of view, is a very imperfect form of emancipation; for its tendency is to prolong the metaphysical stage indefinitely," because it still talks about theological problems "instead of setting aside all inaccessible researches on the ground of their utter inutility." Positivism was about studying *how* instead of *why.* "Now this is wholly incompatible with the ambitious and visionary attempts of Atheism to explain the formation of the Universe, the origin of animal life, etc." In his opinion, if people "persist in attempting to answer the insoluble questions which occupied the attention of the childhood of our race," well, then, "by far the more rational plan is to do as was done then, that is, simply to give free play to the imagination."[116] It's a surprising position.

Comte explained that for his part, if he had to guess, he found the world "far more compatible with the hypothesis of an intelligent Will than with that of a blind mechanism." He believed that it was only "the pride induced by metaphysical and scientific studies" that made atheists "modern or ancient." Comte seems not to have known, or cared, about the history of doubt. He did not like the people he met who called themselves atheists, claiming the doctrine was "generally connected with the visionary but mischievous tendencies of ambitious thinkers to uphold what they call the empire of Reason. . . . Politically, its tendency is to unlimited prolongation of the revolutionary position."[117] Atheism had this political dimension in France. Comte's followers loved him for providing a calm yet modern and rationalist doctrine, devoted to progress and allowing one to skip church.

When people thought of the classic Comtian, many of them had in mind a character in Flaubert's *Madame Bovary* (1856): the village pharmacist

Monsieur Homais. Homais claimed: "I do have a religion, my religion, and I have rather more than that lot with their jiggery-pokery. . . . I believe in the Supreme Being . . . but I don't need to go into a church and kiss a lot of silver plate, paying out for a bunch of clowns who eat better than we do!"[118] He said his was the same God as that of Socrates, of Franklin and Voltaire, and he could not, therefore, "abide an old fogey of a God who walks round his garden with a stick in his hand, lodges his friends in the bellies of whales, dies with a loud cry and comes back to life three days later"; all this was "absurd" and "completely opposed . . . to every law of physics." He added that "priests have always wallowed in squalid ignorance, doing their utmost to engulf the population along with them."[119] At the end of the novel, the village priest and M. Homais argue in a room where a body lies. As we see the young widower look in on his dead wife over and over, we hear these two in the background: "Read Voltaire! said the one; read d'Holbach! read the Encyclopédie," and the priest retorted with a call to read the modern Christian tomes.[120] For Flaubert, both priest and Comtian had lost connection to the mysteries of life, love, and death.

Over the course of the century, the French increasingly saw secularism and democracy as locked in a pitched battle with church and monarchy. Church and monarchy dominated the century. When a lasting French democracy was set up at the end of the nineteenth century, leading republicans—often medical doctors, and often Comtian positivists, if not atheists—zealously secularized the French state. Across the century, in France, baptism and confession declined and civil marriages (by the town mayor) increased. People did not even turn to the Church to handle their deaths: the civil burial became the great banner of republicanism and secularism. Not only did the republicans work to get the priests out of politics, they transformed the school system in 1883 so that there was free, mandatory, secular education for children, to ensure "countless young reserves of republican democracy, trained in the school of science and reason."[121] Historian René Rémond wrote of the schoolteacher as "apostle of the new religion, an officiate of the cult of reason and science."[122] The early Third Republic passed a battery of laws limiting the clergy's place in the government and public institutions and ending some remaining privileges—seminarians, too, now had to perform military service. Ernst Renan's secular, novelized *Life of Jesus* hit France the way Euhemerus's *Sacred History* hit the ancient world. Just as Zeus had been a man who lived and died in Crete, Jesus had been a fellow who had a bit of an adventure in Galilee.

In the fevered pitch of the religion-and-science wars of the last decade of the century, Emile Durkheim picked up the word Comte had coined, *sociology*, and in its name made an observation that calmed things down a good bit. Durkheim said that avid atheists were wrong for arguing that religion is false. Religion is not about knowing the world factually but about feelings and experience. He accepted the skeptical claim that we cannot know reality, and Kant's notion that our minds shape our experience of reality such that we are ignorant of reality itself. Durkheim said that it was society that created the shared dreamworld of likenesses, categories, and meanings, and that it did so, primordially, through religion. As he wrote in an essay of 1898, the moral obligations we feel that seem like they come from outside ourselves, really do—they come from society. "This obligation," he explained, "is the proof that these ways of acting and thinking are not the work of the individual but come from a moral power above him, that which the mystic calls God but which can be more scientifically conceived."[123]

All those strange internal-yet-external forces that had always seemed to be the properties of God were, really, the properties of society. There was no more reason to attack religion as inherently bad, soothed Durkheim. Durkheim would not argue with religion by throwing reason at it, and instead took religion as a real human phenomenon and used reason to understand it. Durkheim's father was the chief rabbi of their region, and his grandfather and great-grandfather had also been rabbis, but the ideas here were especially to do with French Catholicism. Intellectuals in France were able to ease the tension in part because the secularization of the state was well under way. Catholic Italy also went through a period of heightened anticlericalism in which science was a political banner. In Italy a "Bruno mania" was part of the intellectual enthusiasm of cultured Italians. Elsewhere, doubters of other religions had other political concerns.

In Russia, from the 1830s on, German philosophy and French socialism began to dominate intellectual life. By the 1840s the "Westernizers," who followed Hegel, Proudhon, and Fourier, were calling for a far-reaching rejection of all Russian traditions, institutions, and social habits. Soon their central texts included the new European Utilitarian and materialist books. By the 1860s the heat of these issues had produced an intellectual generation called the Nihilists—a term that first appeared in Ivan Turgenev's *Fathers and Sons* in 1861.[124] The fathers in this novel all spoke of progress and humanitarianism in great waves of vague sentiment and optimistic romanticism. The sons all spoke of science, and scolded their fathers for speaking in theories while ignoring the actual suffering going on around them. Whereas the Westernizers had fed on the meaning-laden historical visions of Hegel and idealism, the Nihilists were raised on Feuerbach, Comte's positivism, and the popular German materialists: Büchner, Moleschott, and Vogt. John Stuart Mill also had a big influence. Although always haunted by a Russian preference for Lamarckianism (competition was not the only thing that could run a group or lead to progress), Darwinism still came as a great windfall for doubters, as it did everywhere else. Among the Russian Nihilists, the mood of doubt and emphasis on science were linked with anthropology and sociology. The radicalization of Russian intellectuals took place through the influence of the Nihilists, and it was their materialism and socialism that developed into the Marxist-Leninist doctrines of the twentieth century.

Dostoyevsky was attracted to these ideas in his youth, but later was too pro-Russian to abide the Westernizers and too spiritualist to bear the Nihilists. In his novel translated as *The Devils* or *The Possessed,* he attacks the Nihilists, and in the planning stages, the book was to be called *The Atheist.* In one passage of the book, an army officer goes mad and attacks his commander. In the investigation it comes out that the officer had recently smashed up his landlady's little shrine of Christian icons. In its place he set up works by Vogt, Moleschott, and Büchner, like a trio of Bibles on stands, and then burned a wax church candle in front of each. He turns out to be a political radical, and violent, too. Dostoyevsky doubted much about religion, but he did not like the look of the moral world without God.

The German David Friedrich Strauss got all Europe's attention with his *Life of Jesus* (1835) by discussing the miracles of the Bible not as mistaken perceptions, but as meaningful myth.[125] Strauss cited Spinoza a lot, but his work was also very original and modern: he advanced the scholarly techniques for teasing out the likely historical realities in the Gospels. Secularization advanced so that there was less religion in the other disciplines, and more of the other disciplines in the study of religion.

As for Britain, in 1851 Harriet Martineau translated Comte into English and condensed his six-volume work to two, to his approval and to tremendous success. In 1859 Leslie Stephen was a young tutor at Cambridge and a newly ordained priest, but soon after he began reading Mill, Comte, and Kant he had doubts. In 1862, having declined to take part in the chapel services, he was asked to resign his tutorship. He left the church as well, and married a freethinking woman. In 1873 he published *Essays on Free Thinking and Plain Speaking,* a book that made him a famous doubter longing for real spiritual fulfillment that he said could not come from religion, since religion was impossible to believe.[126] Still, reading Stephen's article "An Agnostic's Apology" in the *Fortnightly Review* was the first time many people ever came across the term *agnostic.* After that, use of it was in high vogue. Stephen's biography of George Eliot (real name Marian Evans and a brilliant doubter—as I'll discuss below) appeared in 1902; that same year he was knighted for his editorship of the *Dictionary of National Biography,* which did not do the status of doubt any harm. He died in 1903, but a posthumous work on Hobbes appeared in 1904. Sir Leslie was also survived by his daughter, the novelist Virginia Woolf. She was less polemical on this issue than her father, but her books are full of searching agnostic and atheist characters. In unpublished autobiographical essays, she wrote things such as "certainly, emphatically, there is no God," and in a private letter: "I read the book of Job last night—I don't think God comes well out of it."[127]

The greatest idol of late-century British doubt was Charles Bradlaugh, a big man with an imposing manner and a lot of power to his speech. His motto was "Thorough!" which meant he was quite an atheist. His lectures on unbelief were among the most popular of the era. Consider this conversation in Bradlaugh's *Doubts in Dialogue:*

CHRISTIAN PRIEST: At least, belief is the safe side. When you die, if your unbelief be right, there is an end of you and of all your heresy; and if it is wrong, there is eternal torment as your sad lot.

UNBELIEVER: Hardly so. If I am right, my unbelief will live after me, in its encouragement to others to honest protest against the superstitions which hinder progress.

PRIEST: But you, at any rate, may be wrong, and belief is, therefore, safest for you.

UNBELIEVER: Which belief? Must I accept alike all creeds?

PRIEST: No; that is not possible. You are asked to accept the true Christian faith.

UNBELIEVER: Why not the true Jewish faith?

PRIEST: A new dispensation was given through Jesus.

UNBELIEVER: Why not the true Mahommedan faith?

PRIEST: Mahommed was an impostor.

UNBELIEVER: About two hundred millions of human beings now believe that he was the prophet of God, and that the Koran is a divine revelation.

PRIEST: He was a false prophet. His pretense that the Koran was revelation was an imposture.

UNBELIEVER: Then it would not be safe for me to believe in Mahommed?

PRIEST: Certainly not; you must believe in Christ and in the Gospels.

UNBELIEVER: Would it not be enough to believe in Buddha, and the blessing of eternal repose in Nirvana?

PRIEST: Buddhism is the equivalent of Atheism. Nirvana is another word for annihilation.

UNBELIEVER: But some four hundred millions are Buddhists, and the character of Buddha is placed very high.

PRIEST: The true faith is that in Jesus, and in him crucified.

UNBELIEVER: Do you mean the man Jesus in whom the Unitarians believe?

PRIEST: Unitarians! Do you not know that there is a special canon of the law-established Church against the damnable and cursed heresy of Socinianism? It is belief in Jesus as God, the second person in the Holy Trinity.[128]

Along with this cosmopolitan critique, Bradlaugh wrote that "the Atheist does not say 'There is no God,'" but says: "'I know not what you mean by God; I am without idea of God; the word *God* is to me a sound conveying no clear or distinct affirmation. I do not deny God, because I cannot deny that of which I have no conception' especially when even those who believe in the thing cannot even define it." He adds that if *God* is defined to mean an existence other than the human kind of existence, "then I deny 'God,' and affirm that it is impossible such 'God' can be. That is, I affirm one existence, and deny that there can be more than one."[129] Bradlaugh wrote that a large part of his atheism was based in having glimpsed Spinoza's *Ethics* in secondary sources and found himself in agreement that the universe is one—it has no other, mystical part.[130]

Bradlaugh's is a good story: As a young man he had been rejected by his family because he doubted the Gospels. In his travels he met free-thought enthusiasts and embraced atheism; he got a job clerking for a lawyer and began a side career as a free-thought lecturer. His audiences were groups that had been followers of Robert Owen but were being retooled by George Holyoake as secular clubs. In 1858 Bradlaugh replaced Holyoake as president of the London Secular Society and was soon made editor of *The National Reformer,* a paper dedicated to atheism, democracy, and birth control. He founded the National Secular Society and became its first president and started a campaign to show the legal injustice of court oaths. He won: in 1869 a law granted the right of atheists to "affirm" in court cases rather than to swear. Meanwhile, he led the movement to get rid of the monarchy and was powerful enough that in 1873 the *New York Herald* proclaimed him "The Future President of England." Queen Victoria, however, rallied and returned to popularity; that ended that, but it was a gesture of democratic zeal by a great doubter. In 1880 Bradlaugh was elected to Parliament. When invited to take the standard oath before taking his seat, he asked to affirm instead—it was his famous issue—but a committee announced that the right to affirm did not extend to Parliament. Bradlaugh said, fine, he would take the oath, but another committee insisted that his well-known atheism prevented this: he should "affirm" and vote "under pain of statute"—meaning subject to penalties for voting without taking the oath. Legal proceedings against him began as soon as he voted the first time and lasted six years. His seat was empty meanwhile, but he was reelected in three by-elections. At each juncture he would try to take the oath (and his seat) but other members would not let him. He was once thrown bodily out of the Palace of Westminster, and he was the last person in history imprisoned in the Clock Tower.

Finally, in 1886 the Speaker helped him take the oath quickly, before anyone could object, and henceforth Bradlaugh was able to participate. He grew to be a very well respected member of Parliament, encouraging public vaccination, ending aristocratic privileges, and arranging systems of support for the working class. Among these wide-ranging concerns, he introduced and successfully championed the 1888 Oaths Act, so that anyone could affirm as an alternative to any oath. He was also famous for his support of India. The year 1880 saw the foundation of Free-Thought International, in Brussels, and its first conference; Bradlaugh was one of its founders, along with Büchner, Spencer, Vogt, Royer, and Moleschott.

In the early 1870s Bradlaugh met Annie Besant, another powerful figure in the history of doubt, and they became close collaborators. Annie Wood had been married off at age nineteen to a preacher named Frank Besant, and they had two children. Annie's increasing religious doubt, culminating in her refusal to take communion, led Frank to kick her out and, in 1873, a legal separation resulted. After her writing caused further scandals, the husband ended up with both children. She became a member of the National Secular Society and the Fabian Society, where Sidney and Beatrice Webb and George Bernard Shaw talked over socialism (leading to the founding of the Labour Party). Besant tells us in her autobiography that while married she'd stood at her husband's pulpit when the church was empty, "to try how it felt," and had liked it. Once on her own, she put her interest in public speaking to use supporting the causes of free thought, national education, women's right to vote, and birth control. She famously organized the matchworkers' union to protect the girls of that trade, and had a major impact on labor power and women's rights. She and Bradlaugh coedited the *National Reformer* now, and they wrote a pamphlet on birth control for which they were brought to trial but acquitted. Together they also wrote *The Freethinker's Text Book,* indicting Christianity for opposing "all popular advancement, all civil and social progress, all improvement in the condition of the masses." Progress has come now, they said, because the "failing creed has lost the power to oppose."[131]

In 1876 Besant wrote *The Gospel of Atheism,* in which she said, "An Atheist is one of the grandest titles . . . it is the Order of Merit of the World's heroes . . . Copernicus, Spinoza, Voltaire, Paine, Priestly." In 1889 Annie Besant was elected to the London School Board with a fifteen thousand majority over the next candidate, and took the landslide as mandate for profound reform of the local schools. Beyond her educational initiatives, she was famous for instating free meals for poor students and free medical exams for every child in elementary school. Later, Besant read Hindu and Buddhist texts and these changed her focus profoundly. She became a leader of the Theosophical Society, which sought to combine the philosophies and religions of the East and the West into something modern people could believe. This all soon took on a spiritualism that included the possibility of communication with the dead. European atheists tend to see her as having abandoned them at that point, but she never betrayed the wider cause of doubt. Besant went to India, became involved in Indian nationalism, and in 1916 established the Indian Home Rule League of which she became president; meanwhile, her *Gospel of Atheism* and *Freethinker's Text Book* were very popular among the intelligentsia in India and Sri Lanka. She wrote a library of theosophist books on atheism, Buddhism, yoga, Hinduism, and the nature of the soul, and in 1923 translated the Bhagavad-Gita, among other key texts of Indian religion. One of her many books was *Giordano Bruno: Theosophy's Apostle in the Sixteenth Century* (1913). She thus brought to India the doubt of the West and brought Eastern thought to the West. Besant lived out much of her life in India, dying there in 1933.

Consider also Hypatia Bradlaugh Bonner (1858–1934), daughter of Charles and named for the great Hypatia of Alexandria. Charles Bradlaugh named his daughter after the ancient philosopher in praise of philosophy. Hypatia Bradlaugh Bonner was a wonderful freethinker: she took over her father's affairs while he was under prosecution (with Annie Besant, for their campaign on birth control), and meanwhile studied inorganic chemistry and animal physiology at London University. She later taught these subjects at her father's Hall of Science, sponsored by his National Secular Society. Hypatia met her husband there (he taught math). Hers was a quiet life, but when her father died in 1891 and there were rumors he'd had a deathbed conversion, Hypatia was riled: she wrote retorts, successfully sued one party, and eventually published a pamphlet titled *Did Charles Bradlaugh Die an Atheist?* She began to campaign against the death penalty, edited a complete edition of Paine's *Rights of Man* and *Age of Reason,* and launched a journal called *The Reformer* with the motto "Heresy makes for progress." She was friends with Ernestine Rose toward the end of the older woman's life. After forty years of public service, Hypatia was appointed Justice of the Peace for London in 1922 and sat on the bench until 1934. When she died, a testament offered a rebuttal to any rumors: "Now, in my seventy-eighth year, being of sane mind, I declare without reserve or hesitation that I have no belief, and never have had any belief, in any of the religions which obsess and oppress the minds of millions of more or less unthinking people throughout the world." She added, "Away with these gods and godlings; they are worse than useless. I take my stand by Truth."[132]

Robert Ingersoll, sometimes called "the Pagan prophet," was the best-known doubting lecturer in the United States. Born in New York, he had been required to attend his minister father's two Sunday sermons; on his own he read Epicurus, Zeno, Voltaire, and Tom Paine. Ingersoll saw his naturalist morality based on happiness as directly indebted to Epicurus rather than to the Utilitarians. He called himself an agnostic, writing, "Let us be honest. . . . Let us have the courage and the candor to say: We do not know."[133] There was American pluck in it: "If by any possibility the existence of a power superior to, and independent of, nature shall be demonstrated, there will be time enough to kneel. Until then let us stand erect."[134] Ingersoll married Eva Parker after having been comforted to learn there were works of Paine and Voltaire on her family's bookshelves, and that Eva's grandmother was an atheist and her parents deists. Ingersoll worked with Mortimer Bennet, another important doubting lecturer, who reported having been converted to atheism as he read Paine's *Age of Reason.* Bennet started the atheist paper *The Truth Seeker.* When he was arrested under the Comstock laws for mailing birth control information, Ingersoll argued in his defense that the Bible was full of sexually explicit passages, so it, too, should be barred from the mail system.

In the United States there were more reforming doubters than could be listed, and many who incorporated doubt in their attack on political wrongs. Frederick Douglass, for example, did not speak much on religion, but here is something he wrote in 1852:

> The church of this country is not only indifferent to the wrongs of the slave, it actually takes sides with the oppressors. . . . For my part, I would say, welcome infidelity! Welcome atheism! Welcome anything! in preference to the gospel, as preached by these Divines! They convert the very name of religion into an engine of tyranny and barbarous cruelty, and serve to confirm more infidels, in this age, than all the infidel writings of Thomas Paine, Voltaire, and Bolingbroke put together have done![135]

Douglas goes on to say that the antislavery movement will cease to be an antichurch movement as soon as the churches join the antislavery movement. So far, he howls, "YOUR HANDS ARE FULL OF BLOOD."[136]

Consider also the atheist, women's rights activist, and eventual anarchist, Voltairine de Cleyre. Her father had named her after his favorite author but apparently forgot to give Voltaire's work to the girl. Instead, she was sent to Catholic boarding school, where she questioned her faith in terrified solitude—stunning when her very name should have been company. She became a renowned atheist lecturer. De Cleyre had a sense of doubt's history, but it was a different one. At a speech on Easter, 1896, she explained, "Whether it be the festival of a risen Christ, or of the passage of Judah from the bondage of Egypt, or the old Pagan worship of light, 'tis ever the same—the celebration of the breaking of bonds. We, too, may allow ourselves the poetic dream."[137] She then gave a list of "resurrected" heroes, dominated by doubters: Hypatia, Frances Wright, Ernestine Rose, Harriet Martineau, and Lucretia Mott. De Cleyre was a marvelous character, famed as strikingly beautiful, brave, and principled. She made her living in the public advocacy of atheism, as well as women's rights and labor rights.

There were many more. Etta Semple and Laura Knox edited *The Free-Thought Ideal* in turn-of-the-century Kansas, and from 1897 Semple was president of the Kansas Freethought Association and later vice president of the American Secular Union. She also did so much good in her community—she opened a hospital and sanitarium for the poor, for example—that when she died the banner headlines mentioned only her humanitarianism. In one of her editorials, "Liberty of Conscience Is All That We Ask," Semple wrote, "If I deny the existence of a God—if I deny the idea of a gold paved city with pearly walls and jasper gates somewhere out of knowledge and space and prefer to die and trust to the unfaltering laws of nature—if, in plain word I don't want to go to heaven, whose business is it but my own?"[138] Consider also Helen Hamilton Gardener, who the *New York Sun* called "Ingersoll done in soprano" and the *Chicago Times* called "the pretty infidel."[139] Her book *Men, Women, and Gods* included lines such as "I do not know the needs of a god or of another world. . . . I do know that women make shirts for seventy cents a dozen in this one."[140] In 1920 she was appointed to the Civil Service Commission—the highest office a woman had yet occupied in the federal government. The great abolitionist and women's rights campaigner Lucy N. Colman wrote that she herself gave up the church "more because of its complicity with slavery than from a full understanding of the foolishness of its creeds."[141] She used to date her articles in *The Truth Seeker* strangely: 1887 was 287. It marked the martyrdom of Bruno, 287 years earlier.

THE POETS

There were doubters in many of the arts: in 1885 the great actress Sarah Bernhardt, for instance, was quoted answering the question *Do you pray?* by saying "Never. I'm an atheist."[142] But of all the doubting artists, poets may be the ones who most directly wrestle their doubt in their art. Most doubting poets of the nineteenth century are as far from polemic as can be; more like Cheshire Cats than Red Queens. That's why the earliest one is such an unusual story. In 1811 one of the most beloved poets in the English language, Percy Bysshe Shelley (1792–1822), was in college at Oxford. Shelly was nineteen. He and his friend T. J. Hogg wrote what has often been described as the first published attestation of atheism in Britain. We have seen that there was precedent, and indeed England knew a handful of isolated atheists and other radical doubters in this same period.[143] Still, there had not been anything like it. It was called *The Necessity of Atheism*. They circulated it to the heads of colleges and to bishops and were summarily kicked out of school. Two years later, Shelley expanded the pamphlet and included the whole thing as a note to his long poem *Queen Mab*.

The book follows through on its titular confidence, but it begins with a note bearing the book's one caveat, tucked weirdly under the section heading

"There Is No God"; it reads: "This negation must be understood solely to affect a creative Deity. The hypothesis of a pervading Spirit coeternal with the universe remains unshaken."[144] The book henceforth ignores that spirit. It begins as Cicero began: "A close examination of the validity of the proofs adduced to support any proposition is the only secure way of attaining truth, on the advantages of which it is unnecessary to descant: our knowledge of the existence of a Deity is a subject of such importance that it cannot be too minutely investigated . . ." For Shelley, the way to do that is to examine the proposition of God while remembering that the senses are our strongest link to truth, that our minds are a somewhat weaker link, and that other people's words are the worst. How does God hold up according to the evidence of the senses? It's not good. "If the Deity should appear to us, if he should convince our senses of his existence, this revelation would necessarily command belief. Those to whom the Deity has thus appeared have the strongest possible conviction of his existence. But the God of Theologians is incapable of local visibility." That last switch is an adroit reminder that the biblical stories and the philosophically argued God (variously distant, unimaginable, and unmoving) do not mesh. It is a major rupture, generally ignored in the history of both belief and doubt.

Our next link to truth, reason, also fails to prove God for Shelley: he dismissed the idea that the "created" universe proves a creator, since "we may reasonably suppose that it has endured from all eternity." Echoing Aristotle here, Shelley went on to echo Cicero, writing, "We must prove design before we can infer a designer"; and then to echo Hume: not only do we not need a first cause, we don't even know what *cause* is. Our weakest link to truth, witnesses other than ourselves, fared worst of all in the attempt to prove God's existence. Anyway, a God who cursed those who dared disbelieve could not *be,* explained Shelley, because it is folly to think you can legislate belief. Shelley also said it is true we do not understand the generative force of life, but it doesn't clarify anything to imagine that this force is eternal, omniscient, and omnipotent. His conclusion: "Hence it is evident that, having no proofs from either of the three sources of conviction, the mind cannot believe the existence of a creative God. . . . Every reflecting mind must acknowledge that there is no proof of the existence of a Deity."

Shelley introduced many familiar arguments: God was invented by those in power to control the masses; belief is based on blind custom and obedience; believers themselves, "who make a profession of adoring the same God," disagree on every aspect of him and deny each other's proofs.

But there were some fresh notions here. Shelley added his century's sense of human mastery: "Is there a country on earth where the science of God is really perfect? Has this science anywhere taken the consistency and uniformity that we see the science of man assume, even in the most futile crafts, the most despised trades?" Shelley had a clever take on believers in the philosophical, distant God: "The being called God by no means answers with the conditions prescribed by Newton; it bears every mark of a veil woven by philosophical conceit, to hide the ignorance of philosophers even from themselves. They borrow the threads of its texture from the anthropomorphism of the vulgar." But, he assures, if ignorance of nature gave birth to gods, knowledge of nature destroys them. "In a word, [man's] terrors dissipated in the same proportion as his mind became enlightened. The educated man ceases to be superstitious." This was 1811; the chutzpah was remarkable.

Shelley paraphrased d'Holbach's French, musing of God: "If he is reasonable, how can he be angry at the blind, to whom he has given the liberty of being unreasonable? If he is immovable, by what right do we pretend to make him change his decrees? If he is inconceivable, why occupy ourselves with him? IF HE HAS SPOKEN, WHY IS THE UNIVERSE NOT CONVINCED?"[145] He quoted Pliny the Elder's *Natural History*, concluding that "The enlightened and benevolent Pliny thus Publicly professes himself an atheist." He ended with a quotation by Spinoza in Latin, saying "God and Nature are one and the same."[146]

Shelley believed that life is a different thing than matter, but where others used this as an argument for immortality, he wondered how you could leap from "different than matter" to immortal. In an essay called "On a Future State," Shelley wrote:

Suppose . . . that the intellectual and vital principle differs in the most marked and essential manner from all other known substances. . . . In what manner can this concession be made an argument for its imperishability? All that we see or know perishes and is changed. Life and thought differ indeed from everything else. But that it survives that period, beyond which we have no experience of its existence, such distinction and dissimilarity affords no shadow of proof, and nothing but our own desires could have led us to conjecture or imagine.[147]

The essay ends with the claim that only "this desire to be forever as we are" and the fear of an "un-experienced change," which all the animated and inanimate "combinations of the universe" undergo, is "the secret persuasion which has given birth to the opinions of a future state."[148]

Shelley advocated free thought his whole short life, fighting for tolerance and for doubt. In his "Refutation of Deism," he wrote that there can be no middle ground between accepting revealed religion and disbelieving in the existence of God. Drawing on Epicurus, Locke, and Hume, he concluded that "the existence of God is a chimera."[149] His *Queen Mab* is one of the great dialog poems. Its epigrams include Lucretius's Epicurean denial of the gods and Voltaire's "Ecrasé l'infame!", and its whole point is to critique revealed religion and champion doubt. As one famous scene begins, Spirit speaks of a burning:

SPIRIT:
I was an infant when my mother went
To see an atheist burned. She took me there.
The dark-robed priests were met around the pile;
The multitude was gazing silently;
And as the culprit passed with dauntless mien,
Tempered disdain in his unaltering eye,
Mixed with a quiet smile, shone calmly forth;
The thirsty fire crept round his manly limbs;
His resolute eyes were scorched to blindness soon;
His death-pang rent my heart! the insensate mob
Uttered a cry of triumph, and I wept.
"Weep not, child!" cried my mother, "for that man
Has said, There is no God." [150]

It is often thought that Shelley was honoring Bruno. The Fairy answers:

FAIRY:
There is no God!
Nature confirms the faith his death-groan sealed.
. . . Let every seed that falls
In silent eloquence unfold its store
Of argument; infinity within,
Infinity without, belie creation;
The exterminable spirit it contains
Is Nature's only God; but human pride

Is skilful to invent most serious names
To hide its ignorance.
The name of God
Has fenced about all crime with holiness,
Himself the creature of his worshippers . . .

The Fairy continues on to say that "priests dare babble of a God of peace, / Even whilst their hands are red with guiltless blood," that as they murder they uproot "every germ / Of truth" and make the earth into "a slaughter-house!" It's worth seeing this particular aspect of what people meant when they said they were reading Shelley. Yet more powerful than his drama about what the universe is not, was his verse about what the universe really was, to his eyes. His sonnet "Ozymandias" is one of doubt's great poems. It begins: "I met a traveler from an antique land / Who said: 'Two vast and trunkless legs of stone / Stand in the desert. Near them, on the sand, / Half sunk, a shattered visage lies . . .'" This huge, broken face sneers its "cold command," unaware that its moment is over. The poem concludes: "And on the pedestal these words appear: / 'My name is Ozymandias, King of Kings: / Look on my works, ye mighty, and despair!' / Nothing beside remains. Round the decay / Of that colossal wreck, boundless and bare / The lone and level sands stretch far away."

The contemporary Romantic poet John Keats (1795–1821) never launched any attack on religion, but he is a darling of the history of doubt—and one of the best poets there ever was. In a letter to his brother, Keats wrote that the poet is no one because the poet takes on the point of view and experience of everyone and everything. That capacity for ambiguity, he asserted, is necessary for *all* greatness: "At once it struck me what quality went to form a Man of Achievement, especially in Literature, and which Shakespeare possessed so enormously—I mean Negative Capability, that is, when a man is capable of being in uncertainties, mysteries, doubts, without any irritable reaching after fact and reason." A few lines later he told his brother, "Shelley's poem is out, and there are words about its being objected to as much as 'Queen Mab' was. Poor Shelley . . . !!" Keats had a hard life, saw several beloved family members die—and then he himself began spitting blood and knew what it meant. He came to believe that the world was a place for "soul making"; this brutal world is the only way for each of us to become a unique identity. Many Romantic poets, and many modern artists of all types, would explicitly take the transcendence and meaning of art as a substitute for religion. Keats's Negative Capability and "soul making" world are among the sublime ideas of that tradition.

In the United States at this time Ralph Waldo Emerson (1803–1882) was developing a movement called Transcendentalism. He, and most of his followers, were former Unitarians who no longer believed even in the distant God with whom they had been raised, and found a replacement in the beauty of nature. Emerson had been a Unitarian minister; he passionately read Plotinus and the Neoplatonist texts that followed therefrom, as well as the Stoics Epictetus and Marcus Aurelius. Whereas Jefferson's library contained lots of Epicurean texts, especially Lucretius, Emerson's had none, but he owned several Stoic texts—it was Velleius and Balbus on the eastern seaboard.[151] Emerson married Lydia Jackson, who was also "a nonconformist Unitarian with Neo-Platonist beliefs," but to his great grief, she died within two years of their wedding. He resigned from the church. Soon after, in 1832, he left for England where he became friends with Thomas Carlyle, Samuel Taylor Coleridge, and William Wordsworth, all Romantic poets who had rejected traditional religion but basically maintained belief in a spirited world. From this and from his reading of ancient Stoicism, Emerson formulated his Transcendental faith. It rejected religion, yet was spiritual in its mood and its vision of the natural world as a humming abundance of beauty, love, and creativity. It would have an important future in modern American thought—up to the present day.

The two other major figures of Transcendentalism were Henry David Thoreau and Margaret Fuller. Both were considerably influenced by Stoicism before they met Emerson. They were all poets of nature. Thoreau lived on Emerson's Walden Pond for two years, basking in Nature, but it was not all leaves and sky. In a paragraph praising the morning, and waking to it, alone in nature, Thoreau managed to cite "the Greeks," Confucius, and the Vedas. Indeed in his solitude in the woods he had time to read Cicero and the Bhagavad-Gita; both he and Emerson read the Samkhyya Karika and the Vishnu Purana as well. What was being advised was a highly informed naturalism. Margaret Fuller, who was editor of the Transcendentalist journal *The Dial,* was more the reformer than either Thoreau or Emerson. They were all abolitionists, but Thoreau and Emerson tended to the removed life of the philosopher. Fuller became famous for her books on women's rights, for her defense of Native Americans against the missionaries who plagued them, and for her "conversation classes" for women, where her paying guests learned to discuss ideas. The first lesson was "Greek mythology." In her *Memoirs* (1852), Fuller wrote that when she read Christian teachings, they made her cry out for her "dear old Greek gods."[152] Here's part of what she meant: "The missionary . . . vainly attempts to convince the red man that a heavenly mandate takes from him his broad lands. He bows his head, but does not at heart acquiesce. He cannot. It is not true." Further, "Let the missionary, instead of preaching to the Indian, preach to the trader who ruins him."[153] In Fuller, doubt was philosophical but also moral.

The future Victorian novelist Marian Evans, mentioned earlier, was brought up with religion in Derbyshire, England, but got hold of a freethinking book on the origins of Christianity and looked up the references. Friends who shared her doubt introduced her to other doubters; she met Emerson once. By the early 1840s she had openly renounced Christianity and set to work on writing the first English translation of Strauss's *Life of Jesus*. It was published anonymously, to much acclaim and uproar, in 1846. In 1854 her translation of Feuerbach's *Essence of Christianity* sent another shock through the English-speaking world and was the only book she ever published in her own name. On its strength, she became a contributing editor of John Stuart Mill's old journal, the *Westminster Review;* she also translated portions of Spinoza, whom she particularly admired, and of Comte. Only later, with *Adam Bede* in 1859 to *Middlemarch* in 1871, did she became England's beloved poet and novelist, under the name George Eliot. Her novels center on the inner life and the mistakes of perception and expectation. Thus she moved from an early role of quiet, brilliant evangelism to a later life as an investigator and poet of the human.

If Keats was the great doubting poet of the first half of the century, Emily Dickinson (1830–1886) took the prize for the second half. She, too, was a Cheshire Cat of a doubter, welcoming ambiguity, playful, but exquisitely serious:

> *Ourselves we do inter with sweet derision.*
> *The channel of the dust who once achieves*
> *Invalidates the balm of that religion*
> *That doubts as fervently as it believes.*[154]

She could do in a four-line poem what other people took a chapter for, but in the best circumstances they take as long to read. Even as she tells us here how it feels to bury a loved one and mouth religious words about an afterlife ("sweet derision"), she keeps the language so tricky that the second line is about the grave accepting the casket but also about the human being, now dust. The poem is also a little psalm to the inner life. And it's got amazing rhythm. It's one of doubt's best.

Dickinson grew up during the period of New England revivalism, but refused to make the public confession of faith that would formally admit her to the church. By the time she was thirty she had stopped going to services entirely. Soon she did not leave the house at all. She stayed in, listening to her inner moods and conversing with herself, in verse, about what she heard there. She sang this song according to a deeply religious melody, the hymns of that church she did not attend. As the critic Dennis Donoghue has written, "of her religious faith virtually anything may be said, with some show of evidence. She may be represented as an agnostic, a heretic, a sceptic, a Christian."[155] She wrote to a friend about her family, "They are religious—except me—and address an Eclipse, every morning—whom they call Father."[156] The woman doubted; that was her whole business.

Those—dying then,
Knew where they went—
They went to God's Right Hand—
That Hand is amputated now
And God cannot be found—

The abdication of Belief
Makes the Behavior small—
Better an ignis fatuus
Than no illume at all—[157]

Dickinson was brilliant at keeping the tension of doubt, and at generating a private religion, of art and inner life, that "doubts as fervently as it believes."

There is more worry in Thomas Hardy's "God's Funeral." In the first five stanzas Hardy describes coming upon a funeral procession and, able to sense that it was God's funeral, he watched with pained awe. He overheard the mourners ask, "O man-projected Figure," how can we outlive you? "Whence came it we were tempted to create / One whom we can no longer keep alive?" The speech explained that "rude reality" had "mangled the Monarch" we invented, until he "quavered, sank; and now has ceased to be." The sadness here echoed other religious losses:

So, toward our myth's oblivion,
Darkling, and languid-lipped, we creep and grope
Sadlier than those who wept in Babylon,
Whose Zion was a still abiding hope.

How sweet it was in years far hied
To start the wheels of day with trustful prayer,
To lie down liegely at the eventide
And feel a blest assurance he was there!

And who or what shall fill his place?
Whither will wanderers turn distracted eyes
For some fixed star to stimulate their pace
Towards the goal of their enterprise?

In the background, the poet could hear some people protest that God still lived. Hardy wished he, too, still believed, but for him, "how to bear such loss I deemed / The insistent question for each animate mind."[158] Some few mourners insisted they could see the glow of something new on the horizon, but the poet lingered with the crowd, aimless between "the gleam and the gloom."

It was doubt's most evangelical century to date. There had been a long history that considered doubt a private club with no other aspirations than truth and enough freedom to debate with one's colleagues in peace. There had never been campaigns for the doubters to convert the believers. Now when the doubters' ranks swelled in response to religious incursions into politics (including persecution and the support of social injustice), people grew convinced that atheism would soon take over. Some were glad for it, some horrified; what is remarkable is that so many expected it. Few remembered that through the millennia, doubt had never been evangelical, but had been passed down, quietly and respectably, in families, had leapt gently from student to teacher, and had appeared de novo now and again—all without looking to take over. In their visions of the future, perhaps Nietzsche was wrong and Etta Semple was right: to many modern doubters, tolerance seems more attractive than uniform disbelief. In any case, nineteenth-century doubt hammered out the basic needs of a mass democracy, achieved a staggering revolution in atoms and anthropolgy, and reenvisioned art as sufficiently transcendent and transformative to enchant modernity.

TEN

Principles of Uncertainty, 1900–

The New Cosmopolitan

If you ran in the right circles, a New Year's Eve party welcoming the twentieth century might have presented you with any of a great variety of doubters. That couple dancing the cakewalk beside you might be an American, freethinking, women's rights lecturer and a British, atheistic, birth control pamphleteer. Or they might be a French anticlerical anthropologist and a Russian nihilist; or a couple of German materialists. It is a fanciful notion, of course, but the cakewalk—a craze in the United States and across Europe beginning in the 1890s—was the first dance to cross over from the African-American community, having originated with slaves mocking the formal strut of their owners' dances. A rising popular culture supported all sorts of rebellion. If we jump forward to 1910, the couples might be doing the turkey trot to Scott Joplin's ragtime. The turkey trot was a silly dance (one hopped four times on each foot) and as good a marker for rebelliousness as the cakewalk had been: it became hugely popular almost entirely as a result of its being denounced by the Vatican. The story of doubt in the twentieth century has been about many things, and kicking it up—doubting authority and custom—is one of them.

The twentieth century has qualities of the peak cosmopolitan moments throughout history: the Hellenistic, Rome, the Tang, the golden age of Baghdad, and the Renaissance. There is deep skepticism about our ability to know the world, to say anything true, to find a universal value. The philosopher Ludwig Wittgenstein set the tone for much modern philosophy when he wrote, "Whereof one cannot speak, thereof one must be silent."[1] Others, such as Wittgenstein's teacher and colleague, Bertrand Russell, have remained confident in science and its brand of doubt. The foundational condition for these doubts and others has been a degree of public secularism. Cosmopolitan doubt has often struggled over public secularism, and over the idea of a cultish version of politics, like that of ancient Rome. Both raise new questions in twentieth-century doubt, and this chapter begins with some flash portraits of the secularization of nations and the religiosity of politics. It then looks at a group of spunky American doubters, at the tricky new philosophies of doubt, and at the specific doubts confronted by Christians, Jews, and Muslims in this tumultuous century. Finally, "Doubt at the New Millennium" will bring us up to date.

SECULAR NATIONS

The secular state became a widespread ideal as governments from various traditions decided to get religion out of the state and to encourage public secularism. Also, the state seemed to be becoming a religion itself: ever since the French Revolution, political forces had spoken of their campaigns as a "new religion" and had devoted that kind of attention to ritual and symbolism. Philosophers explained the history of the modern state as a kind of religious drama, and revolutionaries spoke of the sacred state, its martyrs, and its liturgy. A "cult of personality" grew up around charismatic leaders and was often discussed as a replacement for religion. There were other phenomena that fit into this model, too. States purposefully planned nonreligious festivities and rituals like Thanksgiving, the Fourth of July, or Memorial Day, and such holidays began bringing the population together for something a lot like religion, but secular. Many commented that welfare seemed like a new, secular version of charity. Elementary education soon included all sorts of pledges, shared lore, and anthem singing. Athletics and sporting events, like secular parades and holidays, helped provide unifying communal experiences. Modern doctors in their expanded role (newly empowered by germ theory and vaccination) and the new professionals of psychiatry, sociology, and anthropology were all envisioned as the new clergy, giving advice, defining humanity, and setting standards. There are many levels of this. Thomas Mann wrote during World War I that it is a mistake to think that religion can be separated from politics. "For man is so made that, having lost all metaphysical religion, he transposes the religious into the social, he puts his social life on the altar."[2]

The rise of what Michel Foucault called "the pastoral state"—a secular state that takes on pastoral functions—coincides with the decline of public religion, but one did not simply take the place of the other. A practicing believer may well love secular state rituals, too: sports, parades, and flags. The twentieth century displayed a rise of religious behavior in secular settings (consider also the film *idol* and the fantasy of transformation and ascension implicit in modern stardom), but that does not quite mean that the decline of religion *caused* the rise of these other religiosities. There was clearly a relationship, but here it is my intention not to uphold the idea that modernity has variously replaced religion, but to point out that the popularity of the idea was an important aspect of twentieth-century doubt.

With the separation of church and state in 1905, the heyday of French anticlericalism was over. It had been a movement of the middle class, eager to run a secular, modern democracy. Now political doubt was more likely to be Marxist. By the opening of the century, there were strong socialist and communist parties and associations across Europe. Many members espoused atheism, and all of them were exposed to atheist arguments and atheist pride. Lenin agreed with Marx that religion was bad, but that there need be no attack on it because once the world was rid of wage slavery, religion would disappear. As Lenin put it in 1905, "The modern proletariat ranges itself on the side of Socialism, which, with the help of science, is dispersing the fog of religion and is liberating the workers from their faith in a life after death, by rallying them to the present-day struggle for a better life here upon earth."[3] Heaven was explicitly exchanged for earthly happiness. But consider also this amazingly coercive call for religious freedom: "Everyone must be absolutely free to profess whatever religion he likes, or to profess no religion, i.e., to be an atheist, as every Socialist usually is."[4]

In a 1909 speech Lenin said, "Marxism is materialism. As such it is as relentlessly opposed to Religion as was the materialism of the Encyclopaedists of the eighteenth century, or as was the materialism of Feuerbach." But, said Lenin, Marxism does not attack religion head on, with "purely theoretical propaganda," but by helping real people with their real problems. In this speech Lenin even said the priest should be welcomed in the Socialist movement and so should the person who gets things wrong and says, "Socialism is my religion." In 1917, the Russian Revolution resulted in the first state created and maintained under an empty sky. The United States was the first nation to encourage dissent and ensure the right to doubt; the Soviet Union was the first to maintain a state in decided hostility to religion.

Lenin had wanted the secularization to be unforced, but new revelations from the Russian archives show that in 1922 he responded violently to a community of clergy and their followers who refused to collect church valuables and turn them over to the government. He encouraged the arrest and quick trial of the insurrectionists and then proceeded to murder a large number of the clergy and their supporters. Lenin died in 1924, and Stalin's campaign against religion was worse. His main target over the next two decades was the Russian Orthodox Church, which had the most members. Almost all of its clergy were killed or sent to labor camps, and by 1939 only about five hundred of more than fifty thousand churches were still open. Most organized religions were never outlawed, but were decimated by severe limitations. By 1926 the Roman Catholic Church had no bishops left in the Soviet Union; attacks on Judaism were waged all through the Soviet period, so that the organized practice of Judaism practically disappeared; and Protestant denominations were persecuted. Fearing a pan-Islamic movement, Stalin suppressed Islam with methodical force. As unbelief was imposed, faith became a powerful voice of dissent and freedom. Although much opposition to the regime was in the name of secularist Enlightenment principles, the image of the secret meeting changed from a place where one might confess illicit doubt to a place where one might confess illicit faith. Meanwhile, among the mass of people living without religion, there were many true unbelievers. Open doubt was suddenly the norm.

Turkey secularized fast, too. When the Ottoman Empire fell in 1923, Turkey was proclaimed a republic and Kemal Ataturk became its first president. He ruled as a dictator all his life, but it was Ataturk's aim to democratize and modernize the country quickly, and he did. Of religion, he said, "I have no religion, and at times I wish all religions at the bottom of the sea. He is a weak ruler who needs religion to uphold his government; it is as if he would catch his people in a trap. My people are going to learn the principles of democracy, the dictates of truth, and the teachings of science. Superstition must go." Still, he did not mean to war against faith. "Let them worship as they will," he added, "every man can follow his own conscience, provided it does not interfere with sane reason or bid him act against the liberty of his fellow men."[5]

Ataturk changed every aspect of Turkish life, and most of his modernizing decrees directly attacked Islam. He abolished the caliphate, abolished religious orders, made polygamy illegal, and even forced every man to take off his fez and wear a European hat. In 1926 religious law codes were replaced with new codes borrowed from western Europe. Civil marriage (as opposed to religious) was made compulsory. In 1928 Islam ceased to be the state religion. Ataturk closed the religious schools, which taught in Arabic, and replaced them with secular, Turkish-language schools with modern concerns. He had canvassed the army's support before making these changes, and the army continued to support secularism in the future. Constantinople became Istanbul. Ataturk promulgated a constitution that provided for a parliament elected by universal manhood suffrage, and he let in the women ten years later, in 1934. He is remembered as a hero of democracy by modern Turks. By the time he died in 1938, Ataturk had radically secularized the country on the Western model. His biographer Andrew Mango explains that in Ataturk's world, "Most Turkish officers and gentlemen accepted Islam as a general framework for their lives and the life of their society. Others, like Ataturk and many of his friends, seem to have been freethinkers from their earliest years." This group accepted that Islam was part of other people's lives and had to be taken seriously for that reason, but they did not like it. Writes Mango, "In the eyes of many educated Muslims in Turkey, as in other Mediterranean countries, religion was the province of women; the sincerity of men who showed religious enthusiasm was suspect."[6] One of the few positive things Ataturk said about religion was that since his soldiers thought they were going to heaven, they were conveniently willing to die. It is what doubters had long suspected.

What happened in Italy has been well told by Emilio Gentile in his *The Sacralization of Politics in Fascist Italy* (1996). Early in the century the modernist intelligentsia of Italy thought that a new religion was necessary, an "irreligious religion" to replace Catholicism and energize life.[7] That generation searched for a civic religion, but they did not actually try to hash out a state liturgy so much as mull over the question of what they could possibly believe. What they came up with sounded like a somewhat elitist humanitarian cultural revival. They did not expect a quick solution and spoke of being stranded between grand mythic periods. Some of the Fascist leaders had been associated with this group, but had new rhetoric. Gentile tells us that Benito Mussolini considered himself a militant atheist in these years and (unlike Lenin) specifically described his revolutionary socialism as "religious." Enrico Corradini was one of the major contributors to the cultic rise of Italian nationalism and the creation of Fascism.[8] He admired the "religion of heroes and nature" of Japan, writing that "Japan is the God of Japan" and hoping Italy could copy the mood.[9] In 1932 Giuseppe Bottai wrote that Fascism was, "for my comrades or myself, nothing more than a way of continuing the war, of transforming its values into a civic religion."[10] By 1920 Mussolini spoke of a "religious notion of Italianism."[11] Funerals for Fascist soldiers were full of ornate ritual attaching their memory to "the immortal soul of the nation."[12] A cult of the Italian flag was instituted in the 1920s, with a daily saluting ritual at school, and on Flag Day students were told to receive the flag as the "new eucharist."

As early as 1923, Mussolini was dating letters "Year One of the Fascist Era," and as soon as they gained power, the Fascists added state holidays to the calendar and issued details on how to celebrate them.[13] April 21 was to be the Founding of Rome Day. Secular architecture got as ornate as cathedrals. Il Duce was imaged as the savior, of course, but the real God, as Cordini had hoped, was Italy and its Roman past. Emilio Bodrero wrote that Rome's name was "no longer that of a city but that of a divine entity," and that "being a citizen of Rome meant partaking of that divinity."[14] The patter recalls the ancient world and the polis. Later Mussolini insisted, "There is no need to get all tied up with antireligiousness and give Catholics reason for unease. We need instead to multiply our efforts in education, sports and culture." Some tolerance for religion was all right: "Protestants save their own souls, but we are Catholics and we let priests do their work. On the other hand, when they try to interfere in politics, socially, in sport, then we fight them."[15] Elsewhere Il Duce said that, "The State's duty does not consist in writing a new gospel or other dogmas, in overthrowing old gods, substituting them with others, called 'blood,' 'race,' 'Nordic,' and things of the kind." His point was that, to him, the Nazis were the ones getting religious about politics.

A great deal has been written about the religious qualities of Nazism, with its processionals, symbols, savior, and sacred state, but one does not have to doubt religion to create a religious, mystical nationalism. Nazism held the state above religion in devotion and importance, but celebrated an image of simple Christian piety, with a very blond Christ. The Nazis even argued he hadn't been Jewish. Nazism was described as a replacement religion so widely that the role of doubt cannot be ignored. But neither should it be exaggerated. Like a vacuum, doubt can have brutal power—especially where it has no philosophy, but merely the unspoken absence of belief and religious community—yet there is no evidence that Germany was experiencing more religious doubt than other nations at the time.

Gentile mentions that "the cult of the fallen" in many nationalist movements was an early and key part of the "sanctification of the nation."[16] Historian George Mosse has also shown that a new state religiosity grew up around the bodies of the world-war dead. The memory of the war was "refashioned into a sacred experience which provided the nation with a new depth of religious feeling, putting at its disposal ever-present saints and martyrs, places of worship, and a heritage to emulate."[17] The soldiers' martyrdom became the "all-encompassing civic religion."[18] In the past, soldiers were often not buried at all, but left on the field to be eaten by ravens and dogs.[19] With World War I, thirteen million soldiers died, more than twice as many as in all the major wars between 1790 and 1914. As Mosse shows, the movement to bury every soldier was discussed as a new, nationalist faith.

In China, students in the May Fourth Movement of 1919 embraced Western Utilitarianism as an alternative to the self-sacrifice and social hierarchy of Confucianism. They read the French philosophers and many of them rejected all religions. Historian Vera Schwarz tells of a written debate on "What is the point of human existence after all?" She describes one writer's response as "moving swiftly, and rather superficially, from Feuerbach, to Darwin, to Nietzsche," because while the desire for a great break with the past was not much about God in China, the story of Western civil rights was told as a story of rising religious doubt. Mao Zedong read Voltaire, Diderot, Montesquieu, and d'Holbach in his youth and took on philosophical materialism. By the time Mao became head of the Chinese Communist Party, in 1935, his philosophy of religion was an interpretation of the materialism of Marx and Lenin. He followed them in their contention that religion itself was bad, beyond the idea that science had proved God did not exist: religion drained resources from the state, belonged to the old world, deceived the people about reality, and took up workers' time, bodies, and minds. In the Cultural Revolution many monasteries and churches were destroyed, organizations disbanded, and communities annihilated. Meanwhile, as religion was forced out, the cult of Mao was becoming a vibrant focus of communal identity.

The idea that states were going to have to be secular, and that the secular state was going to have to be emotionally charged, was finding various expressions in the first half of the century. Meanwhile, back in the United States, there were some terrific maverick doubters.

AMERICANA: EDISON, HARRISON, JOHNSON, GOLDMAN, SANGER, AND TWAIN

On October 2, 1910, Thomas Edison told the *New York Times,* "No, all this talk of an existence for us, as individuals, beyond the grave is wrong. It is born of our tenacity of life—our desire to go on living—our dread of coming to an end as individuals. I do not dread it though. Personally, I cannot see any use of a future life."[20] That's a hell of a thing to tell the *New York Times*. Edison was roundly scolded for the statement by all manner of public and private personalities. His investors begged him to tell America that he believed, and after a storm of letters came to his laboratory, he offered this breathtaking claim to faith:

> I have never seen the slightest scientific proof of the religious theories of heaven and hell, of future life for individuals, or of a personal God. . . . I work on certain lines that might be called, perhaps, mechanical. . . . Proof! Proof! That is what I have always been after. I do not know the soul, I know the mind. If there is really any soul, I have found no evidence of it in my investigations. . . . I do not believe in the God of the theologians; but that there is a Supreme Intelligence, I do not doubt.[21]

This critical opinion had been aroused in him by Gibbon's *Decline and Fall of the Roman Empire* and Paine's *Age of Reason*. He wrote of the Paine: "I can still remember the flash of Enlightenment that shone from its pages."[22] He was also devoted to Darwin and Huxley. Edison got away with a lot because he was Edison: God was supposed to have lit up the heavens, and Prometheus brought fire, but neither act was well documented; the historical figure who really let there be light was Edison. Still, in early-twentieth-century America, outspoken doubt seems to have been more publicly acceptable than in most eras, anywhere. Being a dissenter, taking public opposition, had some moral cachet to it as a good unto itself.

The decade of the 1920s saw a burst of creative activity among African-American artists, writers, and social commentators based in Harlem in New York City. Hubert Harrison was a central participant in this and an amazing figure in the history of doubt. He was born on Saint Croix, Danish West Indies, and his working-class mother reared him alone, with few resources. But for those of African descent, Saint Croix had some advantages; there was no formal segregation, no lynching, and some real opportunity for advancement within society. Harrison studied hard and was working as an under-teacher of a school while still a young teenager. His mother died when he was only seventeen and he came to the United States, arriving in New York in 1900. He did menial jobs by day, earned a high school degree at night, read voraciously, found a job with the post office, married, and had five children. Meanwhile, Harrison wrote pro-labor letters to newspapers, which brought him to the attention of a variety of intellectual and workers' movements, including the free-thought movement.

In 1911 Harrison wrote a short essay on Thomas Paine for *The Truth Seeker.* It began: "If you should ask a man in the street who Thomas Paine was, he would say he was an Atheist; and he might probably qualify his statement with an adjective more forceful than polite."[23] That is not what a man in the street would say today. It seems we can either remember Paine was an atheist and forget he was an American hero (as they did a hundred years ago), or remember that he was an American hero but forget that he was an atheist (as we do now). Harrison added that while many hated Paine's atheism, some of those who liked it hated his style: "If you had asked a cultured liberal like Leslie Stephen the same question fifty years ago, he would have said that Paine was one of the cruder kinds of Infidels, fit perhaps for the unlettered minds of the mob."[24] Harrison sided with Paine against the imagined critique of Virginia Woolf's father: for Harrison, Paine's crudeness was part of his success. Harrison, meanwhile, demonstrated more knowledge of the history of doubt than anyone had in a while. He celebrated Galileo, Descartes, Newton, Hume, Spinoza, Diderot, d'Alembert, Voltaire, and d'Holbach. He also said that while most people could not understand this stuff, they could understand Paine. Harrison himself went further than most of his heroes, noting that "These French Deists, however, made certain false premises which we smile at today": they believed in keeping monotheism but fixing it, "which was absurd"; in the worship of nature, "which was foolish"; and "in the origin of religion as conscious fraud," which wasn't true.[25]

Harrison praised Lord Herbert and Edward Gibbon, too. He even mentioned John Toland, the seventeenth-century doubter who taught us about offering a "bouncing compliment" as a misdirection at the beginning of a doubting text, and who died with a copy of Lucretius's *On the Nature of Things* near his bed. The history of doubt had not heard Toland's name much for a long while. In the last three paragraphs of this article on Paine, Harrison returned to his subject, explaining the variety of Paine's biblical criticism, showing all sorts of errors in the Bible, and quipping that everyone knew of these now, "except, perhaps, in America." Because Paine brought the debate to common people: "He was 'the Apostle to the Gentiles' of the Free-thought movement." This meant Paine had brought "the results of that great conflict down to the level of the democracy," just as Paul brought the worship of Jesus from the small world of the Jews and out to the masses of gentiles.[26]

Harrison was friends with the activist and educator Frances Reynolds Keyser, and wrote her a striking letter in 1908. It seems he had said something to her about the attraction of Catholicism, and she asked what he saw in it. In response, he joked that he always wanted to learn Latin, but that really, he was a deep doubter. It was Paine who had awakened him to "certain rationalist results which bore their own proof on their face." On her behalf, he asked himself, "Did it hurt?"

> I said already that I was not one of those who did not care: I suffered. Oh, how my poor wounded soul cried out in agony! I saw the whole fabric of thought and feeling crumbling at its very foundations, and in those first fearful weeks of stern reaction I could not console myself as so many have done with husks of a superior braggadocio.

He then searched desperately for something to believe: "What *had* gone was the authenticity of the Bible," and that was a lot, because "my God was the Bible God": the Hebrew God, "plus the tribune God fused from four centuries of Persian, Babylonian and Hindu teaching and the Alexandrine cobwebs of . . . Plotinus, and the Neo-Platonists. So when my Bible went, my God went also. But I had to get one to worship, and I proceeded to build me a God of what was left."

For a while he believed there was one universal God with no religious dogma. "But in the meanwhile, Time, the great healer, closed the wound and I began again to live—internally. But now I had a new belief— Agnosticism." He had rejected the idea that Jesus was a fraud, as "comparative mythology had more rational explanations to offer," yet since he mentioned it we know it was still a concern. The strongest hold was Jesus: "The power of his personality haunted me for a long time, but in the end that also went. Now I am an Agnostic; not a dogmatic disbeliever nor a bumptious and narrow infidel. I am not at all of Col. Ingersoll's school." Instead, he called himself "such an agnostic as Huxley was."[27] He was a little sad about it:

> I wish to admit here something that most Agnostics are unwilling to admit. I would pay a tribute to the power of that religion which was mine. It is only fair to confess that Reason alone has failed to satisfy all my needs. For there are needs, not merely ethical but spiritual, inspirational—what I would call personal dynamics; and these also must be filled.

Harrison spoke of a spiritual side to humanity as being a real part of human experience, whatever the truth of specific religious ideas. "Rationally," he explained, "I believe the scientific explanation to be the correct one. And yet—Shall we stunt the soul by refusing to develop it in any one direction while conceding the necessity for development in all other directions?" If we can show that a set of beliefs "can develop the spiritual side of man," he asks, why should we "refuse the aid of the belief" just because it doesn't correspond with the facts? For these reasons he had considered Catholicism, with its beautiful rituals and antique institution. "Besides, as I got to know more, I found that Reason was not everything and I admired the sublime courage of the Church which boldly demands the subjection of Reason to faith." The Latin lessons were attractive, too. But "Entre nous, I doubt whether I will ever be anything but an honest Agnostic, because I prefer, as I once told you, to go to the grave with my eyes open."[28]

Harrison did the history of doubt a great service by describing how doubt stood with his fellows. In an article called "The Negro Conservative: Christianity Still Enslaves the Minds of Those Whose Bodies It Has Long Held Bound" (1914), he opined that African-American scholars were essentially stuck in the eighteenth century and wondered why. After all, "It should seem that Negroes, of all Americans, would be found in the Free-thought fold, since they have suffered more than any other class of Americans from the dubious blessings of Christianity." He noted that some said the two main forces of racial prejudice in the United States were the Associated Press and the Christian church. "This is quite true," agreed Harrison. "The church saw to it that the religion taught to slaves should stress the servile virtues of subservience and content." Further, "It was the Bible that constituted the divine sanction of this 'peculiar institution.'" To show his readers "the relation of church and slavery," Harrison referenced a book called *A Short History of the Inquisition;* it is important to see that he saw these abuses as informing on each other. This fine historian of doubt then turned to Nietzsche, wryly commenting that the philosopher's description of Christian ethics as slave ethics "would seem to be justified in this instance."

As for doubters around him, there were "a few Negro Agnostics in New York and Boston," but these were almost always West Indians from the French, Spanish, and English islands. "The Cuban and Porto Rican cigarmakers are notorious Infidels, due largely . . . to their acquaintance with the bigotry, ignorance and immorality of the Catholic priesthood in their native islands." But African-Americans "reputed to have Agnostic tendencies" are rare, and "these are seldom, if ever, openly avowed." Then in an essentially Marxist turn, he said: "Myself, I am inclined to believe that freedom of thought must come from freedom of circumstance," and he did not think it wise to campaign on the subject. In his words: "there is a terrible truth in Kipling's modern version of Job's sarcastic bit of criticism: 'No doubt but ye are the people—your throne is above the king's / Whoso speaks in your presence must say acceptable things.'" Job says the first line only, of course, and he says it in response to his friends' defense of God. Harrison reminds us that tucked in Job are warnings about the tyranny of believers. Popular opinion can support doubt, but it can also be as dangerous to it as any monarch.

In 1920 Harrison was happy to say that after a period of being always drawn as conservatives by reporters, "Today Negroes differ on all those great questions on which white thinkers differ, and there are Negro radicals of every imaginary stripe—agnostics, atheists, . . . and even Bolshevists."[29] By then Harrison had been nicknamed the Black Socrates. He would also be called the father of Harlem radicalism. As historian Jeffrey Perry put it, in the 1910s and 1920s Harrison created or cofounded "almost every important development originating in Negro Harlem—from the Negro Manhood Movement to political representation in public office, from collecting Negro books to speaking on the streets, from demanding Federal control over lynching to agitation for Negroes on the police force."[30] He was also a literary critic and "a pioneer Black activist in the free thought and birth control movement."[31]

Other well-known freethinkers of the Harlem Renaissance include the labor activists Asa Philip Randolph and Chandler Owen; the authors J. A. Rogers and George Schuyler; poets Claude McKay and Walter Hawkins; and activists Cyril Briggs, Richard Moore, and Rothschild Francis. Perry writes that "W. E. B. Du Bois, according to his biographer David Levering Lewis, was 'agnostic and anticlerical.'"[32] Historians have noted more doubting poets of the Harlem Renaissance, including Countee Cullen, Waring Cuney, Georgia Douglas Johnson, and Helene Johnson. Consider Georgia Douglas Johnson's small poem "The Suppliant":

Long have I beat with timid hands upon life's leaden door,
Praying the patient, futile prayer my fathers prayed before,
Yet I remain without the close, unheeded and unheard,
And never to my listening ear is borne the waited word.

Soft o'er the threshold of the years there comes this counsel cool:
The strong demand, contend, prevail; the beggar is a fool!

Doubt, once again, brought strength. This time it had managed to rhyme two of its favorite notions: the fool that does not struggle, and the cool of accepting the real world.

It is no secret that anarchists tended to be doubters. Emma Goldman (1869–1940), the Russian-American anarchist, preached atheism in her magazine *Mother Earth*. In a piece called "The Philosophy of Atheism" (1916), she declared having chosen "the concept of an actual, real world with its liberating, expanding and beautifying possibilities, as against an unreal world," whose spirits, oracles, and "mean contentment" have kept humanity down.[33] Goldman gives a little history of doubt and accounts for her place in it: "I am not interested in the theological Christ. Brilliant minds like Bauer, Strauss, Renan, Thomas Paine, and others refuted that myth long ago." For her, the theological Christ is "less dangerous" than "the ethical and social" one. Science will loosen the hold of theology. "But the ethical and poetical Christ-myth has so thoroughly saturated our lives that even some of the most advanced minds find it difficult to emancipate themselves from its yoke."[34] She hated seeing workers, especially women, still saddled with ideas of self-denial and penance. For her, doubt was a source of happiness. "Atheism in its negation of gods is at the same time the strongest affirmation of man, and through man, the eternal yea to life, purpose, and beauty."[35] In 1932 Goldman wrote a biography of Voltairine de Cleyre.[36] President Hoover had Goldman deported to the Soviet Union, but she left there, disillusioned with Bolshevism, and became a British citizen. When she died, at seventy, her body was brought back to the United States and buried in Chicago, beside the graves of de Cleyre and other Chicago radicals.

When Margaret Sanger opened the first birth control clinic, in 1916, the motto of her journal was "No Gods, No Masters." Sanger left us a nice conversion story: She was given a Christian upbringing by her mother, but soon found herself influenced by her father's freethinking. When Margaret was very young, her father, a monument cutter, arranged for Ingersoll to come speak in their town, and he made it clear to the town that he supported the man. His daughter later confessed to remembering only the excitement and the trees outside the window of the hall, and not a word Ingersoll said. Henceforth, however, Margaret and her siblings were called "devil children" and heretics. She was hurt by this, but it did not touch her belief. It was some years later, after dinner one night, that her father asked her why she had spoken to the bread. She said she was thanking God for it, and he asked if God was a baker. That, she said, was the start of her awakening. In a great pair of lines from her autobiography, Sanger wrote, "It was not pleasant, but father had taught me to think. He gave none of us much peace."[37] He hounded them to take up action for the oppressed workers of the world. "Unceasingly he tried to inculcate in us the idea that our duty lay not in considering what might happen to us after death, but in doing something here and now to make the lives of other human beings more decent."[38] She spent her life working so that women could have some control over their lives and so that people could limit the sizes of their families to fit their resources, but by the end of her life she was also involved in the eugenics aspect of birth control—encouraging the "less fit" to limit reproduction. Some people who pressed for sweeping changes in this century were doubters and took up their causes because they believed that, in the absence of God, humanity must design a better world for itself. Results varied widely.

Of the great American doubters, the most raucous and odd was surely Samuel Clemens—Mark Twain. He left so much material critiquing religion that a recent book has been compiled of his "irreverent writings."[39] Twain's questions on religion were based mostly on Paine's *Age of Reason,* along with some Darwinism and a sense of the late-nineteenth-century science versus religion debate. His work on belief and doubt was usually in the form of allegory or fiction, and unlike most of his contemporaries who were doubters, Twain examined the stories in the Bible. His belief was expressed in his book *Letters from the Earth,* in which God had made earth and universe as an experiment to which he paid little attention. One of God's deputies, Satan, visits earth and writes letters back to the other deputies. He's astounded to find that the human beings think God is watching them, that they talk to him and ask him for things. He marvels that humanity invented heaven and yet kept from it the one thing they love most—sex— and included in it a whole bunch of things they usually avoid: harp playing, endless group singing, and prayer.

As the essay goes on, Twain's indictments become more serious, though always fully engaged in the Christian mythology. Satan describes how the Creator seemed to have come up with distinct, horrible things to infest every little part of a human being. Consider hookworm: "Many poor people have to go barefoot, because they cannot afford shoes. The Creator saw his opportunity. I will remark, in passing, that he always has his eye on the poor. Nine-tenths of his disease-inventions were intended for the poor, and they get them."[40] Twain describes the ravages of African sleeping sickness, and God's "atrocious cruelty" in creating it. Wrote Twain, "his chosen agent was a fly."[41] Then six thousand years go by, science figures out what to do about it, and everyone praises God for having inspired the scientists. Twain could not. "He commits a fearful crime, continues that crime unbroken for six thousand years, and is then entitled to praise because he suggests to somebody else to modify its severities." Twain describes the symptoms of the disease (sleepiness, skin eruptions, convulsions, madness, death), then issues one of the angriest statements of his life: "It is he whom Church and people call Our Father in Heaven who has invented the fly and sent him to inflict this dreary long misery and melancholy and wretchedness, and decay of body and mind, upon a poor savage who has done the Great Criminal no harm." The Great Criminal! Twain said there isn't a man in the world who wouldn't cure the victim of sleeping sickness if he had the power. "To find the one person who has no pity for him you must go to heaven; to find the one person who is able to heal him and couldn't be persuaded to do it, you must go to the same place." Put flatly: "There is only one father cruel enough to afflict his child with that horrible disease—only one. Not all the eternities can produce another one."[42] Twain fully rejected the idea of a just God.

It's strange to hear someone struggling with the details of Christian dogma again—it had been out of favor among doubters for so long. Twain offers complaints about biblical morality, such as the Old Testament's vicarious punishments. Scolding God for never considering that he, as Creator, was to blame for humanity's sins, Twain lamented that just because the people of Shittim had been "committing whoredom with the daughters of Moab," the Lord told Moses to hang the leaders of both peoples. Twain said, "If it was fair and right in that day it would be fair and right to-day," because God's morals are supposed to be eternal and unchanging.[43] "Very well, then," he wrote, "we must believe that if the people of New York should begin to commit whoredom with the daughters of New Jersey, it would be fair and right to set up a gallows in front of the city hall and hang the mayor and the sheriff and the judges and the archbishop on it, although they did not get any of it. It does not look right to me."[44]

In June of 1906, Twain dictated some personal reflections on religion. He wrote to his friend William Dean Howells, "To-morrow I mean to dictate a chapter which will get my heirs and assigns burnt alive if they venture to print it this side of 2006 A.D.—which I judge they won't." He added that "The edition of A.D. 2006 will make a stir when it comes out. I shall be hovering around taking notice, along with other dead pals. You are invited."[45] These confessions rip into religion, and Christianity in particular. For one: "If Christ had really been God, He could have proved it, since nothing is impossible with God."[46] Twain believed in God, but not one that cared for us. He described the size of the universe, our place in it, and then laughed at the idea of God "passing by Sirius to choose our potato for a footstool."[47]

For Twain we are to the creator of the universe as a scientist's vial of microbes would be to the Emperor of China: it's unlikely that he even notices we exist, but impossible that he cares for some of us and is angry at others. Yet Twain had a strong belief in God: "the myriad wonders and glories and charms and perfections of this infinite universe" are all "slave of a system of exact and inflexible law," and when we realize this, "we seem to know—not suppose nor conjecture, but know—that the God that brought this stupendous fabric into being . . . is endowed with limitless power." So he exists. But what more?

> Do we also know that He is a moral being, according to our standard of morals? No. If we know anything at all about it we know that He is destitute of morals—at least of the human pattern. Do we know that He is just, charitable, kindly, gentle, merciful, compassionate? No. There is no evidence that He is any of these things—whereas each and every day as it passes furnishes us a thousand volumes of evidence, and indeed proof, that he possesses none of these qualities.[48]

Twain sounds nothing less than Gnostic in his conviction that there is a God and that this God is cruel: "He proves every day that He takes no interest in man, nor in the other animals, further than to torture them, slay them and get out of this pastime such entertainment as it may afford." The last of Twain's autobiographical dictation is less heated but equally Gnostic in its mood, insisting that "Man is a machine and God made it—without invitation from anyone." The maker of a machine is responsible for the machine's performance; not the machine. Twain goes so far as to say, "In our secret hearts we have no hesitation in proclaiming as an unthinking fool anybody who thinks he believes that he is by any possibility capable of committing a sin against God—or who thinks he thinks he is under obligations to God and owes Him thanks, reverence and worship."[49] It is a nice piece of home-spun American doubt, animated by a conviction that the world is amoral and that therefore there is no good God. It's also beholden to the an-cient argument from design and the Enlightenment idea of knowing God through his natural laws.

EVOLUTION, EINSTEIN, UNCERTAINTY, AND FREUD

When confronted with Darwinism, many American religious leaders were doing what religious leaders all over the world were doing: accepting the idea. A series of conservative Christian works, *The Fundamentals*, was an origin of the term *fundamentalist,* but even here some articles in the series asserted that evolution was "coming to be recognized as but a new name for 'creation.'"[50] It was easy to "assume God—as many devout evolutionists do—to be immanent in the evolutionary process."[51] So what happened? As historian Edward Larson explains in his study of the Scopes trial, the politi-cian William Jennings Bryan was a very particular character. A progressive in many ways, he stood for the vote for women and was enough of a pacifist before World War I that he resigned from his position as secretary of state when President Wilson moved the United States toward war. But he also sided with fundamentalism, and when his resignation ended his political career, he began making speeches for a living, often in support of religious traditionalism. It was he who championed the idea that people shouldn't teach evolution. Education had been the major point in French seculariza-tion, and it would be in the United States, too. It expanded noticeably in the 1920s: for example, in 1910 there were not quite ten thousand students in Tennessee high schools; in 1925 there were more than fifty thousand. This massive change created tension over what the state was teaching, and when Bryan hollered that the state-approved textbooks took evolution for granted, he provided a rallying point for nervous parents.

Spurred by Bryan's campaign, in 1925 Tennessee passed a law against teaching anything that denies the "story of the Divine Creation of man as taught in the Bible" and teaches instead that we descended from a "lower order of animal." In response, the American Civil Liberties Union offered to defend anyone who would challenge the law, and the science teacher John T. Scopes took on the challenge in what became the Scopes "Monkey

Trial." Bryan volunteered his aid to the prosecution and Clarence Darrow was secured for the defense. Darrow is known as a legendary lawyer today, but in his time he was known as an atheist lawyer. While he was alive, Darrow did not let anyone forget he was a doubter. He could argue that you were one, too. His essay "Why I Am an Agnostic" begins:

> An agnostic is a doubter. The word is generally applied to those who doubt the verity of accepted religious creeds of faiths. Everyone is an agnostic as to the beliefs or creeds they do not accept. Catholics are agnostic to the Protestant creeds, and the Protestants are agnostic to the Catholic creed. Any one who thinks is an agnostic about something, otherwise he must believe that he is possessed of all knowledge. And the proper place for such a person is in the madhouse or the home for the feeble-minded. In a popular way . . . an agnostic is one who doubts or disbelieves the main tenets of the Christian faith.[52]

Having reeled in his reader, he pounces:

> I am an agnostic as to the question of God. . . . Since man ceased to worship openly an anthropomorphic God and talked vaguely and not intelligently about some force in the universe, higher than man, that is responsible for the existence of man and the universe, he cannot be said to believe in God. One cannot believe in a force excepting as a force that pervades matter and is not an individual entity.[53]

It's nice to see the reinvention of Hume's point that the "God" that reason leads us to needs another name. Darrow then explains that science builds railroads and bridges, steamships, telegraph lines, cities, and plumbing; science keeps up the food supply and "the countless thousands of useful things that we now deem necessary to life." And science needs doubt: "Without skepticism and doubt, none of these things could have been given to the world." For that alone, "The fear of God is not the beginning of wisdom," but the death of wisdom. "Skepticism and doubt lead to study and investigation, and investigation is the beginning of wisdom."[54]

The final personality in the cast of the Scopes trial was H. L. Mencken, the most prominent American journalist and critic of the time, and a passionate doubter. A recent biography of him is called *The Skeptic,* and this attitude was especially keen when aimed at religion—he coined the phrase

"Bible Belt" and it was not a compliment.[55] Mencken wrote much criticism of Bryan's fundamentalism, so the ACLU took him on as an adviser. He also wrote reports on the trial that shaped the event in national memory. The judge was hostile to evolution and refused to allow scientific witnesses to speak on the value of the theory. On Mencken's advice, the defense put Bryan on the stand. He went willingly, happy for the chance to defend religion, but Darrow had a field day with him. The radio broadcasts of the trial and the reporters' dispatches brought Bryan's literal trust in the Bible to all of America, and America thought it sounded silly. That was clear right away, and the next day the judge struck Bryan's testimony from the record, asking the jury to decide only if Scopes had taught evolution. They found him guilty, but the battle for public opinion had been won by Darrow, Mencken, and doubt. For many believers, the old doubter's doctrine that life and humanity developed naturally was now too well documented to dismiss, so they accepted it and kept believing anyway. For others, rejecting the old doubter's doctrine was a banner issue for rejecting the culture of science and cosmopolitanism. For doubters, the rise of Darwinism was a triumph.

Something similar happened in physics. Einstein's theory of relativity appeared at the very beginning of the century, in 1905, and whatever people understood of it, they could tell the universe was becoming stranger. Energy and mass were now somehow the same; translated versions of one another, convertible back and forth. Time, the great certainty, was shown to slow down in relation to speed! Quantum mechanics and Werner Heisenberg's Uncertainty Principle seemed to make a fact of ancient Skepticism. The old physics assumed we could say where a particle is and what its momentum is, at a given moment. Quantum mechanics holds that this is impossible: the more precisely we measure the position of a particle, the less precisely we can know its momentum. That's the uncertainty. The Skeptics had said our senses and our minds could not know the world because it was not knowable to our senses and minds. Buddhists had come to the same conclusion. The scientists had to admit, and usually did, that the ancients had said all this before. What often got lost was that this conceptual doubt had a historical correlation with religious doubt and the rejection of all dogma.

By the 1930s everyone knew of Albert Einstein and that he was a genius. That made it matter what he believed; his quip "I am convinced that God does not play dice with the universe" has been famous since he said it. The comment, however, was about Einstein's beliefs regarding the role of chance, not the existence of God. When he was talking about religion, he gave a different impression. In 1921 a New York rabbi asked Einstein if he believed in God and he answered, "I believe in Spinoza's God who reveals himself in the orderly harmony of what exists, not in a God who concerns himself with fates and actions of human beings."[56] Thoughtful and judicious, Einstein never seemed to answer the question with the kind of zeal that might have been understood as clear-cut unbelief. In a private letter he lamented the problem: "It was, of course, a lie what you read about my religious convictions, a lie which is being systematically repeated. I do not believe in a personal God and I have never denied this but have expressed it clearly. If something is in me which can be called religious then it is the unbounded admiration for the structure of the world so far as our science can reveal it."[57] His awe was real but should not be misread as mysticism. Consider an early comment and a late one: In 1921 a woman wrote to him asking about his beliefs and Einstein responded that "the mystical trend of our time" evident in "Theosophy and Spiritualism" seemed to him "a symptom of weakness and confusion."[58] He added, "Since our inner experiences consist of reproductions and combinations of sensory impressions, the concept of a soul without a body seems to me to be empty and devoid of meaning." In 1953 a woman wrote to Einstein asking about life after death; he responded: "I do not believe in immortality of the individual, and I consider ethics to be an exclusively human concern with no superhuman authority behind it."[59]

In 1932 he gave a speech entitled "My Credo" to the German League of Human Rights in Berlin. It is a lovely piece. "Our situation on this earth seems strange," began Einstein. We all appear here, "involuntary and uninvited for a short stay," without knowing anything of why. "In our daily lives we only feel that man is here for the sake of others, for those whom we love and for many other beings whose fate is connected with our own." This love was matched by awe:

> The most beautiful and deepest experience a man can have is the sense of the mysterious. It is the underlying principle of religion as well as all serious endeavor in art and science. He who never had this experience seems to me, if not dead, then at least blind. To sense that behind anything that can be experienced there is a something that our mind cannot grasp and whose beauty and sublimity reaches us only indirectly and as a feeble reflection, this is religiousness. In this sense I am religious. To me it suffices to wonder at these secrets and to attempt humbly to grasp with my mind a mere image of the lofty structure of all that there is.[60]

As for so many of those who devote their lives to the study of the universe, it seems to have been for Einstein a school for both reverence and reason.

At about the same time, another conceptual revolution occurred. Sigmund Freud published his early *Studies on Hysteria* in 1895 and *Interpretation of Dreams* in 1899, but it was across the early decades of the twentieth century that his work transformed much of Western culture. Freud's father was a freethinker; he'd been emancipated from the narrow ghetto life of Galicia and had brought up his children in a secular mood in a secular world. Although Freud was Jewish, his critique of religion was aimed at the Roman Catholic Church that surrounded him in Vienna, which he saw as repressive to its flock and anti-Semitic to his.[61] Early on, Freud's work affected the history of doubt in an important, but general way. It showed human morality and civility as a thin covering over a mass of blind hungers and needs. Even noble self-sacrifice was now reducible to the fulfillment of some personal psychological need. Where could good and evil, let alone sin and saintliness, be located in such a schema? There is also in psychoanalysis an impulse toward freeing people from unnecessary bonds, and religious bonds were usually seen as of the unnecessary variety.

Freud also plays an interesting role in the history of Skepticism. On the one hand, his dissecting of compulsion and illusion powerfully suggested that the Skeptics had been right to doubt our ability to see the world. On the other hand, he provided a lot of help in correcting one's vision problems so that the world could be known as clearly as possible. Freud saw both roles, but saw himself much more in the latter role. As he put it privately, he felt that he did for human reason what Copernicus had done for the universe and Darwin had done for our origins.[62] He did not often place himself in the history of doubt, or in any intellectual history. But as Freud scholar Philip Rieff wrote, "Privately he admitted to remote intellectual connections—with Kant, Voltaire, Feuerbach."[63] He also acknowledged that Schopenhauer's "will" was the same as unconscious desire.[64]

In his 1913 *Totem and Taboo*, Freud got more specific about doubt. Here he explained religion as having originated in a primitive patricide carried out by brothers: religious ritual walks us through the killing and the eating of the father's body; the hovering absence of his authoritarian voice serves as the basis for God. The whole notion of sacred and profane, Freud explained, had grown out of the incest taboo. Up till now, the idea that religion had been invented in response to human needs did not really disqualify religion, since believers could say, yes, the need is the indication God gives us that he is there. As Rieff explains it, while other critiques of dogma could be integrated into religion, Freud's work could not be, because here "There is no distinctively religious need—only psychological need."[65] If we believe that we are deeply shaped by our primary family histories, the idea of God as a translation of feelings toward parents is compelling. If we believe there are neuroses, a lot of religious behavior seems psychological.

Still, it was in *The Future of an Illusion* (1927) that Freud entered the ranks of the world-class doubters. Reading *The Future of an Illusion,* it takes a moment, but we recognize in it an old friend. In chapter 4 of this history, we looked at Cicero's dialog *The Nature of the Gods,* and then in chapter 8 we saw that Hume's *Dialog Concerning Natural Religion* borrowed its structure from Cicero and reenvisioned the characters. Both used misdirection in their final assessment of who won their debate. Later, Schopenhauer reimagined the conversation as *Dialogue on Religion* with Philalethes, the philosopher, talking to Demopheles, the people. Schopenhauer's debate was not really on the nature of the gods, but on whether to preach the truth or leave the mass of people the religion they need. In the end, he called it a draw. *The Future of an Illusion* is loosely modeled on the Schopenhauer dialog.[66] After a few chapters, Freud explains that in order to keep his argument rigorous he had decided to "imagine that I have an opponent who follows my arguments with mistrust" and "I shall allow him to interject some remarks."[67] This was the Demopheles character, and the two argue for the rest of the book. Although Freud agrees that for most people religion gives them their only inkling of the philosophical world and their only chance to think about the universe, he thinks they could do better.

Freud calls it maturity. Religion, he says, "has an infantile prototype."[68] It is born of the child's feeling for the parent, and then in adulthood, and the adulthood of humanity, it is often sustained by fear of community rebuke.

> Let no one suppose that what I have said about the impossibility of probing the truth of religious doctrines contains anything new. It has been felt at all times—undoubtedly, too, by the ancestors who bequeathed us this legacy. Many of them probably nourished the same doubts as ours, but the pressure imposed on them was too strong for them to have dared to utter them. And since then countless people have been tormented by similar doubts and have striven to suppress them, because they thought it was their duty to believe.[69]

The book's message is that illusion is different from error; it is willful error. Because we want so much the consolations of God, heaven, purpose, and moral order, we should recognize that religion is a willful error, an illusion. As surprising as it would be if our very wishes turned out true, "it would be more remarkable still if our wretched, ignorant, and downtrodden ancestors had succeeded in solving all these difficult riddles of the universe."[70] Freud added disdain for those who believe we can't know anything about the real nature of things, so we "might as well" believe in God. "If ever there was a case of a lame excuse we have it here. Ignorance is ignorance; no right to believe anything can be derived from it."[71] The fideist God was thus a false idea. The God of the philosophers fared no better: "Philosophers stretch the meaning of words until they retain scarcely anything of their original sense. They give the name of 'God' to some vague abstraction which they have created for themselves; having done so they can pose before all the world as deists, as believers."[72] They even claimed this was a higher concept of God, "notwithstanding that their God is now nothing more than an insubstantial shadow and no longer the mighty personality of religious doctrines."[73] There's that idea again.

Freud's nameless Demopheles begins the penultimate chapter, asking, "Have you learned nothing from history?" He cites the French Revolution and Robespierre and "how short lived and miserably ineffectual the experiment was."[74] Furthermore, "The same experiment is being repeated in Russia at the present time, and we need not feel curious as to its outcome." Then he asks, Isn't it obvious people need religion?[75] Freud answered that people could handle the shock of the truth:

> They will have to admit to themselves the full extent of their helplessness and their insignificance in the machinery of the universe; they can no longer be the center of creation, no longer the object of tender care on the part of a beneficent Providence. . . . But surely infantilism is destined to be surmounted. . . . We may call this "education to reality." Need I confess to you that the sole purpose of my book is to point out the necessity for this forward leap?
>
> You are afraid, probably, that they will not stand up to the hard test? Well, let us at least hope they will. It is something, at any rate, to know that one is thrown upon one's own resources. One learns then to make a proper use of them. And men are not entirely without assistance. Their scientific knowledge has taught them much since the days of the Deluge. . . . As hon-

est smallholders on this earth they will know how to cultivate their plot in such a way that it supports them . . . they will probably succeed in achieving a state of things in which life will become tolerable for everyone and civilization no longer oppressive to anyone. Then, with one of our fellow-unbelievers, they will be able to say without regret: "We leave heaven to the angels and the sparrows."[76]

The *fellow-unbeliever* who wrote that last line was the poet Heine. Heine had coined the word *Unglaubensgenossen*, in reference to Spinoza.

In the final section Freud tries to explain to his imagined opponent, almost apologetically, why he is bothering humanity with his calls to doubt. Listen to the care in Freud's language: "Take my attempt for what it is. A psychologist who does not deceive himself about the difficulty of finding one's bearings in this world, makes an endeavor to assess the development of man" based on a lifetime of studying children and adults. In the process, "the idea forces itself upon him that religion is comparable to a childhood neurosis, and he is optimistic enough to suppose that mankind will surmount this neurotic phase." Could one blame him for saying so? After all, Freud said, he knew the proposed psychologist (himself) might be wrong: "These discoveries derived from individual psychology may be insufficient, their application to the human race unjustified, and his optimism unfounded. I grant you all these uncertainties. But often one cannot refrain from saying what one thinks, and one excuses oneself on the ground that one is not giving it out for more than it is worth."[77] Freud had a method for working toward what he called maturity, and so could imagine that the whole group might develop, and this made it easier to hypothesize that society might someday dispense with religion.

Freud's most dedicated scholars have treated him rather badly in regard to his beliefs about religion. For Rieff, Freud's genius for some reason fails him on the issue of religion:

> Freud's customary detachment fails him here. Confronting religion, psycho-analysis shows itself for what it is: the last great formulation of nineteenth-century secularism, complete with substitute doctrine and cult—capacious, all-embracing, similar in range to the social calculus of the utilitarians, the universal sociolatry of Comte, the dialectical historicism of Marx, the indefinitely expandable agnosticism of Spencer. What first impresses the student of Freud's psychology of religion is its polemical edge.[78]

Peter Gay reminds his readers that "Critics might say that he was merely looking for scientific-sounding reasons to confirm his anti-religious stance."[79] It is true that nineteenth-century secularism was a very particular movement—based on the transition to democracy and identification of religion with unyielding authority and obscurantism—and Freud was alive and working within that secularist movement. He used some of its devices, and his thought was shaped by it. But to read his work is to see that he was no yahoo indicting God out of anger at the church. Freud was part of a long and sober history of doubt. His company is not just Mill, Comte, Marx, and Spencer but also Cicero, Montaigne, Schopenhauer, and Hume.

PHILOSOPHERS OF SCIENCE, OF SILENCE, AND OF SISYPHUS

Twentieth-century philosophy was rich in outspoken and engaged doubters. Bertrand Russell was a central figure of the first half of the century. In his essay "Why I Am Not a Christian" (a 1927 lecture), Russell revived an old approach: he enumerated the proofs of God and why they no longer held. Here's his take on the idea of a first cause:

> I for a long time accepted the argument of the First Cause, until one day, at the age of eighteen, I read John Stuart Mill's *Autobiography*, and I there found this sentence: "My father taught me that the question 'Who made me?' cannot be answered, since it immediately suggests the further question 'Who made god?'" That very simple sentence showed me, as I still think, the fallacy in the argument of the First Cause.[80]

If everything must have a cause, what caused God? "If there can be anything without a cause, it may just as well be the world as God, so that there cannot be any validity in that argument." Next Russell came to "the natural-law argument," which tried to prove God through the existence of his natural laws. That did not wash either, because "Nowadays," with Einstein "you no longer have the sort of natural law that you had in the Newtonian system, where, for some reason that nobody could understand, nature behaved in a uniform fashion." On the argument that the world needs a designer, Russell mentioned Darwin. There were no good reasons to believe, but that did not stop people. "You all know, of course," wrote Russell, "that there used to be in the old days three intellectual arguments for the existence of

God, all of which were disposed of by Immanuel Kant," yet "no sooner had he disposed of those arguments than he invented a new one, a moral argument, and that quite convinced him."[81] Added Russell, "He was like many people: in intellectual matters he was skeptical, but in moral matters he believed implicitly in the maxims that he had imbibed at his mother's knee." "The psycho-analysts" were right, mused Russell: "our very early associations" have an "immensely stronger hold on us" than our later experiences. This modern doubter ticked off philosophical proofs, shook a fist against injustice, and scribbled a Freudian note on the illusions of believers.

Russell is not worried that doubt ought not be sown among the masses. He is certain that doubt is good. In his terms, "We want to stand upon our own feet and look fair and square at the world—its good facts, its bad facts, its beauties, and its ugliness; see the world as it is and be not afraid of it."[82] For him, the idea of God is an idea taken from "ancient Oriental despotisms" and "quite unworthy of free men." In his 1947 "Am I an Atheist or an Agnostic?" Russell says he can't prove "that either the Christian God or the Homeric gods do not exist, but I do not think that their existence is an alternative that is sufficiently probable to be worth serious consideration."[83]

He was awarded the Nobel Prize for Literature in 1950. In 1954 he published a book on ethics and, having seen the horrors of the century, gave this account of modern doubt: "I do not believe that a decay of dogmatic belief can do anything but good. I admit at once that the new systems of dogma, such as those of the Nazis and the Communists, are even worse than the old systems, but they could never have acquired a hold over men's minds if orthodox dogmatic habits had not been instilled in youth." Specifically, Stalin's language "is full of reminiscences of the theological seminary," Russell explained. "What the world needs is not dogma, but an attitude of scientific inquiry, combined with a belief that the torture of millions is not desirable, whether inflicted by Stalin or by a Deity imagined in the likeness of the believer."[84] A doubter could thus blame religion, rather than doubt, for the religiosity of the modern state, but the main point here is that freedom of thought is better than state religion or state atheism.

Russell made major contributions in ethics and epistemology and took logic way beyond the Scholastic logic that had long reigned. He also added a new voice of contented doubt. "I believe that when I die I shall rot, and nothing of my ego will survive," he explained. "I am not young, and I love life. But I should scorn to shiver with terror at the thought of annihilation. Happiness is none the less true happiness because it must come to an end";

thought and love do not lose their value because they are not eternal. As with other doubters, Russell's advice was that after the initial shock, doubt is nice: "Even if the open windows of science at first make us shiver after the cozy indoor warmth of traditional humanizing myths, in the end the fresh air brings vigor, and the great spaces have a splendor of their own."[85] For the enlightened, the universe can be a wonderful home.

It has not always been easy to hear the voice of women doubters, although we have seen that marriages were often made on the basis of mutual religious doubt. Russell's wife, Dora Black Russell (1894–1986), had become a freethinker at a young age and joined a local Heretic's Society while still in high school. After marrying Bertrand, and traveling, she increasingly saw religion as bad rather than merely erroneous. Her first book, *Hypatia* (1925), was about women's reproductive and sexual freedom, but—as suggested by its title—a critique of Christianity ran through it. In her *Religion of the Machine Age,* Black Russell wrote, "When the male of the species, enamored of his stargazing, set up a God outside this planet as arbiter of all events upon it, and repudiated nature, together with sex, for a promised dream of a future life, he turned his back on that creative life and inspiration that lay within himself and his partnership with woman. In very truth he sold his birthright for a mess of pottage."[86] She would not be the last theorist to suggest that religion was a denial of the good real things of life—work, love, children, and play. Others would also champion her claim that while men embraced a sacred fantasy that privileged them, the mundane world left to women was not only the real world, it was the better one.

A very different doubter knew the Russells. Ludwig Wittgenstein did not go around rejecting the Bible, or religion, or God, and yet he is a key doubter. He was a fascinating man. The youngest of eight siblings in an extremely rich family, he gave up his inheritance and lived in austere simplicity. Three of his four brothers committed suicide. As a young man he studied philosophy at Cambridge with Bertrand Russell, and at that time wrote a brilliant book of philosophy, the *Tractatus Logico-Philosophicus.* He then left the discipline for ten years, taught school to children, then returned to philosophy and Cambridge (as a professor now) to write a wealth of brilliant material that began by questioning his own first book, the *Tractatus.* Most of this writing came to light only after his death, but his lectures had already transformed much of philosophy.

Wittgenstein claimed that philosophy was just a matter of conceptual knots that got tied by language. Language is ad hoc; it works in weird ways developed by the community over history, so that a concept like "time" seems to have all sorts of paradoxes in it that are really just problems of language. We all know what time is and how to use all the complex terms for it; only the philosophers had confused things by trying to resolve the paradoxes of language and by insisting on some overall picture of what the world means. For Wittgenstein, meaning is defined by a community playing a language game together—although reality does exist and limits the kinds of games that can be played.[87] Philosophy cannot think outside the communal game, so what it is good for is undoing its own language knots: "Philosophy is a battle against the bewitchment of our intelligence by means of language."[88] He got called the new Socrates and shifted philosophy toward teasing out the rules and sense of language to figure out how we mean things. Lots of people, from Comte, at least, to Russell, had thought that philosophy was going to be about science now, that science was the only approach to investigating the world that had held up. Wittgenstein showed that philosophical problems were still worth talking about because they can be unraveled in language, and since language is our reality, this process has a lot to teach.

The book Wittgenstein was finishing in the days before his death was called *On Certainty* and was an answer to Skepticism. It claimed that doubting, by its nature, is done within the realm of believing something: "If you tried to doubt everything you would not get as far as doubting anything."[89] For Wittgenstein, to doubt everything is like searching for your keys by opening the same drawer over and over: it's missed the point of the game. He defines true doubt as being hinged on certainties, and if we accept that, the Skeptic argument becomes a mistake. As for the Skeptic worry that we are dreaming, Wittgenstein denies the problem: what you think in sleep (even if it is "it is raining" and it so happens that it *is* raining outside) isn't part of the conversation: "And suppose a parrot says: 'I don't understand a word' or a gramophone: 'I am only a machine'?" They are not telling the truth or lying; they do not have the criteria for being in the conversation and neither does a sleeping person. As the Wittgenstein scholar Avrum Stroll has put it, "The background conditions for sensible assertion are, among other things, that a person be fully aware of what his words mean and intend them to make a statement."[90] If we are left with life as a language game, it may be more accurate to say Wittgenstein rewrote the Skeptic problem than that he conquered it. Still, reading the Skepticism of Sextus Empiricus, Montaigne, and Descartes can be kind of annoying—for instance, because we know when we are awake—and there is a way that Wittgenstein's *On Certainty* grabs hold of doubt where Skepticism had only badgered it. Yet, it is not certainty we are left with. Wittgenstein's process leaves us with even more of a sense that we inhabit a world of belief and cannot see out of it.

Wittgenstein did not write much about whether he believed in God or religion. Still, we know he praised mysticism and that there was tension between him and Russell over the latter's idea that science was the route to truth. Wittgenstein was born and died Catholic, although three out of four of his grandparents' families had been Jewish and converted to Catholicism after Napoleon's defeat. We know he was very conflicted about this, and even a bit self-hating. We also know that as a soldier in World War I, he carried Tolstoy's translation of the Bible around so much that he was known as the guy with the Gospels. In 1938 Wittgenstein gave three lectures on religious belief, and his students have left us their notes. Here Wittgenstein did not speak about his own belief but of the separate nests of propositions occupied by the believer and the unbeliever, such that they cannot even contradict each other:

> If you ask me whether or not I believe in a Judgment Day, in the sense in which religious people have belief in it, I wouldn't say: "No. I don't believe there will be such a thing." It would seem to me utterly crazy to say this.
>
> And then I give an explanation: "I don't believe in . . . ," but then the religious person never believes what I describe.
>
> I can't say. I can't contradict that person.
>
> In one sense, I understand all he says—the English words "God," "separate," etc. I understand. I could say: "I don't believe in this" and this would be true, meaning I haven't got these thoughts or anything that hangs together with them. But not that I could contradict the thing.[91]

Meaning is embedded in constructs of beliefs. There is also a sense of Wittgenstein's own religious doubt here. In *On Certainty* we also get hints of his doubt: "I believe that every human being has two human parents; but Catholics believe that Jesus only had a human mother. And other people might believe that there are human beings with no parents, and give no credence to all the contrary evidence."[92] Still, his greatest doubt insists we can speak of the world only in our language game, and otherwise must be silent.

Later in the century the philosophy of doubt was most often expressed in literature, especially that of the existentialists. Existentialism was a mostly German and French movement that started between the world wars and hit its height after the Holocaust and the atom bomb. Pascal and Kierkegaard are understood as forerunners of existentialism, because they both spoke of the choice of faith, and because both were marked by fear and trembling at the thought of the abyss. Existentialists saw pain the way the Buddha did, as the predominate feature of existence. They handled it differently—the Buddha with a triumphant exit from the pain; the existentialist with either a leap of faith or a sneer and a shrug.

Existentialist philosopher Karl Jaspers argued for belief in God in the way that Kierkegaard had discussed: through absurdist irrationalism, a leap. Other well-known existentialists, like Jean-Paul Sartre, did not believe in God. Sartre's father died when he was young; he lived with his mother, a mild Christian, and with his grandparents, one of whom was Protestant and the other Catholic. Although they did go to church occasionally, there wasn't much to it, and as Sartre later attested, "I was led to unbelief not through conflicting dogma but through my grandparents' indifference." For him, belief was good-natured conformism: "In our circle, in my family," he explained, "faith was nothing but an official name for sweet French liberty."

Sartre's atheism was at the heart of his theory of human identity, for in his mind "There is no human nature because there is no God to have a conception of it." Sartre said if there is no idea of us before the fact of us, if there is no blueprint, we are really free. Even the brain itself doesn't dictate who we are. Sartre said "existence precedes essence," which means that a human being "first of all exists, encounters himself, surges up in the world—and defines himself afterwards."[93] Since "existence is prior to essence, man is responsible for what he is."[94] Thus, Sartre explained, if you marry and have children, for all your own reasons, you are also creating human reality.[95] In the way we live, we are the author of what humanity shall be. For Sartre, this was an intense command upon us to be moral, for our actions are the only reality. He was very aware of the recent history of doubt, writing that the "philosophic atheism" of the Enlightenment had the idea that God was gone, but not the idea that essence is prior to existence. "In Diderot, in Voltaire and even in Kant" there is human nature. For Sartre there was heroism in a moral life because we are all free and all responsible.

In Sartre's play *No Exit* the character Garcin provides the cosmology: "Hell is just other people."[96] It is a sour and funny little notion, from the sour and funny faction of the history of doubt. Sartre says more about his childhood in his *War Diaries*. His grandfather (the Protestant) disliked "the whole religious business" on principle and had "a 'dissenters' contempt for clerics."[97] Sartre wrote, "I think he cracked anti-clerical jokes at the table and my grandmother rapped him on the fingers for it, saying, 'Be quiet Dad!'" Sartre's mother made him take First Communion, but more for propriety "than from true conviction." She herself had "no religion, but rather a vague religiosity, which consoles her a bit when necessary and leaves her strictly in peace the rest of the time." He then described his own vague childhood belief, and announced:

> So there you are. It's pretty thin. God existed, but I didn't concern myself with him at all. And then one day at La Rochelle, while waiting for the Machado girls who used to keep me company every morning on my way to lycée, I grew impatient at their lateness and, to while away the time, decided to think about God. "Well," I said, "he doesn't exist." It was something authentically self-evident. . . . I settled the question once and for all at the age of twelve. Much later I studied religious proofs and atheist arguments. . . . I think I ought to say all this because, as I have said, I am affected by moralism, and because moralism often has its source in religion. But with me it was nothing of the kind.[98]

He reminded his readers that for him this was not rebellion: "the truth is I was brought up and educated by relatives and teachers most of whom were champions of secular morality and everywhere sought to replace religious morality by it."[99] These were the anticlerical, secularist teachers we saw at the end of the previous century. Existentialism emerged into the history of doubt with a pedigree.

Simone de Beauvoir, philosopher and central voice of twentieth-century feminism, started out as a religious girl. She stopped believing in God when she was fourteen, as she recounts at some length in the first volume of her autobiography. Reading Balzac one afternoon:

> "I no longer believe in God," I told myself, with no great surprise. . . . That was proof: if I had believed in Him, I should not have allowed myself to offend Him so light-heartedly. I had always thought that the world was a

small price to pay for eternity; but it was worth more than that, because I loved the world, and it was suddenly God whose price was small: from now on His name would have to be a cover for nothing more than a mirage.[100]

For a long time she had had an idea of God that was "purified and refined, sublimated to the point where He no longer had any countenance divine" and was not at all connected to earth or any being on it. Finally, "His perfection cancelled out His reality." It is a now classic observation beautifully expressed. Yet there is something new here, too: "My Catholic upbringing had taught me never to look upon any individual, however lowly, as of no account: everyone had the right to bring to fulfillment what I called their eternal essence. My path was clearly marked: I had to perfect, enrich, and express myself in a work of art that would help others to live."[101] Ever since the Inquisition, doubters who were former Catholics had been generally furious with Catholicism. It's important that Beauvoir's break with the church was not political, not a thing of outrage, but a matter of thought. When she met Sartre, she adopted his notion that in a godless universe there is a desperate need for each of us to be moral and to act for the betterment of life. Yet for her, the original impulse came from Catholicism's inspiring notion of human value and its call to action.

Albert Camus is often associated with the existentialists and was a serious doubter. In his "Myth of Sisyphus" (1942), Camus tells us that Sisyphus is the absurd hero.[102] In his "scorn of the gods" and because he is condemned to push the boulder up the mountain, over and over, with no progress, he represents the image of humanity itself in a godless world. But it's not all bad: "If this myth is tragic," explained Camus, "that is because its hero is conscious. Where would his torture be, indeed, if at every step the hope of succeeding upheld him? The workman of today works every day in his life at the same tasks, and his fate is no less absurd. But it is tragic only at the rare moments when it becomes conscious."[103] Creepy. It comes to seem less creepy to be conscious all the time. That very consciousness is what makes humanity victorious: "The lucidity that was to constitute his torture at the same time crowns his victory. There is no fate that cannot be surmounted by scorn."

Despite the funny idea of world-conquering scorn, Camus keeps coming back to happiness, even when headed downhill with the stone. "If the descent is thus sometimes performed in sorrow, it can also take place in joy. This world is not too much." He mused that, sometimes, "The boundless grief is too heavy to bear. These are our nights of Gethsemane. But crushing truths perish from being acknowledged."[104] Acknowledging the absurdity of the human condition is what saves us. In fact, he writes: "One does not discover the absurd without being tempted to write a manual of happiness." Why? Because it makes fate a human matter. "It drives out of this world a god who had come into it with dissatisfaction and a preference for futile suffering."[105] It's our ball game, such as it is. "All Sisyphus' silent joy is contained therein. His fate belongs to him. His rock is his thing."

> I leave Sisyphus at the foot of the mountain! One always finds one's burden again. But Sisyphus teaches the higher fidelity that negates the gods and raises rocks. . . . This universe henceforth without a master seems to him neither sterile nor futile. Each atom of that stone, each mineral flake of that night filled mountain, in itself forms a world. The struggle itself toward the heights is enough to fill a man's heart. One must imagine Sisyphus happy.[106]

This is graceful-life philosophy and has remarkable echoes of the ancient past.

BACK TO JOB, BACK TO THE GYMNASIUM

Jews in Germany were emancipated in 1871 and had equal rights under the Weimar Constitution, between the world wars. Twentieth-century descendants of "Enlightenment Jews" in Berlin carried on a Spinoza craze culminating in the 1927 celebration of the 250th anniversary of his death, and in the 1932 celebration of the 300th anniversary of his birth. A rationalist intellectualism reigned there between the wars.

When Mordecai Kaplan's *Judaism as a Civilization: Toward a Reconstruction of American-Jewish Life* came out in May 1934, he had already been running a very modern Jewish group for years. Kaplan saw problems with the Orthodox, Reform, and newer, in-between position of Conservatism. Judaism, the book announced, should be seen as a "civilization" with religion as but one part. Jewish poetry, painting, music, and other arts were emphasized. The book consciously followed Emile Durkheim's notion that the sense people have of God is a real sense, but that the feeling emanates from the community. Jewish civilization was a living people that had constantly evolved. As it did so, many of its central ideas, including its "God ideal," shifted. Jews had reached a point at which some of them no longer believed in God, Kaplan explained, but they were still Jewish because the

Jewish people are the definition of Judaism. Kaplan rejected all supernaturalism: revelation was not true, Jews were not the chosen people, and although prayer felt good, it could not work. He saw Jewish rituals as "folkways" that preserved the group's memories, identity, and values. Kaplan hoped that this cultural Judaism would take moral responsibility for the world, encourage social justice, and celebrate the arts. After Kaplan, these ideas grew into Reconstructionism, a fourth Jewish denomination.

Milton Steinberg's *A Driven Leaf* (1939) took as its hero the famous heretic of the Talmud, Elisha ben Abuyah. In the years before the last revolt against ancient Rome, Rabbi Elisha had left Judaism. The Talmud said that one day, as Elisha and a few rabbis were walking, they witnessed a boy die moments after doing a mitzvah said to give long life. Elisha declared, "There is no Justice and there is no Judge." Steinberg drew on his great scholarship of the period and the little information we have on these people and events, and made up the rest. As Steinberg imagined it, Elisha's mother died in childbirth. Elisha's father did not believe in God or practice Judaism, but rather loved Greek wisdom and used Stoic naturalism to argue against miracles.[107] Elisha was circumcised only because of the insistence of his late mother's brother, a rabbi. Later, his father finds the young Elisha a Greek tutor, Nicholas, but the father soon dies and Elisha's uncle steps in, fires Nicholas, and, over time, helps Elisha to become a great rabbi. Years later, Elisha suffers doubts about God. He asks his friends to explore a proof for God, looking to mysticism and Greek rationalism; one friend dies and one goes mad. Elisha cannot stop searching. In a quest for philosophy he meets up with Nicholas the Greek again, and his old tutor helps him choose books, including what he calls "the most blasphemous book ever written, but brilliant": Euhemerus's *Sacred History*.[108] Nicholas reminds Elisha that Judaism has a beauty and a pity for humanity that the philosophers do not know, but Elisha responds that he can't help his doubting. When Job's friends advised submission, Job cried out, "'Wilt Thou harass a driven leaf?' I know how he felt. The great curiosity is like that. It is not a matter of volition."[109] Here the line is a plea for tolerance for doubters: they cannot refuse their questions. The book's title shows the importance, for Steinberg, of this link from the Job author, to Elisha ben Abuyah, to the doubts of modern Jewry.

Steinberg gives Elisha a friendship with the real wife of Elisha's real student, Beruriah, wife of Rabbi Meir, and the death of her young boys grinds at Elisha's faith. Meanwhile, other rabbis worry over the rise of "godlessness" and of "exercising in gymnasiums, sitting in circuses, lounging all night in drunken symposiums and running in pursuit of harlots."[110] Amid this tension it becomes known that Rabbi Elisha has Greek ideas and Jewish doubts. One day, walking with some other rabbis, he sees a father exhorting his son to a mitzvah—go chase away the mother bird before we take the eggs. The rabbis say the boy will have a long life, and the boy falls dead. Steinberg wrote of Elisha, "A great negation crystallized in him. The veil of deception dissolved before his eyes." Then Elisha says the line from the Talmud flanked by some flourish: "It is all a lie. . . . There is no reward. There is no justice. There is no Judge. For there is no God."[111]

Elisha is excommunicated and heads off to cosmopolitan Antioch to devote himself to Greek study. He cuts his hair like a Greek and goes to banquets with brilliant, beautiful women and a cast of Stoics, Epicureans, and even a Cynic who comes uninvited, eats messily, and then stretches out on the floor. One of Elisha's Greek tutors mentions that he does not believe in the gods and, "What is more, I do not know a single educated person who does, except the Stoics who interpret them as allegories for the forces of nature."[112] The tutor reports that he is not bothered by this. Elisha, by contrast, sacrifices everything on his quest for truth, to the point of betraying his own people to the Romans. Steinberg's Elisha chooses the secular, open mind of the Roman state over the Jewish clannish cult. In the end, Elisha concludes that it was a bitter and unbearable mistake: he continues to value Roman freedom of thought and unintrusive secularism but cannot accept the lack of heart in its bloody circuses, its materialism, and its treatment of workers and slaves. His devotion to finding truth fares no better: Elisha concludes that there is no way to find the truth by reason, not even with Euclid, and Steinberg gives him a little non-Euclidean geometry to make the argument.

The truth that Elisha comes to is Steinberg's point: logic fails, but humanism does not. At the very end, alienated from both Rome and the Jews, Elisha sees his old student Rabbi Meir. It is Sabbath and they meet by chance near the road mark past which a Jew could not walk on Sabbath. Meir begs Elisha to return to the Jews, and Elisha admits that he wants what is back there, in the world of belief and community, but sighs that he cannot: "For those who live there insist, at least in our generation, on the total acceptance without reservation of their revealed religion. And I cannot surrender the liberty of my mind to any authority. Free reason, my son, is a heady wine. It has failed to sustain my heart, but having drunk of it, I can never be content with a less fiery draught."[113] Then off he goes, past the boundary marker, and down the road, to disappear into the distance. Who was Steinberg? He was born in 1903, grew up in Harlem, and studied with Mordecai Kaplan at the Jewish Theological Seminary. He was ordained in 1928 and then served as a rabbi in Indiana, becoming famous for siding with the workers during a local strike.

Early-twentieth-century Jewish doubt was deep and melancholy before World War II, but there was much rationalism. The atrocities of the Holocaust were difficult to reconcile with any concept of God, so henceforth, alongside this rationalist Jewish doubt there was a new doubt that came more from the belly; more Job than Spinoza. Elie Wiesel was religious when he was very young, but when he was still a child he was taken to Auschwitz and he lost his faith on his first night there, looking upon the crematoriums to which his mother and sister were headed. In his stunning memoir, *Night*, he wrote, "Never should I forget that nocturnal silence which deprived me, for all eternity, of the desire to live. . . . Never shall I forget these moments which murdered my God and my soul and turned my dreams to dust."[114] He also tells us of a day when the Gestapo took the young boy who was the tiny light of goodness for everyone in the camp and hanged him—he was not heavy enough for his neck to break so it took half an hour for him to slowly suffocate. Watching, someone asked, "Where is God now?" and Wiesel said to himself, "Where is He? Here He is—He is hanging here on this gallows."[115] For many Jews, belief in God ended.

The most famous story about belief and the Holocaust shows that it is a complicated matter. It is said that a group of Jews in Auschwitz put God on trial, indicting him for abuse and evil beyond any argument or mitigation. They found him guilty. The rabbi solemnly spoke the verdict and then announced it was time for evening prayers. For the observant, profound doubt in God did not always mean stopping the practices. For those who did not have practices and did not have God, Jewish identity was growing elusive, just as the attack on Judaism made it seem imperative that Jews remain a vibrant presence in the world, in defiance.

Some secular philosophy came out of the camps. The renowned psychotherapist Viktor E. Frankl lost wife, parents, everyone, and was himself in the death camps for years and suffered endless unspeakable horrors. In a night marked by disease, starvation, and darkness, someone asked Frankl to say something to comfort everyone, to keep them from suicide. He found himself saying that even if you expect nothing more from life, life still expects things from you: to be there for someone else, to use a talent, to bear suffering. In another moment of awakening, on a long, horrible march

with almost no hope of survival in the long run, he suddenly began thinking about the psychology of his experience there, and even imagined giving
a talk on it to a posh crowd in some distant future. It was good. "Both I
and my troubles became the object of an interesting psychoscientific study
undertaken by myself. What does Spinoza say in his Ethics? . . . 'Emotion,
which is suffering, ceases to be suffering as soon as we form a clear and precise picture of it.'"[116] Frankl developed all this into logotherapy, or *meaning
therapy*, which held that Freudian psychoanalysis was wrong to say that
instinctual drives are real while ideas of meaning are just a "secondary
rationalization."[117] Frankl's experience had led him to believe people need
and want meaning more than anything else, and most of them freely
acknowledge that when asked. Therapy must help people find meaning.
There is nothing antireligious about this doctrine, and Frankl encouraged
the believer to see God as his or her meaning, but no more than he encouraged a nonbelieving musician to live for the song. From the most terrible
place, Frankl joined doubt's conversation on how to live in a world like this.

Jewish doubt after the Holocaust was both innovative and traditional. For
many, the liberal Jewish God no longer maintained any coherent meaning,
but the old Gnostic, Cabalistic images of God rang truer than ever. Cabala's
emphasis on religious feelings and the pointedly nonrational character of its
knowledge sources have appealed to many who could not reconcile rationality and religion. Also, not only does it deny providence, it makes human
beings responsible for fixing the world. Our generosity is not judged; rather,
it adds up to something. The great scholar of medieval Jewish mysticism
Gershom G. Scholem explained that mysticism—which seems so full of
belief—actually depends on people feeling separated from God.[118] It was true
of Jewish medieval mysticism and of the modern Jews who read Scholem.
The Holocaust brought the idea of Jewish exile to the fore again, and it also
caused an about-face in the nineteenth-century rejection of Zionism.

Margarete Susman (1874–1966), the brilliant German-Jewish philosopher and poet, offered a moral struggle with God, even in the absence of
God. Susman described her worldview through Job. An essay of 1956
begins with the words "Since the earliest times and down to this day Israel
has not ceased to quarrel with God, to take man's part in his dispute with
God for His justice."[119] The Jew, she explained, has long suffered homelessness, but now the Jews have to share the rest of the Western culture's homelessness as well as their own. "God, for whose sake they have accepted all
this, cannot be found any longer, because . . . the revealed God whom

the Jew has accepted has become, in a manner unimaginable till now, the *Deus absconditus,* the absent God, the God who simply can no longer be found."[120] Then she makes a wonderful leap: "The dispute with God cannot cease even now. . . . Just as He evaded Job in his personal fate, so He evades the modern Jew in his universal fate. For this reason the process against God must assume a new shape . . . a version in which God is all silence and man alone speaks. And yet, though His name is never mentioned, only He is addressed." If God is the idea of true good and true power, we must keep wrestling with him even without him. Susman calls out with Job, "Wilt Thou harass a driven leaf? And wilt Thou pursue the dry stubble?"[121] Like the rabbis who found God guilty and then hurried to evening prayers, Susman issued a total rejection but would not leave it all behind. It is a testament to the poetry of the conversation.

The philosopher Walter Kaufman has likened rationalist believers in God to the friends of Job, suggesting that those whose belief in God is based on life experience "must have led sheltered lives."[122] Kaufman quotes Ecclesiastes, saying that when he sees how the oppressed suffer, and how everyone ignores it, he thinks the dead are better off than the living. For Kaufman, the division between theists and atheists is not as important as the division between those who feel the suffering of the world and those who do not. Hence, "The only theism worthy of our respect believes in God not because of the way the world is made but in spite of that."[123] The only theism as profound as the Buddha's atheism, Kaufman averred, was the theism of people like Job: "born of suffering so intense" that they "must shriek, speak, accuse and argue with God—not about Him—for there is no other human being who would understand."

Richard Rubenstein's *After Auschwitz* (1966) also refused the God of Judaism. Rubenstein spoke of "death of God theology," which had been the purview of the Christian existentialists, as the only way to think about Judaism in the world after the Holocaust.[124] Rubenstein was deeply engaged in the study of psychology and would come to practice Buddhism. Other doubters discussed above were philosophers and poets; Rubenstein was a rabbi, writing for other rabbis.[125] He met with career blocks at first, which amounted to what has been called a bureaucratic excommunication, but now he is seen as having inspired the wealth of post-Holocaust theology that has become a vital component of Jewish culture.

Jews came back to Job, and also back to the gymnasium: the world of the gentiles, "Greek learning," and Mammon, with New York City as the new Alexandria. In many other cities, too, there are Jews who observe as many laws of Moses as did Abraham and Sarah, but who identify strongly as Jews. Does the Jew in the gymnasium doubt? The Jewish voice in mainstream society is often cynical, skeptical, and philosophically materialist. Indeed, the doubt is so thick that wry humor is as much a factor of Jewishness as the beards of the Orthodox. Cultural, secular Judaism has its own claims. As Lenny Bruce said, "Even if you are Catholic, if you live in New York you're Jewish. If you live in Butte, Montana, you are going to be goyish even if you are Jewish." Humor is a good place to look for Jewish doubt in the gymnasium of modern culture. Consider the professed philosophy of an icon of this humor, Jackie Mason:

> Life has no meaning beyond this reality. But people keep searching for excuses. First there was reincarnation. Then refabrication. Now there's theories of life after amoebas, after death, *between* death, *around* death. Now you come back as a shirt, as a pair of pants. . . . People call it truth, religion; I call it insanity, the denial of death as the basic truth of life. "What is the meaning of life?" is a stupid question. Life just exists. You say to yourself, "I can't accept that I mean nothing so I have to find the meaning of life so that I shouldn't mean as little as I know I do." Subconsciously you *know* you're full of shit. I see life as a dance. Does a dance have to have a meaning? You're dancing because you enjoy it.[126]

Mason describes his rejection of religion in part as a rebellion against his strict Orthodox father, but he has dismissed God most resolutely in an ethical critique. The Holocaust and the world's poverty, sickness, and cruelty led him to believe that "If God exists he's an idiot. That's why I don't believe in any God. Because if that's how he behaves, I don't want to know such a person."[127] Few Jewish comedians have offered such analysis, but they speak to the issue in other ways. Henny Youngman said, "I wanted to become an atheist but I gave up. They have no holidays." As for Woody Allen: "Not only is there no God, but try getting a plumber on weekends."[128]

COLD WAR AND POSTMODERN CULTURE

In Cold War America, atheist meant communist. In 1954 a law was passed changing the national motto of the United States from "E Pluribus Unum" to "In God We Trust." In 1955 another law required the new motto to be on all

U.S. currency (it had been there occasionally since 1863), and in 1956 yet another law added the words "under God" to the Pledge of Allegiance. These laws had largely been initiated by Lyndon Johnson when he was Speaker of the House, and were put through by Presidents Nixon and Eisenhower. The *Congressional Record* shows that Congressman Charles G. Oakman supported the laws because: "Our belief in God highlights one of the fundamental differences between us and the Communists."[129]

The well-known Reverend George Docherty preached a sermon in favor of God in the Pledge, and Eisenhower listened. Congressman Louis C. Rabaut, sponsor of the law on the Pledge (House Joint Resolution 243), cited him later, saying: "You may argue from dawn to dusk about differing political, economic, and social systems, but the fundamental issue which is the unbridgeable gap between America and Communist Russia is a belief in Almighty God."[130] Rabaut added, "Unless we are willing to affirm our belief in the existence of God and His creator-creature relationship to man, we drop man himself to the significance of a grain of sand and open the floodgates to tyranny and oppression. An atheistic American, as Dr. Docherty points out, is a contradiction in terms." Rabaut clearly did not know his history of doubt: "This country," he claimed, "was founded on theistic beliefs, on the belief in the worthwhileness of the individual human being which in turn depends solely and completely on the identity of man as the creature and son of God."

The money bill (HR 619) was introduced in the House with the following intent: "Nothing can be more certain than that our country was founded in a spiritual atmosphere and a firm trust in God. . . . At the base of our freedom is our faith in God and the desire of Americans to live by His will and by His guidance. As long as this country trusts in God, it will prevail." The money would bear "a constant reminder of this truth."[131]

Madalyn Murray O'Hair was a strange bird, part Twain, part Paine, part Royall, part Menocchio (the blaspheming miller put to death in 1599). In 1963 students at public schools in Baltimore started every day by reading a chapter of the Christian Bible and/or the Lord's Prayer. William Murray was in high school at the time, and he and his mother, Madalyn, sued. Along with another suit filed by the Schempp family against school Bible reading in Pennsylvania, the Murrays' action brought about two landmark Supreme Court rulings: compulsory prayer and Bible reading were both banned in public schools. Taking on the name O'Hair in a second marriage, Madalyn went on to found American Atheists, which became the largest atheist organization in the nation. Lots of people were put off by O'Hair, and it wasn't just the atheism. She could be very brash, even crass. When *Playboy* ran its famous interview with her in 1965, the magazine called her the "most hated woman in America," and in the article she said things such as: "The 'Virgin' Mary should get a posthumous medal for telling the biggest goddamn lie that was ever told. Anybody who believes that will believe that the moon is made out of green cheese." And furthermore, "I'm sure she played around as much as I have."[132] Adding to the discomfort she aroused, in 1995 she and two family members disappeared, and were found, killed, in 1999—apparently the victims of a crime that had to do with money and not metaphysics. Just before her disappearance, O'Hair was asked by an interviewer what she was proudest of having accomplished. She replied, "Oh, one of the things I'm most proud of is that people can say, 'I am an atheist,' in the United States today, without being called a Communist atheist, or an atheist Communist. I separated the two words. I think that that's probably the best thing that I did."[133] It is funny to think that a lot of doubters are today a little embarrassed by the way atheism is now associated with O'Hair, and to realize how profoundly she thereby achieved her goal. Evangelistic atheism no longer feels treasonous— it just seems, like O'Hair, harsh, a little coarse, and not at all mainstream.

In the second half of the twentieth century, Margaret Knight held a similar position as the most famous British atheist—although with a cleaner image. Reading Bertrand Russell and other philosophers at Cambridge had swept away the last of her already dwindling religious beliefs: "I let them go with a profound sense of relief, and ever since I have lived happily without them."[134] She became famous when she convinced the BBC to let her do a series of atheist broadcasts, offering advice on how to teach morals to children without religion. The shows were avidly antireligious and argued that secular morality was superior to religious morality. The texts were later published in a 1955 book, and it is useful to see how she framed the matter there: "The fundamental opposition is between dogma and the scientific outlook. On the one side, Christianity and communism, the two great rival dogmatic systems; on the other Scientific Humanism."[135] Clearly Knight's project is in part the same as O'Hair's, but not only does she separate communism and atheism, she seeks to link communism and Christianity: all the tired old beliefs were going to have to make way for secular humanism and science.

When the Cold War ended, it became clear that religion had been a big part of the resistance movements that brought down the Berlin Wall in

1989. Yet doubt had a strong hand in this, too. Václav Havel vitalized the revolution with a nonreligious devotion to speaking and living the truth: the simple resistance of not lying or hiding. Havel was brought up Roman Catholic and does not now consider himself a believer, but he wants more than materialism, self-interest, and gadgetry. Havel has said that the principle of human rights emerged from the Enlightenment as "conferred on man by the Creator," but the anthropocentrism of that idea "meant that He who allegedly endowed man with his inalienable rights began to disappear from the world."[136] Human rights have to be anchored in something new. "If it is to be more than just a slogan mocked by half the world, it cannot be expressed in the language of a departing era." But it can't be just science either. Havel has said that the Gaia Hypothesis, which sees the whole earth as a single system, a mega-organism, might be the right kind of idea. Human rights won't be respected until there is respect for "the miracle of the universe." There's a secular, Stoic, mysticism here. Havel tends to use the word *atheism* to decry the emptiness of life in the West, hoping for more cultivation of transcendence, unity, and joy.

After the disasters of the first world war, the shocking horrors of the next world war, and the nerve-wracking threat of the Cold War, for many the value systems of civilization could no longer be trusted. In the 1960s questioning the status quo and challenging the grounds of authority was again championed as a virtue, and intellectual movements based on uncertainty seemed most compelling. The French philosopher Jacques Derrida's Deconstruction appealed because it upended the relationships between established ideas. The fragmented feel of our art and philosophy, often referred to as postmodern, is a response to this. It is also a reflection of the frantic pace of modern culture and its kaleidoscope of images and variety of experts. In taking images from popular culture and advertising, contemporary artists such as Cindy Sherman and, in another way, Robert Rauschenberg reflect a sense of meaning having come unstuck. Both express the pain of that, but also celebrate the fun of it. American artist Jenny Holzer has variously decorated buildings with "truisms," sound-bite claims that seem designed to disrupt the logic of certainty. Artist Matthew Barney's creation of an entire alternate world in his Cremaster Cycle excites the viewer's sense that the rules of our real world are arbitrary conventions. Film and television impart the feeling that the world cannot be quite got hold of, from the fast cuts of music videos to the antihero cowboy. Some feel that the world is meaningless, so why should art bother making meaning? We ought to have fun with nonsense.

Through many sources, the idea arose that the central individual imagined by the Enlightenment does not exist. To approach truth, one must look at things from many vantage points. It is an elegant modern contribution to Skepticism and to cosmopolitan doubt.

ENLIGHTENMENT MEETS ENLIGHTENMENT

The European idea of Chinese beliefs, particularly Confucianism, had included a good dose of atheism for centuries. Later, from the early nineteenth century on, other Eastern productions—the Vedas and Buddhism—also offered new ideas that seemed secular and rational. We have seen how ancient Eastern doubt came into Europe through Schlegel, the champion of Romanticism and husband of Dorothea Mendelssohn, and how Schopenhauer brought attention to these ideas among artists and intellectuals. There was also the Frenchman Eugène Burnouf, whose studies finally made clear to Europeans that Buddhism was not an aspect of Hinduism, and delineated the differences between the two. In all these efforts, little attention was paid to Buddhism as a living tradition; Europeans loved the books and ignored the monks. By the late nineteenth century, there were many texts and lots of readers. The most popular late-century interest in Buddhism was by the Theosophical Society, founded in New York in 1875 by Madame Helena Petrovna Blavatsky and the American Henry Steel Olcott (Annie Besant joined later). In the 1880s, interest was in the Pali Canon, the Pali-language texts that are the foundation of Theravada Buddhism. These were held to be more authentic and pure—and nontheist and rationalist—than the Buddhism commonly practiced. The idea of the Buddhist in Europe now was of a model of restraint and moral goodness, rationalism and detachment—but still, everyone ignored the monks.

The first European to go off and become a Buddhist monk was ordained in 1899. More Europeans followed, and they came back to Europe to set up several famous retreats. There remained an idea of Buddhism as nonmythic. According to historian Martin Baumann, Buddhists in Asia were aware of their European audience and came to emphasize the rationalist scientific character of Buddhism for that reason.[137] In 1903 Karl Seidenstucker created the Society for the Buddhist Mission in Leipzig, Germany, praising Buddhism as the "religion of reason."[138] It seemed a new pragmatic attitude for the new middle class; rationalist Buddhist groups sprang up in Germany and Britain first. Then, in the twentieth century, Buddhism spread through the rest of Europe, usually championed by a single famous enthusiast in each nation. In France in the 1920s it was the American-born Grace Constant Lounsbery. All the proponents were still Western, still devoted to a rationalist Buddhism.

By the 1950s, the movement of Asian Buddhists around the globe brought a variety of Buddhisms, some of which incorporated theism and supernaturalism. Just in these years, meditation was no longer seen as an expendable part of a basically intellectual Buddhism. By the 1960s meditation was growing popular, as were courses in Zen. In the next decades, refugees from Vietnam, Laos, and Cambodia brought more Buddhism to the West. Soon the political tragedies of Tibet and the plight and personality of the Dalai Lama called attention to Tibetan Buddhism, which is now growing most rapidly. Two different Buddhisms in the West were on the rise in the 1960s: that of the so-called European Buddhist community, with small numbers but huge conferences, journals, and societies; and that of the immigrant community, with huge numbers but hardly any formal structure (or anything to which one owes regular attendance). The European version tends to be leery of anything supernatural; many of the immigrant versions are also devoted to practices, community, and identity, and are uninterested in the supernatural.

Buddhism in America is usually described as having three components: the "old-line Asian-Americans"; the "convert" Buddhists (not an accurate term since the second-generation adherents are not converts, but it signifies non-Asian Buddhists whose families came to Buddhism in the 1960s or later); and, last, "ethnic" Buddhists (recent immigrants). The first group brought the dharma to the United States for the first time during the 1849 gold rush, so that by 1875 there were eight Buddhist temples in San Francisco. In 1942 Franklin Roosevelt sent more than a hundred thousand people of Japanese ancestry to internment camps; about 60 percent of them were Buddhists. The experience led this Buddhist community to accelerate in its Americanization and, in 1944, to break all ties with Japan. It is now the quietest, but in some ways most successful, of the three branches, serving as a stable institutionalized form of the dharma that happily coexists with the rest of American life. As for the "converts," between 1910 and the early 1960s twenty-one meditation centers were founded in the United States. Buddhism guides for the year 2003 list more than a thousand. Indeed, sitting meditation is now the chief characteristic of Euro-American Buddhism. Zen is the oldest form of convert Buddhism in America. It came in with a terrific speech by a Rinzai monk, Shaku Soen, at the Chicago

World Parliament of Religions in 1893. His student D. T. Suzuki went on almost single-handedly to popularize and explain Zen philosophy at mid-century. The Beat poets and the rising counterculture brought about the "Zen boom" of the 1950s and 1960s, and Suzuki did much to guide the American understanding of it. Although the counterculture was not against spirituality, the Zen that Suzuki brought to Americans was at home in the history of doubt. Here's Suzuki on religion and God:

> Is Zen a religion? It is not a religion in the sense that the term is popularly understood; for Zen has no God to worship, no ceremonial rites to observe, no future abode to which the dead are destined, and, last of all, Zen has no soul whose welfare is to be looked after by somebody else and whose immortality is a matter of intense concern with some people. Zen is free from all these dogmatic and "religious" encumbrances. When I say there is no God in Zen, the pious reader may be shocked, but this does not mean that Zen denies the existence of God; neither denial nor affirmation concerns Zen. . . . Therefore, in Zen, God is neither denied nor insisted upon; only there is in Zen no such God as has been conceived by Jewish and Christian minds.[139]

Rick Field's study of Buddhism in the United States credits the introduction of Eastern thought to Transcendentalists Emerson, Fuller, and Thoreau, though it was Suzuki who was most responsible for Buddhism in the West as we know it.[140]

Zen, more than Theravada or Tibetan Buddhism, is expressly about doubt.[141] In Suzuki's words, "I discovered that it is necessary, absolutely necessary, to believe in nothing. . . . No matter what god or doctrine you believe in, if you become attached to it, your belief will be based more or less on a self-centered idea."[142] In a recent work, Rick Fields, too, praises doubt, writing: "Doubt is a state of openness and unknowing. It's a willingness to not be in charge, to not know what is going to happen next. The state of doubt allows us to explore things in an open and fresh way."[143] Bernard Glassman explains his notion of "bearing witness": "It's about living a questioning life, a life of unknowing," and just bearing witness to suffering is the path to peace and healing.[144] In 1994 Glassman founded the Zen Peacemaker Order, based on three tenets: "Not-Knowing: and thereby giving up fixed ideas about ourselves and the universe," bearing witness, and loving acts.[145] Glassman has said that Not-Knowing is what his Japanese teacher taught him, but that it also fit well with his own Jewish background, or rather with the Cabalist tradition of it, where all is one, the *Ein Sof.* No knowledge can help one reach it.[146] The Korean master Seung Sahn called his 1982 book *Only Don't Know;* the great Western student of Zen Alan Watts has a book called *The Wisdom of Insecurity.* In each case, the message is doubt.

Another of the most important interpreters of Buddhism today, Stephen Batchelor, has given his books such titles as *The Faith to Doubt* (1992) and *Buddhism Without Beliefs* (1997), and cites the ancient Zen maxim "Great doubt: great awakening. Little doubt: little awakening. No doubt: no awakening."[147] Batchelor explains that we can get to "this condition of unknowing" a number of ways, but for many it comes as the final acceptance that the questions we have are not open to rational answers. "It is the palpable silence which follows the breakdown of an apparatus which has been strained to its limits. The acknowledgement 'I don't know' comes finally not as failure or disgrace but as release."[148] Against critics who might think his call to doubt is a refusal to investigate the world, Batchelor cites Thomas Huxley, saying that agnosticism was about the testing of ideas, not the rejection of all knowledge. Batchelor writes that this agnosticism also describes the Buddha: the program was pragmatic and falsifiable.[149]

In the 1960s, when Americans began visiting the Theravada retreat centers of south and southeast Asia, they brought back "insight meditation," divorcing it from the rituals of its origins. Many of these early students later trained in psychotherapy and gave rise to a movement that combined the two.[150] Recently, Mark Epstein has added his experience working with Buddhism and psychotherapy. He writes that in the case of both specific psychological pain and common existential pain, the trick turns out to be "going *into* the doubt rather than away from it."[151] Batchelor and Epstein both quote the passage from the great Zen master Takasui, and the context of psychotherapy gives Takasui's words new meaning: "Only doubt more and more deeply, gathering together in yourself all the strength that is in you, without aiming at anything or expecting anything in advance, without intending to be enlightened and without even intending not to intend to be enlightened; become like a child in your own breast." The merging of these therapies is a nice story for the history of doubt, especially because psychotherapy has within it so much of the West: the rationalism behind Freud's scowl was Greek, his brave face about it was Roman, and his pity for the forsaken world was Jewish. With the mix of Zen and psychotherapy there is a deep commingling of the great traditions of doubt.

SATANIC VERSES

In the first half of the twentieth century, the greatest scholar of the medieval poet Jalal ad-Din Rumi (1207–1273)—often known as just Rumi—was a professor at Cambridge University, Reynold Alleyne Nicholson. His eight-volume translation and critical commentary on Rumi has been highly regarded in Iran, and the commentary has recently been translated into Persian.[152] He had great credentials. In 1920 Nicholson's *Studies in Islamic Poetry* offered a brief chapter anthologizing earlier poetry, with the rest of the book entirely devoted to one of the great doubters of all time, Abdallah al-Ma'arri (973–1057), the early medieval poet. There had been important work on al-Ma'arri in Europe at the end of the nineteenth century, by Alfred von Kremer, but Nicholson believed that despite its brilliance, Kremer had missed a major point: "He did not examine the language and style with sufficient closeness to detect the subtle manner in which the poet at once disguises and proclaims his unbelief in the Mohammedan or any other revealed religion."[153]

Nicholson told his readers that al-Ma'arri had been a major force of Skepticism. He also explained the poet in terms of the greater history of doubt, finding him at times full of the "pure skepticism of Carneades." Nicholson cited al-Ma'arri lines such as: "Certainty is not to be found in a time whose sagacity brought us no result but supposition. / We said to the lion, 'Art thou a lion?' and he replied doubtfully, 'Perhaps I am' or 'I seem to be.'"[154] It is as delightful an image of Skeptical times as one could ever hope to conjure: the very lions qualify their identity with reservations! Nicholson wrote that al-Ma'arri had sketched "many anecdotes of the zindiqs" and that he expressed himself "in the manner of Lucian."[155] Nicholson saw al-Ma'arri as a freethinker, a deist, and a doubter of creeds, and he not only translated al-Ma'arri's doubt but focused on it. In the 1930s, Nicholson continued to translate al-Ma'arri and write of him as a materialist and disbeliever. Nicholson also said of al-Ma'arri that he "contemplates life with the profound feeling of Lucretius."[156] Thus was a great Islamic doubter returned to the history of doubt.

Later in the century, Islamic doubt drew the attention of the world. The *fatwa* (ruling) on Salman Rushdie is better known than his precise level of doubt. In a 1995 essay, Rushdie wrote that "God, Satan, Paradise, and Hell all vanished one day in my fifteenth year, when I quite abruptly lost my faith," and "afterwards, to prove my new-found atheism, I bought myself a rather tasteless ham sandwich, and so partook for the first time of the forbidden flesh of the swine. No thunderbolt arrived to strike me down." That was it, he wrote. "From that day to this I have thought of myself as a wholly secular person."[157] *The Satanic Verses* (1988) is not an attack on Islam or on religion. It is a work of fiction, of magical realism. Rushdie has written that the book's use of fantasy was intended to faithfully represent the world of Indian believers (there are some 330 million gods), but it also reflects chaotic cosmopolitan uncertainty.[158] The novel is partly based on a story in which Muhammad, not yet established, accepts the idea that some local, female gods can share the pantheon with Allah (the locals will not abandon their gods) and then, later, revokes the verses that okayed the deal, saying they came from Satan rather than God. Hence, the satanic verses. They have been seen as weakening the case for revelation. In the spinning, jumbled adventures of *The Satanic Verses*, figures of the Koran are parodied, but Rushdie's central drama is of the immigrant or displaced individual, wandering the cosmopolitan world. In fact, one main character turns into a hoofed beast in a direct borrowing from that Roman novel of a vast cosmopolitan world, *The Golden Ass.* The book also gave a warty caricature of the Ayatollah Khomeini, and this was perhaps most provocative. With the fatwa, the rise of fundamentalist Islam became visible to a wider world and Rushdie joined the long train of doubters forced into hiding to save their lives.

The headline for Taslima Nasrin's official Web site announces, "I am an atheist. I do not believe in prayers. I believe in work. And my work is that of an author. My pen is my weapon."[159] Nasrin is a Bangladeshi physician and writer—a highly respected poet, essayist, novelist, and journalist. The call for her execution came after the publication of her novel *Shame* (1992), about the suffering of a Hindu family after they are attacked by Muslims. Nasrin's crime was "blasphemy and conspiracy against Islam, the Holy Qur'an, and its prophet." With the fatwa on her, it became clear to the West that Islamic fundamentalism had spread to Bangladesh. In a 1998 interview Nasrin was asked what originally prompted her to be so outspoken in her opposition to Islam.[160] She answered, "When I began to study the Koran, the holy book of Islam, I found many unreasonable ideas. The women in the Koran were treated as slaves. They were nothing but sexual objects." When she set down the Koran and looked at her world, she "realized that religious oppression and injustices are only increasing, especially in Muslim countries," and especially against women. She said that she then began to write against "the crimes of religion, particularly the injustice and oppression against women." Her interviewer asked if she was chiefly critical of fundamentalists, to which Nasrin replied:

> I criticized fundamentalists as well as religion in general. I don't find any difference between Islam and Islamic fundamentalists. I believe religion is the root, and from the root fundamentalism grows as a poisonous stem. If we remove fundamentalism and keep religion, then one day or another fundamentalism will grow again. I need to say that because some liberals always defend Islam and blame fundamentalists for creating problems. But Islam itself oppresses women. Islam itself doesn't permit democracy and it violates human rights. And because Islam itself is causing injustices, so it is our duty to make people alert. It is our responsibility to wake people up, to make them understand that religious scriptures come from a particular period in time and a particular place.

Such statements impugn mainstream Islam in the same way Christians unbelievers have impugned all Christianity along with its crueler excesses. Average moderate Christians never like it, and moderate Muslims do not like it much either. But her point is that just as average moderate believers may support each other across religions, radical doubters and secularists should be able to support each other across religions.

In 1994, while Rushdie was still in hiding, he published an open letter to Taslima Nasrin in the *New York Times,* both to comfort her and to bring attention to her plight.[161] The letter began by saying how tiring it must be to be called the "female Salman Rushdie"—"what a bizarre and comical creature that would be!—when all along you thought you were the female Taslima Nasrin." He said instead the press should say her enemies are "the Bangladeshi Iranians." "What an Islam they have made, these apostles of death, and how important it is to have the courage to dissent from it!" The fundamentalists, he explained, always say they want simplicity, but in fact, they are obscurantists. "What is simple is to agree that if one may say 'God exists' then another may also say 'God does not exist'; that if one may say 'I loathe this book' then another may also say 'But I like it very much.'" Rushdie was at pains to demonstrate that doubt and pluralism have a long history in his part of the world.

In 1991 an interviewer reported that "Mr. Rushdie lamented that people in the rich and powerful West had failed to see that he was the victim of religious persecution, not unlike so-called heretics of generations past who were burned or drowned for their dissenting beliefs."[162] Like Nasrin, Rushdie protests the "multiculturalist" respect some in the West accord to acts that would be seen as simply wrong in their own culture. The West, he asserts, seeking to avoid its old crime of cultural imperialism, now perpetrates a new injustice by denying universal Enlightenment standards for human rights.

To these great Muslim doubters we add two with interesting pseudonyms. First, there is Ibn al-Rawandi—he borrowed the name of the great medieval doubter. His *Islamic Mysticism: A Secular Perspective* (2000) explains that its author was a devoted convert to Sufism and was enchanted by its metaphysics. He studied and worshiped in Cyprus and was content at first, but when doubts emerged for which the traditionalist authors had no answers, and then the Salman Rushdie affair divided Islam, Rawandi left Sufism and sought to offer a critical evaluation of it. His conclusion was that mystical experience is not a trustworthy validation of religion. This new Ibn al-Rawandi lives in London and writes articles for *Philosophy Now* and *New Humanist*. Second, there is the extraordinary author of *Why I Am Not a Muslim* (1995), who took the name Ibn Warraq in honor of that other great medieval Islamic doubter, Muhammad al-Warraq. Ibn Warraq explains that he was raised Muslim, but "As soon as I was able to think for myself, I discarded all the religious dogmas that had been foisted on me. I now consider myself a secular humanist who believes that all religions are sick men's dreams, false—demonstrably false—and pernicious."[163]

He said he would have kept this opinion to himself had it not been for the Rushdie affair and the rise of fundamentalist Islam: "For those who regret not being alive in the 1930s to be able to show their commitment to a cause, there is, first, the Rushdie affair, and, second, the war that is taking place in Algeria, the Sudan, Iran, Saudi Arabia and Pakistan, a war whose principal victims are Muslims, Muslim women, Muslim intellectuals, writers, ordinary decent people. This book is part of my war effort." Ibn Warraq announces, "The present work attempts to sow a drop of doubt in an ocean of dogmatic certainty by taking an uncompromising and critical look at almost all the fundamental tenets of Islam."[164]

Like other Muslim doubters, Ibn Warraq is angry that multiculturalism has made it difficult for an irreligious liberal to be flatly against religion. He quotes John Stuart Mill's encouragement to radicalism and generally calls out to the tradition of doubt: "Muslims cannot hide forever from the philosophical implications of the insights of Nietzsche, Freud, Marx, Feuerbach, . . . and

Renan," he writes. "Hume's writings on miracles are equally valid in the Islamic context." And further: "What of the rise of the critical method in Germany in the nineteenth century, and its application to the study of the Bible and religion in general? When biblical scholars say that Jonah never existed or that Moses did not write the Pentateuch, then, implicitly, the veracity of the Koran is being called into question."[165] He quotes Nicholson as saying "this blasphemous sentence, among others: 'the Koran is an exceedingly human document.'"[166] He cites Xenophanes, Montaigne, Galileo, Spinoza, La Peyrère, Hobbes, Gibbon, Bayle, Voltaire, Kant, Schopenhauer, Paine, and Carlyle, as well as Averroës and Avicenna. Darwin, Huxley, Ingersoll, Russell, and Sartre also give witness. Einstein is quoted as saying that anyone who believes in the law of causation "cannot for a moment entertain the idea of a being who interferes in the course of events. . . . He has no use for the religion of fear."[167] Realizing that many Christians have incorporated science into their faith, Ibn Warraq laments, "Muslims have yet to take even this first step."[168] He also writes that a series of articles by Ibn al-Rawandi in *New Humanist* "somehow gave me enormous moral encouragement and support."[169]

Ibn Warraq tells of a friend, a well-educated Muslim, who saw Russell's *Why I Am Not a Christian* on Ibn Warraq's bookshelf and "pounced on it with evident glee." Later, Ibn Warraq was disappointed to learn that the friend "apparently considered Russell's classic to be a great blow to Christianity; at no time was my friend aware that Russell's arguments applied, *mutates mutandis,* to Islam." Ibn Warraq even suggests that Muslims try reading "Allah" wherever Western doubting texts say "God"; only if they hear Nietzsche saying, "Allah is dead," he states, will they get the point.[170] Ibn Warraq then compiles a great variety of criticisms against Islam. It may be the first Muslim history of doubt. He also issues various critiques of the Koran and of Muhammad's morality.

Popular modern Muslim doubt tends to be ethical; it critiques the behavior of Muhammad as the ancient Greeks critiqued the behavior of the pantheon. One central issue is Muhammad's preteen wife; it is often introduced as a reason a former Muslim has left the faith. The treatment of Islamic women in general is also very important, followed by the treatment of non-Muslims, even, sometimes, Muslim non-Arabs. There is also the issue of democracy. Ibn Warraq cites the founders of the United States on the relationship between religion and politics and concludes that Islam cannot be reconciled to that vision. He also devotes an entire chapter to al-

Ma'arri. Citing many verses from Nicholson's several books, he puts together his own anthology of al-Ma'arri's great doubt. To consider just one conversational passage and a stanza:

> Sometimes you may find a man skillful in his trade, perfect in sagacity and in the use of arguments, but when he comes to religion he is found obstinate, so does he follow the old groove. . . . To the growing child that which falls from his elders' lips is a lesson that abides with him all his life. . . . If one of these had found his kin among the Magians, or among the Sabians, he would have declared himself a Magian, or among the Sabians he would have become nearly or quite like them.[171]

As for the stanza:

> . . . *The creeds of man: the one prevails*
> *Until the other comes; and this one fails*
> *When that one triumphs; ay, the lonesome world*
> *Will always want the latest fairytales.*[172]

The consistent theme of Ibn Warraq's book is distress that Western intellectuals, having won separation of church and state in their own nations, no longer want to fight religion.

DOUBT AT THE NEW MILLENNIUM

When I started writing this book, I did not think there would be much to say about the third millennium CE, but things changed fast in the autumn of 2001. During the Cold War, the idea was that Americans believed more than our opponents did; in the confrontation with fundamentalist Islam, we are faced with an enemy that violently rejects public secularism. It demands a reappraisal of our attitude. In the most immediate terms, consider the response of a few doubting Muslims. Salman Rushdie published an article called "Yes, This Is About Islam" in the *New York Times* on November 2, 2001. He wrote: "'This isn't about Islam.' The world's leaders have been repeating this mantra for weeks." And for good reason, said Rushdie: to keep innocent Muslims from being harassed and to keep the peace with other Muslim countries. But for him, "The trouble with this necessary disclaimer is that it isn't true." Rushdie sighed, "Of course this is 'about Islam.'" But he explained that it was not so much about religion, as religion in politics. "The restoration of religion to the sphere of the personal, its depoliticization, is the nettle that all Muslim societies must grasp in order to become modern." Rushdie concluded, "If terrorism is to be defeated, the world of Islam must take on board the secularist-humanist principles on which the modern is based, and without which Muslim countries' freedom will remain a distant dream."

After September 11, 2001, Ibn Warraq has many times called for politicians and intellectuals to stop trying to "protect" Islam. In October 2001, ABC Radio National devoted an entire program to an interview with Ibn Warraq, and here again he rejected the West's respectful response to Islam.[173] For him, "we will not get anywhere until we emphasize the things that we value, like separation of church and state, liberalism, democracy, the value of rationality, discussing our problems and so on." When asked if he thought Islam capable of a modern transformation, Ibn Warraq replied that it is possible if only we begin to look critically at Islam as we have looked critically at Christianity. "Higher biblical criticism," he encouraged, "has existed since at least the seventeenth century with Spinoza and so on, going on to the nineteenth century in Germany. And yet nobody dares to look at the Koran in the same way."

Ibn Warraq lamented that even in academia there is a taboo about discussing the Koran "scientifically," enough so that excellent work, such as that of Christoph Luxenberg,[174] has been shunned in the West. And with what explanation? "'Well, we do not wish to hurt the sensibilities of the Muslims.' I mean it's incredible." For him the answer is Bible scholarship and lots of it, so "then the Muslims will be forced to look at their own religion in a critical way as well." He added, "As somebody once said, we're not doing Islam any favors by shielding it from Enlightenment values." For him, defusing the present global threat should be understood as dragging Islam through the same process that her older sisters have undergone: separation of church and state, an increase in gender equality, recognition of other religions as partaking in the same truths, and a willingness to have secular standards of conduct applied within their ranks.

A similar critique has been offered by Ramendra Nath, a philosopher who teaches in India and wrote his Ph.D. dissertation on Russell. He has recently published *Why I Am Not a Hindu*. Nath writes: "Though I agree with Buddhism in its rejection of god, soul, infallibility of the Vedas . . . , still I am not a Hindu even in this broad sense of the term 'Hindu,' because as a rationalist and humanist I reject all religions." Nath, too, argues that fundamentalism cannot be extracted from Hinduism and that the religion itself—and all religion—endangers peace, equality, and truth.

The new millennium nurtures its share of doubting scientists. The title alone of Francis Crick's essay "How I Got Inclined Towards Atheism" clues us in to the beliefs of the Nobel Prize–winning biologist who helped decipher the structure of DNA.[175] James Watson, his Nobel Prize–winning partner, has spoken of his religious doubt, as well. Stephen Hawking has presented an attitude toward God very reminiscent of Einstein's (if a little less respectful): willing to muse about "God's mind," but dismissive of a personal God. The Nobel Prize–winning physicist Steven Weinberg has been a vocal proponent of atheism for a long time. He complains that even those, like Havel, who doubt religion, still reject "reductionist materialism." "Thale's ocean had no room for Poseidon. In Hellenistic times the cult leader Epicurus adopted the atomist theory of Democritus as an antidote to belief in the Olympian gods. . . . Scientists are often driven in their work by motives of this sort. Of course, none of this bears on the question of whether the reductionist perspective is correct. And since in fact it is correct, we had all better learn to live with it."[176]

Science fiction, started by the great Latin doubter Lucian, began by laughing at the Olympian gods and never really changed its attitude in this regard. A surprising number of leading science-fiction writers have been open about their atheism. Isaac Asimov was outspokenly critical of religion as early as the 1960s, but in his later years he went further, saying:

> I am an atheist, out and out. It took me a long time to say it. I've been an atheist for years and years, but somehow I felt it was intellectually unrespectable to say that one is an atheist, because it assumed knowledge that one didn't have. Somehow it was better to say one was a humanist or agnostic. I don't have the evidence to prove that God doesn't exist, but I so strongly suspect that he doesn't that I don't want to waste my time.[177]

H. P. Lovecraft said that he gave up belief in God soon after he rejected Santa Claus, before he was ten years old. Harlan Ellison reported himself "so far beyond atheism" that there was no word for his level of disbelief in the English language.[178] In a 1997 profile in the *New York Times,* Arthur C. Clarke mused about the world of 3001, saying, "Perhaps most controversially, religions of all kinds have fallen under a strict taboo, with the citizenry looking back on the religious beliefs and practices of earlier ages as products of ignorance that caused untold strife and bloodshed. But the concept of a God, known by the Latin word Deus, survives, a legacy of man's continuing wonder at the universe." In 1998 Douglas Adams said he called himself a "radical atheist" because if he said just "atheist," people assumed he meant he was agnostic and he would have to explain, "I really do not believe that there is a god, in fact I am convinced that there is not a god (a subtle difference), I see not a shred of evidence to suggest that there is one . . . etc., etc."[179] The *Time* magazine columnist Barbara Ehrenreich, a self-avowed fourth-generation atheist herself,[180] wrote a very funny line on the phenomenon: "What gets me is all the mean things people say about Secular Humanism, without even taking the time to read some of our basic scriptures, such as the Bill of Rights or *Omni* magazine."[181]

In 1995 the Freedom from Religion Foundation made the social commentator Katha Pollitt their "Freethought Heroine" for her willingness to publicly call herself an atheist. She wrote about their 1995 convention in her column in *The Nation,* and the piece is called "No God, No Master"—an adjusted version of Sanger's phrase.[182] In 2001 Natalie Angier, a Pulitzer Prize–winning science writer for the *New York Times,* published an essay entitled "Confessions of a Lonely Atheist," pointing to how illicit atheism feels to her today. Angier wrote, "So, I'll out myself. I'm an Atheist. I don't believe in God, Gods, Godlets or any sort of higher power beyond the universe itself, which seems quite high and powerful enough to me. I don't believe in life after death, channeled chat rooms with the dead, reincarnation, telekinesis or any miracles but the miracle of life and consciousness, which again strike me as miracles in nearly obscene abundance."[183] Angier also writes that public atheists are either too wacky, like O'Hair, or more concerned with secularism than with godlessness. Angier quotes Katha Pollitt here as saying that if someone believes, she is "not interested in trying to persuade that person there is no intelligent design to the universe." Rather, Pollitt explained, "Where I become interested and wake up is about the temporal power of religion, things like prayer in schools, or Catholic–secular hospital mergers." Angier understands, but writes, "And yet there is something to be said for a revival of pagan peevishness and outspokenness."

Churches are dwindling in Europe. They are dwindling more slowly in the United States, but here there is a rather loud buzz of doubt from the world of culture and entertainment. Many sports heroes, artists, and others have come forward as doubters, from comedians such as George Carlin, who made a fabulous career of laughing at the "invisible man living in the sky" who always needs money, to a great range of people who quietly mention their atheism when asked. In an interview in *Ladies' Home Journal*, Katharine Hepburn said, "I'm an atheist, and that's it. I believe there's nothing we can know except that we should be kind to each other and do what we can for each other."[184] Doubt speaks in many voices, from actor Richard Gere promoting Tibetan Buddhism to Karen Armstrong, the historian of religion and former nun, musing with a good deal of open doubt about the future of God.[185] Author and gay rights activist Quentin Crisp wrote of having shocked an audience in Ireland without much mentioning his irreligion. He described their tension as sectarian rather than moral and quipped that when you tell them you are an atheist they ask, "But is it the God of the Catholics or the God of the Protestants in whom you do not believe?"[186] Paul Geisert and Mynga Futrell recently coined the word "Brights" as a new umbrella term for those who have "a naturalist worldview," and early promoters of the new term include the evolution scientist Richard Dawkins and the philosopher Daniel Dennet. The list of doubters now in the public eye, or recent icons, includes musicians Frank Zappa and John Lennon, actors Christopher Reeve and Jodie Foster, financier Warren Buffett, writer Camille Paglia, filmmaker Ingmar Bergman, historian Gerda Lerner, Nobel Prize–winning author Nadine Gordimer, Linus Torvalds (creator of Linux), fashion designer Bill Blass, and both Penn and Teller—magicians.

We are in an age of intellectual uncertainty and we are in an age of science. We are in an age of cosmopolitan secularism and an age of ardent, doubt-conscious faith. We are marked by moral ambiguity. We investigate graceful-life philosophies and various transcendentalist and therapeutic meditations. Nowadays, all the classic forms of doubt run wild. There is uncertainty in modern society, modern art, and modern cosmology. In politics, there is doubt of moral absolutes, but also uncertainty about moral relativism. The last hundred years have nurtured every aspect of the great history of doubt, so that Skepticism, rationalism, and cultural relativism were redefined and ancient doubting practices found new audiences. At moments like this one, for those who love doubt, it seems that a culture is not only held together by shared beliefs, but also by shared dedication to inquiry, and defense of a secular public sphere.

CONCLUSION

The Joy of Doubt

Ethics, Logic, Mood

The history of doubt exists, has galloped across some twenty-six centuries, and has been very conscious of itself for much of that time. From Cicero to Schopenhauer, from Fanny Wright to Hubert Harrison, from Socrates to Wittgenstein, the long strange story of the history of doubt has loved its heroes. It seems crucial that this history be known, if only so that its theorists, poets, comedians, and martyrs may be understood in their proper context. People should be able to speak to each other about doubt without having to establish all the old arguments every time the conversation begins again. Doubters and believers alike should know that Epicurus and Lucretius, the books of Job and Ecclesiastes, and the teachings of the Buddha have been remarkably constant resources in the history of unbelief. So has the whole history of Skepticism and doubt in our ability to know the world from the Carvaka, Socrates, Pyrrho, Sextus Empiricus, Montaigne, Charron, Hume, Bayle, through to all the modern skeptics. Since I began writing this book, well before September 2001, the significance of its subject has redoubled. The book is now offered as a way to contextualize the struggle over religion and secularism that is at the heart of the crisis. There are limits to any specific parallel, but it helps to know something of the historical course of secularization and the responses it has drawn. Most notably, one sees how slowly ideas change and how many heroines and heroes of the cause are required.

The quiz that introduced this book was geared toward teasing out the issues of doubt in general, and was interpreted in terms of the categories that make sense to our times: *believers, agnostics,* and *atheists.* Now that we have run through the story of doubt, the categories *believer, agnostic,* and *atheist* stand out as very recent ways of dividing thought on this issue. They may not be the most interesting or useful ways anymore.

According to common usage, the term *agnosticism* holds that we cannot reasonably make an assessment on the question of whether God exists. Why not then extend Skepticism to all knowledge; that is, why are agnostics supposed to be Skeptics only on this question? Agnosticism often ends up being a catchall term for those who do not think there is a God, but who harbor a tiny allowance that there might be some force that creates meaning and makes possible an afterlife.

What of the difference between belief and atheism? There have been mystics and philosophers who said they believed in God but who did not believe anything about the universe that was different than how the atheists described it. They just called something about it "God." If your idea of God is a being that thinks, does things, or even *exists,* you would have to re-classify a great many self-titled believers as atheists. If, instead, what divides belief and atheism is that believers have a taste for religion and atheists think it's dangerous bunk, then what of the great atheist religions? Believer mystics and believer philosophers have more in common with atheist mystics and atheist philosophers than with those who accept a Creator God who is aware of us and does things.

I think politics drives a lot of clinging to the three terms, but I also think it is easier to force yourself to be clear if you avoid using *believer, agnostic,* and *atheist* and just try to say what you think about what we are and what's out there.

Divisions that seem more historically stable might include: the *sectarian,* who accepts the stories, rituals, and rules of his or her own religion as true; the *"one-of-many" religionist,* who believes all religions are equally true and relate to a thinking, creative force; the *meaning and science spiritualist,* who interprets the universe as having some force that unites life and perhaps gives it meaning; the *Skeptic,* who doesn't believe we can know anything about anything; the *perplexed,* who believes knowledge is possible but who identifies him- or herself as personally unresolved; the *ritualist,* who thinks the universe is a natural phenomenon and we should celebrate our humanity in the ritual and allegory of traditional religion; and the *science secularist,* who thinks the universe is a natural phenomenon and that religion adds more bad than good. All these people can be doubters—open to the idea that they do not know everything. Whatever the terms, knowing the history of doubt seems to open up the conversation.

This history also allows broad assessment of doubt. For instance, is it good or bad to find out you are living next door to it? It seems to have a knack for generating and popularizing very useful theories. In atomism, anthropology, and cosmology, in politics and neurology, we now hold doubters' doctrines. It is not a coincidence: doubters have wanted to know how the world works and expected to find answers in the world around them. Doubt has been a disproportionately industrious and dynamic stream of human culture and cosmology, espoused by such productive figures as Thomas Edison and Albert Einstein, Frederick Douglass and Elizabeth Cady Stanton, Socrates and Sigmund Freud—it gets a lot done. The story of doubt has run alongside the religions all along, and I think it always will. There will always be individuals and groups who doubt the religious faith in which they were brought up, and parts of the world in which cosmopolitanism encourages the mixing of ideas and the growth of doubt. When religion stays out of politics (in the broad sense of the term), atheism quiets down to a calm chat: we hear about it from contemplative people, with little or no attack on the mythic aspects of religion, because in these circles, none is necessary.

In the ancient world a tremendous range of doubt emerged early on: Anaxagoras guessed that instead of a god, the sun was a hot rock; Democritus came up with atoms as an explanation of a completely self-functioning universe; Epicurus worked out a graceful-life philosophy to live well in this world, without religion; the Carvaka rejected God, gods, inference, causality, karma, and life after death, and declared religion a trap by leaders and priests for power and money; Siddhartha Gautama, the Buddha, counseled a nontheistic transcendence program; there was a grand and nuanced tradition of philosophical Skepticism. The great cosmopolitan doubt started with Xenophanes considering the gods of Ethiopians and red-headed Thracians, and in the same breath the relativism this inspired was extended to animals, so that oxen and horses and lions all imagined gods with tails. Also in the ancient history of doubt, the author of the Book of Job wailed at the injustice of the world; Miriam called the Temple a wolf that devoured time, energy, and intellect, and smacked the building with her sandal; and the essentially secular Roman Empire encouraged a bread-and-circuses materialism, along with a good deal of state ritual. From the ancient world, seven key doubting projects emerged:

- science, materialism, and rationalism
- nontheistic transcendence programs

- cosmopolitan relativism
- graceful-life philosophies
- the moral rejection of injustice
- philosophical skepticism
- the doubt of the believer

They are all very different, but many people involved in one are also involved in others. Christian doubt arose as the struggle to believe in God, in oneself, in one's strength to bear the challenges of renunciation and in some cases martyrdom: Jesus' last words, in two out of the three synoptic Gospels, asked God why he had forsaken him; it seems he was expecting something and now doubted it was coming. Augustine gnashed his teeth and rolled around in his garden yearning to devote himself to God. The doubt of believers is one of the beauties of religions that stress belief; Jesus admonished Thomas to believe without proof, and it is a central image of that faith. Eastern asceticism and the emerging notions of sin and hell and, much later, Protestant justification by faith, further intensified believers' doubt. It is an aspect of the howl of Job, through the looking glass where belief itself is the test rather than life as the test for belief.

People tend to think that the doubt of believers shut down the others, but that is true in only the most limited geographical terms. There was a "dark age" of European doubt, which can be described as beginning with Hypatia's murder in 415 CE; or in 529 when the emperor Justinian outlawed paganism and closed the Epicurean Garden, the Skeptic Academy, and the Lyceum. It took a while before the philosophies of doubt were forgotten, but let's consider the sixth century as the time the lights dimmed, and the twelfth century as when doubt begins to start up again in Europe. But that was just Europe. Although the period was marked by a great movement toward theism all over the world, there was still a lot of room for doubt. Theravada Buddhism continued to doubt the supernatural throughout the period, as did the nontheist traditions of Confucianism and Taoism. In the late fifth century CE Zen gave doubt new life, concentrating on doubt itself, staring at it, making doubt its only goal. The Buddha had suggested that we deny the reality of this world. Zen really loved doubt for its own sake: it *liked* answering questions by holding up lotus flowers or tweaking a nose, celebrating the development of doubt as a highly charged experience. Still this doubt was productive: "Great doubt: great awakening." There were Indian rationalist philosophies in this period, too: in the seventh century, the sage Purandara added probability to the Carvaka's denial of all inference, and the eighth-century Sankara argued that as milk, without intelligence, flows when it is needed, so the entire world can self-generate according to natural relationships.

Europe had its dark age, but around the world, and even on the other coasts of the Mediterranean Sea, the missing six centuries of doubt were actually lively with doubt. Al-Rawandi died about 860 having rejected almost every aspect of Islam, God, and religion in general. The older al-Warraq had taught him much of his doubt. If we ask, then, how long the record of doubt goes dark around the Mediterranean Sea, we are down to about three centuries. Al-Rawandi doubted that the Koran was miraculously beautiful, argued that the prophets had tricked believers on purpose, and that Muhammad himself showed the weakness of religion when he criticized Judaism and Christianity. Al-Razi lived around the same time (854–925) and was as beloved as al-Rawandi was hated, but he doubted as fiercely. He wondered what kind of God would use prophets instead of just telling everyone what they need to know. He suggested, instead, the contemplation of philosophy, and al-Razi's Baghdad joined the ranks of the world's great doubting cities.

By the twelfth century, doubt was back in Europe. It had swept like an hour hand down through the Mideast, across North Africa, up into Spain, and soon, with Saadia ben Joseph (882–942), into France. Averroës and Maimonides were doubting in Spain in the tenth century and Gersonides in France between 1288 and 1344. Doubt must have been in Paris a good while before the Condemnation of 1277 said it was no longer permissible to teach "that theological discussions are based on fables"; "that there is no higher life than philosophical life"; "that Christian Revelation is an obstacle to learning"; and "that man could be adequately generated from putrefaction," among other doubting propositions. In 1417 a manuscript of Lucretius's *On the Nature of Things* was discovered, and by 1435 Lorenzo Valla's celebration of Epicurus almost got Valla burned at the stake.

By the mid-thirteenth century a movement of Latin Averroism was centered in Paris and Padua. Averroism got really big in the West with Pietro Pomponazzi. At Padua in the early sixteenth century, Pomponazzi denied the existence of immortal souls and argued that people do not need threats of heaven and hell in order to be moral. Epicurus, Lucretius, and Diagoras were Pomponazzi's heroes. Pomponazzi, in turn, was hero to the entire Libertine group in France—those who also drew upon Montaigne and Charron as evidence that we cannot know anything about religion or God.

The Inquisition let us hear common doubt, and it was remarkably fertile and imaginative, doubting all aspects of religion; doubting God because of personal tragedy, because the myths of the Old and New Testaments did not ring true, and because of the injustice of the world. From Menocchio to Bruno, the history of doubt saw many of its martyrs in this horror. That claims to certainty were enforced by fire helped drive many to bold expressions of skepticism. Montaigne had lines from Ecclesiastes and Sextus Empiricus carved into the walls of his home, and his self-designed symbol featured a scale balance, in perfect, level indifference, underlined by the motto "What do I know?" He also concluded that since Copernicus had overturned cosmology, we should ignore science, for the whole thing might be overturned again. Not everyone took it this way: Gassendi and Galileo headed off in the direction of experimental science—transforming atomism into a religiously neutral idea along the way. Modern thought still follows both these divergent paths.

Maimonides and Gersonides, Newton and Galileo, all worked on the problem of finding naturalist ways to understand miracles. Spinoza shocked everyone by saying they did not happen. His biblical criticism and disbelief got him excommunicated and made him one of the most consistently important doubters of all time. His near contemporary Hobbes had a similar effect; indeed, for a while it seemed all doubt was Hobbist, but in the long run Spinoza remained more significant to doubters through to the twentieth century. Perhaps if today's politically conservative doubters realized Hobbes's philosophical position, he would again enjoy a vogue.

Several authors explained in writing that it was well known that doubting authors used recognizable tricks to fool the censors. Charron, in *Of Wisdom* (1601), discussed hidden messages in pious texts. Toland did the same, in more detail, teaching us about the "bouncing compliment" and other forms of subterfuge. Doubt was already powerful and popular by the time the English deists rejected the mythology and dogma of religion and the French Enlightenment philosophers popularized deism and traced out its consequences. When Hume sat down to dinner with d'Holbach and met fifteen atheists around one table, the voice of doubters got bolder, and when Diderot wrote up the anecdote, that voice got loud.

The most striking innovation of modern doubt is its political nature. American doubters had unique new reasons for anger at religion: slavery, the Calvinist witch trials, missionaries gouging money from people and using the money to further harass the Native Americans. America also had a new kind of doubter, the kind in petticoats: reforming women who generally supported their families with their books and speeches. Reformist doubters did not usually read widely in the history of doubt, citing only Thomas Paine of all critics, and noting only Galileo of all the harassed. The other new factor of modern doubt was the evangelism of nineteenth-century atheism. Much of the avant-garde of the century expected the end of religion and the end of belief in God, and many of them hoped for it. They thought universal atheism was best because it was true and because it would force us to live better, to realize how entirely responsible we are for what we do and what we fail to do.

Throughout the history of doubt, Epicurus and Lucretius were probably the most consistently visible figures. To take a quick tour: The Mishnah mentioned the Epicureans alone of all Greek philosophers and warned against them. Augustine wrote, "To my mind Epicurus would have been awarded the palm of victory, had I not believed that after death the life of the soul remains . . . a belief which Epicurus rejected." Maimonides was well aware of Epicurus and through him of atheism, and of the idea of a world run entirely by chance. Bruno praised Democritus and Epicurus in the preface to his *On the Infinite Universe and Worlds* of 1591, and many future doubters would find Democritus and Epicurus through Bruno. In the early drafts for an edition of *Principia,* Newton included ninety lines from Lucretius's *On the Nature of Things.* In the late seventeenth century, the philosopher Pierre Bayle wrote that "Atheism does not necessarily lead to corruption of morals," and praised the ethical lives of Epicurus and his followers. Hume demanded: "Epicurus' old questions are yet unanswered. Is He willing to prevent evil, but not able? then is He impotent. Is He able, but not willing? then is He malevolent. Is He both able and willing? whence then is evil?" Jefferson wrote to a friend, "As you say of yourself, I too am an Epicurean," and in another private letter he recommended: "Question with boldness even the existence of a God; because, if there be one, he must more approve the homage of reason than of blindfolded fear. . . . If it end in a belief that there is no God, you will find incitements to virtue in the comfort and pleasantness you feel in its exercise and in the love of others which it will procure for you." Marx wrote his dissertation on Epicurus, lauding the Greek's effort to free humanity from fear of gods. Fanny Wright gave a lecture about Epicurus, including a discussion of Leontium, the courtesan and Epicurean philosopher; Ingersoll saw his naturalist morality based on happiness as indebted to Epicurus more than anyone else. Epicurean ethics based on pleasure and pain inspired Utilitarians and Transcendentalists.

And then there is the modern triumph of ideas Epicurus championed: atomic theory and the self-creation of the world and life.

Doubt has had some other towering figures. Cicero called to order a doubters' debate that was reconvened in works by Montaigne, Hume, Schopenhauer, and Freud. In Petrarch's hands Cicero was the start of the Renaissance—a grand honor in the history of doubt. Also, the Job author started a conversation about justice in the universe that has been joined by doubters throughout the centuries. In the Jewish tradition, there was Menelaus and Miriam; Koheleth, the Preacher of Ecclesiastes. The psalms speak of the godless, and the fool who "in his heart, says there is no God." Rabbi Elisha ben Abuyah, in the early second century, became Aher, "the Other," saying that there is no justice and there is no judge. It was a moral rage that precipitated this, but we know Elisha never tired of singing Greek songs: this was also a cosmopolitan doubt. These issues would come up again throughout the centuries, in Spinoza, Maimonides, Mendelssohn, and into the new millennium.

Doubt has come in dialogs, scholastic questions, cryptic books, anonymous compilations, ribald novels, essays, astronomy lessons, Inquisition trial reports, treatises on political science, speeches, poetry, interviews, and open letters to the *New York Times*. There are left-wing doubters such as those of the Enlightenment tradition, including the French Revolution and much of communism, and there are authoritarian doubters such as Hobbes, Scarpi, Naudé, Napoleon, and Mussolini. There are forgotten doubters such as Miriam and Julius Caesar Vanini; there are hated doubters such as Elisha and al-Rawandi; widely beloved doubters such as al-Razi and Twain. Among the drinking doubters are Socrates, Rabelais, and Ikkyu Sojun. There are also teetotaling doubters, as well as sexual adventurers and phlegmatic types who are never much tempted to look away from their books.

People throughout the ancient world had argued that a thinking person could be happy and moral without God or gods, but most of them worried about what the average man or woman would do, and feel, without religion. This issue was in the background of the debate Cicero started. Bayle decided people could handle it. This conversation continued along without resolve. Something dramatic changed when Harriet Taylor and John Stuart Mill thought to encourage doubting and nonconformism, as such. Now the healthy state needs doubt. In the next century, Freud said the healthy psyche needs doubt, indeed should embrace disbelief as a part of maturity.

There have been many people who have seen the modern world of doubt as a matter of the translation of services and relationships, calendars and allegiances, from religious to secular. The Romantics made art a religion; a whole range of people made politics a religion; and the practitioners of medicine and psychotherapeutics are often described as modern confessors. All of these have added their nontheistic counsel to a great tradition. That counsel could only benefit from being understood within its wider context. Most crucially, the murderous tension surrounding fundamentalism right now demands that the history of doubt be understood, and that secularists, arguing for cosmopolitan tolerance, be conversant with its history. It is worth mentioning again that the last "great" enemy of the United States was more secularist than the United States. That is no longer the case, and we might want to revise our approach to the matter.

Some of my favorite items in the history of doubt are that Ben Franklin was converted to deism in a book intended to argue against deism; that Emma Goldman wrote a biography of Voltairine de Cleyre; that Bradlaugh was the last prisoner of the Clock Tower; and that he named his daughter Hypatia. I love that Paine picked up Doubting Thomas as a hero for doubters. I love the claim by the first century CE philosopher Wang Ch'ung that he would believe men fly as soon as one grows feathers; and that Nagarjuna, the Buddhist philosopher of the early third century CE, claimed that everything we could think of about reality, even the doctrine of no-self, is equally wrong. I love that Lucian, a doubting cutup in late ancient Rome, was so well known as a doubter in Rabelais and Dolet's time that they were both called "apes of Lucian"; and that in Rushdie's *Satanic Verses,* there are shadows of Lucian's ancient transformation. I also love Freud's mentioning Heinrich Heine as an *Unglaubensgenossen,* a "fellow-unbeliever," and that Heine himself had coined the unwieldy word in reference to Spinoza; that Margaret Sanger's first introduction to doubt was when her father brought Ingersoll to town. More favorites are the weird compilation texts of doubt published in the seventeenth century, like the *Theophrastus Redivivus;* and that modern Muslim doubters, in search of protective pen names, have made it so that once again al-Warraq and al-Rawandi labor over the problems of Islam. Marcus Aurelius's warm advice also stands out; as does the fact that George Eliot translated Strauss and Feuerbach; and that Pliny the Elder was able to believe that sometimes rain comes down as blood, but he still laughed at the idea of life after death. Finally, I love that Albert Einstein doubted a personal God of any kind, but said he believed in Spinoza's God, recognizing Spinoza's God as the awesome universe itself.

Cicero ended his study *The Nature of the Gods* by picking a winner among the doubters, and the tradition is worth keeping up. The contest I want to judge, however, is not between various doubters but between the great doubting tradition and all the other traditions of religious thought. Theistic religions all have in them an amazing human ability: belief. Belief is one of the best human muscles; it can be very good. The religions are all beautiful and horrible, filled with feasts, sacrifices, miracles, wars, songs, lamentations, stained glass, onion matzos, and intense communal joy: everyone kneeling, everyone rocking, everyone silent, everyone nose to the floor. The religions have also been the energy behind much generosity, compassion, and bravery. The story of doubt, however, has all this, too. It also has a relationship to truth that is rigorous, sober, and, when necessary, resigned—and it prizes this rigorous approach to truth above the delights of belief. Doubt has its own version of comforts and challenges. From doubt's beginnings, it has advised that if you create your own desires and model them after what you actually experience, you can be happy. Accept that we are animals, but ones with special problems, and that the world is natural, but natural is just an idea that we animals have in our heads. Devote yourself to wisdom, self-knowledge, friends, family, and give some attention to community, money, politics, and pleasure. Know that none of it brings happiness all that consistently. It's best to stay agile, to keep an open mind. Anyway, if you live long enough, you will likely find yourself believing something that you'd never believe today. Or disbelieving. In a funny way, the one thing you can really count on is doubt. Expect change. Accept death. Enjoy life. As Marcus Aurelius explained, the brains that got you through the troubles you have had so far will get you through any troubles yet to come.

Throughout history, many great thinkers have argued that the study of these questions could give life meaning, grace, and happiness. Many heartily suggest, indeed insist, that doubters should do some practices, some therapy, some art, to tune themselves to a manageable relationship with a universe that very possibly has no humanness at all. Doubters in the modern world have all sorts of philosophies and communal experiences in which to engage and participate, and it is not uncommon for doubters to compose a sacred-but-secular world for themselves out of reading philosophy of some sort, taking part in psychotherapy, art and poetry, meditation, dance, secular solemnities, and festivals. The only thing such doubters really need, that believers have, is a sense that people like themselves have always been around, that they are part of a grand history. I hope it is clear now that doubt has such a history of its own, and that to be a doubter is a great old allegiance, deserving quiet respect and open pride. For its longevity, its productivity, its pluck, its warmth, its service to friend and foe, and its sometimes ruthless commitment to demonstrative truth, I give the palm to the story of doubt.

Notes

CHAPTER ONE

1. Two good histories of ancient doubt, both of which informed the present chapter, are James Thrower, *The Alternative Tradition: Religion and the Rejection of Religion in the Ancient World* (The Hague: Mouton, 1980), 197–200; and A. B. Drachmann, *Atheism in Pagan Antiquity* (London: Ares, 1922).
2. John Burnet, *Early Greek Philosophy* (London: Black, 1930), 14.
3. Burnet, 49–50.
4. Heraclitus B 5, as cited in Walter Burkert, *Greek Religion* (Cambridge: Harvard Univ. Press, 1985), 309.
5. As cited in Burkert, 181.
6. Burkert, 315.
7. Burkert, 311.
8. Xenophanes, "Fragment 11," in G. S. Kirk, J. E. Raven, and M. Schofield, *The Presocratic Philosophers,* 2nd ed. (Cambridge: Cambridge Univ. Press, 1957), 168.
9. Xenophanes, "Fragment 15," in Kirk, Raven, and Schofield, 168–169.
10. Xenophanes, "Fragment 16," in Kirk, Raven, and Schofield, 168–169.
11. Burkert, 314.
12. Plato, *Phaedra,* in *The Collected Dialogues,* ed. Edith Hamilton and Huntington Cairns (Princeton: Princeton Univ. Press, 1961), 94–95.
13. Aristophanes, *The Clouds,* William Arrowsmith, trans. (New York: Mentor, 1962), 43–44.
14. Michael Morgan, "Plato and Greek Religion," in *The Cambridge Companion to Plato,* ed. Richard Kraut (Cambridge: Cambridge Univ. Press, 1992), 227–247.
15. Plato, *The Laws* X, in *The Collected Dialogues,* 1445.
16. Plato, *The Laws* XII:1512.
17. Plato, *The Laws* X:1446.
18. Plato, *The Laws* XII:1512.
19. Etienne Gilson, *God and Philosophy* (New Haven: Yale Univ. Press, 1941, 2002), 26.
20. Constance C. Meinwald, "Good-bye to the Third Man," in *Companion to Plato,* ed. Kraut, 365–396, esp. 367.
21. Plato, *Timaeus and Critias,* Desmond Lee, trans. (London: Penguin Classics, 1977), 123.
22. Sextus Empiricus, *Selections from the Major Writings on Scepticism, Man, and God,* ed. Philip P. Hallie and trans. Sanford G. Etheridge (Indianapolis: Hackett, 1985), 189.
23. For a discussion of this, see Edward Grant, *God and Reason in the Middle Ages* (Cambridge: Cambridge Univ. Press, 2001), 160–164.
24. As cited in Jonathan Barnes, "Metaphysics," in *The Cambridge Companion to Aristotle,* ed. Jonathan Barnes (Cambridge: Cambridge Univ. Press, 1995), 104.

25. See A. A. Long, *Hellenistic Philosophy: Stoics, Epicureans, Sceptics* (Berkeley: Univ. of California Press, 1986); and Luther H. Martin, *Hellenistic Religions: An Introduction* (Oxford: Oxford Univ. Press, 1987).

26. Lucius Apuleius, *The Golden Ass* (New York: Horace Liveright, 1927), 258.

27. Drachmann, 109.

28. Epicurus, "Letter to Menoeceus," in Diogenes Laertius, *Lives of the Eminent Philosophers,* Robert Drew Hicks, trans. (Cambridge: Harvard Univ. Press, 1972), 651.

29. Epicurus, "Letter to Menoeceus," 649–650.

30. Epicurus, "Letter to Pythocles," in Diogenes Laertius, 623.

31. Epicurus, "Letter to Menoeceus," 649.

32. Epicurus, "Letter to Menoeceus," 659.

33. Epicurus, "Letter to Herodotus," in Diogenes Laertius, 598–599.

34. On Hellenistic Skepticism, see introductory chapters in Richard H. Popkin, *The History of Scepticism from Erasmus to Spinoza* (Berkeley: Univ. of California Press, 1979); and introduction by Philip P. Hallie to Sextus Empiricus, *Selections.*

CHAPTER TWO

1. The scholarship and judgments found in three superb works on Hellenistic Jews much informed this section: Elias J. Bickerman, *The Jews in the Greek Age* (Cambridge: Harvard Univ. Press, 1988); Arnaldo Momigliano, *Alien Wisdom: The Limits of Hellenization* (Cambridge: Cambridge Univ. Press, 1971); and Menahem Stern, "The Period of the Second Temple," in *A History of the Jewish People,* ed. Hayim Ben-Sasson (Cambridge: Harvard Univ Press., 1976), 183–306.

2. Paul Johnson, *A History of the Jews* (New York: HarperCollins, 1988), 147.

3. Bickerman, 254.

4. Bickerman, 92–93.

5. Bickerman, 254–256.

6. Maccabees III 1:2.

7. Momigliano, 92–93.

8. Momigliano, 93.

9. Stern, in Ben-Sasson, *A History of the Jewish People,* 195–204.

10. Maccabees II 13.

11. As cited in Johnson, 100.

12. Ben Zion Wacholder, *Eupolemus: A Study of Judaeo-Greek Literature* (Cincinnati: Hebrew Union College Press, 1974). See also Martin Hengel, *Judaism and Hellenism: Studies in Their Encounter in Palestine During the Early Hellenistic Period,* trans. John Bowden, 2 vols. (Philadelphia: Fortress, 1974).

13. Bickerman, 161–176.

14. Bickerman, 93; Maccabees II 8:20.

15. Sukkot 56b. See Johnson, 103. In Genesis Rabbi 99:3, the altar is referred to as a wolf and the commentary is that the altar "seizes" its sacrifices. (Correspondence with Rabbi Eric A. Silver.)

16. Johnson, 102–105.

17. Johnson, 103.

18. Bickerman, 226, 227.

19. Josephus, *Jewish Antiquities* (Ant. Jud.) 12.260, as cited in Momigliano, 108–109.

20. Maccabees I 13–43 to 13–58.

21. Maccabees II 4:40.

22. Johnson, 99.
23. Momigliano, 91.
24. Johnson, 101.
25. Bickerman, 254.
26. For a discussion of the Book of Job and a compendium of responses to it, see Nahum N. Glatzer, ed., *The Dimensions of Job: A Study and Selected Readings* (New York: Schocken, 1969).
27. Job 1:8 (KJV).
28. Job 2:9.
29. Job 3:11.
30. Job 19:7.
31. Job 16:2.
32. Job 21:7–11.
33. Job 29:5–23.
34. Job 24:2–12.
35. Job 24:14–21.
36. Job 7:15.
37. Job 7:9–11.
38. Job 38–39.
39. Job 41:27–29.
40. Job 42:3–6.
41. Martin Buber, *The Prophetic Faith* (New York: Harper, Torchbooks, 1949), 191.
42. Buber, 188.
43. Jack Miles, *God: A Biography* (New York: Vintage, 1995), 329.
44. On Ecclesiastes, see Tremper Longman III, *The Book of Ecclesiastes* (Cambridge: Eerdmans, 1998); James Crenshaw, *Urgent Advice and Probing Questions: Collected Writings on Old Testament Wisdom* (Macon, GA: Mercer Univ. Press, 1995); Crenshaw, "The Silence of Eternity: Ecclesiastes," in *A Whirlpool of Torment: Israelite Traditions of God as an Oppressive Presence* (Philadelphia: Fortress, 1984), 77–92; Robert Gordis, *Koheleth—The Man and His World* (New York: Schocken, 1978).
45. Eccles. 7:16–17 (KJV).
46. Eccles. 2:2.
47. Eccles. 2:16–17.
48. Eccles. 7:23–24.
49. Eccles. 2:18–19.
50. Eccles. 2:23.
51. Eccles. 3:1–9.
52. Eccles. 3:12.
53. Eccles. 3:19–22.
54. Eccles. 4:1.
55. Eccles. 4:2–4.
56. Eccles. 4:8–11.
57. Eccles. 9:9.
58. Eccles. 1:1–7.
59. Eccles. 1:9–11.
60. Eccles. 7:10.
61. Walter Kaufman wrote that we do not know "whether the author of Ecclesiastes retained any faith in God." Kaufman, *The Faith of a Heretic* (Garden City, NY: Anchor, 1968).
62. Eccles. 5:18–19.

63. Eccles. 5:1.
64. Eccles. 5:7.
65. Eccles. 9:4.
66. Eccles. 8:15.
67. Eccles. 9:14–15.
68. Eccles. 11:1.
69. Eccles. 11:5.
70. Eccles. 7:1–3.
71. Eccles. 7:4–5.
72. Eccles. 11:8.
73. Eccles. 9:5–6.
74. Crenshaw, *Urgent Advice*, 509.

CHAPTER THREE

1. Donald S. Lopez Jr., *Asian Religions in Practice* (Princeton, NJ: Princeton Readings in Religion, 1999), 48–50.
2. James Thrower, *The Alternative Tradition: Religion and the Rejection of Religion in the Ancient World* (The Hague: Mouton, 1980), 63. Thrower offers a far-ranging history of Ancient Asian atheism and rationalism.
3. See "Carvaka," in *A Sourcebook in Indian Philosophy*, ed. Sarvepalli Radhakrishnan and Charles A. Moore (Princeton: Princeton Univ. Press, 1973), 227–249.
4. "Prabodha-Candrodaya," in *Sourcebook*, ed. Radhakrishnan and Moore, 247.
5. Sarvasiddhantasasgraha, in *Sourcebook*, ed. Radhakrishnan and Moore, 234.
6. "Prabodha-Candrodaya," 247.
7. Madhava Acarya's *Sarvadarsanasamgraha*, in *Sourcebook*, ed. Radhakrishnan and Moore, 229.
8. *Sarvasiddhantasasgraha*, 235.
9. "Prabodha-Candrodaya," 248.
10. "Prabodha-Candrodaya," 248.
11. "Prabodha-Candrodaya," 248.
12. *Sarvasiddhantasasgraha*, 233.
13. *Sarvasiddhantasasgraha*, 235.
14. *Sarvasiddhantasasgraha*, 233–234.
15. Ninian Smart, as cited in Thrower, *The Alternative Tradition*, 77.
16. See Thrower, 85.
17. Huston Smith, *The World's Religions* (San Francisco: HarperSanFrancisco, 1991), 82.
18. The use of Reich to illustrate the Buddhist concept is from Mark Epstein, *Thoughts Without a Thinker: Psychotherapy from a Buddhist Perspective* (New York: Basic Books, 1995), 19.
19. Nyanaponika Thera and Bhikkhu Bodi, trans. and eds., *Numerical Discourses of the Buddha: An Anthology of Suttas from the Anguttara Nikaya* (Oxford: Altamira, 1999), 72–73.
20. Numerical Discourses of the Buddha, 58.
21. Numerical Discourses of the Buddha, 194–195.
22. Numerical Discourses of the Buddha, 47.
23. Arthur C. Danto, *Mysticism and Morality: Oriental Thoughts and Moral Philosophy* (New York: Columbia Univ. Press, 1987), 82–83.
24. Alan Watts, *The Way of Zen* (New York: Vintage, 1989), 66.
25. Wing Tsit Chan, "On Nature," as cited in Thrower, 119.

26. Wang Ch'ung, *Lun-Hêng: Philosophical Essays of Wang Ch'ung*, trans. and intro. Alfred Forke (New York: Paragon, 1962, reprint of 1907 ed.). Language here is slightly cleaned up following the guidance of a version of this text on the Web at: www.humanistictexts.org/wangchung.htm. On Wang Ch'ung see Forke's introduction to the above work, 4–44; see also: "The Sceptical Philosophy of Wang Chhung," in Joseph Needham, *Science and Civilisation in China* (Cambridge: Cambridge Univ. Press, 1991, reprint of 1956 ed.), vol. 2, 368–395.

27. Wang Ch'ung, 92–93.

28. Wang Ch'ung, 166. Some editions translate this chapter as "Wrong Notions of Unhappiness."

29. Wang Ch'ung, 349.

30. Wang Ch'ung, 228.

31. Wang Ch'ung, 182.

32. Wang Ch'ung, 335.

33. Wang Ch'ung, 346–347.

CHAPTER FOUR

1. Werner Jaeger, *Early Christianity and Greek Paideia* (Cambridge: Harvard Univ. Press, 1961), 32.

2. See for instance: James Frazer, *The Golden Bough: A Study in Comparative Religion*. (London: Macmillan, 1890); and Arthur Darby Nock, *Conversion: The Old and the New in Religion from Alexander the Great to Augustine of Hippo* (Oxford: Clarendon, 1933).

3. This revision began perhaps with: Keith Hopkins, *Conquerors and Slaves* (Cambridge: Cambridge Univ. Press, 1977); and Simon Price, *Rituals and Power: The Roman Imperial Cult in Asia Minor* (Cambridge: Cambridge Univ. Press, 1984). See also: Mary Beard, John North, and Simon Price, *Religions of Rome* (Cambridge: Cambridge Univ. Press, 1998).

4. Arnaldo Momigliano, *On Pagans, Jews, and Christians* (Middletown, CT: Wesleyan Univ. Press, 1987), 95.

5. Momigliano, 95.

6. Momigliano, 100.

7. Momigliano, 107.

8. Cicero, *The Nature of the Gods*, Horace C. P. McGregor, trans. (London: Penguin, 1972), 69, 70.

9. Cicero, 75.

10. Cicero, 77.

11. Cicero, 87.

12. Cicero, 91–92.

13. Cicero, 94.

14. Cicero, 108.

15. Cicero, 117.

16. Cicero, 120.

17. Cicero, 124.

18. Cicero, 129.

19. Cicero, 145.

20. Cicero, 130.

21. Cicero, 124.

22. Cicero, 152.

23. Cicero, 161.

24. Cicero, 162.

25. Cicero, 187.

26. Cicero, 132.
27. Cicero, 193.
28. Cicero, 194.
29. Cicero, 195.
30. Cicero, 197.
31. Cicero, 197.
32. Cicero, 201–202.
33. Cicero, 202.
34. Cicero, 206.
35. Cicero, 207–208.
36. Cicero, 209.
37. Cicero, 232.
38. Cicero, 234.
39. Cicero, 234.
40. To cite a master of the period who is of this opinion: "The essential point is that Cotta, as an Academic, finds himself in greater sympathy with the Epicureans than with the Stoics when it comes to deciding whether gods intervene in human life. Cotta is really as uncertain about the existence of the gods as he is about the immortality of the soul." This scholar notes that Cicero has Cotta attribute Rome's greatness to the observance of her rituals. "This is good enough, but very short and followed by a close argument for the impossibility of proving the existence of gods, to which there is no reply. The inescapable conclusion a reader was bound to draw from the end of the De natura deorum was that Cicero, with all due precautions, intended to be negative." Momigliano, 69.
41. Momigliano, 72.
42. Cicero, 234.
43. Lucretius, On the Nature of Things, trans. and ed. Anthony M. Esolen (Baltimore: Johns Hopkins Univ. Press, 1995), 26–27.
44. Lucretius, 112.
45. Lucretius, 82.
46. Lucretius, 114–116.
47. Lucretius, 121.
48. Lucretius, 159.
49. Lucretius, 192.
50. Lucretius, 75–76.
51. Lucretius, 202.
52. Lucretius, 203.
53. Lucretius, 77.
54. Lucretius, 86–87.
55. Lucretius, 88.
56. Pliny the Elder, Natural History (London: Penguin, 1991), 13.
57. Pliny, 12.
58. Pliny, 13.
59. Pliny, 14.
60. Pliny, 103.
61. Pliny, 14.
62. Pliny, 103–104.
63. Marcus Aurelius, The Meditations of Marcus Aurelius, trans. George Long (New York: Collier, 1909–1914), 71–72.

64. Marcus Aurelius, 77.
65. Marcus Aurelius, 23.
66. Marcus Aurelius, 73.
67. Marcus Aurelius, 77–78.
68. Marcus Aurelius, 22–23.
69. Marcus Aurelius, 72.
70. Marcus Aurelius, 79.
71. Marcus Aurelius, 80.
72. Marcus Aurelius, 82.
73. Marcus Aurelius, 22.
74. Marcus Aurelius, 73–74.
75. Marcus Aurelius, 61.
76. Marcus Aurelius, 24–25.
77. Marcus Aurelius, 22.
78. Marcus Aurelius, 72.
79. Marcus Aurelius, 81.
80. Marcus Aurelius, 72–73.
81. Sextus Empiricus, *Selections from the Major Writings on Scepticism, Man, and God,* ed. Philip P. Hallie and trans. Sanford G. Etheridge (Indianapolis: Hackett, 1985), 188.
82. Sextus Empiricus, 190.
83. Sextus Empiricus, 185.
84. Sextus Empiricus, 206.
85. Sextus Empiricus, 207.
86. Sextus Empiricus, 209.
87. Sextus Empiricus, 211.
88. Sextus Empiricus, 213.
89. Lucian, *True History and Lucius or the Ass,* trans. Paul Turner (Bloomington: Indiana Univ. Press, 1974).
90. Lucian, 17.

CHAPTER FIVE

1. Hayim Ben-Sasson, ed., *A History of the Jewish People* (Cambridge: Harvard Univ. Press, 1976), 218–225.
2. Ben-Sasson, 288.
3. Ben-Sasson, 288–289.
4. Jerusalem Talmud Haggigah 1:7.
5. See E. P. Sanders, *The Historical Figure of Jesus* (London: Penguin, 1993), esp. 131. See also Michael Grant, *Jesus: An Historian's Review of the Gospels* (New York: Macmillan, 1977).
6. Sanders, 176, 230–234, 255–257.
7. John Dominic Crossan, *Jesus: A Revolutionary Biography* (San Francisco: HarperSanFrancisco, 1989), 122. See: Burton Mack, *The Lost Gospel: The Book of Q and Christian Origins* (San Francisco: HarperSanFrancisco, 1994) and John Dominic Crossan, *The Historical Jesus: The Life of a Mediterranean Jewish Peasant* (San Francisco: HarperSanFrancisco, 1993).
8. Matt. 26:38–39. Unless otherwise indicated, citations in this chapter refer to the New International Version.
9. Matt. 24:9–14; Mark 13:9–13; Luke 21:12–19.
10. Matt. 14:25–31.

11. Mark 9:21–24.
12. Matt. 17:20.
13. Mark 6:5–6 (KJV).
14. John 20:24–29.
15. Luke 24:36–43 (KJV).
16. Rom. 4:11.
17. Rom. 4:13–18.
18. Rom. 7:14–25.
19. Rom. 9:30–32.
20. Rom. 10:4.
21. Werner Jaeger, *Early Christianity and Greek Paideia* (Cambridge: Harvard Univ. Press, 1961), 11.
22. Jaeger, 26–35.
23. On Plotinus, see Karen Armstrong, *A History of God* (New York: Knopf, 1993), esp. 101–102, 171, 175, 181.
24. Elaine Pagels, *The Gnostic Gospels* (New York: Vintage, 1979), xix–xx.
25. Peter Brown, *The Rise of Western Christendom* (Oxford: Basil Blackwell, 1996), 73.
26. Augustine, *Confessions,* trans. Henry Chadwick (Oxford: Oxford Univ. Press, 1992), 109–110.
27. Augustine, *Confessions,* 130.
28. Augustine, *Confessions,* 131–132.
29. Augustine, *Confessions,* 145.
30. Augustine, *Confessions,* 147.
31. Augustine, *Confessions,* 150.
32. Augustine, *Confessions,* 151.
33. Augustine, *Confessions,* 152.
34. Augustine, *Confessions,* 153.
35. Augustine, *De Trinitate,* trans. E. Hill (New York: New City, 1991), 10.10.14, as cited in Simon Harrison, "Do We Have a Will?: Augustine's Way into the Will," in *The Augustinian Tradition,* ed. Gareth B. Matthews (Berkeley: Univ. of California Press, 1999), 201.
36. Augustine, *The City of God,* trans. D. S. Wiesen (Cambridge: Harvard Univ. Press, 1988), 11.26, as cited in Harrison, 201.
37. Augustine, *Teaching Christianity (On Christian Doctrine),* ed. Edmund Hill (Hyde Park, NY: New City, 1996), 159–160, sec. 60, as cited in Edward Grant, *God and Reason in the Middle Ages* (Cambridge: Cambridge Univ. Press, 2001), 37.
38. Augustine, *Teaching Christianity,* 153–154, sec. 48, as cited in Grant, 39.
39. Cited in Jonathan Barnes, "Boethius and the Study of Logic," in *Boethius: His Life, Thought and Influence,* ed. Margaret Gibson (Oxford: Basil Blackwell, 1981), 73.
40. Grant, 41.
41. Yer. Meg. i. 9.
42. 2:6, 7.
43. Babylonian Talmud, Tractate Eruvin 53b.
44. Socrates Scholasticus, *Ecclesiastical History* (London: S. Bagster and Sons, 1853), vol. 7, 15.
45. Brown, 46.
46. As cited in Brown, 97.
47. Armstrong, 131, 305–306.
48. Stephen Batchelor, *The Faith to Doubt: Glimpses of Buddhist Uncertainty* (Berkeley, CA: Paralax, 1990).
49. Batchelor, 40.

50. Keiji Nishitani, *Religion and Nothingness,* trans. Jan Van Bragt (Berkeley: Univ. of California Press, 1983), 16.
51. S. Dasgupta, *A History of Indian Philosophy* (London: Cambridge Univ. Press, 1922), 536.
52. James Thrower, *The Alternative Tradition* (The Hague: Mouton, 1980), 83.

CHAPTER SIX

1. Peter Brown, *The Rise of Western Christendom* (Oxford: Basil Blackwell, 1996), 181.
2. Joel Kraemer, *Humanism in the Renaissance of Islam: The Cultural Revival During the Buyid Age* (Leiden: E. J. Brill, 1986), vi, 189–190.
3. Karen Armstrong, *A History of God* (New York: Knopf, 1993), 173.
4. Sarah Stroumsa, *Freethinkers of Medieval Islam* (Leiden: Brill, 1999).
5. Stroumsa, 145–150.
6. Ignaz Goldziher, *Muslim Studies,* S. M. Stern, ed., C. R. Barber and S. M. Stern, trans., 2 vols. (London, 1967–71), 363–364.
7. Cited in D. S. Margoliouth, "Atheism (Muhammadan)," in *Encyclopedia of Religion and Ethics,* as cited in Ibn Warraq, *Why I Am Not a Muslim* (New York: Prometheus, 1995), 255.
8. Stroumsa, 43.
9. Stroumsa, 63.
10. Stroumsa, 83.
11. Stroumsa, 73.
12. Stroumsa, 47.
13. Stroumsa, 131–132.
14. Stroumsa, 90.
15. Stroumsa, 96.
16. Stroumsa, 96.
17. Stroumsa, 96.
18. Stroumsa, 98.
19. P. Kraus and S. Pines, "al-Razi," *The Encyclopaedia of Islam,* 3.II (Leiden, 1913–1938), 1136, as cited in Stroumsa, 98.
20. As cited in Stroumsa, 99–100.
21. Stroumsa, 100.
22. Stroumsa, 103.
23. Stroumsa, 113.
24. As cited in Armstrong, 180.
25. Warraq, *Why I Am Not a Muslim,* 286.
26. Reynold Alleyne Nicholson, *Studies in Islamic Poetry* (Cambridge: Cambridge Univ. Press, 1920), 176.
27. Nicholson, 110.
28. Nicholson, 173.
29. Nicholson, 190.
30. Nicholson, 191.
31. Nicholson, 167.
32. Nicholson, 174.
33. Nicholson, 174.
34. Nicholson, 177.
35. Nicholson, 177.

36. Al-Ghazzali, *Deliverance from Error,* in *The Faith and Practice of al-Ghazali,* trans. W. M. Watt (London: George Allen, 1951), 20.

37. Armstrong, 187.

38. Al-Ghazzali, *Deliverance from Error,* 21. See also the new translation Al-Ghazzali, *Deliverance from Error and Mystical Union with the Almighty,* George F. McLean, ed., Muhammad Abulaylah, trans. (Washington, DC: Council for Research in Values & Philosophy, 2002), esp. 65–96.

39. Al-Ghazzali, *Deliverance from Error,* 23.

40. Al-Ghazzali, *Deliverance from Error,* 25.

41. Al-Ghazzali, *Deliverance from Error,* 26–27.

42. Al-Ghazzali, *Deliverance from Error,* 30.

43. Abulaylah, 74, translates this: "According to them, the animal issued from the sperm, and the sperm from the animal continuously. These are atheist (zanadiqa)."

44. Al-Ghazzali, *Deliverance from Error,* 33.

45. Al-Ghazzali, *Deliverance from Error,* 42.

46. Al-Ghazzali, *The Incoherence of the Philosophers,* in *Philosophy in the Middle Ages,* Arthur Hyman and James J. Walsh, eds. (Indianapolis: Hackett, 1973), 298.

47. Al-Ghazzali, *Incoherence of the Philosophers,* 301.

48. Al-Ghazzali, *Incoherence of the Philosophers,* 302.

49. Al-Ghazzali, *Incoherence of the Philosophers,* 302.

50. Al-Ghazzali, *Incoherence of the Philosophers,* 310.

51. Saadia ben Joseph, *The Book of Doctrines and Beliefs,* in *Philosophy in the Middle Ages,* 344.

52. Ben Joseph, 348.

53. Moses Maimonides, *Guide for the Perplexed* (New York: Dover, 1956), 314–315 (III, 28).

54. Maimonides, 173 (II, 13).

55. Maimonides, 172.

56. Maimonides, 87.

57. Maimonides, 89.

58. Maimonides, 87.

59. Maimonides, 85.

60. Maimonides, 85.

61. Maimonides, 384.

62. As cited in Marvin Fox, *Interpreting Maimonides* (Chicago: Univ. of Chicago Press, 1990), 331.

63. Strousma, 222–238.

64. Gershom G. Scholem, *Major Trends in Jewish Mysticism* (New York: Schocken, 1941), 7–9.

65. Scholem, 139.

66. Granted by the International Astronomical Union at the beginning of the twentieth century.

67. Gersonides, *The Wars of the Lord,* ch. 3, in *Philosophy in the Middle Ages,* 434.

68. Gersonides, 434.

69. Edward Grant, *God and Reason in the Middle Ages* (Cambridge: Cambridge Univ. Press, 2001), 87.

70. Grant, 101.

71. G. E. Hughes, intro. to his trans. of *John Buridan on Self-Reliance* (Cambridge: Cambridge Univ. Press, 1982), 4, as cited in Grant, 127.

72. These are taken from Grant, 125–127, who takes them from several sources.

73. Grant, 179.

74. John E. Murdoch, "The Anayltic Character of Late Medieval Learning: Natural Philosophy Without Nature," in *Approaches to Nature in the Middle Ages,* ed. Lawrence D. Roberts (Binghamton, NY: CMES, 1982).

75. Grant, 192.
76. Etienne Gilson, *Reason and Revelation in the Middle Ages* (New York: Scribner, 1938), 64.
77. Grant, 190.
78. Grant wrote that out of 310 questions he investigated, in 217 of them one could not tell if the author was an atheist, a Christian, a Jew, or a Muslim. Only 10 actually discussed God or the faith. Grant, 186.
79. Grant, 199.
80. Nicholas of Autrecourt, "Letters to Bernard of Arezzo," in *Philosophy in the Middle Ages,* 707–708. I've used "of your body" where this translator had "inside you."
81. Richard H. Popkin, *The History of Scepticism from Erasmus to Spinoza* (Berkeley: Univ. of California Press, 1979), 19.

CHAPTER SEVEN

1. Alan Watts, *The Way of Zen* (New York: Vintage, 1989), 106, 170.
2. Ikkyu, as cited in R. H. Blyth, *Zen and Zen Classics,* ed. Frederick Franck (New York: Vintage, 1978), 144. See also Ikkyu, *Crow with No Mouth,* trans. Stephen Berg (Port Townsend, WA: Copper Canyon, 2000).
3. James Thrower, *The Alternative Tradition: Religion and the Rejection of Religion in the Ancient World* (The Hague: Mouton, 1980), 88.
4. Diané Collinson, Kathryn Plant, and Robert Wilkinson, eds., *Fifty Eastern Thinkers* (London: Routledge, 2000), 279–280.
5. Samkhya Sutra Vrtti, as cited in Thrower, 88.
6. Edward Grant, *God and Reason in the Middle Ages* (Cambridge: Cambridge Univ. Press, 2001), 294.
7. Grant, 295.
8. Grant, 295.
9. Petrarch, "An Averroist Visits Petrarca," in *The Renaissance Philosophy of Man,* ed. Ernst Cassirer, Paul Oskar Kristeller, and John Herman Randall Cassirer Jr. (Chicago: Univ. of Chicago Press, 1948), 141.
10. Don Cameron Allen, *Doubt's Boundless Sea* (Baltimore: Johns Hopkins Univ. Press, 1964; New York: Arno Press, 1979), 32.
11. Allen, 35.
12. David Wootton, "New Histories of Atheism," in *Atheism from the Reformation to the Enlightenment,* ed. Michael Hunter and David Wootton (Oxford: Clarendon, 1992), 32.
13. Nicholas Davidson, "Unbelief and Atheism in Italy, 1500–1700," in *Atheism,* ed. Hunter and Wootton, 68.
14. Erasmus, *Praise of Folly,* trans. Betty Radice (London: Penguin, 1993), 87.
15. Erasmus, 93.
16. Erasmus, 94.
17. Erasmus, 71.
18. As cited in Grant, 303.
19. Luther, *De Servo Arbitrio,* in *Luther and Erasmus: Free Will and Salvation,* trans. and ed. Philip S. Watson (Philadelphia: Westminster, 1969), 109.
20. Luther, 109.
21. As cited in Allen, 7.
22. Lucien Febvre, *The Problem of Unbelief in the Sixteenth Century: The Religion of Rabelais,* trans. Beatrice Gottlieb (Cambridge: Harvard Univ. Press, 1947), 192.
23. Febvre, translator's preface, xi.

24. Febvre, 53–54.
25. Febvre, 56.
26. There were some exceptions, of course, but they tended to be maverick performances, like the odd and interesting *Doubt's Boundless Sea* by Don Cameron Allen, noted above.
27. Karen Armstrong, *A History of God* (New York: Knopf, 1994), 286–287.
28. Susan Reynolds, "Social Mentalities and the Case of Medieval Scepticism," *Transactions of the Royal Historical Society,* 6th ser., 1 (1991), 25.
29. Reynolds, 35.
30. That book's eleven authors all discuss writing the history of atheism with regard to Febvre's nondoubting "mentalities," and their convictions against his theory have helped to create the context I present here.
31. François Rabelais, *Gargantua and Pantagruel,* trans. J. M. Cohen (London: Penguin, 1955), 37.
32. Rabelais, 544–545,
33. Rabelais, 193.
34. Rabelais, 194–195.
35. Febvre, 98.
36. Reynolds, 36.
37. Davidson, in *Atheism,* ed. Hunter and Wootton, 64.
38. Davidson, 57.
39. Davidson, 66–79.
40. Davidson, 79.
41. As cited in Richard H. Popkin, *The History of Scepticism from Erasmus to Spinoza* (Berkeley: Univ. of California Press, 1979), 27–28, 29.
42. Davidson, 63.
43. Davidson, 13–14.
44. Davidson, 44.
45. Davidson, 48.
46. Davidson, 52.
47. Davidson, 65.
48. Davidson, 67.
49. Davidson, 80.
50. Davidson, 75.
51. Davidson, 83.
52. Wootton, in *Atheism,* ed. Hunter and Wootton, 29.
53. Carlo Ginzburg, *The Cheese and the Worms: The Cosmos of a Sixteenth-Century Miller* (Baltimore: Johns Hopkins Univ. Press, 1980).
54. Ginzburg, 2.
55. Ginzburg, 4.
56. Ginzburg, 10.
57. Ginzburg, 6.
58. Ginzburg, 21.
59. Ginzburg, 47.
60. Ginzburg, 46.
61. Ginzburg, 42.
62. Ginzburg, 50.
63. Ginzburg, 11.
64. Ginzburg, 69.

65. Ginzburg, 65.
66. As cited in Davidson, in Hunter and Wootton, 58.
67. Giordano Bruno, *The Ash Wednesday Supper* (1584). La cena de le ceneri, Dial. I, pp. 1–2.
68. As cited in Davidson, 61.
69. Bruno, *On Cause, Primary Origin, and the One.* See Epicurus, "Letter to Herodotus," in Diogenes Laertius, *Lives of Eminent Philosophers,* vol. 2 (Cambridge: Harvard Univ. Press, 1972), 591.
70. Davidson, 73–74.
71. As cited in Davidson, 77–78.
72. Michel de Montaigne, *The Complete Essays of Montaigne,* trans. Donald M. Frame (Stanford: Stanford Univ. Press, 1958), 320–321.
73. Montaigne, 324–325.
74. Montaigne, 394.
75. Montaigne, 258.
76. Montaigne, 359.
77. Montaigne, 361.
78. Montaigne, 400.
79. Montaigne, 397.
80. Montaigne, 355, 356.
81. Montaigne, 438.
82. Montaigne, 355.
83. Montaigne, 377.
84. Montaigne, 413.
85. Montaigne, 424.
86. Montaigne, 428.
87. Montaigne, 455.
88. Montaigne, 456.
89. Montaigne, 423.
90. Montaigne, 426.
91. Montaigne, 429.
92. Montaigne, 373.
93. Montaigne, 375n, 393.
94. Montaigne, 428.
95. Montaigne, 436.
96. Montaigne, 429–430.
97. Montaigne, 430–431.
98. Montaigne, 362.
99. Montaigne, 445.
100. Montaigne, 457.
101. Popkin, 82.
102. Wootton, in Hunter and Wootton, 38.
103. Tullio Gregory, "Pierre Charron's 'Scandalous Book,'" in Hunter and Wootton, 87; Pierre Charron, *Petit traicté de sages* (Paris, 1635, first published 1606), 223–224, as cited by Gregory in Hunter and Wootton, 89.
104. As cited in Popkin, 79.
105. Popkin, 90.

106. Popkin, 214–228.
107. Popkin, 88.
108. Wootton, in Hunter and Wootton, 22.

CHAPTER EIGHT

1. René Descartes, *Philosophical Works I* (Cambridge: Cambridge Univ. Press, 1975), 144.
2. Descartes, 145.
3. Descartes, 148.
4. Descartes, 151.
5. T. M. Rudavsky, "Galileo and Spinoza: Heroes, Heretics, and Hermeneutics," *Journal of the History of Ideas* 62 (1951): 611–631.
6. Rudavsky, 613–615.
7. As cited in Nicholas Davidson, "Unbelief and Atheism in Italy, 1500–1700," in *Atheism from the Reformation to the Enlightenment,* ed. Michael Hunter and David Wootton (Oxford: Clarendon, 1992), 61.
8. Davidson, 62.
9. Rudavsky, 629–630.
10. Benedict de Spinoza, *Ethics,* James Gutmann, ed. (New York: Hafner, 1949), 76.
11. As cited in Don Garrett, "Spinoza's Ethical Theory," in *The Cambridge Companion to Spinoza,* ed. Don Garrett (Cambridge: Cambridge Univ. Press, 1996), 278.
12. As cited in W. N. A. Klever, "Spinoza's Life and Work," in *Companion to Spinoza,* ed. Garrett, 44.
13. Thomas Hobbes, *Leviathan* (New York: Penguin, 1968), 483.
14. Hobbes, 167–168.
15. Hobbes, 169.
16. Hobbes, 170.
17. Hobbes, 172–173.
18. David Berman, *A History of Atheism in Britain: From Hobbes to Russell* (London: Routlage, 1988), 49–50.
19. As cited in Berman, 59.
20. Blaise Pascal, *Pensées* (London: Penguin, 1995),122–123.
21. Pascal, 124.
22. Pascal, 129.
23. Richard H. Popkin, *The History of Scepticism from Erasmus to Spinoza* (Berkeley: Univ. of California Press, 1979), 237.
24. Davidson, in *Atheism,* ed. Hunter and Wootton, 83–84.
25. David Wootton, "New Histories of Atheism," in *Atheism,* ed. Hunter and Wootton, 30.
26. Peter Harrison, "Newtonian Science, Miracles, and the Laws of Nature," *Journal of the History of Ideas* 56 (October 1995): 531–553.
27. Harrison, 540.
28. Thomas M. Lennon, *Reading Bayle* (Toronto: Univ. of Toronto Press, 1999), 7.
29. Pierre Bayle, *Historical and Critical Dictionary: Selections,* trans. Richard Popkin (Indianapolis: Hackett, 1991), 119.
30. Bayle, 272.
31. Bayle, 288.
32. Bayle, 289–290.
33. Bayle, 290.

34. Bayle, 290.
35. Wootton, in Hunter and Wootton, 43.
36. Bayle, 340.
37. Popkin, Translator's Introduction, in Bayle, xxi–xxii.
38. Bayle, 195.
39. Bayle, Translator's Introduction, xxiii.
40. Bayle, *The Dictionary Historical and Critical of Mr. Peter Bayle* (London: J. J. and P. Knapton, 1734–38), vol. V, 440–441; as cited in Wootton in Hunter and Wootton, 13–15.
41. See Silvia Berti, "The First Edition of the *Traité des trios imposteurs* and Its Debt to Spinoza's *Ethics*," in *Atheism,* ed. Hunter and Wootton, 182–220.
42. Berti, 218.
43. Berti, 214.
44. Berti, 214.
45. Berti, 186.
46. Popkin, *History of Scepticism,* 151–161.
47. David Berman, "Disclaimers as Offence Mechanisms in Charles Blount and John Toland," in *Atheism,* ed. Hunter and Wootton, 269.
48. Berman, in Hunter and Wootton, 255.
49. Toland, *Tetradymus* (London, 1720), 72, as cited in Berman, *Atheism in Britain,* 75.
50. Toland, *Tetradymus,* 95, as cited in Berman, *Atheism in Britain,* 76.
51. See Wootton, in Hunter and Wootton, 40.
52. Berman, *Atheism in Britain,* 72.
53. Berman, *Atheism in Britain,* 82.
54. Berman, *Atheism in Britain,* 81.
55. Berman, *Atheism in Britain,* 81.
56. Michael Hunter, "Aikenhead the Atheist," in Hunter and Wootton, 221.
57. Hunter, 224–225.
58. Hunter, 225.
59. Joanna Whaley-Cohen, *The Sextants of Beijing: Global Currents in Chinese History* (New York: Norton, 1999), 110.
60. Whaley-Cohen, 110.
61. "Decree of Emperor Kangxi," in Dan Li, trans., *China in Transition, 1517–1911* (New York: Van Nostrand, 1969), 22–24.
62. Berman, *Atheism in Britain,* 80.
63. Dena Goodman, *The Republic of Letters: A Cultural History of the French Enlightenment* (Ithaca, NY: Cornell Univ. Press, 1994), 125.
64. Goodman, 74–75.
65. Bernard Fontenelle, *A Conversation on the Plurality of Worlds* (1686), *The Origin of Fables* (1724), *Of the Island of Borneo* (1686).
66. Goodman, 86.
67. Goodman, 89.
68. Voltaire, *The Philosophical Dictionary* (New York: Coventry House, 1932), 397.
69. Voltaire, 197.
70. Goodman, 77.
71. Denis Diderot, *Diderot: Interpreter of Nature* (Westport, CT: Hyperion, 1937), 218.
72. Diderot, 225.
73. Diderot, 226.
74. Diderot, 228.

75. Diderot, 227.
76. Diderot, "Dithyrambe sur la fête de rois."
77. David Hume, *An Enquiry Concerning Human Understanding: A Letter from a Gentleman to His Friend in Edinburgh* (Indianapolis: Hackett, 1977), 93.
78. Hume, 97.
79. Berman, *Atheism in Britain,* 102.
80. Alan Charles Kors, "The Atheism of d'Holbach and Naigeon," in Hunter and Wootton, 292.
81. Baron d'Holbach, *System of Nature,* vol. 2, ch. 2, excerpted in *Varieties of Unbelief: From Epicurus to Sartre,* ed. J. C. A. Gaskin (Upper Saddle River, NJ: Prentice-Hall, 1989), 92.
82. Holbach, ch. 8, as cited in Gaskin, 95.
83. Edward Gibbon, *The Decline and Fall of the Roman Empire* (New York: Modern Library, 1954), 1:406.
84. Gibbon, 1:406.
85. Gibbon, 1:407.
86. J. Y. T. Grieg, *The Letters of David Hume,* 2 vols. (Oxford: Clarendon, 1969), 2:310.
87. Gibbon, 1:504.
88. Thomas Paine, *The Age of Reason* (Amherst, NY: Prometheus, 1984), 9.
89. Paine, 13.
90. Paine, 13.
91. Paine, 11–12.
92. Paine, 12.
93. Thomas Jefferson, "To William Short, October 31, 1819," in *A Jefferson Profile: As Revealed in His Letters,* ed. Saul K. Padover (New York: John Day, 1956), 306.
94. Jefferson, "To William Short, August 4, 1820," in *The Works of Thomas Jefferson,* H. A. Washington, ed. (New York: Townsend, 1884), vol. 2, 217.
95. Thomas Jefferson, "To Peter Carr, August 10, 1787," in *Basic Writing of Thomas Jefferson,* Philip S. Foner, ed. (New York: Wiley, 1944), 561. Carr was one of Jefferson's favorite nephews.
96. Jefferson, "To Peter Carr," 562.
97. Jefferson, "To William Short, August 4, 1820," 2:217.
98. It was written by Joel Barlow, U.S. Consul.
99. From *The American Enlightenment: The Shaping of the American Experiment and a Free Society,* ed. Adrienne Koch (New York: G. Braziller, 1965), 234.
100. Peter Mercer-Taylor, *The Life of Mendelssohn* (Cambridge: Cambridge Univ. Press, 2000), 8.
101. Mercer-Taylor, 206.
102. Mercer-Taylor, 12.
103. Moses Mendelssohn, *Jerusalem, or On Religious Power and Judaism,* Allan Arkush, trans., Alexander Altmann, intro. and comment. (Hanover: published for Brandeis Univ. Press by University Press of New England, 1983), 35.
104. Mendelssohn, 59.
105. Mendelssohn, 61.
106. Mendelssohn, 63.
107. Mendelssohn, 73.
108. Mendelssohn, 74.
109. Mendelssohn, 89.
110. Mendelssohn, 90.
111. Mendelssohn, 90.
112. Mendelssohn, 94.
113. Mendelssohn, 95.

114. Mendelssohn, 97–98.
115. Mendelssohn, 100.
116. Mendelssohn, 111, 115, 117.
117. Mendelssohn, 133.
118. Jerusalem Talmud Haggigah 1:7.
119. Mendelssohn, 137, 138.
120. Mendelssohn, 139.
121. There is some confusion as to his dates, but this seems right; see Keiji Nishitani, *Religion and Nothingness*, trans. Jan Van Bragt (Berkeley: Univ. of California Press, 1983), 20.
122. Nishitani, 20–21.

CHAPTER NINE

1. Ludwig Feuerbach, *The Essence of Christianity*, trans. George Eliot (New York: Harper, Torchbooks, 1957), 277.
2. David Berman, "Disclaimers as Offence Mechanisms in Charles Blount and John Toland," in *Atheism from the Reformation to the Enlightenment*, ed. Michael Hunter and David Wootton (Oxford: Clarendon, 1992), 234.
3. Peter Mercer-Taylor, *The Life of Mendelssohn* (Cambridge: Cambridge Univ. Press, 2000), 16.
4. As cited in Mercer-Taylor, 31.
5. As cited in Mercer-Taylor, 32.
6. Heinrich Heine, *Gedanken und Einfälle*, vol. 10.
7. Heine, as cited in Joseph Ratner, "Introduction," in *The Philosophy of Spinoza: Selections from His Works* (New York: Modern Library, 1927).
8. See W. Gunther Plaut, *The Rise of Reform Judaism: A Sourcebook of Its European Origins* (New York: World Union for Progressive Judaism, 1963), 22.
9. "Metz, 1841, A Public Appeal (31 Jewish Households)," in Plaut, 45.
10. "Metz, 1841," in Plaut, 45.
11. "Metz, 1841" in Plaut, 44.
12. "The Status of Women—No Spiritual Minority, Abraham Geiger," in Plaut, 253.
13. "Status of Women," in Plaut, 254.
14. See Plaut, 243–255.
15. See Plaut, 99.
16. See Plaut, 205.
17. Cited in full in David Philipson, *The Reform Movement in Judaism* (Jersey City, NJ: KTAV, 1967), 354–357.
18. Anne Newport Royall, *Letters from Alabama, 1817–1822* (Birmingham: Univ. of Alabama Press, 1969), 176.
19. Royall, *Black Book* (Washington, DC, 1828), vol. 1, 163.
20. Royall, *Black Book*, 165.
21. Royall, *Black Book*, 240–241.
22. Royall, *Letters from Alabama*, 218–219.
23. Royall, *Black Book*, 195.
24. Royall, *Black Book*, 194.
25. George Stuyvesant Jackson, *Uncommon Scold: The Story of Anne Royall* (Boston: B. Humphries, 1937).
26. John Stuart Mill, *The Autobiography of John Stuart Mill* (New York: Columbia Univ. Press, 1924), 30.

27. Mill, *Autobiography*, 28.
28. Mill, *Autobiography*, 33.
29. Mill, *On Liberty*, ed. Edward Alexander (Toronto: Broadview, 1999), 108.
30. Mill, *On Liberty*, 113.
31. Harriet Martineau, *Autobiography*, ed. Maria Weston Chapman (Boston: James R. Osgood, 1877), 1:28.
32. Harriet Martineau, with Henry George Atkinson, *On the Laws of Man's Nature and Development* (Boston: Josiah P. Mendum, 1851), 222–223.
33. Martineau, *Autobiography*, 2:110.
34. Martineau, *Autobiography*, 1:30.
35. Martineau, *Autobiography*, 1:89.
36. Martineau, *Autobiography*, 2:45–46.
37. Martineau, *Autobiography*, 2:46.
38. Cited in *Women Without Superstition, "No Gods—No Masters": The Collected Writings of Women Freethinkers of the Nineteenth and Twentieth Centuries,* ed. Annie Laurel Gaylor (Madison, WI: Freedom from Religion Foundation, 1997), 49.
39. Martineau, *Autobiography*, 2:104–107.
40. Peter M. Rinaldo, *Atheists, Agnostics, and Deists in America: A Brief History* (New York: DorPete, 2000), 55.
41. Fanny Wright, *A Few Days in Athens* (New York: Arno Press, 1972), 211.
42. Wright, *Views of Society and Manners in America*, ed. Paul R. Baker (Cambridge: Harvard Univ. Press, 1963), 224.
43. Robert Owen, *The Life of Robert Owen, Written by Himself* (New York: A. M. Kelly, 1967), 1:206–207.
44. Wright, "Divisions of Knowledge," in her *Life, Letters, and Lectures, 1834/1844* (New York: Arno, 1972).
45. Wright, "Divisions of Knowledge," 73–74.
46. Wright, "Religion," in *Life, Letters, and Lectures, 1834/1844.*
47. "Ernestine Rose," in *The History of Women's Suffrage,* ed. Elizabeth Cady Stanton, Susan B. Anthony, and Matilda Joslyn Gage (Rochester, NY: Source Book Press, 1971), 1:661–663.
48. "Ernestine Rose," 1:663.
49. Ernestine Rose, "A Defense of Atheism," in *Women Without Superstition,* ed. Gaylor, 77.
50. Rose, 81.
51. Rose, 85.
52. Rose, 85.
53. Karl Heinrich Marx, "The Difference Between the Democritean and Epicurean Philosophy of Nature," in Norman D. Livergood, *Activity in Marx's Philosophy* (The Hague: Martinus Nijhoff, 1967), 61.
54. Marx, "The Difference Between the Democritean and Epicurean Philosophy of Nature," 62.
55. From Marx's introduction to "Contribution to the Critique of Hegel's Philosophy of Right" (1844), as cited in *Varieties of Unbelief: From Epicurus to Satre,* ed. J. C. A. Gaskin (Upper Saddle River, NJ: Prentice-Hall, 1989), 158.
56. As cited in Gaskin, 158.
57. Owen Chadwick, *The Secularization of the European Mind* (Cambridge: Cambridge Univ. Press, 1975), 65.
58. "Stanton to Anthony," January 24, 1856, Library of Congress; as cited in *Women Without Superstition,* ed. Gaylor, 106.

59. Stanton, "Antislavery" (1860), as cited in *Women's Suffrage,* ed. Stanton, Anthony, and Gage, 2:187.

60. Stanton, "Women's Position in the Christian Church" (1888), as cited in Gaylor, 113.

61. Stanton, "The Universal Creed" (1869), in Gaylor, 137.

62. Stanton, "The Pleasures of Age" (1885), in Gaylor, 150.

63. Stanton, Anthony, and Gage, 1:100.

64. Stanton, Anthony, and Gage, 1:422.

65. "Stanton to Antoinette Brown Blackwell, June 10, 1873," in *Elizabeth Cady Stanton as Revealed in Her Letters, Diary and Reminiscences,* ed. Theodore Stanton and Harriot Stanton Blatch (New York: Harper, 1922), 2:142.

66. *Elizabeth Cady Stanton, Susan B. Anthony, Correspondence, Writings, Speeches,* ed. Ellen Carol DuBois (New York: Schocken, 1981), 246–255.

67. Stanton, "Women's Position in the Christian Church" (1888), 119.

68. Stanton, "Women's Position," 124.

69. Stanton, "Women's Position," 107.

70. Arthur Schopenhauer, *The World as Will and Representation,* trans. E. F. J. Payne, 2 vols. (New York: Dover, 1969), 2:161–162.

71. Schopenhauer, *The World,* 1:420.

72. Schopenhauer, *The World,* 2:357.

73. Schopenhauer, *The World,* 2:354.

74. Schopenhauer, *The World,* 1:326.

75. Schopenhauer, *The World,* 1:419.

76. Schopenhauer, *The World,* 1:357.

77. Schopenhauer, *The World,* 2:586.

78. Schopenhauer, *The World,* 1:375.

79. Schopenhauer, *The World,* 1:422.

80. Schopenhauer, *The World,* 1:361–362.

81. Schopenhauer, *The World,* 1:407.

82. Schopenhauer, *The World,* 1:367.

83. Schopenhauer, *The World,* 2:47.

84. Bryan Magee, *The Philosophy of Schopenhauer* (Oxford: Oxford Univ. Press, 1983), 288.

85. Schopenhauer, *Religion: A Dialogue and Other Essays,* 3d ed., trans. T. Bailey Saunders (Westport, CT: Greenwood, 1973), 4.

86. Schopenhauer, *Religion,* 4.

87. Schopenhauer, *Religion,* 4–5.

88. Schopenhauer, *Religion,* 5.

89. Schopenhauer, "On the Sufferings of the World," in his *Studies in Pessimism* (New York: Boni and Liveright, 1925), 34.

90. Schopenhauer, *Religion,* 5–6.

91. Schopenhauer, *Religion,* 21.

92. Schopenhauer, *Religion,* 51.

93. Søren Kierkegaard, *Fear and Trembling* (London: Penguin Books, 1985), 103.

94. Kierkegaard, 63.

95. Kierkegaard, 67.

96. Kierkegaard, 133.

97. Kierkegaard, 135.

98. Friedrich Nietzsche, *The Gay Science,* trans. Walter Kaufmann (New York: Vintage, 1974), 125.

99. Nietzsche, *Daybreaks*, trans. R. J. Hollingdale (Cambridge: Cambridge Univ. Press, 1982), 89.

100. Adrian Desmond, *The Politics of Evolution* (Chicago: Univ. of Chicago Press, 1989), 9.

101. Desmond, 46.

102. Stephen J. Gould, *Ever Since Darwin* (New York: Norton, 1977), 25.

103. Gould, 26.

104. Chadwick, 176.

105. Chadwick, 166.

106. Chadwick, 171.

107. Jennifer Michael Hecht, *The End of the Soul: Scientific Modernity, Atheism, and Anthropology in France* (New York: Columbia Univ. Press, 2003).

108. Thomas Huxley, "Mr. Balfour's Attack on Agnosticism I," *The Nineteenth Century* 217 (March 1895): 534.

19. Huxley, "Mr. Balfour's Attack," 530.

110. Huxley, "On Descartes' *Discourse Touching the Method of Using One's Reason Rightly and of Seeking Scientific Truth*" (1870), in his *Lay Sermons, Addresses, and Reviews* (New York: Appleton, 1910), 322–323.

111. Huxley, "Mr. Balfour's Attack," 529.

112. James Strick, ed., *Evolution and the Spontaneous Generation Debate* (Bristol: Thoemmes, 2001).

113. Auguste Comte, *A General View of Positivism*, trans. J. H. Bridges (London: Routledge, 1957), 35–36.

114. Comte, 36.

115. Comte, 50.

116. Comte, 51.

117. Comte, 52.

118. Gustave Flaubert, *Madame Bovary*, trans. Geoffrey Wall (London: Penguin, 1992), 61.

119. Flaubert, 61.

120. Flaubert, 270.

121. Jules Ferry, *Revue pédagogique* (1882), as cited in Hecht, *End of the Soul*, 52.

122. René Rémond, *L'anticlericalism en France de 1815 à nos jours* (Paris: Fayard, 1976), 173.

123. Emile Durkheim, "Représentations individuelles et représentations collectives," *Revue de métaphysique et de morale* 5 (1898): 273–302.

124. James M. Edie, James Scanlan, and Mary-Barbara Zeldin, eds., *Russian Philosophy: The Nihilists, the Populists, Critics of Religion and Culture* (Chicago: Quadrangle, 1965), 2:3–10.

125. David Friedrich Strauss, *The Life of Jesus, Critically Examined*, trans. Marian Evans (New York: Blanchard, 1860).

126. Leslie Stephen, *Essays on Free Thinking and Plain Speaking* (London: Longmans, 1873).

127. Virginia Woolf, *Moments of Being*, ed. Jeanne Schulkind (London: Hogarth, 1985), 72; and "Letter to Lady Robert Cecil, 12 November 1922," in *The Question of Things Happening: The Letters of Virginia Woolf*, ed. Nigel Nicolson (London: Hogarth, 1982), 2:585.

128. Charles Bradlaugh, *Doubts in Dialogue* (London: Watts, 1909), 9–10.

129. Bradlaugh, *A Plea for Atheism* (London: Freethought, 1880), 4.

130. C. Cohen, *Bradlaugh and Ingersoll* (London: Pioneer, 1933), 57, cited in David Berman, *A History of Atheism in Britain* (London: Routledge, 1988), 215.

131. Annie Besant and Charles Bradlaugh, *The Freethinker's Text Book* (London: Freethought, 1881), 426, 476.

132. Arthur Bonner and Charles Bradlaugh Bonner, *Hypatia Bradlaugh Bonner: The Story of Her Life* (London: Watts & Co., 1942), 117, 119.

133. Robert G. Ingersoll, "Why I Am an Agnostic," in Ingersoll, *Works*, ed. Clinton P. Farrell (New York: Dresden, 1902), 4:45.

134. Ingersoll, *On Gods and Other Essays* (Buffalo: Prometheus, 1990), 54–55.

135. Frederick Douglass, "The Meaning of Fourth of July for the Negro" (1852), in *Oration, Delivered in Corinthian Hall, Rochester* (Rochester, NY: Lee, Mann, 1852), 29.

136. Douglass, 30.

137. *Boston Investigator*, September 19, 1896, as cited in *Women Without Superstition*, 376.

138. Etta Semple, "Liberty of Conscience Is All That We Ask," *The Free-thought Ideal*, ca. 1898, as cited in *Women Without Superstition*, 325.

139. As cited in *Women Without Superstition*, 295.

140. As cited in *Women Without Superstition*, 295.

141. Lucy N. Colman, *The Truth Seeker*, February 19, 1887.

142. A. Carel, *Histoire anecdotique des contemporains* (Paris, 1885), 46.

143. They included such figures as Richard Carlile, Matthew Turner, Thomas Cooper, and Samuel Francis. Berman, 110–134.

144. Percy Bysshe Shelley, "The Necessity of Atheism," in *The Necessity of Atheism and Other Essays* (Buffalo: Prometheus, 1993), 31.

145. Shelley, "Necessity of Atheism," 43.

146. *The Necessity of Atheism*, 44.

147. Shelley, "On a Future State," in *Necessity of Atheism and Other Essays*, 56–60.

148. Shelley, "On a Future State," 58.

149. Shelley, "Refutation of Deism," in *Necessity of Atheism and Other Essays*, 88.

150. Shelley, *Queen Mab*, in *Complete Poems* (New York: Book of the Month Club, 1993), 21–22.

151. Rinaldo, 69.

152. Margaret Fuller, *Memoirs of Margaret Fuller Ossoli* (Boston: Phillips, Sampson, 1852), 90.

153. Fuller, *Summer on the Lakes* (Boston: Charles Little & James Brown, 1844), 194.

154. Emily Dickinson, *The Complete Poems of Emily Dickinson* (Boston: Little, Brown, 1951), (#1144) 512.

155. Dennis Donoghue, *Emily Dickinson* (Minneapolis: Univ. of Minnesota Press, 1969), 14.

156. Donoghue, 15.

157. Dickinson, *The Complete Poems of Emily Dickinson* (Boston: Little, Brown, 1951), 646. Poem quoted is #1551, ca. 1882.

158. Thomas Hardy, "God's Funeral," in *The Collected Poems of Thomas Hardy* (New York: Macmillan, 1974).

CHAPTER TEN

1. Ludwig Wittgenstein, *Tractatus Logico-Philosophicus*, trans. D. F. Pears and B. F. McGuinness (London: Routledge, 1974).

2. Thomas Mann, *Thoughts of an Unpolitical Man* (New York: F. Unger, 1983).

3. V. I. Lenin, *Socialism and Religion* (Moscow: Foreign Languages Publishing, 1954), 6.

4. Lenin, 7.

5. Grace Ellison's *Turkey Today* (London: Hutchinson, 1928), 24, includes statements from Ataturk that seem to have been made to Ellison; as cited in Andrew Mango, *Atatürk: The Biography of the Founder of Modern Turkey* (New York: Overlook, 2000), 463.

6. Mango, 45.

7. Emilio Gentile, *The Sacralization of Politics in Fascist Italy* (Cambridge: Harvard Univ. Press, 1996), 13.

8. Ernst Nolte, *Three Faces of Fascism: Action Francaise, Italian Fascism, National Socialism* (New York: Holt, Rinehart & Winston, 1965), 147, 150.
9. Nolte, 14.
10. Nolte, 20.
11. Nolte, 21.
12. Nolte, 27.
13. Nolte, 36.
14. Nolte, 80.
15. Nolte, 70.
16. Gentile, 17.
17. George Mosse, *Fallen Soldiers: Reshaping the Memory of the World Wars* (New York: Oxford Univ. Press, 1990), 7.
18. Mosse, 7.
19. Mosse, 45.
20. Thomas Edison interview, *New York Times,* October 2, 1910, sec. 5, p. 1.
21. M. C. Nerney, *Edison, Modern Olympian,* 252, as cited in Matthew Josephson, *Edison: A Biography* (New York: McGraw-Hill, 1959), 438.
22. As cited in Josephson, 438.
23. Hubert Harrison, "Paine's Place in the Deistical Movement," *The Truth Seeker* 38, no. 6 (February 11, 1911): 87–88; in *A Hubert Harrison Reader,* ed. Jeffrey R. Perry (Middletown, CT: Wesleyan Univ. Press, 2001), 40.
24. Harrison, "Paine's Place," 40.
25. Harrison, "Paine's Place," 41.
26. Harrison, "Paine's Place," 42.
27. Harrison, "Paine's Place," 38.
28. Harrison, "Paine's Place," 39.
29. Harrison, *When Africa Awakes* (New York: Porro, 1920), 76.
30. Jeffrey Perry, "Introduction," in *Hubert Harrison Reader,* 2.
31. Perry, 2.
32. Perry, 36.
33. Emma Goldman, "The Philosophy of Atheism" (*Mother Earth,* 1916) in Goldman, *The Philosophy of Atheism and the Failure of Christianity* (New York: Mother Earth, 1916), 5.
34. Goldman, "The Failure of Christianity" (*Mother Earth,* 1913), in Goldman, *The Philosophy of Atheism and the Failure of Christianity,* 9.
35. Goldman, "The Philosophy of Atheism," 7.
36. Goldman, *Voltairine De Cleyre* (Berkeley Heights, NJ: Oriole, 1932).
37. Margaret Sanger, *Margaret Sanger: An Autobiography* (New York: Dover, 1971), 22.
38. Sanger, 23.
39. Mark Twain, *The Bible According to Mark Twain: Irreverent Writings on Eden, Heaven, and the Flood by America's Master Satirist* (New York: Simon & Schuster, Touchstone, 1996).
40. Twain, 240.
41. Twain, 243.
42. Twain, 244.
43. Twain, 252.
44. Twain, 252.
45. Twain, 318.
46. Twain, 328.
47. Twain, 323.
48. Twain, 323.

49. Twain, 330–331.
50. As cited in Edward J. Larson, *Summer for the Gods: The Scopes Trial and America's Continuing Debate Over Science and Religion* (Cambridge: Harvard Univ. Press, 1997), 20–21 and see also 32.
51. Larson, 20.
52. Clarence Darrow, "Why I Am an Agnostic" in *Why I Am an Agnostic and Other Essays* (Amherst, NY: Prometheus, 1995), 11.
53. Darrow, 12.
54. Darrow, 19.
55. Terry Teachout, *The Skeptic: A Life of H. L. Mencken* (New York: HarperCollins, 2002).
56. Ronald W. Clark, *Einstein: The Life and Times* (New York: Avon, 1984), 502.
57. Helen Dukas and Banesh Hoffman, eds., *Albert Einstein, the Human Side: New Glimpses from His Archives* (Princeton: Princeton Univ. Press, 1979), 43.
58. Dukas and Hoffman, 40.
59. Dukas and Hoffman, 39.
60. Michael White and John Gribbin, *Einstein: A Life in Science* (New York: Dutton, 1994), 262.
61. Philip Rieff, *Freud: The Mind of the Moralist* (Chicago: Univ. of Chicago Press, 1959), 258.
62. Rieff, 70.
63. Rieff, 24.
64. Rieff, 36.
65. Rieff, 271.
66. Rieff (295) says, "I should guess that Freud had read it, so closely does his own dialogue in *The Future of an Illusion* follow it."
67. Sigmund Freud, *The Future of an Illusion*, ed. Peter Gay (New York: Norton, 1961), 26.
68. Freud, 21.
69. Freud, 34.
70. Freud, 42.
71. Freud, 41.
72. Freud, 41.
73. Freud, 41.
74. Freud, 59.
75. Freud, 59.
76. Freud, 62–63.
77. Freud, 68.
78. Rieff, 257.
79. Gay, "About This Book," in Freud, *Future of an Illusion*, xxiii.
80. Bertrand Russell, *Why I Am Not a Christian and Other Essays*, ed. Paul Edwards (London: Unwin, 1957), 15.
81. Russell, *Not a Christian*, 19.
82. Russell, *Not a Christian*, 26.
83. Russell, "Am I an Atheist or an Agnostic?" in Russell, *Atheism: Collected Essays* (New York: Arno & New York Times, 1972), 5.
84. Russell, *Human Society in Ethics and Politics* (New York: Simon & Schuster, 1954), 208.
85. Russell, *What I Believe* (New York: Dutton, 1925), 13.
86. Dora Black Russell, *The Religion of the Machine Age*, 236, as cited in *Women Without Superstition, "No Gods—No Masters": The Collected Writings of Women Freethinkers of the Nineteenth and Twentieth Centuries*, ed. Annie Laurel Gaylor (Madison, WI: Freedom from Religion Foundation, 1997), 427.

87. The formulation is from Hans Sluga, "Ludwig Wittgenstein: Life and Work," in *The Cambridge Companion to Wittgenstein,* ed. Hans Sluga and David G. Stern (Cambridge: Cambridge Univ. Press, 1996), 22.

88. Ludwig Wittgenstein, *Philosophical Investigations* (Oxford: Basil Blackwell, 1958), 109.

89. Wittgenstein, *On Certainty* (New York: Harper, Torchbooks, 1972), 18.

90. Avrum Stroll, *Wittgenstein* (Oxford: One World, 2002), 145.

91. Wittgenstein, "Lectures on Religious Beliefs," in *Lectures and Conversations on Aesthetics, Psychology and Religious Belief,* ed. Cyril Barrett (Oxford: Basil Blackwell, 1966), 55.

92. Wittgenstein, *On Certainty,* 32.

93. Jean-Paul Sartre, "Existentialism," in *Existentialism from Dostoevsky to Sartre,* ed. Walter Kaufman (New York: Meridian, 1975), 349.

94. Sartre, "Existentialism," 349.

95. Sartre, "Existentialism," 350.

96. Sartre, *No Exit,* trans. Paul Bowles (New York: Samuel French, 1958), 52.

97. Sartre, *War Diaries: Notebooks from a Phoney War,* trans. Quintin Hoare (London: Verso, 1984), 70.

98. Sartre, *War Diaries,* 71–72.

99. Sartre, *War Diaries,* 72.

100. Simone de Beauvoir, *Memoirs of a Dutiful Daughter* (Cleveland: World, 1959), 1:144–145.

101. Beauvoir, 1:202.

102. Albert Camus, "The Myth of Sisyphus," in Kaufman, *Existentialism,* 377.

103. Camus, 377.

104. Camus, 377.

105. Camus, 378.

106. Camus, 378.

107. Milton Steinberg, *A Driven Leaf* (Springfield, NJ: Behrman House, 1939), 21, 23.

108. Steinberg, 168.

109. Steinberg, 200, 202.

110. Steinberg, 136.

111. Steinberg, 250.

112. Steinberg, 304.

113. Steinberg, 474.

114. Elie Wiesel, *Night,* trans. Stella Rodway (Harmondsworth, England: Penguin, 1981), 45.

115. Wiesel, 76–77.

116. Viktor E. Frankl, *Man's Search for Meaning* (New York: Washington Square, 1959), 95.

117. Frankl, 121.

118. Gershom G. Scholem, *Major Trends in Jewish Mysticism* (New York: Schocken, 1941), 7.

119. Margarete Susman, "God the Creator," in *The Dimensions of Job: A Study and Selected Readings,* ed. Nahum N. Glatzer (New York: Schocken, 1969), 86–92.

120. Susman, 92.

121. Susman, 88.

122. Walter Kaufman, *The Faith of a Heretic* (Garden City, NY: Anchor, 1968), 167.

123. Kaufman, 168.

124. Richard L. Rubenstein, *After Auschwitz: Radical Theology and Contemporary Judaism* (Indianapolis: Bobbs-Merrill, 1966).

125. Rubenstein had first published many of the essays in *After Auschwitz* in the journal *The Reconstructionist.*

126. Used by permission of Jackie Mason. For more on Mason's rejection of religion see: Mason, with Ken Gross, *Jackie, Oy!: Jackie Mason from Birth to Rebirth* (New York: Little, Brown, 1988), 25–42.

127. Interview transcript by Mason and Elli Wohlgelernter, July–August 1988, *Jackie Mason, Oral History Library* (American Jewish Committee, New York Public Library), tape 3, 199.

128. Woody Allen, "My Philosophy," *The New Yorker*, December 27, 1969; see also his *Getting Even* (1971).

129. Rep. Charles G. Oakman, *Congressional Record*, Appendix, p. A2527.

130. *Congressional Record*, House, February 12, 1954, p. 1700.

131. *Congressional Record*, House, June 7, 1955, pp. 7795–7796.

132. "Madalyn Murray Interview," by Richard Tregaskis, *Playboy*, October 1965.

133. "Interview: O'Hair on Church and State," *Freedom Writer*, March–May 1989.

134. Margaret Knight, *Morals Without Religion: And Other Essays* (London: Dobson, 1955), 14.

135. Knight, 35.

136. As cited in Edward E. Ericson Jr., "Living Responsibly: Václav Havel's View," *Religion and Liberty* 8 (September–October 1998).

137. Martin Baumann, "Buddhism in Europe," in *Westward Dharma: Buddhism Beyond Asia*, eds. Charles S. Prebish and Martin Baumann (Berkeley: Univ. of California Press, 2002), 88.

138. Charles S. Prebish and Martin Baumann, eds., *Westward Dharma* (Berkeley: Univ. of California Press, 2002), 87.

139. D. T. Suzuki, *An Introduction to Zen Buddhism* (New York: Grove, 1991), Preface.

140. Rick Fields, *How the Swans Came to the Lake: A Narrative History of Buddhism in America* (Boston & London: Shambhala, 1992), 83–166.

141. See Stephen Batchelor, *The Faith to Doubt: Glimpses of Buddhist Uncertainty* (Berkeley, CA: Paralax, 1990), 14–17.

142. Suzuki, *Zen Mind, Beginner's Mind: Informal Talks on Zen Meditation and Practice* (New York: Weatherhill, 1970), 21.

143. Bernard Glassman and Rick Fields, *Instructions to the Cook: A Zen Master's Lessons in Living a Life That Matters* (New York: Bell Tower, 1996), 51.

144. Bernard Glassman, *Bearing Witness: A Zen Master's Lessons in Making Peace* (New York: Bell Tower, 1998), xiv.

145. The last two, in full, are "Bearing witness to the joy and suffering of the world" and "Loving action toward ourselves and others."

146. *Westward Dharma*, 333.

147. Batchelor, *Faith to Doubt*, 44.

148. Batchelor, *Faith to Doubt*, 45.

149. Stephen Batchelor, *Buddhism Without Beliefs: A Contemporary Guide to Awakening* (New York: Riverhead, 1997), xiii, as cited in *Westward Dharma*, 332.

150. Its founders included Ruth Denison, Sharon Salzberg, and Jack Kornfield.

151. Mark Epstein, *Thoughts Without a Thinker: Psychotherapy from a Buddhist Perspective* (New York: Basic Books, 1995), 57.

152. By Hasan Lâhûtî in 1995.

153. Reynold Alleyne Nicholson, *Studies in Islamic Poetry* (Cambridge: Cambridge Univ. Press, 1920), viii.

154. Nicholson, *Islamic Poetry*, 145.

155. Nicholson, *Islamic Poetry*, 166. Nicholson again likens al-Ma'arri to Lucian in his *Literary History of the Arabs* (Cambridge: Cambridge Univ. Press, 1930), 318–319.

156. Nicholson, *Islamic Poetry*, 44.

157. Salman Rushdie, "In God We Trust," in his *Imaginary Homelands* (London: Granta, 1991), 376–432.

158. Rushdie, "In God We Trust," 380.

159. taslimanasrin.com, consulted December 9, 2002.

160. Taslima Nasrin interview, by Matt Cherry and Warren Allen Smith, *Free Inquiry* 19 (Winter 1998/1999).

161. "Open Letter from Salman Rushdie to Taslima Nasrin," *New York Times*, July 14, 1994.

162. Esther B. Fein, "Rushdie, Defying Death Threats, Suddenly Appears in New York," *New York Times*, December 12, 1991.

163. Ibn Warraq, *Why I Am Not a Muslim* (New York: Prometheus, 1995), xiii.

164. Ibn Warraq, xiv.

165. Ibn Warraq, 33.

166. Ibn Warraq, 14.

167. Ibn Warraq, 141.

168. Ibn Warraq, 33.

169. Ibn Warraq, xv.

170. Ibn Warraq, 32.

171. Ibn Warraq, 283.

172. Ibn Warraq, 283.

173. "Ibn Warraq: Why I Am Not a Muslim," interviewed by Lyn Gallacher, October 10, 2001, ABC Radio National Website.

174. Also important in this work are John Wansbrough of the School of Oriental and African Studies in London; Patricia Crone, professor of history at the Institute for Advanced Study in Princeton; Michael Cook, professor of Near Eastern history at Princeton University; and Fred M. Donner of the University of Chicago.

175. Steven Weinberg, *Facing Up: Science and Its Cultural Adversaries* (Cambridge: Harvard Univ. Press, 2001), 119–120.

176. Steven Weinberg, 119–120.

177. Isaac Asimov, in *Free Inquiry*, Spring 1982, p. 9.

178. Harlan Ellison, on the Tom Snyder show *Tomorrow*.

179. Douglas Adams, interview by David Silverman, *The American Atheist* 37, no. 1, in Adams, *The Salmon of Doubt: Hitchhiking the Galaxy One Last Time* (New York: Ballantine Books, 2003), 96.

180. Barbara Ehrenreich, "My Family Values Atheism," article adapted from the acceptance speech for the 1999 Freethought Heroine Award. *Freethought Today*, April 2000. On the Web, see www.ffrf.org/fttoday/april2000/ehrenreich.html

181. Barbara Ehrenreich, *The Worst Years of Our Lives: Irreverent Notes from a Decade of Greed* (New York: Pantheon Books, 1990), 5. This is a fanciful piece; for more serious notes on her background, see pp. 7–11.

182. Katha Pollitt, "No God, No Master" in her *Subject to Debate: Sense and Dissents on Women, Politics, and Culture* (New York: Random House, 2001), 87–90.

183. Natalie Angier, "Confessions of a Lonely Atheist," *New York Times Magazine*, January 14, 2001, 36.

184. Katharine Hepburn, interview, *Ladies' Home Journal*, October 1991, 215.

185. Karen Armstrong, *A History of God* (New York: Knopf, 1993), 377–399, esp. 397.

186. Quentin Crisp, *How to Become a Virgin* (New York: Saint Martin's Press, 1981), 102.

Bibliography

Adams, Douglas. *The Salmon of Doubt: Hitchhiking the Galaxy One Last Time.* New York: Ballantine Books, 2003.

Allen, Don Cameron. *Doubt's Boundless Sea.* Baltimore: Johns Hopkins Univ. Press, 1964; New York: Arno, 1979.

Allen, Woody. "My Philosophy." *The New Yorker,* December 27, 1969.

Angier, Natalie. "Confessions of a Lonely Atheist." *New York Times Magazine,* January 14, 2001.

Apuleius, Lucius. *The Golden Ass.* New York: Horace Liveright, 1927.

Armstrong, Karen. *A History of God.* New York: Knopf, 1993.

Augustine. *The City of God.* D. S. Wiesen, trans. Cambridge: Harvard Univ. Press, 1988.

_____. *Confessions.* Henry Chadwick, trans. and intro. Oxford: Oxford Univ. Press, 1992.

_____. *De Trinitate.* E. Hill, trans. New York: New City, 1991.

_____. *Teaching Christianity (On Christian Doctrine).* Edmund Hill, ed. and intro. Hyde Park, NY: New City, 1996.

Barnes, Jonathan, ed. *The Cambridge Companion to Aristotle.* Cambridge: Cambridge Univ. Press, 1995.

Barrett, Cyril, ed. *Lectures and Conversations on Aesthetics, Psychology and Religious Belief.* Oxford: Basil Blackwell, 1966.

Batchelor, Stephen. *Buddhism Without Beliefs: A Contemporary Guide to Awakening.* New York: Riverhead, 1997.

_____. *The Faith to Doubt: Glimpses of Buddhist Uncertainty.* Berkeley, CA: Paralax, 1990.

Baumann, Martin. "Buddhism in Europe." In *Westward Dharma: Buddhism Beyond Asia.* Charles S. Prebish and Martin Baumann, eds. Berkeley: Univ. of California Press, 2002.

Bayle, Pierre. *Historical and Critical Dictionary: Selections.* Richard Popkin, trans. Indianapolis: Hackett, 1991.

Beauvoir, Simone de. *Memoirs of a Dutiful Daughter.* Cleveland: World, 1959.

Ben-Sasson, Hayim, ed. *A History of the Jewish People.* Cambridge: Harvard Univ. Press, 1976.

Berman, David. *A History of Atheism in Britain: From Hobbes to Russell.* London: Routledge, 1988.

Besant, Annie, and Charles Bradlaugh. *The Freethinker's Text Book.* London: Freethought, 1881.

Bickerman, Elias J. *The Jews in the Greek Age.* Cambridge: Harvard Univ. Press, 1988.

Blyth, R. H. *Zen and Zen Classics.* Frederick Franck, comp. New York: Vintage, 1978.

Bradlaugh, Charles. *Doubts in Dialogue.* London: Watts, 1909.

_____. *A Plea for Atheism.* London: Freethought, 1880.

Brown, Peter. *The Rise of Western Christendom.* Oxford: Basil Blackwell, 1996.

Bruno, Giordano. *The Ash Wednesday Supper.* 1584.

Buber, Martin. *The Prophetic Faith.* New York: Harper, Torchbooks, 1949.

Burkert, Walter. *Greek Religion.* Cambridge: Harvard Univ. Press, 1985.

Burnet, John. *Early Greek Philosophy.* London: Black, 1930.

Carel, A. *Histoire anecdotique des contemporains.* Paris, 1885.

Cassirer, Ernst; Paul Oskar Kristeller; and John Herman Randall Cassirer Jr., eds. *The Renaissance Philosophy of Man.* Chicago: Univ. of Chicago Press, 1948.

Chadwick, Owen. *The Secularization of the European Mind.* Cambridge: Cambridge Univ. Press, 1975.

Ch'ung, Wang. *Lun-Hêng: Philosophical Essays of Wang Ch'ung.* Alfred Forke, trans. and intro. (New York, 1962, reprint of 1907 ed.).

Cicero. *The Nature of the Gods.* H. C. McGregor, trans. New York: Viking, 1985.

Clark, Ronald W. *Einstein: The Life and Times.* New York: Avon, 1984.

Cohen, C. *Bradlaugh and Ingersoll.* London: Pioneer, 1933.

Collinson, Diané, Kathryn Plant, and Robert Wilkinson, eds. *Fifty Eastern Thinkers.* London: Routledge, 2000.

Comte, Auguste. *A General View of Positivism.* J. H. Bridges, trans. London: Routledge, 1957.

Congressional Record. House. February 12, 1954, and June 7, 1955.

Crenshaw, James. "The Silence of Eternity: Ecclesiastes." In *A Whirlpool of Torment: Israelite Traditions of God as an Oppressive Presence.* Philadelphia: Fortress, 1984.

_____. *Urgent Advice and Probing Questions: Collected Writings on Old Testament Wisdom.* Macon, GA: Mercer Univ. Press, 1995.

Crick, Francis. *What Mad Pursuit: A Personal View of Scientific Discovery.* New York: Basic Books, 1988.

Crossan, John Dominic. *The Historical Jesus: The Life of a Mediterranean Jewish Peasant.* San Francisco: HarperSanFrancisco, 1993.

_____. *Jesus: A Revolutionary Biography.* San Francisco: HarperSanFrancisco, 1989.

Danto, Arthur C. *Mysticism and Morality: Oriental Thoughts and Moral Philosophy.* New York: Columbia Univ. Press, 1987.

Dasgupta, S. *A History of Indian Philosophy.* London: Cambridge Univ. Press, 1922.

Descartes, René. *Philosophical Works I.* Cambridge: Cambridge Univ. Press, 1975.

Desmond, Adrian. *The Politics of Evolution.* Chicago: Univ. of Chicago Press, 1989.

Dickinson, Emily. *The Complete Poems of Emily Dickinson.* Boston: Little, Brown, 1951.

Diderot, Denis. *Diderot: Interpreter of Nature.* Westport, CT: Hyperion, 1937.

Donoghue, Dennis. *Emily Dickinson.* Minneapolis: Univ. of Minnesota Press, 1969.

Douglass, Frederick. "The Meaning of Fourth of July for the Negro" (1852). In *Oration, Delivered in Corinthian Hall, Rochester.* Rochester, NY: Lee, Mann, 1852.

Drachmann, A. B. *Atheism in Pagan Antiquity.* London: Ares, 1922.

DuBois, Ellen Carol, ed. *Elizabeth Cady Stanton, Susan B. Anthony, Correspondence, Writings, Speeches.* Gerda Lerner, Foreword. New York: Schocken, 1981.

Dukas, Helen, and Banesh Hoffman, eds. *Albert Einstein, the Human Side: New Glimpses from His Archives.* Princeton: Princeton Univ. Press, 1979.

Durkheim, Emile. "Représentations individuelles et représentations collectives." *Revue de métaphysique et de morale* 5 (1898): 273–302.

Edie, James M., James Scanlan, and Mary-Barbara Zeldin, eds. *Russian Philosophy: The Nihilists, the Populists, Critics of Religion and Culture.* Chicago: Quadrangle, 1965.

Edison, Thomas. Interview by *New York Times,* October 2, 1910.

Ehrenreich, Barbara. *The Worst Years of Our Lives: Irreverent Notes from a Decade of Greed.* New York: Pantheon Books, 1990.

Ellison, Grace. *Turkey Today.* London: Hutchinson, 1928.

Epstein, Mark. *Thoughts Without a Thinker: Psychotherapy from a Buddhist Perspective.* New York: Basic Books, 1995.

Erasmus. *Praise of Folly.* Betty Radice, trans. London: Penguin, 1993.

Ericson, Edward E. Jr. "Living Responsibly: Václav Havel's View." *Religion and Liberty* 8 (September–October 1998).

Febvre, Lucien. *The Problem of Unbelief in the Sixteenth Century: The Religion of Rabelais.* Beatrice Gottlieb, trans. Cambridge: Harvard Univ. Press, 1947.

Fein, Esther B. "Rushdie, Defying Death Threats, Suddenly Appears in New York." *New York Times,* December 12, 1991.

Feuerbach, Ludwig. *The Essence of Christianity.* George Eliot, trans. New York: Harper, Torchbooks, 1957.

Fields, Rick. *How the Swans Came to the Lake: A Narrative History of Buddhism in America.* Boston and London: Shambhala, 1992.

Flaubert, Gustave. *Madame Bovary.* Geoffrey Wall, trans. London: Penguin, 1992.

Foucault, Michel. *Religion and Culture.* Jeremy R. Carrette, ed. New York: Routledge, 1999.

Fox, Marvin. *Interpreting Maimonides.* Chicago: Univ. of Chicago Press, 1990.

Frankl, Viktor E. *Man's Search for Meaning.* New York: Washington Square, 1959.

Freud, Sigmund. *The Future of an Illusion.* Peter Gay, ed. and intro. New York: Norton, 1961.

Fuller, Margaret. *Memoirs of Margaret Fuller Ossoli.* Boston: Phillips, Sampson, 1852.

_____. *Summer on the Lakes.* Boston: Charles Little & James Brown, 1844.

Garrett, Don, ed. *The Cambridge Companion to Spinoza.* Cambridge: Cambridge Univ. Press, 1996.

Gaskin, J. C. A., ed. *Varieties of Unbelief: From Epicurus to Sartre.* Upper Saddle River, NJ: Prentice-Hall, 1989.

Gaylor, Annie Laurel, ed. *Women Without Superstition, "No Gods—No Masters": The Collected Writings of Women Freethinkers of the Nineteenth and Twentieth Centuries.* Madison, WI: Freedom from Religion Foundation, 1997.

Gentile, Emilio. *The Sacralization of Politics in Fascist Italy.* Cambridge: Harvard Univ. Press, 1996.

Al-Ghazzali. *The Faith and Practice of Al-Ghazali.* W. M. Watt, trans. London: George Allen, 1951.

_____. *Deliverance from Error and Mystical Union with the Almighty.* George F. McLean, ed., Muhammad Abulaylah, trans. Washington, D.C.: Council for Research in Values & Philosophy, 2002.

Gibbon, Edward. *The Decline and Fall of the Roman Empire,* vol. 1. New York: Modern Library, 1954.

Gibson, Margaret, ed. *Boethius: His Life, Thought and Influence.* Oxford: Basil Blackwell, 1981.

Gilson, Etienne. *God and Philosophy.* New Haven: Yale Univ. Press, 1941, 2002.

_____. *Reason and Revelation in the Middle Ages.* New York: Scribner, 1938.

Ginzburg, Carlo. *The Cheese and the Worms: The Cosmos of a Sixteenth-Century Miller.* Baltimore: Johns Hopkins Univ. Press, 1980.

Glassman, Bernard. *Bearing Witness: A Zen Master's Lessons in Making Peace.* New York: Bell Tower, 1998.

Glassman, Bernard, and Rick Fields. *Instructions to the Cook: A Zen Master's Lessons in Living a Life That Matters.* New York: Bell Tower, 1996.

Glatzer, Nahum N., ed. and intro. *The Dimensions of Job: A Study and Selected Readings.* New York: Schocken, 1969.

Goldman, Emma. *The Philosophy of Atheism and the Failure of Christianity.* New York: Mother Earth, 1913.

———. *Voltairine De Cleyre.* Berkeley Heights, NJ: Oriole, 1932.

Goldziher, Ignaz. *Muslim Studies.* S. M. Stern, ed.; C. R. Barber and S. M. Stern, trans. 2 vols. London, 1967–71.

Goodman, Dena. *The Republic of Letters: A Cultural History of the French Enlightenment.* Ithaca, NY: Cornell Univ. Press, 1994.

Gordis, Robert. *Koheleth—The Man and His World.* New York: Schocken, 1978.

Gould, Stephen J. *Ever Since Darwin.* New York: Norton, 1977.

Grant, Edward. *God and Reason in the Middle Ages.* Cambridge: Cambridge Univ. Press, 2001.

Grant, Michael. *Jesus: An Historian's Review of the Gospels.* New York: Macmillan, 1977.

Grieg, J. Y. T. *The Letters of David Hume.* 2 vols. Oxford: Clarendon, 1969.

Hamilton, Edith, and Huntington Cairns, eds. *Plato: The Collected Dialogues.* Princeton: Princeton Univ. Press, 1989.

Hardy, Thomas. *The Collected Poems of Thomas Hardy.* New York: Macmillan, 1974.

Harrison, Hubert. *When Africa Awakes.* New York: Porro, 1920.

Harrison, Peter. "Newtonian Science, Miracles, and the Laws of Nature." *Journal of the History of Ideas* 56 (October 1995): 531–553.

Hecht, Jennifer Michael. *The End of the Soul: Scientific Modernity, Atheism, and Anthropology in France.* New York: Columbia Univ. Press, 2003.

Heine, Heinrich. *Gedanken und Einfalle,* vol. 10.

Hengel, Martin. *Judaism and Hellenism: Studies in Their Encounter in Palestine During the Early Hellenistic Period.* 2 vols. John Bowden, trans. Philadelphia: Fortress, 1974.

Hepburn, Katharine. Interviewed by *Ladies' Home Journal,* October 1991.

Hume, David. *An Enquiry Concerning Human Understanding: A Letter from a Gentleman to His Friend in Edinburgh.* Indianapolis: Hackett, 1977.

Hunter, Michael, and David Wootton, eds. *Atheism from the Reformation to the Enlightenment.* Oxford: Clarendon, 1992.

Huxley, Thomas. "Mr. Balfour's Attack on Agnosticism I." *The Nineteenth Century* 217 (March 1895).

———. "On Descartes' *Discourse Touching the Method of Using One's Reason Rightly and of Seeking Scientific Truth*" (1870). In *Lay Sermons, Addresses, and Reviews.* New York: Appleton, 1910.

Hyman, Arthur, and James J. Walsh, eds. *Philosophy in the Middle Ages.* Indianapolis: Hackett, 1973.

Ibn Warraq. *Why I Am Not a Muslim.* New York: Prometheus, 1995.

"Ibn Warraq: Why I Am Not a Muslim." Interviewed by Lyn Gallacher, October 10, 2001. ABC Radio National Website.

Ikkyu Sojun. *Crow with No Mouth.* Stephen Berg, trans. Port Townsend, WA: Copper Canyon, 2000.

Ingersoll, Robert G. *On Gods and Other Essays.* Buffalo: Prometheus, 1990.

———. *Works.* Clinton P. Farrell, ed. New York: Dresden, 1902.

"Interview: O'Hair on Church and State." *Freedom Writer,* March–May 1989.

Jackson, George Stuyvesant. *Uncommon Scold: The Story of Anne Royall.* Boston: B. Humphries, 1937.

Jaeger, Werner. *Early Christianity and Greek Paideia.* Cambridge: Harvard Univ. Press, 1961.

Jefferson, Thomas. *The Writings of Thomas Jefferson.* Washington, D.C.: Jefferson Memorial Association, 1904–05.

_____. "To Peter Carr, August 10, 1787." In *Basic Writing of Thomas Jefferson.* Philip S. Foner, ed. New York: Wiley, 1944.

_____. "To William Short, August 4, 1820." In *The Works of Thomas Jefferson,* H. A. Washington, ed. New York: Townsend, 1884.

_____. "To William Short, October 31, 1819." In *A Jefferson Profile: As Revealed in His Letters.* Saul K. Padover, ed. New York: John Day, 1956.

Johnson, Paul. *A History of the Jews.* New York: HarperCollins, 1988.

Josephson, Matthew. *Edison: A Biography.* New York: McGraw-Hill, 1959.

Kaufman, Walter. *The Faith of a Heretic.* Garden City, NY: Anchor, 1968.

_____, ed. *Existentialism from Dostoevsky to Sartre.* New York: Meridian, 1975.

Kierkegaard, Søren. *Fear and Trembling.* London: Penguin Books, 1985.

Knight, Margaret. *Morals Without Religion: And Other Essays.* London: Dobson, 1955.

Koch, Adrienne, ed. *The American Enlightenment: The Shaping of the American Experiment and a Free Society.* New York: G. Braziller, 1965.

Kraemer, Joel. *Humanism in the Renaissance of Islam: The Cultural Revival During the Buyid Age.* Leiden: E. J. Brill, 1986).

Kraut, Richard, ed. *The Cambridge Companion to Plato.* Cambridge: Cambridge Univ. Press, 1992.

Larson, Edward J. *Summer for the Gods: The Scopes Trial and America's Continuing Debate Over Science and Religion.* Cambridge: Harvard Univ. Press, 1997.

Lenin, V. I. *Socialism and Religion.* Moscow: Foreign Languages Publishing, 1954.

Lennon, Thomas M. *Reading Bayle.* Toronto: Univ. of Toronto Press, 1999.

Li, Dan, trans. *China in Transition, 1517–1911.* New York: Van Nostrand, 1969.

Long, A. A. *Hellenistic Philosophy: Stoics, Epicureans, Sceptics.* Berkeley: Univ. of California Press, 1986.

Longman, Tremper, III. *The Book of Ecclesiastes.* Cambridge: Eerdmans, 1998.

Lopez, Donald S., Jr. *Asian Religions in Practice.* Princeton, NJ: Princeton Readings in Religion, 1999.

Lucian. *True History and Lucius of the Ass.* Paul Turner, trans. Bloomington: Indiana Univ. Press, 1974.

Lucretius. *On the Nature of Things.* Anthony M. Esolen, trans. and ed. Baltimore: Johns Hopkins Univ. Press, 1995.

Mack, Burton. *The Lost Gospel: The Book of Q and Christian Origins.* San Francisco: HarperSanFrancisco, 1994.

"Madalyn Murray Interview." Interviewed by Richard Tregaskis. *Playboy,* October 1965.

Magee, Bryan. *The Philosophy of Schopenhauer.* Oxford: Oxford Univ. Press, 1983.

Maimonides, Moses. *Guide for the Perplexed.* New York: Dover, 1956.

Mango, Andrew. *Atatürk: The Biography of the Founder of Modern Turkey.* New York: Overlook, 2000.

Mann, Thomas. *Thoughts of an Unpolitical Man.* New York: F. Unger, 1983.

Marcus Aurelius. *The Meditations of Marcus Aurelius.* George Long, trans. New York: Collier, 1909–1914.

Martin, Luther H. *Hellenistic Religions: An Introduction.* Oxford: Oxford Univ. Press, 1987.

Martineau, Harriet. *Autobiography.* Maria Weston Chapman, ed. Boston: James R. Osgood, 1877.

Martineau, Harriet, with Henry George Atkinson. *Letters on the Laws of Man's Nature and Development.* Boston: Josiah P. Mendum, 1851.

Marx, Karl Heinrich. "The Difference Between the Democritean and Epicurean Philosophy of Nature." In Norman D. Livergood. *Activity in Marx's Philosophy.* The Hague: Martinus Nijhoff, 1967.

Mason, Jackie, with Ken Gross. *Jackie, Oy!: Jackie Mason from Birth to Rebirth.* New York: Little, Brown, 1988.

Matthews, Gareth B., ed. *The Augustinian Tradition.* Berkeley: Univ. of California Press, 1999.

Mendelssohn, Moses. *Jerusalem, or On Religious Power and Judaism.* Allan Arkush, trans., Alexander Altmann, intro. and comment. Hanover: Brandeis Univ. Press, 1983.

Mercer-Taylor, Peter. *The Life of Mendelssohn.* Cambridge: Cambridge Univ. Press, 2000.

Mill, John Stuart. *The Autobiography of John Stuart Mill.* New York: Columbia Univ. Press, 1924.

_____. *On Liberty.* Edward Alexander, ed. Toronto: Broadview, 1999.

Momigliano, Arnaldo. *Alien Wisdom: The Limits of Hellenization.* Cambridge: Cambridge Univ. Press, 1971.

_____. *On Pagans, Jews, and Christians.* Middletown, CT: Wesleyan Univ. Press, 1987.

Montaigne, Michel de. *The Complete Essays of Montaigne.* Donald M. Frame, trans. Stanford: Stanford Univ. Press, 1958.

Mosse, George. *Fallen Soldiers: Reshaping the Memory of the World Wars.* New York: Oxford Univ. Press, 1990.

Nasrin, Taslima. Interviewed by Matt Cherry and Warren Allen Smith. *Free Inquiry* 19 (Winter 1998/1999).

Needham, Joseph. *Science and Civilisation in China,* 7 vols. (Cambridge: Cambridge Univ. Press, 1991, reprint of 1956 ed.), vols. 1–7.

Nicholson, Reynold Alleyne. *Literary History of the Arabs.* Cambridge: Cambridge Univ. Press, 1930.

_____. *Studies in Islamic Poetry.* Cambridge: Cambridge Univ. Press, 1920.

Nicolson, Nigel, ed. *The Question of Things Happening: The Letters of Virginia Woolf.* London: Hogarth, 1982.

Nietzsche, Friedrich. *Daybreaks.* R. J. Hollingdale, trans. Cambridge: Cambridge Univ. Press, 1982.

_____. *The Gay Science.* Walter Kaufmann, trans. New York: Vintage, 1974.

Nishitani, Keiji. *Religion and Nothingness.* Jan Van Bragt, trans. Berkeley: Univ. of California Press, 1983.

Nolte, Ernst. *Three Faces of Fascism: Action Francaise, Italian Fascism, National Socialism.* New York: Holt, Rinehart & Winston, 1965.

"Open Letter from Salman Rushdie to Taslima Nasrin." *New York Times,* July 14, 1994.

Owen, Robert. *The Life of Robert Owen, Written by Himself.* New York: A. M. Kelly, 1967.

Pagels, Elaine. *The Gnostic Gospels.* New York: Vintage, 1979.

Paine, Thomas. *The Age of Reason.* Amherst, NY: Prometheus, 1984.

Pascal, Blaise. *Pensées.* London: Penguin, 1995.

Perry, Jeffrey R., ed. *A Hubert Harrison Reader.* Middletown, CT: Wesleyan Univ. Press, 2001.

Philipson, David. *The Reform Movement in Judaism.* Jersey City, NJ: KTAV, 1967.

Plato. *Timaeus and Critias,* Desmond Lee, trans. London: Penguin Classics, 1977.

Plaut, W. Gunther. *The Rise of Reform Judaism: A Sourcebook of Its European Origins.* New York: World Union for Progressive Judaism, 1963.

Pliny the Elder. *Natural History.* London: Penguin Classics, 1991.

Pollitt, Katha. *Subject to Debate: Sense and Dissents on Women, Politics, and Culture.* New York: Random House, 2001.

Popkin, Richard H. *The History of Scepticism from Erasmus to Spinoza.* Berkeley: Univ. of California Press, 1979.

Prebish, Charles S., and Martin Baumann, eds. *Westward Dharma.* Berkeley: Univ. of California Press, 2002.

Radhakrishnan, Sarvepalli, and Charles A. Moore, eds. *A Sourcebook in Indian Philosophy.* Princeton: Princeton Univ. Press, 1973.

Ratner, Joseph. *The Philosophy of Spinoza: Selections from His Works.* New York: Modern Library, 1927.

Rémond, René. *L'anticlericalism en France de 1815 à nos jours.* Paris: Fayard, 1976.

Renan, Ernest. *Averroès et l'averroïsme: Essai historique.* Paris: Michel Lévy Frères, 1861.

Reynolds, Susan. "Social Mentalities and the Case of Medieval Scepticism." *Transactions of the Royal Historical Society,* 6th ser., 1 (1991): 21–41.

Rieff, Philip. *Freud: The Mind of the Moralist.* Chicago: Univ. of Chicago Press, 1959.

Rinaldo, Peter M. *Atheists, Agnostics, and Deists in America: A Brief History.* New York: DorPete, 2000.

Roberts, Lawrence D., ed. *Approaches to Nature in the Middle Ages.* Binghamton, NY: CMES, 1982.

Royall, Anne Newport. *Black Book,* vol. 1. Washington, DC, 1828.

_____. *Letters from Alabama, 1817–1822.* Birmingham: Univ. of Alabama Press, 1969.

Rubenstein, Richard L. *After Auschwitz: Radical Theology and Contemporary Judaism.* Indianapolis: Bobbs-Merrill, 1966.

Rudavsky, T. M. "Galileo and Spinoza: Heroes, Heretics, and Hermeneutics." *Journal of the History of Ideas* 62 (1951): 611–631.

Rushdie, Salman. *Imaginary Homelands.* London: Granta, 1991.

_____. *The Satanic Verses.* New York: Picador, 1988.

Russell, Bertrand. "Am I an Atheist or an Agnostic?" In Russell, *Atheism: Collected Essays.* New York: Arno & New York Times, 1972.

_____. *Human Society in Ethics and Politics.* New York: Simon & Schuster, 1954.

_____. *What I Believe.* New York: Dutton, 1925.

_____. *Why I Am Not a Christian and Other Essays.* Paul Edwards, ed. London: Unwin, 1957.

Samkhya Sutra Vrtti. Sinha, 1915.

Sanders, E. P. *The Historical Figure of Jesus.* London: Penguin, 1993.

Sanger, Margaret. *An Autobiography.* New York: Dover, 1971.

Sartre, Jean-Paul. *No Exit.* Paul Bowles, trans. New York: Samuel French, 1958.

_____. *War Diaries: Notebooks from a Phoney War.* Quintin Hoare, trans. London: Verso, 1984.

Scholasticus, Socrates. *Ecclesiastical History.* London: S. Bagster and Sons, 1853.

Scholem, Gershom G. *Major Trends in Jewish Mysticism.* New York: Schocken, 1941.

Schopenhauer, Arthur. *Religion: A Dialogue and Other Essays.* 3d ed. T. Bailey Saunders, select. and trans. Westport, CT: Greenwood, 1973.

_____. *Studies in Pessimism.* New York: Boni & Liveright, 1925.

_____. *The World as Will and Representation.* 2 vols. E. F. J. Payne, trans. New York: Dover, 1969.

Sextus Empiricus. *Selections from the Major Writings on Scepticism, Man, and God.* Philip P. Hallie, ed. and intro.; Sanford G. Etheridge, trans. Indianapolis: Hackett, 1985.

Shelley, Percy Bysshe. *Complete Poems.* New York: Book of the Month Club, 1993.

_____. *The Necessity of Atheism and Other Essays.* Buffalo: Prometheus, 1993.

Sluga, Hans, and David G. Stern, eds. *The Cambridge Companion to Wittgenstein.* Cambridge: Cambridge Univ. Press, 1996.

Smith, Huston. *The World's Religions*. San Francisco: HarperSanFrancisco, 1991.

Spinoza, Benedict de. *Ethics*. James Gutmann, ed. New York: Hafner, 1949.

Stanton, Elizabeth Cady; Susan B. Anthony; and Matilda Joslyn Gage, eds. *The History of Women's Suffrage*, vol. 1. Rochester, NY: Source Book Press, 1971.

Stanton, Theodore, and Harriot Stanton Blatch, eds. *Elizabeth Cady Stanton as Revealed in Her Letters, Diary and Reminiscences*. New York: Harper and Brothers, 1922.

Steinberg, Milton. *A Driven Leaf*. Springfield, NJ: Behrman House, 1939.

Stephen, Leslie. *Essays on Free Thinking and Plain Speaking*. London: Longmans, 1873.

Strauss, David Friedrich. *The Life of Jesus, Critically Examined*. Marian Evans, trans. New York: Blanchard, 1860.

Strick, James, ed. *Evolution and the Spontaneous Generation Debate*. Bristol: Thoemmes, 2001.

Stroll, Avrum. *Wittgenstein*. Oxford: One World, 2002.

Stroumsa, Sarah. *Freethinkers of Medieval Islam*. Leiden: Brill, 1999.

Suzuki, D. T. *An Introduction to Zen Buddhism*. New York: Grove, 1991.

_____. *Zen Mind, Beginner's Mind: Informal Talks on Zen Meditation and Practice*. New York: Weatherhill, 1970.

Teachout, Terry. *The Skeptic: A Life of H. L. Mencken*. New York: HarperCollins, 2002.

Thera, Nyanaponika, and Bhikkhu Bodi, trans. and eds. *Numerical Discourses of the Buddha: An Anthology of Suttas from the Anguttara Nikaya*. Oxford: Altamira, 1999.

Thrower, James. *The Alternative Tradition: Religion and the Rejection of Religion in the Ancient World*. The Hague: Mouton, 1980.

Twain, Mark. *The Bible According to Mark Twain: Irreverent Writings on Eden, Heaven, and the Flood by America's Master Satirist*. Baetzhold, Howard G., and Joseph B. McCullough, eds. New York: Simon & Schuster, Touchstone, 1996.

Voltaire. *The Philosophical Dictionary*. New York: Coventry House, 1932.

Wacholder, Ben Zion. *Eupolemus: A Study of Judaeo-Greek Literature*. Cincinnati: Hebrew Union College Press, 1974.

Watson, Philip S., trans. and ed. *Luther and Erasmus: Free Will and Salvation*. Philadelphia: Westminster, 1969.

Watts, Alan. *The Way of Zen*. New York: Vintage, 1989.

Weinberg, Steven. *Facing Up: Science and Its Cultural Adversaries*. Cambridge, MA: Harvard Univ. Press, 2001.

Whaley-Cohen, Joanna. *The Sextants of Beijing: Global Currents in Chinese History*. New York: Norton, 1999.

White, Michael, and John Gribbin. *Einstein: A Life in Science*. New York: Dutton, 1994.

Wiesel, Elie. *Night*. Stella Rodway, trans. Harmondsworth, England: Penguin, 1981.

Wittgenstein, Ludwig. *On Certainty*. New York: Harper, Torchbooks, 1972.

_____. *Philosophical Investigations*. Oxford: Basil Blackwell, 1958.

_____. *Tractatus Logico-Philosophicus*. D. F. Pears and B. F. McGuinness, trans. London: Routledge, 1974.

Woolf, Virginia. *Moments of Being*. Jeanne Schulkind, ed. London: Hogarth, 1985.

Wright, Fanny. *A Few Days in Athens*. New York: Arno, 1972.

_____. *Life, Letters, and Lectures, 1834/1844*. New York: Arno, 1972.

_____. *Views of Society and Manners in America*. Paul R. Baker, ed. Cambridge: Harvard Univ. Press, 1963.

Acknowledgments

I want to thank my husband, John Chaneski, for seeing me through this huge project and for always having a sectarian zeal for it even though the subject matter, about which I talked constantly, might have encouraged equivocation. Mary Keller and Amy Allison Hecht both read many chapters and provided keen advice and crucial encouragement. My editor, John Loudon, offered insight and a broad sense of the task at hand. Thanks also go to the book's production editor, Lisa Zuniga; the copyeditor, Anne Collins; the designer, Kris Tobiassen; Henning Gutmann, Virginia Vitzthum, Gil Schrank, and Jeffrey Levine, who read parts of the book and helped me see it, or just cheered for it. Thanks to Jean-Pierre Trebot of the New York Friars Club for help with a reference. For conversations over the years, and for their friendship, I thank Meema Spadola, Jack Stoller, Stephen Hull, Yvany Peery, Tanya Elder, Amy Holman, Joshua Swanson, Mindy Roseman, Paul Kelly, Toby Baker, Christine Krol, Stephan Shaw, Liz and David Goldstein, Melissa Hotchkiss, Harry Williams, Michael White, Chad Jones, Phil Stevenson, Mollie Hecht, Dolly Brivic, Jamey and Kerry Hecht, Claudia R. Ferrer, Victoria R. Ferrer, Thomas R. Ferrer, Arthur R. Ferrer, Angela Rose Altomare, Frankie Altomare, Angela Marie Altomare, and all the Chaneskis: Bob and Marietta, Bobby and Alison, Chris and Mary Ann, and Nannie. Finally, thanks to my parents, Carolyn and Gene, to whom this book is dedicated, for their belief, and for their enthusiasm for the subject of doubt.

Index